AUTHORS

Robert Alexander   **Brendan Kelly**

Contributing Writers

Paul Atkinson
Superintendent of Instruction
Waterloo Board of Education
Waterloo, Ontario

Maurice Barry
Vista School District
Newfoundland

Fred Crouse
Annapolis Valley Regional
   School Board
Berwick, Nova Scotia

Garry Davis
Saskatoon Public Board
   of Education
Saskatoon, Saskatchewan

George Gadanidis
Durham Board of Education
Whitby, Ontario

Liliane Gauthier
Saskatoon Board of Education
Saskatoon, Saskatchewan

Florence Glanfield
Mathematics Education
   Consultant
Edmonton, Alberta

Jim Nakamoto
Sir Winston Churchill
   Secondary School
Vancouver, British Columbia

Linda Rajotte
Georges P. Vanier
   Secondary School
Courtenay, British Columbia

Elizabeth Wood
National Sport School
Calgary Board of Education
Calgary, Alberta

Rick Wunderlich
Shuswap Junior
   Secondary School
Salmon Arm, British Columbia

PEARSON

Addison
Wesley

Toronto

DEVELOPMENTAL EDITORS

Claire Burnett
Lesley Haynes
Sarah Mawson

EDITORS

Mei Lin Cheung
Julia Cochrane
David Gargaro
Tony Rodrigues
Rajshree Shankar
Anita Smale

RESEARCHER

Lynne Gulliver

DESIGN/PRODUCTION

Pronk&Associates

ART DIRECTION

Pronk&Associates/Joe Lepiano

ELECTRONIC ASSEMBLY & TECHNICAL ART

Pronk&Associates

Acknowledgements appear on page 615.

ClarisWorks is a trademark of Claris Corporation.
Claris is a registered trademark of Claris Corporation.
Microsoft and Windows are either registered trademarks or trademarks of Microsoft Corporation.
Macintosh is a registered trademark of Apple Computer, Inc. Graphmatica is a trademark of kSoft, Inc.

Canadian Cataloguing in Publication Data

Alexander, Bob, 1941 –
        Addison-Wesley mathematics 11

Western Canadian ed.
Includes index.
Previous ed. written by Brendan Kelly, Bob Alexander, Paul Atkinson.
Includes index.
ISBN 0–201-34624–9

1. Mathematics.   I.Kelly, B. (Brendan), 1943 – .   II. Davis, Garry, 1963 – .   III. Title.

QA39.2.K447 1998   510   C97-932421-1

ISBN 0–201–34624–9

Printed and bound in Canada

7 –FP– 08

REVIEWERS/CONSULTANTS

Lana Rinn
Mathematics Consultant
Transcona-Springfield School Division #12
Winnipeg, Manitoba

Darryl Smith
Mathematics Department Head
Austin O'Brien High School
Edmonton, Alberta

Don Smith
Aden Bowman Collegiate
Saskatoon, Saskatchewan

Mila Stout
Swan Valley Regional Secondary School
Swan River, Manitoba

Margaret Warren
Program Coordinator
Peel Board of Education
Mississauga, Ontario

Mike Wierzba
Mathematics Coordinator
Etobicoke Board of Education
Etobicoke, Ontario

CONTENTS

Welcome to Addison-Wesley Mathematics 11

Western Canadian Edition

This book is about mathematical thinking, mathematics in the real world, and using technology to enhance mathematical understanding. We hope it helps you see that mathematics can be useful, interesting, and enjoyable.

These introductory pages illustrate how your student book works.

Mathematical Modelling

Each chapter begins with a provocative problem. Once you have explored relevant concepts, you use mathematical modelling to illustrate and solve the problem.

Consider This Situation is the first step in a four-stage modelling process. This page introduces the chapter problem. You think about the problem and discuss it in general terms.

FYI Visit refers you to our web site, where you can connect to other sites with information related to the chapter problem.

You revisit the chapter problem in a Mathematical Modelling section later in the chapter.

Develop a Model suggests a graph or table, a formula or pattern, or some approximation that can represent the situation. Related exercises help you construct and investigate the model.

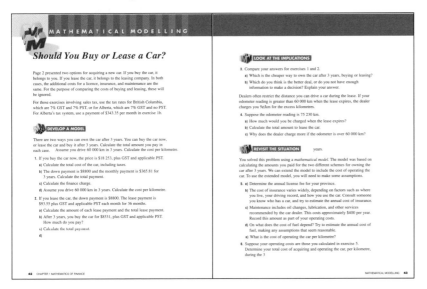

Look at the Implications encourages you to relate your findings to the original problem. What is the significance of your results?

Revisit the Situation invites you to criticize the model you developed. Does it provide a reasonable representation of the situation? Can you refine the model to obtain a closer approximation to the situation? Can you develop another problem that fits a similar model?

Short **Mathematical Modelling** boxes echo the fourth stage of the modelling process. Each box leads you back to an earlier problem to reflect on the validity of the solution, and to consider alternative ways to model the problem.

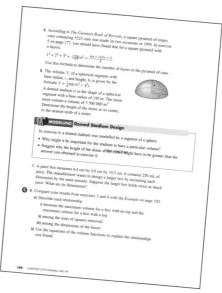

Concept Development

Here's how a typical lesson in your student book works.

By completing **Investigate**, you discover the thinking behind new concepts.

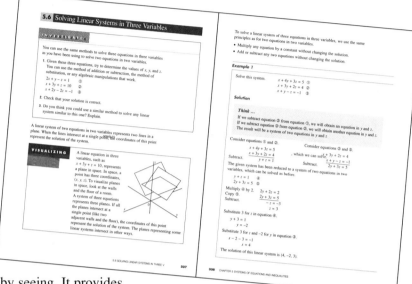

Visualizing helps you learn by seeing. It provides another way to understand and remember concepts.

Examples with full solutions provide you with a model of new methods.

Discussing the Ideas helps you clarify your understanding by talking about the preceding explanations and examples.

Communicating the Ideas helps you confirm that you understand important concepts. If you can explain it or write about it, you probably understand it!

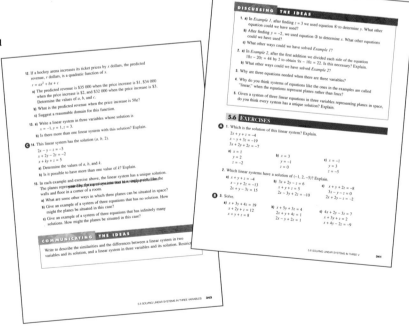

Exercises, including end-of-chapter **Review** and **Cumulative Review** exercises, help you reinforce your understanding.

There are three levels of **Exercises** in the sections in this book.

A exercises involve the simplest skills of the lesson.

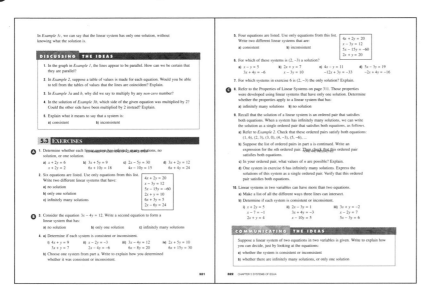

B exercises usually require several steps, and they may involve applications or problem solving.

C exercises are more thought-provoking. They may call on previous knowledge or foreshadow upcoming work.

Technology is incorporated into exercises in several ways. Special logos tell you when an exercise requires the use of technology.

This logo tells you that you need a graphing calculator to complete this exercise.

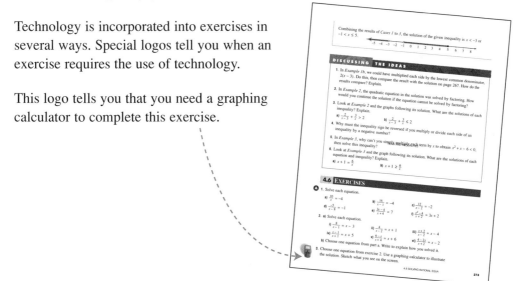

Technology

The graphing calculator and the computer are tools for learning and doing mathematics in ways that weren't possible a few years ago.

For some computer activities, you use computer applications such as spreadsheets.

This computer logo tells you that the exercise involves a spreadsheet.

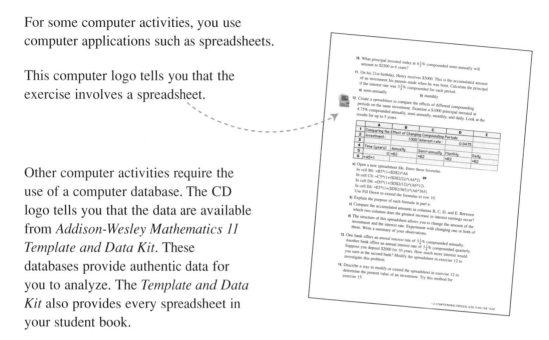

Other computer activities require the use of a computer database. The CD logo tells you that the data are available from *Addison-Wesley Mathematics 11 Template and Data Kit*. These databases provide authentic data for you to analyze. The *Template and Data Kit* also provides every spreadsheet in your student book.

You need ClarisWorks™, Microsoft Works®, or Microsoft® Office 97 to use *Addison-Wesley Mathematics 11 Template and Data Kit*.

Computer applications are also featured in **Linking Ideas: Mathematics & Technology**. Completing these activities will show you an efficient and effective use of technology.

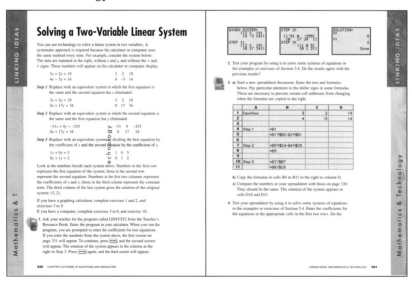

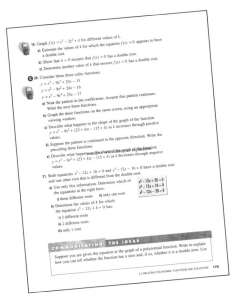

The graphing calculator is a powerful mathematical tool. It does not replace the need to develop good graphing skills, but it can enhance your mathematical understanding.

In selected Investigates and exercises, the calculator helps you explore patterns graphically, numerically, and symbolically. You develop skills to connect these different ways of looking at a situation.

Exploring with a Graphing Calculator presents a series of exercises that involve the graphing calculator and relate to a specific mathematical concept.

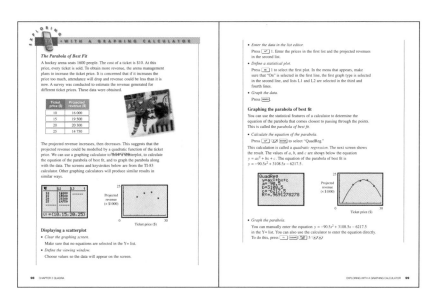

Here are some other sections you will encounter as you work through your student book.

Linking Ideas

These sections show how mathematics topics relate to other school subjects or to the world outside school.

This linking feature shows how you can use circle properties to explain eclipses.

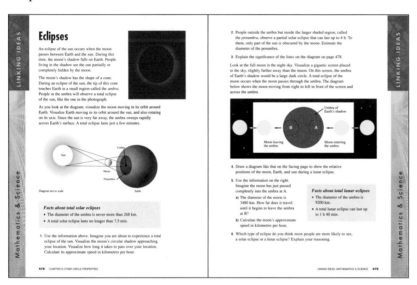

Other links

- Mathematics and Sports
- Mathematics and Art
- Mathematics and History

Problem Solving

Special **Problem Solving** sections introduce classic problems and puzzles, and their solutions. The accompanying exercises offer problems that may involve extensions to the introductory problem.

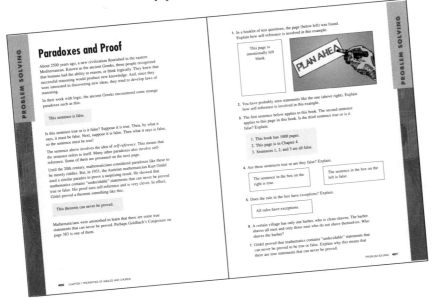

Mathematics Files

These short sections present key mathematical concepts that form part of the foundation for upcoming lessons. Selected **Mathematics Files** may present an alternative method to one already taught, or a specialized method for solving a problem.

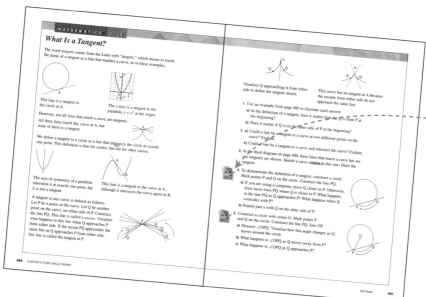

This logo tells you to use either a computer with dynamic geometry software, or geometric instruments.

Should You Buy or Lease a Car?

When you're ready to own a car, you may have to borrow money to buy the car, or you can lease the car from the dealer. Both decisions have different effects, so it's useful to think about the consequences of each.

 CONSIDER THIS SITUATION

Suppose you've saved $8800 toward a new car. You've chosen a Saturn SL1 for $18 253 plus tax. If you use your savings as a down payment to buy the car, the monthly cost over 3 years is $365.81. You will own the car at the end of 3 years.

The dealer suggests leasing as an alternative. You pay $8800 down, and your monthly cost for 3 years is $93.55 plus tax. At the end of 3 years, you must return the car, or you can buy it for $8531.

• Do you have enough information to make a decision?

• If so, would you buy or lease the car? Why?

• If not, what additional information do you need? Why?

On pages 42 and 43, you will use a mathematical model to analyze the costs involved in buying, leasing, and operating a car.

 FYI Visit www.awl.com/canada/school/connections

For information related to the above problem, click on <u>MATHLINKS</u> followed by *Mathematics 11*. Then select a topic under Should You Buy or Lease a Car?

In this chapter, you will examine some current banking and business practices, and explore how these practices can affect you, the customer. For simplicity, we shall assume that interest rates remain constant in each problem although, in practice, they change frequently.

When you deposit money in a bank account, you lend the money to the bank. In return, the bank pays you *interest*, which is money paid for the use of money.

INVESTIGATE | Simple Interest Calculations

1. Suppose you make a $100 investment that earns interest at a rate of 8% per year. How much interest would you earn in each time period?

   a) 1 year   b) 2 years   c) 3 years

   d) 5 years   e) 10 years   f) 1.5 years

   g) 6 months   h) 1 month   i) 30 days

2. Write a general method for calculating interest over any time period.

Simple interest, I dollars, can be calculated using the formula $I = Prt$, where P dollars is the principal, t is time in years, and r is the annual interest rate expressed as a decimal. In practice, simple interest usually applies over relatively short time periods of 1 year or less. In the examples and exercises in this chapter, round interest calculations to the nearest cent.

Example 1

Marsha deposits $360 on May 10 to open a daily interest savings account. The annual interest rate is 2%. Interest is calculated each day on the closing balance but credited to the account on the last day of each month. Calculate the interest Marsha earns from May 10 to June 30. Her only other transaction is a further deposit of $170 on June 5.

Solution

On May 31, Marsha will receive interest for 21 days.

The number of days is $31 - 10 = 21$

Since Marsha's daily balance is $360, the interest can be determined in a single calculation.

$I = Prt$ ← **Principal**

$= 360 \times 0.02 \times \frac{21}{365}$ ← **Annual interest rate expressed as a decimal**

$\doteq 0.41$ ← **Time in years**

On May 31, $0.41 is credited to Marsha's account. Her new balance is $360.41. This is also called the *accumulated amount*.

From June 1 to June 5, her balance is $360.41.

$I = 360.41 \times 0.02 \times \frac{5}{365}$

$\doteq 0.10$

From June 6 to June 30, her balance is $360.41 + $170.00, or $530.41.

$I = 530.41 \times 0.02 \times \frac{25}{365}$

$\doteq 0.73$

The interest earned is $0.41 + $0.10 + $0.73, or $1.24.

In *Example 1*, Marsha's bankbook will display the balance following each transaction, including the interest earned each month.

Date	Item	Withdrawal	Deposit	Balance
MAY 10-97	DEP		***360.00	***360.00
MAY 31-97	ICR		*****0.41	***360.41
JUN 5-97	DEP		***170.00	***530.41
JUN 30-97	ICR		*****0.83	***531.24

Example 2

Calculate the missing information.

	Principal, P ($)	Annual interest rate, r (%)	Time, t (days)	Simple interest, I ($)
a)	500.00	$4\frac{1}{2}$	25	
b)	38.50		149	1.65
c)	1750.00	$5\frac{3}{4}$		39.85
d)		$7\frac{1}{8}$	64	3.07

Solution

Use the formula $I = Prt$.

a) $P = 500$, $r = 0.045$, and $t = \frac{25}{365}$

$$I = 500 \times 0.045 \times \frac{25}{365}$$
$$\doteq 1.54$$

The simple interest is $1.54.

b) $I = 1.65$, $P = 38.50$, and $t = \frac{149}{365}$

$$1.65 = 38.50 \times r \times \frac{149}{365}$$
$$r = \frac{1.65 \times 365}{38.50 \times 149}$$
$$\doteq 0.105$$

The interest rate is 10.5%.

c) $I = 39.85$, $P = 1750$, and $r = 0.0575$

$$39.85 = 1750 \times 0.0575 \times t$$
$$t = \frac{39.85}{1750 \times 0.0575}$$
$$\doteq 0.396$$

The time is 0.396 years, or
0.396×365 days, which is 145 days.

d) $I = 3.07$, $r = 0.071\,25$, and $t = \frac{64}{365}$

$$3.07 = P \times 0.071\,25 \times \frac{64}{365}$$
$$P = \frac{3.07 \times 365}{0.071\,25 \times 64}$$
$$\doteq 245.73$$

The principal is about $245.73.

When a bank or a business provides money, goods, or services on the understanding that payment will be made at a later date, it often charges interest. This arrangement is called *credit*. Utility companies extend credit to their customers every month on the understanding that customers will pay their bills by the due date. If a customer fails to pay the total amount by the due date, the company charges a late payment fee on the outstanding balance.

Example 3

Tony's gas bill for December was $153.93. His payment is due January 14; Tony paid $80 on January 12. The gas company adds a 1.5% late payment charge on any overdue balance. During January, Tony used an additional $103.54 in gas. How much will Tony owe the gas company when his January bill arrives?

Solution

Since Tony did not pay the full balance, there will be a late payment charge.

The unpaid balance from December is: $153.93 − $80.00 = $73.93

The late payment charge is assessed at 1.5%: $0.015 \times \$73.93 = \$1.11$

Tony's new balance = Previous balance + Late payment charge + New charges
$$= \$73.93 + \$1.11 + \$103.54, \text{ or } \$178.58$$

Tony will owe $178.58.

1.1 EXERCISES

A 1. Some banks offer a regular savings account in which interest is credited at the end of the month. The interest calculation is based on the minimum monthly balance. A regular savings account usually has a higher interest rate than a daily interest savings account. Under what circumstances would it be better to have each account?

2. Calculate the simple interest on each investment.

a) $210.00 at 4% annually for 91 days

b) $465.00 at $5\frac{1}{2}$% annually for 150 days

c) $78.50 at $3\frac{1}{4}$% annually for 240 days

d) $1245.00 at $4\frac{3}{4}$% annually for 450 days

3. Determine the missing information.

	Principal ($)	Annual interest rate (%)	Time (days)	Simple interest ($)
a)	627.00	$4\frac{1}{2}$	27	
b)		$3\frac{1}{4}$	58	5.72
c)	265.00		120	16.50
d)	575.00	$16\frac{1}{2}$		8.06
e)		$2\frac{1}{8}$	215	126.25
f)	183.12	1.25	47	

4. Choose one part of exercise 3. Write to explain how you calculated the missing information.

5. The balance in a daily interest savings account on January 31 is $437.25. The annual interest rate is $1\frac{1}{4}$%. No deposits or withdrawals are made. Determine the accumulated amount on February 28.

B 6. On May 7, Reid opens a daily interest account with $2178.65, earning annual interest at $1\frac{3}{4}$%. Determine the accumulated amount at the end of May for each situation.

a) There are no further deposits and no withdrawals.

b) There is a further deposit of $345.18 on May 23.

7. At 5% annual simple interest, how long would it take $500 to grow to $510?

8. Three hundred fifty dollars grows to $362.76 in 5 months. What is the annual rate of simple interest?

9. Dave was charged $1.52 on an unpaid balance of $109.70 from April's telephone bill. What was the rate of the late payment charge?

10. Christina's telephone company charges a 1.5% late fee if the balance due is not paid. The company charged $10.24 for an unpaid balance. What was the balance on which this charge was based?

11. Select one of exercise 8, 9, or 10. Write to explain how you completed the exercise.

12. Determine each accumulated amount. Use the formula following *Example 5*.
    a) $500 earns 5% annual interest for 1 year.
    b) $1500 earns 3.5% annual interest for 3 years.
    c) $2450 earns 4.5% annual interest for 2 years.
    d) $3675 earns 5.5% annual interest for 6 months.
    e) $2585 earns 4% annual interest for 31 days.

13. This table summarizes Marcel's charge account at a department store. Monthly credit charges are $2\frac{1}{4}\%$ of the unpaid balance.

Month	Previous balance ($)	Payment ($)	Unpaid balance ($)	Credit charge ($)	New purchases ($)	New balance ($)
January	782.50	250.00			531.87	
February		300.00			425.76	
March		300.00	919.58		723.05	
April		500.00			673.12	
May		400.00			358.29	
June		375.00		33.27	206.38	

Total credit charge [　]

a) Copy and complete the table.

b) Is Marcel using this charge account wisely? Explain.

c) Suppose Marcel stopped all purchases on this account and paid $300 each month. Predict whether he could reduce his balance to zero by the end of the year. Explain your answer.

d) Create a spreadsheet to determine an exact answer to part c. A possible spreadsheet is started below. You may need to format cells B4 through G8 to display currency.

	A	B	C	D	E	F	G
1	Continuing Marcel's Account Record, Systematic Payment						
2	Month	Prev. bal.	Payment	Unpaid balance	Credit charge	Purchases	New balance
3	July	$1718.46	$300.00	=B3-C3	=D3*0.0225	$0.00	=D3+E3
4	August	=G3	$300.00	=B4-C4	=D4*0.0225	$0.00	=D4+E4

e) What will Marcel's total credit charge be for the year?

f) Use your spreadsheet from part d. Determine the monthly amount Marcel should pay to ensure the account is paid in full by the end of the year.

Many credit card companies follow this policy: if the amount due is not paid in full, interest is charged on each purchase from the day the purchase is processed (posting date) to the due date on the statement. In the next statement, interest is charged on the amount due from the previous statement. Interest is not charged on new purchases if the bill is paid in full by the due date. Use this information to complete exercises 14 and 15.

14. Cal wants to buy a stereo system. It costs $1708.86, including taxes. He expects a large tax refund in April, but he wants to buy the system now. Cal purchases the stereo on March 14, using his credit card. This is his only purchase during this billing period. The transaction is processed on March 16. Cal receives his credit card statement on April 15 and has until April 30 to pay the amount due. On April 30, Cal makes a payment of $1500.

a) The annual interest rate is 18%. Calculate the interest on Cal's purchase.

b) Calculate the unpaid amount.

c) Cal wants to pay off the amount due on May 31. He has not made any additional purchases with the credit card. How much will he have to pay?

d) Based on Cal's experience, what advice would you give someone purchasing an item with a credit card?

15. An executive used one credit card for all her expenses as she travelled around the world on business. Her balance was often over $30 000 a month. One month, unable to access her credit card statement, she estimated her bill to be $35 000 and paid this amount. She later found she owed $41 000. The credit card company charged an annual interest rate of 12%. Assume the executive's purchases were made 30 days before the payment date.

a) Calculate the interest charge.

b) Explain why the interest charge on the next statement might differ from the answer you calculated in part a.

C 16. A payment of $650 is due in 10 months. What principal should be invested now at $5\frac{1}{2}\%$ annual interest to meet the payment?

COMMUNICATING THE IDEAS

Banks compete for your money. They offer many different types of accounts. Investigate two accounts at a bank. Create some numerical examples, or a list of pros and cons, to determine which account would be better for you.

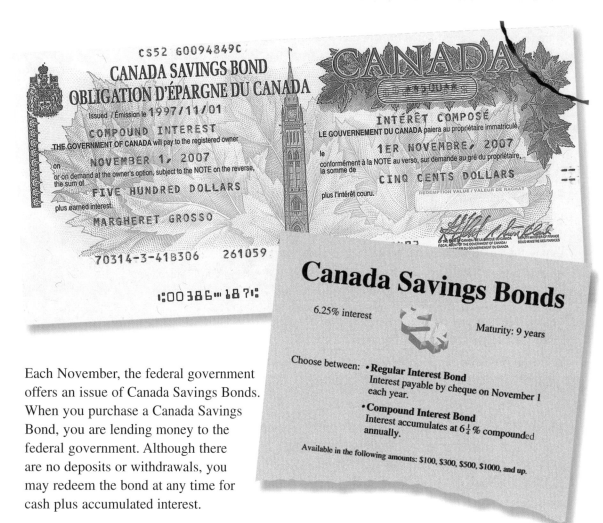

Each November, the federal government offers an issue of Canada Savings Bonds. When you purchase a Canada Savings Bond, you are lending money to the federal government. Although there are no deposits or withdrawals, you may redeem the bond at any time for cash plus accumulated interest.

A savings bond usually has a higher interest rate than a savings account. Unlike a savings account, the interest rate is guaranteed never to be lower than the interest rate at the time of issue; it may even increase if rates rise generally.

There are two types of Canada Savings Bonds—Regular Interest Bonds and Compound Interest Bonds. When you purchase Regular Interest Bonds, you receive annual interest payments. The value of the bond is the initial value plus accumulated interest. When you purchase a Compound Interest Bond, the interest you earn each year is added to the bond. The following year this interest also earns interest. Interest calculated in this manner is *compound interest*.

Graphing Simple and Compound Interest

1. Suppose you purchase a $500 Regular Interest Bond at the annual interest rate of 6.25%. Copy and complete this table.

Year	0	1	2	3	4	5	6	7	8	9
Value ($)	500.00	531.25	562.50							

2. Graph the data from exercise 1. Plot Year on the horizontal axis.

3. Describe how the value of a Regular Interest Bond changes with time.

4. Suppose you repeated exercises 1 and 2 for a higher annual rate of interest. Visualize how the graph would change. Describe the graph.

5. Suppose you purchase a $500 Compound Interest Bond at 6.25%. Copy and complete the table. Remember that the interest earned each year is added to the principal before the next interest calculation.

Year	0	1	2	3	4	5	6	7	8	9
Value ($)	500.00	531.25	564.45							

6. Graph the data from exercise 5. Use the same grid as for exercise 2.

7. Suppose you repeated exercises 5 and 6 for a different annual rate of interest. Visualize how the graph would change for higher interest rates. Visualize how it would change for lower interest rates. Describe each graph.

8. Describe how the value of a Compound Interest Bond changes with time. Write an equation that expresses the value of the bond in dollars as a function of the time in years.

Consider a $500 Compound Interest Bond earning 6.25% interest. Since the interest each year is added to the principal, we can determine the new value of the bond in a single calculation.

1st year: Value of bond on November 1: $\$500.00 \times 1.0625 = \$531.25$
2nd year: Value of bond on November 1: $\$531.25 \times 1.0625 = \$564.45$
3rd year: Value of bond on November 1: $\$564.45 \times 1.0625 = \$599.73$

We look for a pattern to derive a general expression.

Value, in dollars, at end of:

1st year: 500×1.0625
2nd year: $500 \times 1.0625 \times 1.0625$, or $500(1.0625)^2$
3rd year: $500(1.0625)^2 \times 1.0625$, or $500(1.0625)^3$
nth year: $500(1.0625)^n$

The Compound Interest Bond matures in 9 years. After that time, no further interest is paid and the bond should be redeemed.

The *maturity value* will be: $\$500(1.0625)^9 = \$862.84$

This graph shows how the value of the Compound Interest Bond grows over 9 years. Also shown is the growth of the investment in the Regular Interest Bond, assuming that the interest payments are not reinvested. The pattern of the values of the Compound Interest Bond for several years can be generalized.

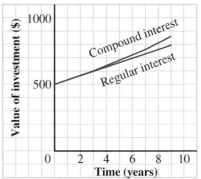

Growth of Canada Savings Bonds

When an amount of money, P (the principal), is invested at an interest rate, i, compounded annually, the accumulated amount, A, after n years is given by the formula $A = P(1 + i)^n$. This is the *compound interest formula*.

Example 1

Determine the accumulated amount of $5000 invested at 5.25% compounded annually for 10 years.

Solution

Use the formula $A = P(1 + i)^n$.
Substitute $P = 5000$, $i = 0.0525$, and $n = 10$.

$A = 5000(1.0525)^{10}$
$\doteq 8340.48$

The accumulated amount is $8340.48.

In some situations, you may want to determine how much you need to invest today to obtain a specific amount in the future. This starting principal is the *present value* of an investment.

The present value may be found by solving the formula $A = P(1 + i)^n$ for P:

$$P = \frac{A}{(1 + i)^n}$$

Example 2

What is the present value of an investment that will yield $3000 after 5 years at 4.5% compounded annually?

Solution

Use the formula $P = \frac{A}{(1 + i)^n}$.

In this case, $A = 3000$, $i = 0.045$, and $n = 5$.

$$P = \frac{3000}{(1.045)^5}$$
$$\doteq 2407.35$$

The present value is $2407.35.

VISUALIZING

To remember the formulas for present value and accumulated amount, picture a time diagram representing the growth of $100 earning compound interest at an annual interest rate i.

Time in years

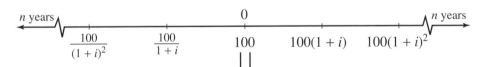

Present value

- Move to the left.

- Divide by $(1 + i)$ for each year.

- The present value of $100 due in n years is $\frac{100}{(1 + i)^n}$.

Accumulated amount

- Move to the right.

- Multiply by $(1 + i)$ for each year.

- The accumulated amount of $100 after n years is $\$100(1 + i)^n$.

We use the formula for accumulated amount to determine the interest rate at which an investment grows to a certain amount in a given number of years.

Example 3

A $200 investment grows to $250 in 4 years, with interest compounded annually. What is the interest rate?

Solution

Let i represent the interest rate.

Use the formula $A = P(1 + i)^n$.

Substitute $A = 250$, $P = 200$, and $n = 4$.

$250 = 200(1 + i)^4$

$(1 + i)^4 = 1.25$ ————— Take the fourth root of each side.

$1 + i = 1.25^{\frac{1}{4}}$

$1 + i \doteq 1.057\ 371$

$i \doteq 0.057\ 371$

5.74% is very close to $5\frac{3}{4}\%$.

$200 will grow to $250 in 4 years if invested at about $5\frac{3}{4}\%$ compounded annually.

For a given interest rate, we can determine the time for an investment to grow to a certain amount.

Example 4

A student has $800 to invest at 5.5% compounded annually. How long will it take for the investment to grow to $1200?

Solution

Use the formula $A = P(1 + i)^n$.
Substitute $A = 1200$, $P = 800$, and $i = 0.055$.

$1200 = 800(1.055)^n$

$1.055^n = 1.5$

The investment grows to 1.5 times its original amount.

Use a graphing calculator to estimate the value of n for which $1.055^n = 1.5$. Enter the function $y = 1.055^x$. Display the corresponding table of values. Scroll down until you see a value of y that is close to 1.5.

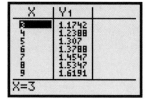

The screen display shows that it takes about 8 years for the investment to grow to 1.5 times its original amount.

1. List some advantages and disadvantages of Canada Savings Bonds as forms of investment.

2. What other equation could you use to solve *Example 4* with a graphing calculator? Which function do you think displays the solution more clearly? Explain.

1.2 EXERCISES

A 1. Make a table of values to show the growth of each investment.

   a) a $500 bond invested at $5\frac{1}{4}$% compounded annually for 5 years

   b) a $300 bond invested at $4\frac{3}{4}$% compounded annually for 6 years

   c) a $1000 bond invested at $5\frac{1}{2}$% compounded annually for 4 years

2. Determine each accumulated amount.

   a) $1000 invested for 6 years at 7% compounded annually

   b) $500 invested for 20 years at $5\frac{1}{2}$% compounded annually

   c) $215 invested for 3 years at $6\frac{3}{4}$% compounded annually

   d) $720 invested for 8 years at $7\frac{1}{2}$% compounded annually

3. Determine the value of a $1000 Canada Savings Compound Interest Bond at $8\frac{1}{4}$% after each time.

   a) 3 years   **b)** 5 years   **c)** 8 years

4. The Canada Savings Compound Interest Bonds issued one year earned $8\frac{1}{4}$% interest compounded annually. They matured in 7 years. Determine the maturity value of each bond.

   a) a $100 bond   **b)** a $500 bond   **c)** a $2500 bond

B 5. A debenture is a certificate entitling its owner to receive both the interest and the principal after a specified time. Use the information in the advertisement. Determine the maturity value of each debenture.

   a) a $5000 one-year debenture

   b) a $2000 three-year debenture

   c) a $10 000 five-year debenture

   d) a $7500 four-year debenture

Debentures
Interest Paid or Compounded Annually

5 years	$5\frac{3}{4}$%
2–4 years	$4\frac{1}{2}$%
1 year	4%

Shoppers
Mortgage and Loan Corporation

6. Select one part of exercise 5. Write to explain how you determined the maturity value.

7. Use the information in the advertisement for exercise 5. How much would you have to invest today to earn each amount?

a) $3980 in 5 years   **b)** $2850 in 3 years

c) $10 400 in 1 year   **d)** $5460 in 2 years

8. Select one part of exercise 7. Write to explain how you determined the present value.

9. A donor gave $75 000 to a town council. The money was to be invested for 10 years, and the accumulated amount used to expand the public library. The money earned 10% interest compounded annually. What amount was available to spend on the library?

10. Suppose $100 is invested for 25 years with interest compounded annually.

a) Copy and complete this table. You could use a spreadsheet or the List feature on a graphing calculator.

Years	Accumulated amount for 5% interest ($)	Accumulated amount for 10% interest ($)
0	100.00	100.00
5		
10		
15		
20		
25		

b) Plot the data for 5% and 10% interest rates on the same grid.

c) About how many years does it take $100 to grow to $300 at each rate?

   i) 5%   **ii)** 10%

d) When the interest rate on an investment is doubled, does the accumulated amount double? Explain.

11. An investor deposits $500 at an interest rate of 7% compounded annually. In about how many years will the investment grow to $1000?

12. What interest rate, compounded annually, will yield each accumulated amount?

a) $350 accumulates to $637.15 in 6 years.

b) $3500 accumulates to $6195.50 in 7 years.

13. Interest is compounded annually. How much does each investment earn?

    a) \$462.50 invested for 3 years at $5\frac{1}{2}\%$

    b) \$1500 invested for 8 years at $4\frac{1}{4}\%$

    c) \$600 invested for 5 years at $3\frac{3}{4}\%$

    d) \$1435 invested for 11 years at $6\frac{7}{8}\%$

14. About how much would you have to invest today at 6% compounded annually to have an accumulated amount of \$5000 in 5 years?

15. This table shows the estimated world population over the past 300 years.

    a) Plot the data, placing Year on the horizontal axis.

    b) From the graph, estimate the year when the world population reached 4 billion.

    c) Use the graph to predict the world's present population.

Year	World population (millions)
1650	500
1850	1100
1930	2000
1950	2500
1970	3600
1990	5730

    d) Compare this graph with a graph of the growth of a Compound Interest Bond. Describe similarities and differences between the graphs, and explain why you think they occur.

16. The 50¢ Bluenose is one of Canada's most famous postage stamps. In 1930, it could be bought at the post office for 50¢. In 1987, a superb copy was sold at an auction for \$500. What rate of interest, compounded annually, corresponds to an investment of 50¢ in 1930 that grows to \$500 in 1987?

Stamp reproduced courtesy of Canada Post Corporation

17. A parent wants to invest money to accumulate to \$8000 in 4 years when her son starts university. Assume the interest is compounded annually. What is the present value for each interest rate?

    a) 3%   **b)** 4%   **c)** 5%   **d)** 6%   **e)** 7%

18. At what interest rate compounded annually would a sum of money double in 7 years?

19. About how many years will it take a sum of money to double when invested at 12% compounded annually?

20. This graph shows the number of years required for money to double when invested at different interest rates.

a) About how long does it take money to double when invested at each rate?

i) 10%   ii) 15%

b) What interest rate compounded annually would double an investment for each time?

i) 6 years   ii) 10 years

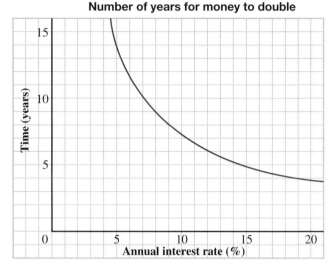

Number of years for money to double

Time (years) / Annual interest rate (%)

 21. Create a spreadsheet to show the growth of a $100 investment, compounded annually, at different interest rates. Use the spreadsheet to confirm the information on the graph in exercise 20. A possible spreadsheet is started below.

	A	B	C	D
1	Growing $100 with Compound Interest (compounded annually)			
2	Years	0.04	=B2+0.01	=C2+0.01
3	1	=100*(1+B$2)	=100*(1+C$2)	=100*(1+D$2)
4	=A3+1	=B3*(1+B$2)	=C3*(1+C$2)	=D3*(1+D$2)
5				
6				

Enter the first 4 rows. Fill Down from row 4 to display year 3 and beyond. Fill Right from column D to display interest rates greater than 6%.

22. A $6 billion issue of Canada Savings Regular Interest Bonds earns interest at $6\frac{3}{4}$%. To remain competitive with other investments, the government increases the interest rate to $7\frac{1}{2}$% on March 1. How much additional interest will the government have to pay that year?

COMMUNICATING THE IDEAS

Write to explain the difference between simple interest and compound interest.

1.3 Compounding Periods Less than One Year

Semi-Annual and Monthly Compounding Periods

Interest is often charged or credited more than once a year. The two most common compounding periods are semi-annual and monthly.

1. A regular savings account pays interest at 2% per annum that is credited to the account every 6 months on April 30 and October 31. Suppose there is $1000 in the account on November 1. No further withdrawals or deposits are made over the next 12 months.

   a) What does 2% per annum mean? What percent interest is credited on April 30 and on October 31?

   b) What is the interest earned on April 30? What is the balance or accumulated amount?

   c) What is the interest earned on October 31? What is the balance?

2. In exercise 1, the principal of $1000 earned 2% interest, compounded semi-annually. Explain why the following expression could have been used to determine the balance on October 31: $1000(1.01)^2$

3. Suppose the savings account in exercise 1 paid interest monthly instead of semi-annually. A balance of $1000 on May 1 is earning an annual interest of 2%, compounded monthly. There are no other transactions to the account from May 1 to April 30. Explain why the following expression can be used to calculate the balance on April 30:

   $$A = 1000\left(1 + \frac{0.02}{12}\right)^{12}$$

4. Evaluate the expression in exercise 3. Compare your results for this calculation and exercise 1c. Why does monthly compounding yield a greater accumulated amount?

Compounding periods of less than 1 year are common in banking and investing. Here are some terms you should know.

Term Deposit: Available from banks and trust companies, a term deposit offers a higher interest rate than a savings account because it involves a larger amount than a typical personal account. The rate of interest is guaranteed for a fixed term or period of investment. Customers who withdraw money before the end of the term forfeit all or part of the interest, depending on the terms.

Guaranteed Investment Certificate (GIC): Like a term deposit, a GIC offers a higher interest rate than a personal savings account. It requires a minimum deposit, typically thousands of dollars. It cannot be redeemed before its maturity date. A GIC is available from banks, trust companies, and other financial institutions such as life insurance companies.

Registered Retirement Savings Plan (RRSP): The major advantage of an RRSP is the tax break it offers. You do not pay income tax on any income contributed to an RRSP; nor do you pay tax on the interest it earns until you withdraw RRSP funds. An RRSP is not a single type of investment. Banks and trust companies offer RRSP plans. Some people set up personalized RRSPs that might hold savings, term deposits, GICs, mutual funds, bonds, stocks, or any combination of investments.

Mortgages: Most people who buy a home, store, or office require a mortgage, (a long-term loan). A mortgage can be complex because banks and trust companies compete by offering customers many options. You will study mortgages in Section 1.9.

For a compounding period of less than 1 year, we adapt the compound interest formula $A = P(1 + i)^n$. When interest is compounded semi-annually, there are two interest calculations per year, so n is double the number of years. The interest rate for each half-year period is half the annual interest rate.

Example 1

Élise put $2000 into an RRSP earning $9\frac{1}{2}\%$ interest compounded semi-annually. Determine the accumulated amount after 7 years.

Solution

Use the formula $A = P(1 + i)^n$.
Substitute $P = 2000$, $i = \frac{0.095}{2}$, $n = 7 \times 2$.

$$A = 2000\left(1 + \frac{0.095}{2}\right)^{7 \times 2}$$
$$= 2000(1.0475)^{14}$$
$$\doteq 3829.89$$

The accumulated amount after 7 years is $3829.89.

When interest is compounded monthly, there are 12 interest calculations each year. The interest rate for each compounding period is $\frac{1}{12}$ the annual interest rate. In the formula $A = P(1 + i)^n$, n is the number of months, and i is $\frac{1}{12}$ the annual interest rate.

Example 2

Pat is the treasurer for her community day care centre. Because the day care centre starts the school year with a surplus, Pat puts $5000 into a short-term deposit. She obtains a nine-month term deposit at 6% interest compounded monthly. The interest earned each month remains in the term deposit. What is the maturity value of the term deposit?

Solution

Use the formula $A = P(1 + i)^n$.
Substitute $P = 5000$, $i = \frac{0.06}{12}$, and $n = 9$.

$$A = 5000\left(1 + \frac{0.06}{12}\right)^9$$
$$A \doteq 5229.55$$

The maturity value of the term deposit is $5229.55.

Accumulated amounts and present values can be found for any compounding periods, including weekly, daily, or even hourly. These periods are rarely used in business transactions.

For the general compound interest formula $A = P(1 + i)^n$:

A is the accumulated amount, in dollars.
P is the principal, in dollars.
i is the interest rate per compounding period.
n is the number of compounding periods.

DISCUSSING THE IDEAS

1. Instead of using one-half of a year for time when calculating semi-annual interest, banks use fractions corresponding to the actual number of days in each interest period. How does this affect the calculations? Why do you think banks do this?

2. All the examples and exercises in this section involve savings and investments. Discuss ways in which loans are different from savings.

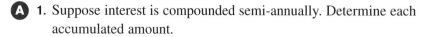

1.3 EXERCISES

A 1. Suppose interest is compounded semi-annually. Determine each accumulated amount.

   a) $480 for 9 years at 4.5%

   b) $100 for 10 years at 5.25%

   c) $260 for 4 years at $3\frac{1}{2}\%$

   d) $1200 for 15 years at $5\frac{1}{4}\%$

2. Suppose interest is compounded monthly. Determine each accumulated amount.

   a) $150 for 3 years at $4\frac{3}{4}\%$

   b) $800 for 5 years at 6%

   c) $325 for 2 years at $3\frac{1}{2}\%$

   d) $2740 for 7 years at $5\frac{1}{2}\%$

3. Determine the interest on $1000 after 1 year at 4.5% compounded for each period.

   a) semi-annually

   b) monthly

4. Calculate the accumulated amount of $250 invested for 3 years at 5% compounded for each period.

   a) monthly

   b) semi-annually

5. Select one part of exercise 3 or 4. Write to explain how you calculated the answer.

6. Determine the principal that should be deposited today to accumulate to $1000 in 3 years at 4.5% compounded for each period.

   a) semi-annually

   b) monthly

B 7. Determine the interest rate necessary for $150 to accumulate to $275 in 8 years with interest compounded semi-annually.

8. On May 1, 1996, Marcie deposited $500 in a term deposit paying $4\frac{3}{4}\%$ compounded semi-annually. On May 1, 1997, she deposited another $500. There are no other deposits and no withdrawals. Determine Marcie's accumulated amount on May 1, 1998.

9. This table shows the growth of a $6000 RRSP.

   a) Plot the data.

   b) Describe the growth of this RRSP investment.

   c) Estimate the time required for the RRSP to accumulate to $15 000.

   d) Estimate the accumulated amount after 7 years.

Time (years)	Accumulated amount ($)
0	6000.00
1	6375.00
2	6773.44
3	7196.78
4	7646.58
5	8124.49

10. What principal invested today at $6\frac{1}{2}\%$ compounded semi-annually will amount to $2500 in 6 years?

11. On his 21st birthday, Henry receives $5000. This is the accumulated amount of an investment his parents made when he was born. Calculate the principal if the interest rate was $5\frac{3}{4}\%$ compounded for each period.

a) semi-annually   **b)** monthly

12. Create a spreadsheet to compare the effects of different compounding periods on the same investment. Examine a $1000 principal invested at 4.75% compounded annually, semi-annually, monthly, and daily. Look at the results for up to 5 years.

	A	B	C	D	E
1	Comparing the Effect of Changing Compounding Periods				
2	Investment:	1000	Interest rate:	0.0475	
3					
4	Time (years)	Annually	Semi-annually	Monthly	Daily
5	0	=B2	=B2	=B2	=B2
6	=A5+1				

a) Open a new spreadsheet file. Enter these formulas.
In cell B6: =$B$5*(1+$D$2)^A6
In cell C6: =$C$5*(1+($D$2/2))^(A6*2)
In cell D6: =$D$5*(1+($D$2/12))^(A6*12)
In cell E6: =$E$5*(1+($D$2/365))^(A6*365)
Use Fill Down to extend the formulas to row 10.

b) Explain the purpose of each formula in part a.

c) Compare the accumulated amounts in columns B, C, D, and E. Between which two columns does the greatest increase in interest earnings occur?

d) The structure of this spreadsheet allows you to change the amount of the investment and the interest rate. Experiment with changing one or both of these. Write a summary of your observations.

13. One bank offers an annual interest rate of $5\frac{1}{4}\%$ compounded annually. Another bank offers an annual interest rate of $5\frac{1}{4}\%$ compounded quarterly. Suppose you deposit $2000 for 10 years. How much more interest would you earn at the second bank? Modify the spreadsheet in exercise 12 to investigate this problem.

14. Describe a way to modify or extend the spreadsheet in exercise 12 to determine the present value of an investment. Try this method for exercise 15.

15. Determine the present value that will yield the accumulated amount for each GIC investment.

   a) $800 in 4 years at $3\frac{1}{2}\%$ compounded monthly

   b) $750 in 6 years at 5% compounded semi-annually

   c) $1260 in 9 years at $4\frac{1}{2}\%$ compounded monthly

16. Select one part of exercise 15. Write to explain how you determined the present value.

17. Determine, to the nearest half year, how long it will take $100 to accumulate to $500 at $6\frac{1}{2}\%$ compounded semi-annually.

18. Determine, to the nearest month, how long it will take $300 to accumulate to $1000 at 5% compounded monthly.

19. Shabir makes regular contributions to his RRSP, starting on January 1. Which investment schedule would you recommend and why?

   a) On the first day of each month, invest $100 at 8.5% interest compounded monthly.

   b) On January 1 and June 1, invest $600 at 8.5% interest compounded semi-annually.

20. Determine how long it takes money to double in each situation.

   a) 7% interest compounded semi-annually

   b) $5\frac{1}{4}\%$ interest compounded semi-annually

   c) $8\frac{1}{2}\%$ interest compounded monthly

MODELLING Saving Situations

The preceding exercises each give a single rate of interest. In practice, interest rates can change daily. Also, banks offer higher interest rates when customers commit their money for longer time periods or accept fewer compounding periods.

• Select one of exercise 11, 12, 15, 19, or 20. Explain how the information above might affect the situation.

• Some banks guarantee a minimum interest rate on some investments but increase that rate if rates in general rise. How might this affect your analysis above?

21. Select one part of exercise 20. Write to explain how you determined the doubling time.

22. Banks use computers to calculate daily interest. They could use the same technology to calculate interest for compounding periods of 1 hour or even 1 second. Use a graphing calculator to investigate whether there would be any advantage to this. Enter these formulas, which calculate interest for 1 day, 1 hour, and 1 second, for $100 invested at 6% interest.

Y1 = 100(1+.06/365)^(365X)
Y2 = 100(1+.06/365/24)^(365*24X)
Y3 = 100(1+.06/365/24/3600)^(365*24*3600X)

Select values of X that are multiples of 5. Display the tables of values corresponding to these functions.

a) Compare the accumulated amounts after 35 years when interest is calculated by the day, the hour, and the second.

b) How do the accumulated amounts in part a compare with the accumulated amount for $100 invested at 6% compounded monthly?

c) Suggest reasons why financial institutions rarely offer interest compounded hourly or by the minute.

C 23. What equal deposits, one made now and another made 6 months later, will accumulate to $1000 in 1 year at $7\frac{1}{2}\%$ compounded semi-annually?

24. A merchant buys $3000 worth of goods from a supplier. She pays a certain amount down and agrees to pay $1000 in 6 months, and another $1000 after a further 6 months. The supplier charges $9\frac{1}{2}\%$ interest compounded semi-annually. Determine the down payment.

COMMUNICATING THE IDEAS

Create an example to show the effect of compounding interest more frequently than once a year. Write to explain the significance of the results.

Investigating Financial Calculations on the TI-83

Some calculators can carry out financial calculations. The TI-83 is one of these. It has a feature called the TVM (Time Value of Money) Solver.

To access the TVM Solver, press [2nd] [x^{-1}], then [ENTER]. This screen appears. The variables represent the following quantities:

N: total number of payment (compounding) periods
I%: annual interest rate as a percent
PV: present value
PMT: payment each period
FV: future value, or accumulated amount
P/Y: number of payments per year
C/Y: number of compounding periods per year

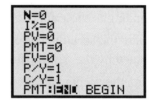

The calculator displays either positive or negative quantities for PV, PMT, and FV. A positive value indicates the amount is earned. A negative value indicates the amount is invested.

When you input four of the first five quantities, the calculator can provide the fifth. Use the calculator to solve this problem.
You invest $5000 at 7.25% interest compounded annually for 10 years. What is the accumulated amount of the investment?

Input the known quantities using the arrow and enter keys. In the case of an investment, the payment each period is 0. Check that both P/Y and C/Y have values of 1.

To find the future value, move the cursor to the FV line and press [ALPHA] [ENTER].

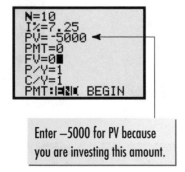

Enter −5000 for PV because you are investing this amount.

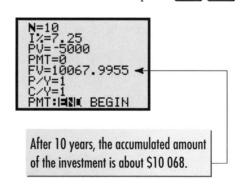

After 10 years, the accumulated amount of the investment is about $10 068.

Use the TVM Solver to solve these problems.

1. How much money would you have to invest today, at 6.5% compounded semi-annually, for the investment to accumulate to $7500 in 3 years?

2. Marge wants to buy a $14 700 car (price includes tax). She plans to finance the car over a four-year period. She estimates that she can afford monthly car payments of $350.

   a) What interest rate allows her to purchase this car?

   b) Is this realistic? Explain. Discuss how Marge can modify her plans so that she can afford the car.

3. How long will it take $400 to accumulate to $1000 if it is invested at 6.75% compounded semi-annually?

4. Jolene puts $100 a month into an account that pays 6% interest compounded monthly. This type of investment is an *annuity*.

   a) Investigate the growth of Jolene's annuity. Copy and complete this table.

Investment period (years)	1	5	10	15	20	25	30	35	40
Accumulated amount ($)									

   b) Graph these data with Investment period on the horizontal axis.

   c) Copy and complete this table.

Investment period (years)	1	5	10	15	20	25	30	35	40
Amount invested ($)									

   d) Plot these data on the graph for part b.

   e) Compare the two sets of data on the graph. Explain why it is advantageous to begin investing early.

5. A company plans to replace its computers in 3 years. It estimates it will cost $50 000. To prepare for the expense, the company invests monthly in an account that pays 4.5% interest compounded monthly. What monthly payment should the company make to ensure it has $50 000 in 3 years?

INVESTIGATE **Which Is the Greater Rate?**

Guaranteed Investment Certificates (GICs) often offer two different rates for the same period of investment. For example, the rates for a two-year certificate might be:

7% compounded semi-annually $7\frac{1}{4}$% compounded annually

Compare these two options by considering the interest on $100 for one year.

1. **a)** What is the accumulated amount on an investment of $100 after one year at 7% compounded semi-annually?

   b) Repeat part a for an interest rate of $7\frac{1}{4}$% compounded annually.

   c) Which GIC offers the greater return?

2. The difference between the two returns may seem insignificant for a $100 investment. However, for corporations, investors, or governments dealing with large amounts of money, compounding periods and rates are very important.

   a) Determine the interest on $1 000 000 invested at 7% compounded semi-annually.

   b) Determine the interest on $1 000 000 invested at $7\frac{1}{4}$% compounded annually.

   c) Which investment (part a or b) earns more? How much more?

   d) Suppose the $1 000 000 is invested at 7% compounded monthly. What effect does this have on your answer to part c?

3. Why do you think banks and other institutions offer investments at different interest rates? How can customers be sure they are selecting the best option? Explain.

To compare interest rates with different compounding periods, determine the accumulated amount of the same principal after one year at each rate.

Example 1

Which is the better rate, 6% compounded semi-annually or 6.09% compounded annually?

Solution

Assume that $100 is invested at each rate.

The accumulated amounts at the end of one year are:

$$\$100(1.03)^2 = \$106.09$$
$$\$100(1.0609) = \$106.09$$

The two rates yield equal amounts.

Example 1 shows that two rates of interest with different compounding periods can have the same effect on a sum of money. Such rates are *equivalent rates*.

Although interest rates are often compounded semi-annually or monthly, it is customary to call them annual rates. A rate may be quoted as 6% compounded semi-annually but, as *Example 1* shows, this is equivalent to 6.09% compounded annually. This second rate is called the *effective annual interest rate* because it is the rate that, with annual compounding, has the same effect as 6% compounded semi-annually.

Example 2

An investment pays interest at 9% compounded semi-annually. What is the effective annual interest rate?

Solution

The accumulated amount of $1 in one year at 9% compounded semi-annually is:

$$1\left(1 + \tfrac{0.09}{2}\right)^2 = (1.045)^2$$
$$= 1.092\ 025$$

In one year, $1 would grow to $1.092 025. Subtract the principal of $1. The difference is the interest earned on $1. This is the interest rate expressed as a decimal. Therefore, the effective annual interest rate is 9.2025%.

Example 3

A department store charges 1.5% interest per month on overdue accounts.

a) What is the annual rate of interest the store charges?

b) Assume that any interest charged is added to the next month's balance. Determine the effective annual interest rate.

Solution

a) Since $1.5\% \times 12 = 18\%$, the annual rate of interest is 18%.

b) The accumulated amount of $1 in one year at 1.5% compounded monthly is:
$1(1.015)^{12} \doteq 1.1956$

The effective annual interest rate is $1.1956 - 1 = 0.1956$, or about 19.6%.

 MODELLING **Department Store Interest Rates**

For the situation in *Example 3* to apply to a customer, the customer would maintain an account for a whole year without making any purchases or paying off the balance. This may not seem realistic, but the assumption is necessary to calculate the store's effective annual interest rate.

• Although customers seldom leave their accounts inactive for one year, is it correct to say that the effective annual interest rate is 19.6%? Explain.

• What arguments might the store management present to justify its high effective annual interest rate?

Banks and companies offer credit under different schemes.
This *Investigate* explores one example.

INVESTIGATE   Selecting the Best Borrowing Scheme

A family wants to buy a new computer that costs $2507 (including tax). It cannot pay the full amount but wants to buy the computer today and pay for it over 2 years. Here are two options.

1. The store dealer offers to sell the computer if the family pays $500 today and 24 monthly installments of $98.

a) Determine the total amount the family pays if it accepts the dealer's offer.

b) How much more than $2507 is this offer? This is the *finance charge* associated with this payment scheme.

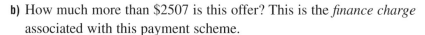

2. A bank lends $2007 at 10% per annum, compounded monthly, with monthly payments. How much will the family pay in total?

a) Construct a spreadsheet to investigate this situation. The monthly payment is not known. Start with $98 to compare the bank loan with the dealer's offer in exercise 1.

	A	B	C	D	E	F
1	Computer loan					
2		Interest rate:		0.1		
3		Principal:		$2007.00		
4		Monthly payment:		$98.00		
5		Amount			Less	Revised
6	Month	owing	Interest	Payment	interest	amount owing
7	1	=$D$3	=B7*($D$2/12)	=$D$4	=D7-C7	=B7-E7
8	=A7+1	=F7	=B8*($D$2/12)	=$D$4	=D8-C8	=B8-E8

Use Fill Down to extend the spreadsheet from row 8, to display the results for 24 months. Format cells D3 and D4 and cells B7 through F30 to display currency.

b) Look at the formulas across row 7. Describe the purpose of each formula. Why is the interest subtracted from the monthly payment before the revised amount owing is calculated?

c) The $98 payment in cell D4 is the payment required by the dealer in exercise 1. Change the number in cell D4 to determine, within $1, the required bank payment on this loan. If necessary, adjust the last payment to reduce the amount owing to $0.00.

d) Add an appropriate cell to calculate the total amount paid on the loan.

e) What is the finance charge on the bank loan? How does this compare to the finance charge in exercise 1?

3. The spreadsheet in exercise 2 represents any two-year loan. Use the spreadsheet to determine the effective annual interest rate offered by the dealer in exercise 1. Set the monthly payments at $98. Leave the loan value at $2007. Change the interest rate in cell D2 until the spreadsheet reflects that the family has paid off the loan in full, within $1. What effective annual interest rate did the dealer offer?

4. What advice would you give to the family making this purchase?

1. Suggest why a bank might offer a slightly lower interest rate on savings accounts with more frequent compounding periods.

2. What advantage is there in having a savings account in which interest is compounded monthly rather than semi-annually?

1.4 EXERCISES

A 1. Show that the rates in each pair are equivalent.

   a) 10% compounded semi-annually

   10.25% compounded annually

   b) 13% compounded semi-annually

   13.4225% compounded annually

   c) 12% compounded semi-annually

   12.36% compounded annually

   d) 9% compounded semi-annually

   9.2025% compounded annually

2. What is the accumulated amount of $100 after one year at each rate? Which is the greater rate?

   a) i) 9% compounded semi-annually   **ii)** $9\frac{1}{4}$% compounded annually

   b) i) $2\frac{1}{4}$% compounded semi-annually   **ii)** $2\frac{1}{2}$% compounded annually

   c) i) $10\frac{1}{2}$% compounded monthly   **ii)** $10\frac{1}{4}$% compounded annually

3. Which is the greater rate?

   a) i) $11\frac{3}{4}$% compounded semi-annually   **ii)** 12% compounded annually

   b) i) $6\frac{1}{2}$% compounded annually   **ii)** $6\frac{1}{4}$% compounded semi-annually

   c) i) $9\frac{1}{4}$% compounded monthly   **ii)** $9\frac{3}{4}$% compounded semi-annually

   d) i) $11\frac{1}{4}$% compounded annually   **ii)** $10\frac{1}{2}$% compounded monthly

B 4. Suppose you invest $2500 for 5 years in one of each pair of GICs. Which is the better investment? How much greater are your earnings?

   a) i) $12\frac{1}{2}$% compounded semi-annually   **ii)** $11\frac{3}{4}$% compounded monthly

   b) i) $10\frac{1}{4}$% compounded annually   **ii)** $9\frac{3}{4}$% compounded monthly

   c) i) 16% compounded monthly   **ii)** 17% compounded semi-annually

5. Select one part of exercise 4. Write to explain how you determined which investment was better.

6. Determine each effective annual interest rate.

   a) 8% compounded semi-annually

   b) $12\frac{1}{4}$% compounded semi-annually

   c) 15% compounded monthly

   d) $9\frac{1}{2}$% compounded monthly

7. Choose one part of exercise 6. Write to explain how you calculated the effective annual interest rate.

8. Po Ling plans to make regular contributions to her RRSP, starting January 1. Which of these two plans would you recommend and why?

   a) Invest $150 at the beginning of each month at 6.25% compounded monthly.

   b) Invest $900 on January 1 and June 1 at 6.5% compounded semi-annually.

9. A family's home renovations cost $5000. A contractor suggests that the family pays $1000 down, followed by 24 monthly payments of $198.

   a) Determine the finance charge for this payment scheme.

   b) Modify the spreadsheet from *Investigate* on page 33. Determine the effective annual rate of interest that the contractor is charging.

10. a) Modify the spreadsheet on page 33. Calculate the monthly payment for a two-year loan of $4000 at $12\frac{1}{4}$% compounded monthly.

   b) Describe any adjustments you made to calculate the last payment.

   c) Compare your results with those for exercise 9. Compare the finance charge and the interest rate for each situation. Write to explain what you notice.

11. Krista borrows money to buy a car. She has $4500 as a down payment and wants to buy a car that costs $12 200.

   a) Modify the spreadsheet on page 33. Krista obtains a two-year loan at 9.5% compounded monthly. Determine her monthly payment and the finance charge.

   b) Suppose Krista obtains a four-year loan at 9% compounded monthly. Modify the spreadsheet on page 33. Determine the monthly payment and the finance charge.

   c) Which option would you recommend for Krista? Write to explain how her personal situation might influence her choice.

12. Ashley applies for a car loan of $16 500. The bank offers financing at 8.5% compounded monthly, for a term of 3, 4, or 5 years, payable monthly.

  a) Determine Ashley's monthly payment for each of the three-year, four-year, and five-year terms.

  b) Calculate the total cost of each loan in part a.

  c) Which term would you recommend for Ashley? Explain your choice.

C **13.** Express as an equivalent semi-annual rate.

  a) 12% compounded monthly   **b)** 12% compounded annually

14. These credit terms appeared on a bank-card statement.

> **Interest Charges**
>
> No interest is charged on purchases that appear on your statement for the first time provided full payment is received by the Due Date. Purchases not paid in this manner incur interest from the date the transaction was posted until full payment is received. Interest is calculated at a daily rate of 0.05094% (18.6% per annum).

Suppose the daily rate of 0.05094% is correct. Is the rate of 18.6% per annum correct? What effective annual interest rate is being charged? Assume monthly compounding.

15. On an investment, interest compounded semi-annually requires a slightly lower rate than interest compounded annually to yield the same amount. Suppose the semi-annual rate is i percent and the annual rate is r percent.

  a) Express r as a function of i.

  b) Graph the function for reasonable values of i.

  c) Use the graph to determine each estimate.

   i) the annual rate equivalent to 10% compounded semi-annually

   ii) the semi-annual rate equivalent to 18% compounded annually

COMMUNICATING THE IDEAS

Two equal investments can be made for the same length of time at different interest rates. They result in the same accumulated amount after one year. Write to explain how this can happen.

INVESTIGATE | Interpreting a Wage Graph

This graph shows the gross income that a fast food restaurant pays an employee for 1 week's work.

1. At what point does the slope of this graph change? Suggest some reasons why the slope changes.

2. Calculate the slope of each part of the graph. What does each slope represent?

3. Suppose an employee worked a total of 70 h during the week. Determine his gross income. Explain how you calculated your answer.

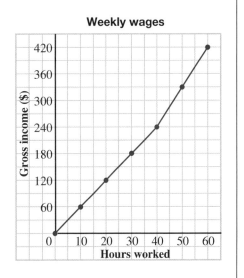

Weekly wages

The money an employee earns before deductions is *gross income* or *gross earnings*. Businesses pay their employees at regular time intervals, which are called *pay periods*. Common pay periods include weekly and monthly intervals, as well as biweekly (once every 2 weeks) and semi-monthly (twice each month) intervals.

Employers offer earnings in different ways.

- *Salary*: This is a fixed amount of money earned over a specified time period, such as monthly or yearly.

- *Wage*: This is money earned hourly, daily, or by piecework. Overtime and gratuities may also be included in a person's wage.

- *Commission*: This is earnings based on a percent of an employee's sales during the pay period and may be paid in combination with a salary or wage.

- *Graduated commission*: This is earnings based on commission, in which the rate of commission increases when sales reach a certain goal or goals.

Example 1

Crystal works full time at a gas station. She is paid $7.50/h plus time-and-a-half for overtime, which is any time over 40 h a week. One week she worked 46 h. Calculate her gross earnings for the week.

Solution

Crystal worked 46 h in total: 40 of these were regular hours and the remaining 6 were overtime.

Regular earnings: Hours worked × hourly wage = 40 × $7.50
$$= \$300$$

Overtime earnings: Hours worked × hourly wage × overtime rate = 6 × $7.50 × 1.5
$$= \$67.50$$

Total earnings: Regular earnings + overtime earnings = $300.00 + $67.50
$$= \$367.50$$

Crystal's gross earnings for the week were $367.50.

Example 2

Dale works at a furniture store where she earns a graduated commission. She earns 9% commission on the first $10 000 worth of furniture she sells, and 12% commission on all sales exceeding $10 000. In May, Dale's sales were $21 675. Calculate her gross earnings.

Solution

Since Dale earns a graduated commission, and her sales exceeded $10 000, calculate her earnings in two parts. She will earn 9% on the first $10 000 and 12% on the remaining $11 675.

First $10 000: 9% of $10 000 = 0.09 × $10 000
$$= \$900$$

Remaining $11 675: 12% of $11 675 = 0.12 × $11 675
$$= \$1401$$

Total earnings: $900 + $1401 = $2301

Dale's gross earnings for May were $2301.

Example 3

Troy has job offers from two clothing stores. Store A will pay $7.25/h. Store B will pay a base salary of $800 plus 4% commission on sales. Both stores want Troy to work 160 h per month. Which store offers higher potential earnings?

Solution

Determine the gross earnings that Troy might expect from each store.

Store A pays an hourly wage with no other remuneration.

Monthly earnings for Store A = Hours worked × hourly wage
$$= 160 \times \$7.25$$
$$= \$1160$$

Store B pays a salary plus 4% commission. This means that Troy's earnings will vary, depending on his sales.

Monthly earnings for Store B = Base salary + 0.04 × monthly sales

From this equation, create a table of values and graph to show how Troy's earnings depend on his sales.

Sales ($)	Earning ($)
0	800
3 000	920
6 000	1040
9 000	1160
12 000	1280

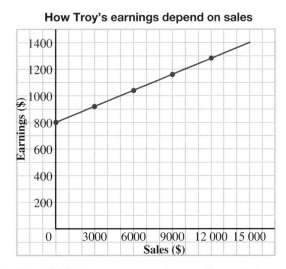

How Troy's earnings depend on sales

Monthly sales over $9000 will yield higher gross earnings at Store B than at Store A. If Troy can regularly exceed $9000 in monthly sales, he will earn more at Store B; otherwise he should work at store A.

1. Explain the difference between earning a salary and earning a wage.

2. List some jobs where employees get paid:

   a) a wage   **b)** a commission   **c)** a salary

3. *Example 2* shows how a graduated commission works. Why do you think a company would structure its commission payment this way?

4. In *Example 3*, Troy needs to consider more than just income to make his decision. What other factors might affect his choice?

1.5 EXERCISES

A 1. What are an employer's advantages of paying commission to her salespeople instead of a salary? What are the advantages to her salespeople?

2. What is the difference between being paid biweekly and semi-monthly?

3. A company pays all its employees weekly. Is it possible that it ever issues pay cheques more than 52 times a year? Fewer than 52 times? Explain.

B 4. Geri earns a base salary of $900 a month plus 5% of total sales. Last month she had a total of $26 324 in sales. Estimate her gross earnings. Write to describe how you determined your estimate.

5. Caroline is hired to arrange flowers. She is paid $2.50 for every arrangement or $12/h. Caroline estimates she could create about four arrangements an hour. Which payment scheme would you recommend and why?

6. Calculate the gross earnings for each person. Identify each payment method.

   a) Joe works in a restaurant 27 h one week. He is paid $7.50/h and receives $134 in gratuities.

   b) Sze Mun worked 46 h last week. She was paid $11.50/h and double time for overtime. Her normal work week is 40 h.

   c) Jack, a salesperson at an appliance store, earns 12% commission. Last week his sales totalled $5675.

   d) Megan plants trees. She is paid 25¢ for every tree she plants. One week she planted 4350 trees.

   e) Marlene works in a computer store. She earns a graduated commission of 3% on her first $15 000 in sales and 5% on sales over $15 000. Last week she sold $32 150 worth of merchandise.

f) Duncan, a salesperson, receives $200 a week plus 4% commission. Last week he sold $4780 worth of merchandise.

g) Frank, a salesperson, is paid $1800 a month and is expected to have monthly sales of $25 000. He is paid 8% commission on any sales beyond this amount. Last month, Frank's total sales were $36 740.

7. Select one part of exercise 6. Write to explain how you calculated the gross earnings.

8. May works as a server. She earns $6.25/h and keeps 70% of the gratuities she receives. The other 30% is shared among the hostess, bus people, and cooks. May worked 34 h last week and received $254 in gratuities, before sharing. What was her gross pay?

9. Eric is offered a job as a server at two restaurants. The Fish House pays $6.50/h and a server averages $34 a day in tips. The Tea House pays $7.25/h and a server averages $28 a day in tips. If Eric considers the gross pay, which job should he take? What other factors might he want to consider?

10. Select exercise 8 or 9. Write to explain how you calculated the gross pay.

11. Jay has just accepted a job in computer sales. The company offers a base salary of $900 plus 6% of his monthly sales, or a straight commission of 9%. Which option should Jay choose?

    a) Draw a graph to show gross earnings against sales. Plot both situations on the same grid. You could use a graphing calculator or a spreadsheet.

    b) Use the graph to advise Jay.

    c) What other information might help Jay make a decision?

12. Pauline earns $9.50/h plus extra for overtime. Her normal work week is 40 h. Last week she worked 51 h. Her gross earnings were $641.25. What is Pauline's overtime rate of pay?

C 13. Raud earns a base salary plus commission. At the end of his first month, he was paid $1440 based on monthly sales of $11 500. At the end of his second month, he made $17 900 in sales and was paid $1824. Determine his base salary and his commission.

COMMUNICATING THE IDEAS

Suppose you are planning to open an up-scale clothing store. Your partners want to pay staff a fixed hourly rate. How might you convince them to accept a commission scheme?

Should You Buy or Lease a Car?

Page 2 presented two options for acquiring a new car. If you buy the car, it belongs to you. If you lease the car, it belongs to the leasing company. In both cases, the additional costs for a licence, insurance, and maintenance are the same. For the purpose of comparing the costs of buying and leasing, these will be ignored.

For those exercises involving sales tax, use the tax rates for British Columbia, which are 7% GST and 7% PST, or for Alberta, which are 7% GST and no PST. For Alberta's tax system, use a payment of $343.35 per month in exercise 1b.

DEVELOP A MODEL

There are two ways you can own the car after 3 years. You can buy the car now, or lease the car and buy it after 3 years. Calculate the total amount you pay in each case.

1. If you buy the car now, the price is $18 253, plus GST and applicable PST.

   a) Calculate the total cost of the car, including taxes.

   b) The down payment is $8800 and the monthly payment is $365.81 for 3 years. Calculate the total payment.

   c) Calculate the finance charge.

   d) Assume you drive 60 000 km in 3 years. Calculate the cost per kilometre.

2. If you lease the car, the down payment is $8800. The lease payment is $93.55 plus GST and applicable PST each month for 36 months.

   a) Calculate the amount of each lease payment and the total lease payment.

   b) After 3 years, you buy the car for $8531, plus GST and applicable PST. How much do you pay?

   c) Calculate the total payment.

   d) Assume you drive 60 000 km in 3 years. Calculate the cost per kilometre.

3. Compare your answers for exercises 1 and 2.

   a) Which is the cheaper way to own the car after 3 years, buying or leasing?

   b) Which do you think is the better deal, or do you not have enough information to make a decision? Explain your answer.

Dealers often restrict the distance you can drive a car during the lease. If your odometer reading is greater than 60 000 km when the lease expires, the dealer charges you 9¢/km for the excess kilometres.

4. Suppose the odometer reading is 75 230 km.

   a) How much would you be charged when the lease expires?

   b) Calculate the total amount to lease the car.

   c) Why does the dealer charge more if the odometer is over 60 000 km?

You solved this problem using a *mathematical model*. The model was based on calculating the amounts you paid for the two different schemes for owning the car after 3 years. We can extend the model to include the cost of operating the car. To use the extended model, you will need to make some assumptions.

5. a) Determine the annual license fee for your province.

   b) The cost of insurance varies widely, depending on factors such as where you live, your driving record, and how you use the car. Consult someone you know who has a car, and try to estimate the annual cost of insurance.

   c) Maintenance includes oil changes, lubrication, and other services recommended by the car dealer. This costs approximately $400 per year. Record this amount as part of your operating costs.

   d) On what does the cost of fuel depend? Try to estimate the annual cost of fuel, making any assumptions that seem reasonable.

   e) What is the cost of operating the car per kilometre?

6. Suppose your operating costs are those you calculated in exercise 5. Determine your total cost of acquiring and operating the car, per kilometre, during the 3 years.

1.6 Taking Home a Pay Cheque

Here is a typical pay stub.

Company Name: One Better		For the period ending: **15/1/97**		
Employee's Name	**Total Hours**		**Deductions**	**To date**
Louis Robitaille	82	Income Tax	213.15	213.15
Regular Pay	**To date**	EI	32.91	32.91
$75 \times 12.75 = 956.25$	956.25	CPP	29.25	29.25
Overtime		RRSP	100.00	100.00
$7 \times 25.50 = 178.50$	178.50	Union Dues	17.02	17.02
Total Gross		Deductions	392.33	392.33
1134.75	1134.75	Net Pay	742.42	742.42

On the left, the pay stub indicates Louis' gross earnings. On the right, the pay stub shows a number of *payroll deductions* that are subtracted from his earnings. Louis' take-home pay is his *net earnings.* An employer is responsible for making these deductions from each employee's 1997 gross earnings:

- *Canada Pension Plan Contributions (CPP)*: Any employee between 18 and 70 years of age must contribute to CPP, up to $944.78 per year. The employer must also contribute an amount equal to that made by the employee. The CPP enables the Canadian government to create a government-run pension plan.

- *Employment Insurance Premiums (EI)*: In almost all employment situations, the worker has *insurable earnings* and pays into the Employment Insurance fund up to $1131 per year. The employer must pay 1.4 times the amount contributed by the employee. The fund insures the worker against unemployment situations such as disability, job loss, and maternity leave.

- *Income Tax:* Both the federal and provincial governments tax personal income, based on the employee's claim code as indicated on her or his *TD1 form.* The TD1 form is completed by anyone receiving employment income.

    For the situations in this text, insurable earnings are equal to gross earnings.

Other possible payroll deductions include union dues, professional dues, life insurance premiums, parking fees, alimony or child support payments, Registered Pension Plan (RPP) contributions, and RRSP contributions. Some of these deductions are *tax-exempt:* the government does not tax any income directed toward these contributions.

Taxable income is gross income minus any of the following tax-exempt deductions: union dues, RPP, RRSP, and child support payments.

Example 1

Joseph owns a convenience store in Saskatchewan. He pays his manager, Angie, every 2 weeks. She has gross earnings of $1160. Her TD1 claim code is 1, and she pays $125 every 2 weeks into an RPP. How will Joseph calculate Angie's deductions and her net earnings?

Solution

Use the 1997 payroll deduction tables for Saskatchewan. A portion of these tables is on pages 78 to 81. Businesses may use these printed tables, or they may get a computer program from Revenue Canada to calculate these deductions.

To determine the CPP deduction, go to section B of the payroll deduction tables; and the pages that specify biweekly pay periods. Look for the pay range that includes Angie's $1160 gross income. The CPP deduction is indicated to the right, in grey.

From - De To - A	CPP	From - De To - A	CPP	From - De To - A
1134.44 - 1134.78	29.25	1159.06 - 1159.39	29.97	1183.67 - 1184.01
1134.79 - 1135.12	29.26	1159.40 - 1159.73	29.98	1184.02 - 1184.35
1135.13 - 1135.46	29.27	1159.74 - 1160.08	29.99	1184.36 - 1184.69
1135.47 - 1135.80	29.28	1160.09 - 1160.42	30.00	1184.70 - 1185.03

Angie's CPP deduction is $29.99, assuming she has not yet paid $944.78 for 1997.

Joseph must match her CPP deduction and pay the federal government $29.99.

To determine the EI deduction, turn to section C in the payroll deduction tables. There is one table that covers all pay periods. Find the appropriate pay range, and read the corresponding deduction.

From - De To - A	EI premium	From - De To - A	EI premium
1159.14 - 1159.48	33.62	1183.97 - 1184.31	34.34
1159.49 - 1159.82	33.63	1184.32 - 1184.65	34.35
1159.83 - 1160.17	33.64	1184.66 - 1184.99	34.36
1160.18 - 1160.51	33.65	1185.00 - 1185.34	34.37
1160.52 - 1160.86	33.66	1185.35 - 1185.68	34.38

Angie's EI deduction is $33.64, provided she has not yet paid $1131 for 1997.

Joseph must pay 1.4 times this amount, or $47.10.

To determine the income tax deduction, first determine Angie's taxable income. Remember that her RPP contribution is tax-exempt.

Taxable income = Gross income − RPP
$$= \$1160 - \$125$$
$$= \$1035$$

Now go to the pages in section D of the tables that correspond to biweekly pay periods. Look for the pay range that includes $1035, then locate the appropriate claim code in that row. This is Angie's deduction.

Pay							
	0	1	2	3	4	5	6
From Less than							
995 - 1011	271.05	204.35	196.20	179.85	163.50	147.20	131.00
1011 - 1027	275.45	208.75	200.60	184.25	167.90	151.55	135.25
1027 - 1043	279.85	213.15	205.00	188.65	172.30	155.95	139.65
1043 - 1059	284.25	217.55	209.40	193.05	176.70	160.35	144.05

Angie's income tax deduction is $213.15.

Calculate Angie's net earnings by adding all the payroll deductions and subtracting this amount from her gross earnings.

Total deductions = $125.00 + $213.15 + $33.64 + $29.99
$$= \$401.78$$

Net earnings = Gross earnings − total deductions
$$= \$1160 - \$401.78$$
$$= \$758.22$$

Employees do not normally have access to the payroll deduction tables, so it may not always be possible to calculate net earnings. However, to plan your spending, you need some idea of what your net earnings will be. To estimate net earnings, we can use the following approximations.

EI	3% of gross earnings, to a maximum of $1131 annually
CPP	3% of gross earnings, to a maximum of $945 annually
Income tax*	18% for annual taxable income from $7000 to $30 000
	29% for annual taxable income from $30 000 to $60 000
	39% for annual taxable income above $60 000

*In practice, if taxable income is closer to the lower end of the range, the estimate will be high. If taxable income is closer to the higher end of the range, the estimate will be low.

Example 2

Suppose Angie wanted to estimate her net earnings before receiving her pay cheque. Estimate her net earnings using the information from *Example 1*.

Solution

EI and CPP deductions are each approximately 3% of $1160.

$0.03 \times \$1160 = \$34.80$

To calculate income tax, use Angie's annual taxable income. From *Example 1*, her taxable income is $1035 every 2 weeks.

$$\text{Annual taxable income} = 26 \times \$1035$$
$$= \$26\ 910$$

Angie's annual taxable income is between $7000 and $30 000. She can expect to pay approximately 18% income tax in each pay period.

$0.18 \times \$1035 = \$186.30$

Since Angie's annual taxable income is close to $30 000, this estimate may be low.

$$\text{Total deductions} = \$125 + \$34.80 + \$34.80 + \$186.30$$
$$= \$380.90$$

$$\text{Estimated net earnings} = \text{Gross earnings} - \text{total deductions}$$
$$= \$1160 - \$380.90$$
$$= \$779.10$$

Angie's net earnings for the biweekly pay period can be estimated at $780 or less. This estimate is confirmed by the calculated result in *Example 1*.

Example 3

Naz earns a monthly salary of $3425. Each month, in addition to the mandatory deductions, she pays $45 in union dues, $225 into an RPP, and $20 for parking. Estimate her net earnings.

Solution

Naz's CPP and EI deductions are each approximately 3% of her gross income.

$0.03 \times \$3425 = \$102.75$

To calculate her taxable income, subtract RPP contributions and union dues, which are tax-exempt.

Taxable monthly income = Gross monthly earnings − (union dues + RPP)

$$= \$3425 - (\$45 + \$225)$$

$$= \$3155$$

Taxable annual income = Taxable monthly income × 12

$$= \$3155 \times 12$$

$$= \$37\ 860$$

Since her taxable income is between $30 000 and $60 000, Naz's income tax deduction will be approximately 29% of her taxable monthly income.

Income tax = 0.29 × $3155

$$= \$914.95$$

This estimate may be high.

Total deductions = $20.00 + $45.00 + $225.00 + $102.75 + $102.75 + $914.95

$$= \$1410.45$$

Net earnings = Gross earnings − total deductions

$$= \$3425 - \$1410.45$$

$$= \$2014.55$$

Naz's net earnings will be about $2014.55.

DISCUSSING THE IDEAS

1. Explain the differences among gross earnings, taxable earnings, and net earnings.

2. Suppose you obtained a job today. Which payroll deductions would you have to pay?

3. How could a person reduce the amount of income tax he or she pays without changing the gross pay or the claim code?

1.6 EXERCISES

A 1. Look at Louis Robitaille's pay stub on page 44. What will Louis' employer pay for CPP and EI?

2. What is a TD1 form and why is it important that an employee fill out one?

3. Why do the payroll deduction tables differ from province to province?

4. Examine the payroll deduction tables on pages 78 to 81. Describe how each deduction changes.

   a) CPP deduction, as income increases

   b) income tax deduction, as income increases

   c) EI contributions, as gross income increases

   d) income tax deduction, as TD1 claim code increases

5. Use the payroll deduction tables. Determine the deduction for each income.

   a) CPP, monthly gross income $1160.00

   b) EI, monthly gross income $1130.45

   c) CPP, monthly gross income $1117.35

   d) income tax deduction, taxable income $1156, claim code 1

   e) income tax deduction, claim code 10, taxable income $1601.98

   f) EI, monthly gross income $1165.32

6. Explain how contributing to an RRSP or RPP affects CPP, EI, and income tax deductions.

B **7.** Jackie and Keith have identical gross earnings. However, Jackie's gross earnings are for 2 weeks of work whereas Keith's are for 1 month.

   a) Which person will have higher CPP deductions over the year?

   b) Which person will have higher EI deductions over the year?

   c) Suppose Jackie and Keith have the same claim code. Which person will pay more income tax?

   d) Consider your answers to parts a to c. Do you think this is fair? Why?

8. **a)** Use the payroll deduction tables. Calculate the net earnings for Sandra whose gross biweekly earnings are $1122. Each pay period, she pays $50 into an RRSP as well as a life insurance premium of $10. Her claim code is 3.

   b) When Sandra's family situation changes, she changes her claim code from 3 to 8. Use her new claim code to recalculate her net earnings.

9. Tim earns a biweekly salary of $1169. He has a claim code of 1 and pays $20 union dues each pay period.

   a) Calculate Tim's biweekly net earnings.

   b) Tim receives a raise of $5 a week. Use Tim's new biweekly salary to recalculate Tim's net earnings.

   c) Write to explain why Tim's net earnings changed by less than $10.

10. Alina earns $1320 a week. Each week she pays $100 into an RRSP and $10 for parking.

   a) Estimate Alina's weekly net earnings.

   b) Will this estimate be higher or lower than the actual net earnings? Why?

   c) Some employees at Alina's workplace have asked to increase their RRSP contribution to $200 a week. Would this be beneficial for Alina? Why?

11. Jim earns a salary of $2325 semi-monthly. He pays $150 every pay period into an RPP, as well as a medical premium of 1.5% of his gross income.

   a) Estimate Jim's net earnings.

   b) Would you expect Jim's actual net earnings to be higher or lower than your estimate? Explain your answer.

12. Select one of exercise 9, 10, or 11. Write to explain how you estimated the net earnings.

13. You have a job offer from a grocery store. The job pays $8.25/h, and you will work 25 h a week. You will be paid every 2 weeks.

   a) Calculate your gross earnings for one pay period.

   b) Estimate your net earnings.

   c) What percent of your gross earnings will you receive in your pay cheque?

14. Copy this table.

   a) Refer to pages 78 to 81. Complete the table for a Saskatchewan resident who earns a biweekly taxable income of $1740.

   b) Plot a graph of Income tax deducted against Claim code. Place the Claim code on the horizontal axis.

   c) Describe how the income tax deducted changes as the claim code increases.

   d) Draw a line of best fit, then write a function to describe this line.

Claim code	Income tax deducted ($)	Percent of taxable income deducted (%)
1		
2		
3		
4		
5		
6		
7		
8		
9		
10		

COMMUNICATING THE IDEAS

A friend writes you a letter stating, "I've been offered a great job that pays $15/h. They want me to work 80 h a month, which means I will receive $1200 a month." Write a paragraph responding to your friend's statement.

The Future Is Now!

When you start your first full-time job, saving for retirement may be the farthest thing from your mind. However, if you can start a small RRSP in your twenties, it can make a huge difference when you retire at 65.

Let's examine two ways of contributing $40 000 to an RRSP. One worker contributes $1000 every year, from the age of 25. We'll call this worker the ant. The other, making a late start at age 45, contributes $2000 every year. We'll call this worker the grasshopper.

Create a spreadsheet to display the RRSP savings for both the ant and the grasshopper up to the retirement age of 65. Although interest rates vary from year to year, we simplify the situation. Assume an average 8% rate with annual compounding.

	A	B	C
1	Invest each year from age 25; from age 45		
2	Fixed Interest Rate:		0.08
3		Amount:	Amount:
4	Age	The Ant	The Grasshopper
5	25	$1000.00	$0.00
6	=A5+1	=$B$5+B5*(1+$C$2)	$0.00

Fill Down to extend the spreadsheet to row 24, where the ant and the grasshopper both turn 44. Format columns B and C to display currency.

In cell C25, enter $2000.00
In cell C26, enter the formula =$C$25+C25*(1+$C$2)

Fill Down to extend the spreadsheet to row 45, when both workers turn 65.

1. Compare the accumulated amount earned by each worker. How many times as great as the grasshopper's accumulated amount is that of the ant?

2. Investigate whether the rate of interest affects the results. Reduce the interest rate in cell C2, and compare the accumulated amounts. Repeat several times. Increase the interest rate, and compare again. Make a general conclusion about your observations.

3. Suppose the grasshopper could afford higher RRSP contributions. What should the grasshopper contribute each year to obtain the same accumulated amount as the ant?

4. The federal government imposes a ceiling on RRSP contributions, according to each employee's annual earnings. Investigate this. Determine whether the RRSP ceiling might affect the grasshopper's potential savings in exercise 3.

Mathematics & Technology

1.7 Determining Expenses

Often when people experience financial difficulties, they have not planned their spending. A *budget* is a written plan outlining how you will spend your money. A budget allows you to analyze your spending in light of your financial goals.

To develop a budget, you must identify your *expenses*. Expenses include all items or services for which you pay. When you identify expenses, it's helpful to distinguish between necessities and desires.

Most people plan personal budgets based on monthly spending. However, you must also budget for expenses that come up once or twice a year, such as car and home insurance, residential taxes, or holiday plans.

Example 1

Ric is planning a three-week vacation 10 months from now to visit his relatives in Norway. The air travel will cost $1200. His relatives have suggested he bring 6000 k (kroner) for spending money. The current exchange rate is 1 k for $0.19 Can.

a) Determine the total cost of the trip.

b) How much should Ric save each month to plan for this vacation?

Solution

a) Since Ric is visiting relatives, assume he will have expenses for transportation and spending money only. His relatives suggested 6000 k spending money. Convert this amount to Canadian dollars.

   1 k = $0.19 Can.
   6000 k = 6000 × $0.19 Can.
   6000 k = $1140 Can.

   The total cost of the trip is: $1200 + $1140 = $2340

b) Ric has 10 months to save $2340. He should save $234 each month for the trip.

When you first leave home, you'll have new expenses — groceries, furniture, kitchenware, linens, and so on. Homeowners need to budget for these items and more. Some responsibilities that come with home ownership are mortgage payments, property taxes, and utilities.

Property owners pay taxes to their local government. These taxes help support education, libraries, and other services.

Property Tax

Property tax depends on the *assessed value* of the property and the local *mill rate*. The assessed value of a property is a percent of its fair market value. The mill rate is set locally; one mill is equal to $\frac{1}{1000}$ of $1. This formula is used.

$$\text{Property tax} = \frac{\text{Assessed value} \times \text{mill rate}}{1000}$$

Example 2

The Meyers' property has a fair market value of $124 000. In their area, the assessed value of a property is 75% of the fair market value. The current mill rate is 28 mills.

a) Determine the Meyers' property tax for the year.

b) How much should the Meyers budget each month for this expense?

Solution

a) The assessed value is 75% of the fair market value.
   Assessed value = $0.75 \times \$124\ 000$, or $93 000

$$\begin{aligned}
\text{Property tax} &= \frac{\text{Assessed value} \times \text{mill rate}}{1000} \\
&= \frac{\$93\ 000 \times 28}{1000} \\
&= \$2604
\end{aligned}$$

   The Meyers' property tax is $2604.

b) The Meyers have 12 months to budget for this expense.
   $2604 ÷ 12 = $217

   They should budget $217 monthly.

DISCUSSING THE IDEAS

1. Discuss some advantages of planning a budget.

2. Discuss some reasons why people may not develop a budget.

1.7 EXERCISES

A 1. When shopping for groceries or other household items, we often make decisions about which items are most economical. Estimate which item is the better buy. Check your estimate using a calculator.

   a) tomato soup: $0.54 for 284 mL or $1.39 for 907 mL

   b) grated cheese: $2.98 for 400 g or $1.98 for 255 g

   c) breakfast cereal: $2.19 for 450 g or $6.99 for 1.4 kg

   d) macaroni dinner: 12 for $9.99 or the first 6 for $0.69 each and the next 6 for $0.96 each

2. A department store is selling a box of 12 video cassettes for $49.99. The same tapes can also be purchased in packages of 3 for $13.99.

   a) Which purchase provides the better value?

   b) Is it always wise to purchase the more economical item? Explain.

B 3. An electrical company offers an equalization plan that charges a customer a fixed amount every month. At the end of the year, the company calculates the difference between customer usage and the amount paid, and it either charges or rebates the difference to the customer. This chart summarizes the electrical bills for the Dirksen family over 1 year. Near the end of this year, the family learned that electrical rates would increase by 4% effective January 1st.

January	$155.43	July	$ 75.64
February	$137.78	August	$ 74.73
March	$ 34.31	September	$ 75.56
April	$ 99.56	October	$ 87.77
May	$ 87.11	November	$ 95.69
June	$ 79.42	December	$119.79

   a) Compare the amounts for February, March, and April. Suggest some reasons for the differences.

   b) The Dirksens decide to use the equalization plan. What monthly amount should the electrical company charge?

   c) Explain how joining this plan would assist in budgeting.

4. The Lees own two cars and their own home. This table summarizes the non-monthly expenses associated with their cars and home.

March 30	Car licence	$ 150.00
April 15	Car insurance	$ 625.00
May 20	Home insurance	$ 520.00
June 30	Property taxes	$2060.00
November 15	Car licence	$ 170.00
November 30	Car insurance	$ 870.00

a) The Lees decide to save money each month to pay for these expenses. How much money do they need to set aside each month?

b) Suppose the Lees save the monthly amount you calculated in part a. Will they be able to pay for all these expenses if they begin to put money aside January 1? Explain your answer.

5. Miki's property has an assessed value of $73 900. Last year's mill rate was 31.8 mills.

a) Determine Miki's property tax for last year.

b) Suppose the mill rate is increased by 3 mills. Determine the increase in Miki's property tax.

c) How much should Miki set aside each month to budget for the increased property taxes?

d) Many banks allow customers with a mortgage to include property tax payments with their mortgage payments. Suggest some advantages to this option.

6. A farmer owns land that has a fair market value of $225 000. The assessed value is 70% of the fair market value. The mill rate is 14.3 mills. How much should the farmer budget each month for property taxes?

7. Two friends are planning a vacation in Alaska. They want to include a one-day coastal cruise that costs $139 U.S. per person. The friends estimate that this cruise will cost approximately $350 Can., and they budget accordingly. When they leave for their trip, $1.00 Can. is worth $0.72 U.S.

a) Without calculating the cost of the cruise, have the two friends budgeted enough money? Explain your answer.

b) Calculate the cost of the one-day cruise.

8. The Sants are planning a 4-day ski vacation at Whitefish, Montana. They estimate daily costs of $125 U.S. for lift tickets and $120 U.S. for accommodation and food. They plan an additional $500 Can. for travel and incidental expenses. When they plan their budget, $1.00 U.S. is worth $1.40 Can. The family has 6 months to save for the vacation. How much money should it set aside each month?

9. Suppose the value of the Canadian dollar increases relative to the American dollar. Is this beneficial for Canadians planning an American vacation? Explain.

10. A person travelling from Canada to the U.S. received an exchange rate of 71%. The same day, a person travelling from the U.S. to Canada received an exchange rate of 140.8%. Write to explain how this is possible.

11. A woman is travelling from the Netherlands to Britain. She knows that one Dutch guilder is equivalent to $0.71 Can., whereas one British pound is equivalent to $2.24 Can. She has 200 guilders and wants to exchange them for British pounds. How many pounds should she receive?

12. Suppose the mill rate is increased by 2 mills. Will all tax payers experience an equal increase in their property taxes? Justify your answer.

C **13.** The inflation rate measures how the cost of living changes. If the inflation rate one year is 2%, it means that it costs 2% more than the preceding year to maintain an equivalent standard of living. Given an average annual inflation rate of 4%, estimate the cost of each item 60 years from now.

a) $0.60 for 1 L of gas

b) $8.00 for admission to a movie

c) $22 000 for a new car

d) $100 000 for a new home

14. Laurette and Lucien have a new grandchild. As their gift, they want to invest in an education fund to cover the cost of the child's first year of post-secondary education, 18 years from now. Currently, the cost of attending university for 1 year is about $5000. Assume an annual inflation rate of 3% and an interest rate of 7% per annum, compounded annually. How much money should they invest today?

15. Use the information from exercise 14. Assume that Laurette and Lucien budget a yearly amount to invest in an education fund. You may treat each deposit as a separate investment.

a) Construct a spreadsheet to determine how much should be invested each year. A possible spreadsheet has been started. The $250 payment in cell A7 is an initial trial.

b) Modify the spreadsheet to see what happens if the interest rate is 6%. How much should be invested each year?

c) How much would have to be invested if Laurette and Lucien wanted to cover 4 years of university?

	A	B	C
1	An Endowment for a Grandchild		
2			
3	Interest rate:		
4	0.07		
5			
6	Payment:		
7	$250.00		
8			
9		Compounding	Future
10	Age	years	value
11	0	18	$844.98
12	1	17	$789.70
13	2	16	$738.04
14	3	15	$689.76
15	4	14	$644.63
16	5	13	$602.46
17	6	12	$563.05
18	7	11	$526.21
19	8	10	$491.79
20	9	9	$459.61
21	10	8	$429.55
22	11	7	$401.45
23	12	6	$375.18
24	13	5	$350.64
25	14	4	$327.70
26	15	3	$306.26
27	16	2	$286.23
28	17	1	$267.50
29			
30		Total	$9094.74

COMMUNICATING THE IDEAS

List the expenses related to owning a car. Describe how you would plan for these expenses when creating a budget.

Keeping Track of Your Money

If you open a chequing account, you may not receive a passbook. The bank sends you a monthly statement showing its record of your account.

Chris Haddad
49 Spire Hillway,
Cooksville, MB, M7B 2S3

Account no. 4325193
Number of enclosures 2
Period ending 15/10/98

Account/description	Debits	Credits	Date	Balance
Balance forward			15/09	523.14
Cash withdrawal	50.00		17/09	473.14
Cheque 034	62.18		20/09	410.96
Deposit		341.60	29/09	752.56
Maintenance fee	9.75		01/10	742.81
Withdrawal	40.00		04/10	702.81
Cheque 035	137.64		05/10	565.17
Withdrawal	60.00		10/10	505.17
Withdrawal	40.00		12/10	465.17

The bank's arithmetic will always be correct, but there could be incorrect charges. If you don't report errors promptly, the bank's statement will be taken as correct. This is why it's a good idea to record transactions in a personal register. Record every cheque, deposit, and withdrawal at the time it is made. This is your only record of your transactions.

Date	No.	Description	Credit	Debit	✔	Balance
		Balance forward				523.14
Sept. 17		Withdrawal		50.00		473.14
Sept. 18	34	The Electronic Store		62.18		410.96
Sept. 29		Deposit	341.60			751.56
Oct. 3	35	Clothing Express		137.46		614.10
Oct. 4		Withdrawal		40.00		574.10
Oct. 12		Withdrawal		40.00		534.10
Oct. 15	36	Bike and Skate		83.12		450.98

When you receive the bank statement, it will show your account's activity up to the date it was printed, which could be a week or more before you receive it. You may have recorded later transactions in your personal register. You need to check whether both records agree up to the date of the bank statement. This process is called *reconciling* your account.

There are different ways to reconcile a statement. Here is one method:

Step 1. Work from the bank statement. Look for each item in your register, and check it off if you have recorded it.

Step 2. The bank statement may show a transaction that you forgot to record. It may report bank fees or interest for the month. Add any further transactions to your register.

Step 3. Start with the final balance on the bank statement. Account for transactions on your personal register that have not yet cleared at the bank: add uncleared deposits and subtract uncleared cheques or withdrawals.

Step 4. The result from Step 3 should agree with the balance in your register. If it doesn't, check for the following errors.

  • Have you recorded each amount correctly?

  • Have you made a mathematical error?

  When you have identified any errors, adjust your personal register and repeat Step 3.

1. Follow the steps above to find three errors in the personal register on page 58. Check your work against the bank statement on page 58.

2. Use the personal register on page 60 and the bank statement below. Reconcile Chris's chequing account for November. If the balance is incorrect, identify any errors and describe how to correct them.

Chris Haddad 49 Spire Hillway, Cooksville, MB, M7B 2S3			Account no. 4325193 Number of enclosures 3 Period ending 15/11/98	

Account/description	Debits	Credits	Date	Balance
Balance forward			15/10	465.17
Cheque 036	83.12		17/10	382.05
Withdrawal	20.00		22/10	362.05
Cheque 037	58.72		23/10	303.33
Withdrawal	40.00		28/10	263.33
Deposit		283.95	30/10	547.28
Maintenance fee	9.75		01/11	537.53
Withdrawal	60.00		05/11	477.53
Cheque 038	33.15		09/11	444.38

Date	No.	Description	Credit	Debit	✔	Balance
Oct. 12		Withdrawal		40.00		534.10
Oct. 15	36	Bike and Skate		83.12		450.98
Oct. 1		Maintenance fee		9.75		441.23
Oct. 10		Withdrawal		60.00		381.23
		Math error (Ch. 35)		0.18		381.05
		Math error (Sept. 29)	1.00			382.05
Oct. 21	37	Clothing Express		58.72		333.33
Oct. 22		Withdrawal		20.00		313.33
Oct. 30		Deposit	283.95			597.28
Nov. 1		Maintenance fee		9.75		587.53
Nov. 5		Withdrawal		60.00		527.53
Nov. 8	38	Record Bin		30.15		497.38
Nov. 14		Withdrawal		40.00		457.38
Nov. 16		Withdrawal		20.00		437.38

3. Why is it important to keep your personal register updated?

4. Do you have enough information from the personal registers to describe Chris' spending habits? Explain.

5. Retail stores reconcile cash register tallies with daily receipts at the end of every business day. Their procedures may differ, depending on the size of the store and the number of cashiers. Contact the manager of a store and ask what procedures are used to reconcile cash register tallies with daily receipts. Determine what procedures are followed when the two amounts do not balance. Write to explain what you learned.

6. Set up a spreadsheet to serve as a personal register. How would you use your spreadsheet to reconcile a bank statement?

1.8 | Preparing a Budget

What plans do you have for the future? What lifestyle do you hope to achieve once you're living on your own? Preparing a budget can help you plan your spending and achieve your overall goals.

Goals may be short-term, such as a summer vacation, or long-term, such as early retirement. When you recognize your goals, you can begin to prepare a budget that works with your income, spending habits, and goals.

You can prepare a budget on paper. A budget worksheet, like the one on page 62, provides a guide. Use a pencil since you may need to change initial estimates. You can also prepare a budget using a computer. The computer will adjust amounts as you revise estimates. The worksheet on page 62 is available as a spreadsheet from *Addison-Wesley Mathematics 11 Template and Data Kit*.

Governments and businesses typically develop yearly budgets. However, for personal expenses, it's a good idea to plan monthly.

Follow these steps to prepare a budget.

Step 1. Calculate and record your monthly net income. Include all sources of income. Use averages if your hours of work vary from month to month.

Step 2. Record all your expenses on a budget worksheet.

Step 3. Subtract your expenses from your monthly income. A positive amount can go toward your financial goals. A negative amount indicates there is not enough income to cover all your expenses. Revise your worksheet based on your goals and your income.

When you have prepared a written budget, monitor it. Make appropriate adjustments based on your actual spending and earning patterns.

Monthly Budget Worksheet for _____

Average Monthly Net Income _____

Projected Expenses

1. HOUSING and UTILITIES

   Rent or Mortgage _____

   Utilities _____

   Cable _____

   Insurance _____

   Taxes _____

   Repairs _____

   _____ _____

   TOTAL _____

   % of Net Income _____

2. FOOD and CLOTHING

   Groceries _____

   Eating Out _____

   Clothing _____

   Footwear _____

   _____ _____

   TOTAL _____

   % of Net Income _____

3. HEALTH and PERSONAL CARE

   Prescriptions _____

   Dental _____

   Other Medical _____

   Skin and Hair Care _____

   _____ _____

   TOTAL _____

   % of Net Income _____

4. TRANSPORTATION

   Public transit _____

   Taxis _____

   Car Payments _____

   Car Licence _____

   Car Insurance _____

   Gas, Oil, etc. _____

   Repairs _____

   _____ _____

   TOTAL _____

   % of Net Income _____

5. RECREATION and EDUCATION

   Entertainment _____

   Hobbies _____

   Vacations _____

   Lessons _____

   School Expenses _____

   _____ _____

   TOTAL _____

   % of Net Income _____

6. SAVINGS

   Short Term _____

   Long Term _____

   _____ _____

   TOTAL _____

   % of Net Income _____

7. MISCELLANEOUS

   Gifts _____

   Donations _____

   _____ _____

   _____ _____

   TOTAL _____

   % of Net Income _____

Summary

Net Income _____

1. Housing and Utilities _____

2. Food and Clothing _____

3. Health and Personal Care _____

4. Transportation _____

5. Recreation and Education _____

6. Savings _____

7. Miscellaneous _____

Total Monthly Expenses _____

Discretionary Income _____

A budget is a personal document; it depends on your financial situation and goals. A restrictive budget may be impossible to follow. One that is too lax may delay your financial goals. A good budget should be realistic, comprehensive, and flexible. Although there are no strict rules for developing a budget, this table provides suggested guidelines from the finance industry.

Expense category	Portion of monthly net income (%)
Housing and utilities	27 to 33
Food and clothing	20 to 26
Health and personal care	3 to 5
Transportation	12 to 14
Recreation and education	6 to 8
Savings	6 to 10
Miscellaneous	12 to 18

Before you prepare personal budgets for yourself or for specific scenarios, review the examples and complete some of the exercises that follow.

Example 1

Kelly is apartment hunting. She earns a monthly net income of $2100. According to the guidelines above, approximately how much can she afford for accommodation?

Solution

The guidelines recommend spending 27% to 33% of net income on housing and utilities. This means that Kelly can probably afford between $567 and $693.

Before she commits to any new expense, Kelly should check her spending in all categories. She should also check whether monthly rent covers all expenses in the apartment or whether there are additional expenses related to maintenance fees, parking, utilities, and so on.

Creating a new budget, or modifying a budget to accommodate new goals, requires analysis, reflection, and care. *Example 2* models a process for analyzing and modifying a budget.

Example 2

This table summarizes Mel's monthly budget.

Mel wants to purchase a house and expects to increase his Housing and utilities costs to $1000 a month. He still has monthly car payments of $475 a month. Describe the decisions Mel could make to be able to afford the house.

Expense category	Amount ($)
Housing and utilities	750.00
Food and clothing	550.00
Health and personal care	75.00
Transportation	700.00
Recreation and education	200.00
Savings	75.00
Miscellaneous	250.00

Solution

Compare Mel's monthly expenditures to the recommended guidelines. Add the amounts: $2600

Express each amount as a percent of this sum.

For example, Housing and utilities $= \frac{750}{2600} \times 100\%$

$\doteq 28.8\%$

Expense category	Actual percent (%)	Recommended percent (%)
Housing and utilities	28.8	27 to 33
Food and clothing	21.2	20 to 26
Health and personal care	2.9	3 to 5
Transportation	26.9	12 to 14
Recreation and education	7.7	6 to 8
Savings	2.9	6 to 10
Miscellaneous	9.6	12 to 18

Mel is spending twice the recommended amount on transportation. If he purchases the house, his housing expense will increase to $\frac{1000}{2600} \times 100\%$ or 38.5%, which also exceeds the recommended guideline. Savings and miscellaneous expenses are significantly lower than recommended. There are many ways Mel could modify his budget. Here are some possibilities.

- Sell or trade in his car for a cheaper model so that he can purchase the house now.

- Delay the purchase of the house, continue to save toward a down payment, and pay off the car loan.

There are other solutions Mel may be tempted to try, but they may not be effective.

- Mel may cut expenses for food, health, and recreation. This could lead to problems if he loses his job or if the car requires major repair.

- Mel might investigate options for car insurance to see if he can lower his monthly $700 transportation expense. However, this may not work. He needs an additional $250 monthly for housing, but there is only $225 per month that is not taken up by the car payments.

In making his final decision, Mel needs to recognize that the car has put him in debt. He already is over-extended in his transportation expenses. Purchasing a house will increase his debt: he will be carrying both a car loan and a mortgage. If emergencies arise, it's unlikely that a bank will grant him a third loan, which means he may miss payments on the car or the house.

Mel may need to acknowledge that he can't have everything at once. He has either to defer the house purchase or to sell the car then buy something more modest. He won't receive the original value of the car when he sells, due to depreciation, so he may prefer to make the house purchase a long-term goal, working toward it while paying off the loan for the car.

DISCUSSING THE IDEAS

1. Explain the difference between long-term and short-term goals. Give one example of each. Why are goals important?

2. The Miscellaneous category might include gifts. What other items do you think would be in this category?

3. Discuss the solution to *Example 2* with a classmate. Is there a correct solution? Explain.

1.8 EXERCISES

A 1. Use the information in the table.

a) What percent of the monthly income is spent in each category?

b) Compare the spending to the recommended guidelines. In which categories is the person overspending? Underspending? Spending within the guidelines?

Monthly Budget

Expense category	Amount ($)
Housing and utilities	650.00
Food and clothing	350.00
Health and personal care	50.00
Transportation	600.00
Recreation and education	150.00
Savings	50.00
Miscellaneous	150.00

2. Eric's annual net income is $32 000. Use the spending guidelines. Determine the range of money Eric should budget monthly in each category.

a) Housing and utilities   b) Food and clothing

c) Transportation   d) Savings

B 3. This circle graph represents Jodi's monthly budget of $1500. Jodi plans to enroll part-time at a local business school. She can continue working while she attends school at night and on weekends. The school costs will be approximately $200 a month for 2 years. Construct a new budget for Jodi that allows her to attend school. Explain the decisions you made.

Monthly Budget

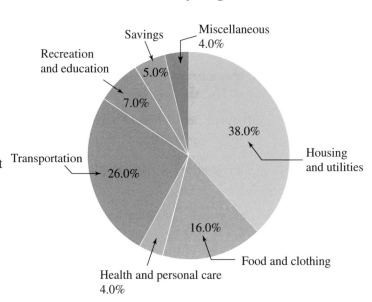

4. Federal and provincial governments often talk about the *deficit* and the *debt*. Investigate then write to explain what these words mean and how they relate to budgets.

5. Use the table and the scenario to complete this exercise.

The Zunigas have three children. Both parents work outside the home. The table shows their monthly budget. The Housing category includes the monthly mortgage payment on their home. Because of increased interest rates, their monthly mortgage payment will increase by $200.

The Zunigas' Monthly Budget

Expense category	Amount ($)
Housing and utilities	1200.00
Food and clothing	800.00
Health and personal care	150.00
Transportation	700.00
Recreation and education	100.00
Savings	50.00
Miscellaneous	100.00

a) Construct a new budget for the Zunigas. Explain your changes.

b) Draw a graph to illustrate your revised budget.

6. Use the table and the scenario to complete this exercise.

Pat and Steve were both working when they created this budget. The transportation expense reflects their purchase of a new car, with monthly loan payments of $750. Steve earns a net income of $1900/month but has just learned that his job is terminated. He will receive 2 months' severance pay. Steve predicts that he can find a new job with a comparable salary within 6 months. Currently, Pat and Steve have $1000 in savings.

Pat and Steve, Monthly Budget

Expense category	Amount ($)
Housing and utilities	1500.00
Food and clothing	700.00
Health and personal care	150.00
Transportation	1000.00
Recreation and education	200.00
Savings	100.00
Miscellaneous	100.00

a) Construct two budgets for Pat and Steve. Base the first budget on the next 2 months, in which monthly net income will not change. Base the second budget on the following 4 months, when only Pat is earning a salary. Explain the decisions you make to get them through the period of unemployment.

b) Financial planners recommend people have 3 months' net earnings in their savings. Does this seem reasonable? Explain.

7. The *Canadian Households* database contains information about average household expenses in Canadian cities from 1982 to 1992.

a) Choose a city and find its records for 1982 and 1992. For each year, record the average household before tax income and personal taxes. Calculate the percent of income paid in taxes. How did this percent change over the 10-year-period?

b) Compare your answers to part a with those of other students. Was there much variation in the percent paid in taxes for a particular year from city to city?

c) Each record also provides the percent of households that had at least two full time earners. Look at this information for the city you selected. Recall the information on page 46. How might the number of earners affect the tax rate?

d) In the database, what other information is provided about factors that might affect a person's tax rate?

8. If you have access to *Addison-Wesley Mathematics 11 Template and Data Kit,* use file Page 62. Otherwise, use the budget worksheet on page 62. This exercise is based on the following scenario.

You decide to work for 1 year after high school before you begin post-secondary training. Starting September 1, you work 40 h/week at a grocery store earning $12.50/h. Your furnished apartment costs $350/month, plus telephone and cable. Your goal is to save $5000 to attend school next year. You purchase a used car and assume payments of $234/month for 1 year. Your parents paid $425 to license and insure the car for the first year; this cost will be due by August 31. You are paid biweekly and have a net claim code of 1.

a) Estimate your monthly net income.

b) Create a monthly budget. Allow for all the expenses above. If you prefer to make other choices, note your ideas for a follow-up assignment.

c) Create a table, then draw a graph to summarize your monthly budget.

d) Compare your spending in each category to the recommended guidelines.

e) On November 1, you must spend $500 to winterize your car. Adjust your budget to accommodate this expense, without missing any bills or relinquishing your goal.

9. Suppose you are a financial planner. Select one of these clients. Describe the client, and determine all the possible expenses that might arise. Estimate the income. Create a monthly budget for your client.

a) a single parent

b) a famous personality

c) a school

d) a non-profit shelter for teen runaways

e) a parks and recreation department for a municipality

COMMUNICATING THE IDEAS

Write a paragraph to explain what a budget is and why it is important to create a budget. Describe how personal goals are related to budgeting.

For many people, buying a home is an important long-term budgeting goal that requires obtaining a mortgage. A mortgage is similar to other loans but is usually paid off over a longer period of time called the *amortization period*. Since a mortgage extends over a long period of time, the mortgage terms may be changed periodically.

Before approving a mortgage, a bank, lender, or financial institution will investigate how much you earn and your credit history. It will also expect you to pay a portion of the house price as a down payment, usually 25% of the sale price.

Banks use computers to handle customers' mortgage inquiries. You can use the TVM Solver on the TI-83, on page 28. If you do not have a calculator that performs financial calculations, use the table on page 76.

Example 1

Daniel decides to purchase a home and requires a mortgage of $85 000. His bank approves a mortgage, amortized over 20 years, with a three-year term at 6.75% interest. Determine the monthly mortgage payment.

Solution

Using the table on page 76

Locate the row corresponding to an interest rate of 6.75%, then read across to the column with the "20 Years" heading. The payment of $7.60 is the monthly payment required for each $1000 of the principal. Since Daniel's mortgage is $85 000, multiply to find his monthly payment.

$85 \times \$7.60 = \$646$

Daniel's monthly mortgage payment is $646.

Using the TVM Solver on the TI-83

Enter the amounts shown on the calculator display. The monthly payment is $646.31.

The calculator result reveals that the calculation from the table is not accurate. This is because the table reflects rounded numbers.

```
N=240
I%=6.75
PV=85000
PMT=-646.30940...
FV=0
P/Y=12
C/Y=12
PMT:END BEGIN
```

Example 2

The Arcands have just assumed a $70 000 mortgage, amortized over 25 years, at 7.25% interest.

a) Determine the monthly mortgage payment.

b) Assume the terms of the mortgage stay the same. Determine the total cost of the mortgage at the end of 25 years. How does this cost compare with the amount of the mortgage?

Solution

a) Using the table: $70\,000 \times \frac{\$7.23}{\$1000} = \$506.10$

The monthly mortgage payment is $506.10.

b) The total cost of the mortgage is the total amount the Arcands would have paid at the end of 25 years.

In 25 years, they will make 300 monthly payments of $506.10.

$300 \times \$506.10 = \$151\,830$

The total cost of the mortgage is $151 830. The Arcands will have paid more than double the original amount owed.

Example 3

Recalculate the Arcands' mortgage in *Example 2* with an amortization period of 20 years. How do the two total costs compare?

Solution

Determine the monthly mortgage payment.

From the table: $70\,000 \times \frac{\$7.90}{\$1000} = \$553$

In 20 years, they will make 240 monthly payments of $553.

$240 \times \$553 = \$132\,720$

The total cost of the mortgage is $132 720.

By paying about $50 more each month, the Arcands can reduce the total cost of their mortgage by almost $20 000.

1. Discuss the similarities and differences between a car loan and a mortgage.

2. Even though the total cost of a mortgage is relatively high, suggest some reasons people choose to take out a mortgage.

3. Discuss the difference between the amortization period of a mortgage and the term of a loan.

4. A bank offers both open and closed mortgages. Investigate to find out the differences between these mortgages. Discuss your findings with others in your class.

1.9 EXERCISES

A 1. Determine the monthly payment and the total cost of each mortgage.

   a) a $120 000 mortgage at 6.5% interest amortized over 20 years

   b) a $60 000 mortgage at 5.75% interest amortized over 10 years

   c) a $90 000 mortgage at 8% interest amortized over 25 years

2. Select one part of exercise 1. Write to explain how you determined the monthly payment.

3. A family moves into its new home. The $90 000 mortgage is amortized over 25 years at 7% for a two-year term.

   a) Calculate the monthly mortgage payment.

   b) At the end of the two-year term, interest rates increased by 2%. The family renews its mortgage at current rates. Calculate the new monthly payment.

4. Consider your results for exercise 3. List reasons why some people keep track of current mortgage rates.

B **5.** Three years ago, Sariah took out a $75 000 mortgage over a five-year term at 9.5% interest. The mortgage was amortized over 25 years.

a) Calculate Sariah's monthly payment.

b) At the end of the third year (after 36 payments), mortgage rates drop to 6.75%. Sariah considers renegotiating her mortgage. She would pay off the balance then remortgage at the lower rate. The bank informs her that a balance of $72 448.65 remains on her mortgage and if she renegotiates, she will have to pay a penalty of $1500. This amount could be added to her present balance.

　i) Calculate Sariah's monthly payment if she renegotiates a mortgage amortized over 20 years.

　ii) Would you encourage her to renegotiate her mortgage? Explain.

6. Kyo applies for a mortgage of $82 000, amortized over 20 years. The bank is currently offering these mortgage rates.
One-year term: 5.25%
Two-year term: 6.00%
Three-year term: 6.50%
Four-year term: 6.75%
Five-year term: 7.00%

a) Calculate the monthly payment for each term.

　i) one-year

　ii) three-year

　iii) five-year

b) Suggest some reasons for choosing a five-year term instead of a one-year term.

7. In the early 1980s, interest rates soared to over 20%. To see the devastating effect this had on some homeowners, consider a family with a mortgage of $67 000 amortized over 20 years. Calculate the monthly payment for each interest rate.

a) 7% **b)** 12% **c)** 22%

8. Maria estimates that she can afford a monthly mortgage payment of $575. Current interest rates are 6.75%. Calculate the mortgage she could assume for each amortization period.

a) 15 years

b) 20 years

c) 25 years

d) What other factors should Maria consider before assuming the mortgage?

9. When deciding on the amortization period of a mortgage, a consumer should consider the total cost of the mortgage, not just the monthly payment. Consider a mortgage of $80 000 at 7% interest. Assume the interest rate does not change.

a) Copy and complete the table.

Amortization period (years)	Monthly payment ($)	Total cost ($)	Percent of original repaid (%)
5			
10			
15			
20			
25			

b) Write a statement to describe what happens to the monthly payment as the amortization period increases.

c) Write a statement to describe what happens to the total cost as the amortization period increases.

10. Consider a $90 000 mortgage amortized over 15 years.

a) Copy and complete the table.

Interest rate (%)	Monthly payment ($)	Total cost ($)	Percent of original repaid (%)
4			
6			
8			
10			
12			

b) Describe what happens to the monthly payment as the interest rate increases.

c) Describe what happens to the total cost of the mortgage as the interest rate increases.

d) If the interest rate doubles, will the monthly payment double? Explain how you know.

11. Some mortgages involve payments other than monthly. Consider a $100 000 mortgage amortized over 20 years at 6.5% interest. Assume the interest rate does not change. Assume the number of compounding periods per year does not change as the number of payments per year does change.

a) Copy and complete the table.

Payment period	Payment ($)	Total cost ($)	Percent of original repaid (%)
Monthly			
Semi-monthly			
Biweekly			
Weekly			

b) Write a statement to describe what happens to the payment as the payment period changes.

c) Write a statement to describe what happens to the total cost of the mortgage as the payment period changes.

12. Consider a $100 000 mortgage amortized over 25 years at 8% interest.

a) Calculate the monthly payment for this mortgage.

b) Use the bal(feature on the TI-83. Calculate the balance owing on this mortgage after 1 year (12 payments).

c) What percent of the money paid during the first year of this mortgage went to reduce the principal?

d) Copy and complete this table.

Number of payments	12	36	60	120	180	240	300
Unpaid balance ($)							

e) Plot a graph of the Unpaid balance against the Number of payments. Place the Number of payments on the horizontal axis.

f) Describe the shape of the graph. Suggest some reasons why the unpaid balance does not decrease linearly.

C 13. Many mortgages offer customers the option of making one additional payment each year. This additional payment is applied directly against the principal. Consider a $65 000 mortgage amortized over 25 years at 6.75% interest.

a) Calculate the monthly payment.

b) What is the balance of the mortgage after 1 year of payments?

c) Suppose you make an additional payment of $3000 at the end of the first year. What is the new balance owing on the mortgage?

d) Assume the monthly payment is not changed. How many more payments are needed to pay off the mortgage? How many years will this take?

e) Suppose one additional payment of $3000 is paid on the mortgage after 10 years of making regular payments. How many more payments are required to pay off the mortgage?

f) Use your results in parts d and e. Is it more beneficial to make additional payments early in your mortgage term or later? Suggest some possible reasons.

MODELLING the Full Amortization of a Mortgage

In this section, we have assumed that interest rates remain constant during the full period during which a mortgage is paid. In reality, interest rates change frequently. Mortgages are affected by these changes.

- What is the advantage of considering a mortgage during its full amortization period at a fixed interest rate?

- Suppose a mortgage comes up for renewal when interest rates are relatively high and have been increasing for the past 12 months. Is it better to select a short term (6 months) or a longer term (3 years) at a slightly lower interest rate? Explain.

- Suppose a mortgage comes up for renewal when interest rates are relatively low and have been dropping for the past 12 months. Is it better to select a long term (3 years) or a short term (6 months) at a slightly higher interest rate? Explain.

COMMUNICATING THE IDEAS

A person has many factors to consider when deciding on a mortgage. Suggest two or three things that you should look for or can do to reduce the total amount you pay on a mortgage.

Amortization Table

This table gives the monthly payments, per $1000, at a given interest rate for the amortization periods shown.

%	5 Years	10 Years	15 Years	20 Years	25 Years
4.00	$18.42	$10.12	$ 7.40	$ 6.06	$ 5.28
4.25	$18.53	$10.24	$ 7.52	$ 6.19	$ 5.42
4.50	$18.64	$10.36	$ 7.65	$ 6.33	$ 5.56
4.75	$18.76	$10.48	$ 7.78	$ 6.46	$ 5.70
5.00	$18.87	$10.61	$ 7.91	$ 6.60	$ 5.85
5.25	$18.99	$10.73	$ 8.04	$ 6.74	$ 5.99
5.50	$19.10	$10.85	$ 8.17	$ 6.88	$ 6.14
5.75	$19.22	$10.98	$ 8.30	$ 7.02	$ 6.29
6.00	$19.33	$11.10	$ 8.44	$ 7.16	$ 6.44
6.25	$19.45	$11.23	$ 8.57	$ 7.31	$ 6.60
6.50	$19.57	$11.35	$ 8.71	$ 7.46	$ 6.75
6.75	$19.68	$11.48	$ 8.85	$ 7.60	$ 6.91
7.00	$19.80	$11.61	$ 8.99	$ 7.75	$ 7.07
7.25	$19.92	$11.74	$ 9.13	$ 7.90	$ 7.23
7.50	$20.04	$11.87	$ 9.27	$ 8.06	$ 7.39
7.75	$20.16	$12.00	$ 9.41	$ 8.21	$ 7.55
8.00	$20.28	$12.13	$ 9.56	$ 8.36	$ 7.72
8.25	$20.40	$12.27	$ 9.70	$ 8.52	$ 7.88
8.50	$20.52	$12.40	$ 9.85	$ 8.68	$ 8.05
8.75	$20.64	$12.53	$ 9.99	$ 8.84	$ 8.22
9.00	$20.76	$12.67	$10.14	$ 9.00	$ 8.39
9.25	$20.88	$12.80	$10.29	$ 9.16	$ 8.56
9.50	$21.00	$12.94	$10.44	$ 9.32	$ 8.74
9.75	$21.12	$13.08	$10.59	$ 9.49	$ 8.91
10.00	$21.25	$13.22	$10.75	$ 9.65	$ 9.09
10.25	$21.37	$13.35	$10.90	$ 9.82	$ 9.26
10.50	$21.49	$13.49	$11.05	$ 9.98	$ 9.44
10.75	$21.62	$13.63	$11.21	$10.15	$ 9.62
11.00	$21.74	$13.78	$11.37	$10.32	$ 9.80
11.25	$21.87	$13.92	$11.52	$10.49	$ 9.98
11.50	$21.99	$14.06	$11.68	$10.66	$10.16
11.75	$22.12	$14.20	$11.84	$10.84	$10.35
12.00	$22.24	$14.35	$12.00	$11.01	$10.53

1. Calculate the simple interest on each investment.

    a) $1600.00 at 5% annually for 80 days

    b) $440.00 at $3\frac{1}{2}$% annually for 210 days

    c) $2125.00 at $7\frac{1}{4}$% annually for 400 days

    d) $3610.00 at $6\frac{3}{4}$% annually for 175 days

2. Determine each accumulated amount.

    a) $500 invested for 4 years at 4% compounded annually

    b) $1250 invested for 2 years at 7% compounded annually

    c) $775 invested for 6 years at $5\frac{1}{2}$% compounded annually

3. Select one part of exercise 2. Write to explain how you determined the accumulated amount.

4. Determine each effective annual interest rate.

    a) 7% compounded quarterly

    b) $8\frac{1}{2}$% compounded semi-annually

    c) 9% compounded monthly

    d) 12% compounded daily

5. Pui Yan is offered two jobs as a chef. He is expected to work 60 h each week. The first pays $17/h plus time-and-a-half for overtime (over 40 h). The second pays $15/h and double time for overtime. Which job should Pui Yan accept? Explain.

6. Suppose you receive an offer for a job that pays $12.50/h plus double time for overtime (over 40 h). Your boss wants you to work 50 h each week. You have union dues of $15 a week and a weekly parking fee of $10. Estimate your weekly net pay. Use the information on page 46.

7. A property has a fair market value of $320 000. The assessed value is 79% of the fair market value. The mill rate is 19 mills.

    a) Determine the property tax.

    b) How much should the property owner set aside each month to pay the property taxes?

8. The *Canadian Households* database contains information about average household expenses in Canadian cities from 1982 to 1992. Open a database and select a city. Find the record with information about 1992.

    a) Group and sum the expenses according to the seven categories on page 63. Calculate the savings by subtracting the taxes and all other expenses from the income before tax.

    b) Draw a circle graph to illustrate the yearly budget.

    c) Write a question for a classmate to answer using your graph. Provide a sample answer.

Canada Pension Plan Contributions

Biweekly (26 pay periods a year)

Cotisations au Régime de pensions du Canada

Aux deux semaines (26 périodes de paie par année)

Pay / Rémunération From - De	To - À	CPP RPC	Pay / Rémunération From - De	To - À	CPP RPC	Pay / Rémunération From - De	To - À	CPP RPC	Pay / Rémunération From - De	To - À	CPP RPC
1119.06	1119.39	28.80	1143.67	1144.01	29.52	1168.29	1168.62	30.24	1192.91	1193.24	30.96
1119.40	1119.73	28.81	1144.02	1144.35	29.53	1168.63	1168.96	30.25	1193.25	1193.58	30.97
1119.74	1120.08	28.82	1144.36	1144.69	29.54	1168.97	1169.31	30.26	1193.59	1193.92	30.98
1120.09	1120.42	28.83	1144.70	1145.03	29.55	1169.32	1169.65	30.27	1193.93	1194.26	30.99
1120.43	1120.76	28.84	1145.04	1145.37	29.56	1169.66	1169.99	30.28	1194.27	1194.60	31.00
1120.77	1121.10	28.85	1145.38	1145.72	29.57	1170.00	1170.33	30.29	1194.61	1194.95	31.01
1121.11	1121.44	28.86	1145.73	1146.06	29.58	1170.34	1170.67	30.30	1194.96	1195.29	31.02
1121.45	1121.78	28.87	1146.07	1146.40	29.59	1170.68	1171.02	30.31	1195.30	1195.63	31.03
1121.79	1122.13	28.88	1146.41	1146.74	29.60	1171.03	1171.36	30.32	1195.64	1195.97	31.04
1122.14	1122.47	28.89	1146.75	1147.08	29.61	1171.37	1171.70	30.33	1195.98	1196.31	31.05
1122.48	1122.81	28.90	1147.09	1147.43	29.62	1171.71	1172.04	30.34	1196.32	1196.66	31.06
1122.82	1123.15	28.91	1147.44	1147.77	29.63	1172.05	1172.38	30.35	1196.67	1197.00	31.07
1123.16	1123.49	28.92	1147.78	1148.11	29.64	1172.39	1172.72	30.36	1197.01	1197.34	31.08
1123.50	1123.84	28.93	1148.12	1148.45	29.65	1172.73	1173.07	30.37	1197.35	1197.68	31.09
1123.85	1124.18	28.94	1148.46	1148.79	29.66	1173.08	1173.41	30.38	1197.69	1198.02	31.10
1124.19	1124.52	28.95	1148.80	1149.13	29.67	1173.42	1173.75	30.39	1198.03	1198.37	31.11
1124.53	1124.86	28.96	1149.14	1149.48	29.68	1173.76	1174.09	30.40	1198.38	1198.71	31.12
1124.87	1125.20	28.97	1149.49	1149.82	29.69	1174.10	1174.43	30.41	1198.72	1199.05	31.13
1125.21	1125.55	28.98	1149.83	1150.16	29.70	1174.44	1174.78	30.42	1199.06	1199.39	31.14
1125.56	1125.89	28.99	1150.17	1150.50	29.71	1174.79	1175.12	30.43	1199.40	1199.73	31.15
1125.90	1126.23	29.00	1150.51	1150.84	29.72	1175.13	1175.46	30.44	1199.74	1200.08	31.16
1126.24	1126.57	29.01	1150.85	1151.19	29.73	1175.47	1175.80	30.45	1200.09	1200.42	31.17
1126.58	1126.91	29.02	1151.20	1151.53	29.74	1175.81	1176.14	30.46	1200.43	1200.76	31.18
1126.92	1127.25	29.03	1151.54	1151.87	29.75	1176.15	1176.49	30.47	1200.77	1201.10	31.19
1127.26	1127.60	29.04	1151.88	1152.21	29.76	1176.50	1176.83	30.48	1201.11	1201.44	31.20
1127.61	1127.94	29.05	1152.22	1152.55	29.77	1176.84	1177.17	30.49	1201.45	1201.78	31.21
1127.95	1128.28	29.06	1152.56	1152.90	29.78	1177.18	1177.51	30.50	1201.79	1202.13	31.22
1128.29	1128.62	29.07	1152.91	1153.24	29.79	1177.52	1177.85	30.51	1202.14	1202.47	31.23
1128.63	1128.96	29.08	1153.25	1153.58	29.80	1177.86	1178.19	30.52	1202.48	1202.81	31.24
1128.97	1129.31	29.09	1153.59	1153.92	29.81	1178.20	1178.54	30.53	1202.82	1203.15	31.25
1129.32	1129.65	29.10	1153.93	1154.26	29.82	1178.55	1178.88	30.54	1203.16	1203.49	31.26
1129.66	1129.99	29.11	1154.27	1154.60	29.83	1178.89	1179.22	30.55	1203.50	1203.84	31.27
1130.00	1130.33	29.12	1154.61	1154.95	29.84	1179.23	1179.56	30.56	1203.85	1204.18	31.28
1130.34	1130.67	29.13	1154.96	1155.29	29.85	1179.57	1179.90	30.57	1204.19	1204.52	31.29
1130.68	1131.02	29.14	1155.30	1155.63	29.86	1179.91	1180.25	30.58	1204.53	1204.86	31.30
1131.03	1131.36	29.15	1155.64	1155.97	29.87	1180.26	1180.59	30.59	1204.87	1205.20	31.31
1131.37	1131.70	29.16	1155.98	1156.31	29.88	1180.60	1180.93	30.60	1205.21	1205.55	31.32
1131.71	1132.04	29.17	1156.32	1156.66	29.89	1180.94	1181.27	30.61	1205.56	1205.89	31.33
1132.05	1132.38	29.18	1156.67	1157.00	29.90	1181.28	1181.61	30.62	1205.90	1206.23	31.34
1132.39	1132.72	29.19	1157.01	1157.34	29.91	1181.62	1181.96	30.63	1206.24	1206.57	31.35
1132.73	1133.07	29.20	1157.35	1157.68	29.92	1181.97	1182.30	30.64	1206.58	1206.91	31.36
1133.08	1133.41	29.21	1157.69	1158.02	29.93	1182.31	1182.64	30.65	1206.92	1207.25	31.37
1133.42	1133.75	29.22	1158.03	1158.37	29.94	1182.65	1182.98	30.66	1207.26	1207.60	31.38
1133.76	1134.09	29.23	1158.38	1158.71	29.95	1182.99	1183.32	30.67	1207.61	1207.94	31.39
1134.10	1134.43	29.24	1158.72	1159.05	29.96	1183.33	1183.66	30.68	1207.95	1208.28	31.40
1134.44	1134.78	29.25	1159.06	1159.39	29.97	1183.67	1184.01	30.69	1208.29	1208.62	31.41
1134.79	1135.12	29.26	1159.40	1159.73	29.98	1184.02	1184.35	30.70	1208.63	1208.96	31.42
1135.13	1135.46	29.27	1159.74	1160.08	29.99	1184.36	1184.69	30.71	1208.97	1209.31	31.43
1135.47	1135.80	29.28	1160.09	1160.42	30.00	1184.70	1185.03	30.72	1209.32	1209.65	31.44
1135.81	1136.14	29.29	1160.43	1160.76	30.01	1185.04	1185.37	30.73	1209.66	1209.99	31.45
1136.15	1136.49	29.30	1160.77	1161.10	30.02	1185.38	1185.72	30.74	1210.00	1210.33	31.46
1136.50	1136.83	29.31	1161.11	1161.44	30.03	1185.73	1186.06	30.75	1210.34	1210.67	31.47
1136.84	1137.17	29.32	1161.45	1161.78	30.04	1186.07	1186.40	30.76	1210.68	1211.02	31.48
1137.18	1137.51	29.33	1161.79	1162.13	30.05	1186.41	1186.74	30.77	1211.03	1211.36	31.49
1137.52	1137.85	29.34	1162.14	1162.47	30.06	1186.75	1187.08	30.78	1211.37	1211.70	31.50
1137.86	1138.19	29.35	1162.48	1162.81	30.07	1187.09	1187.43	30.79	1211.71	1212.04	31.51
1138.20	1138.54	29.36	1162.82	1163.15	30.08	1187.44	1187.77	30.80	1212.05	1212.38	31.52
1138.55	1138.88	29.37	1163.16	1163.49	30.09	1187.78	1188.11	30.81	1212.39	1212.72	31.53
1138.89	1139.22	29.38	1163.50	1163.84	30.10	1188.12	1188.45	30.82	1212.73	1213.07	31.54
1139.23	1139.56	29.39	1163.85	1164.18	30.11	1188.46	1188.79	30.83	1213.08	1213.41	31.55
1139.57	1139.90	29.40	1164.19	1164.52	30.12	1188.80	1189.13	30.84	1213.42	1213.75	31.56
1139.91	1140.25	29.41	1164.53	1164.86	30.13	1189.14	1189.48	30.85	1213.76	1214.09	31.57
1140.26	1140.59	29.42	1164.87	1165.20	30.14	1189.49	1189.82	30.86	1214.10	1214.43	31.58
1140.60	1140.93	29.43	1165.21	1165.55	30.15	1189.83	1190.16	30.87	1214.44	1214.78	31.59
1140.94	1141.27	29.44	1165.56	1165.89	30.16	1190.17	1190.50	30.88	1214.79	1215.12	31.60
1141.28	1141.61	29.45	1165.90	1166.23	30.17	1190.51	1190.84	30.89	1215.13	1215.46	31.61
1141.62	1141.96	29.46	1166.24	1166.57	30.18	1190.85	1191.19	30.90	1215.47	1215.80	31.62
1141.97	1142.30	29.47	1166.58	1166.91	30.19	1191.20	1191.53	30.91	1215.81	1216.14	31.63
1142.31	1142.64	29.48	1166.92	1167.25	30.20	1191.54	1191.87	30.92	1216.15	1216.49	31.64
1142.65	1142.98	29.49	1167.26	1167.60	30.21	1191.88	1192.21	30.93	1216.50	1216.83	31.65
1142.99	1143.32	29.50	1167.61	1167.94	30.22	1192.22	1192.55	30.94	1216.84	1217.17	31.66
1143.33	1143.66	29.51	1167.95	1168.28	30.23	1192.56	1192.90	30.95	1217.18	1217.51	31.67

Employee's maximum CPP contribution for the year 1997 is **$944.78**

La cotisation maximale de l'employé au RPC pour l'année 1997 est de **$944.78**

B-19

Canada Pension Plan Contributions
Monthly (12 pay periods a year)

Cotisations au Régime de pensions du Canada
Mensuel (12 périodes de paie par année)

Pay Rémunération From - De	To - À	CPP RPC	Pay Rémunération From - De	To - À	CPP RPC	Pay Rémunération From - De	To - À	CPP RPC	Pay Rémunération From - De	To - À	CPP RPC
1079.19	1079.52	23.04	1103.80	1104.13	23.76	1128.42	1128.75	24.48	1153.03	1153.36	25.20
1079.53	1079.86	23.05	1104.14	1104.48	23.77	1128.76	1129.09	24.49	1153.37	1153.71	25.21
1079.87	1080.20	23.06	1104.49	1104.82	23.78	1129.10	1129.43	24.50	1153.72	1154.05	25.22
1080.21	1080.54	23.07	1104.83	1105.16	23.79	1129.44	1129.77	24.51	1154.06	1154.39	25.23
1080.55	1080.89	23.08	1105.17	1105.50	23.80	1129.78	1130.12	24.52	1154.40	1154.73	25.24
1080.90	1081.23	23.09	1105.51	1105.84	23.81	1130.13	1130.46	24.53	1154.74	1155.07	25.25
1081.24	1081.57	23.10	1105.85	1106.18	23.82	1130.47	1130.80	24.54	1155.08	1155.42	25.26
1081.58	1081.91	23.11	1106.19	1106.53	23.83	1130.81	1131.14	24.55	1155.43	1155.76	25.27
1081.92	1082.25	23.12	1106.54	1106.87	23.84	1131.15	1131.48	24.56	1155.77	1156.10	25.28
1082.26	1082.60	23.13	1106.88	1107.21	23.85	1131.49	1131.83	24.57	1156.11	1156.44	25.29
1082.61	1082.94	23.14	1107.22	1107.55	23.86	1131.84	1132.17	24.58	1156.45	1156.78	25.30
1082.95	1083.28	23.15	1107.56	1107.89	23.87	1132.18	1132.51	24.59	1156.79	1157.13	25.31
1083.29	1083.62	23.16	1107.90	1108.24	23.88	1132.52	1132.85	24.60	1157.14	1157.47	25.32
1083.63	1083.96	23.17	1108.25	1108.58	23.89	1132.86	1133.19	24.61	1157.48	1157.81	25.33
1083.97	1084.30	23.18	1108.59	1108.92	23.90	1133.20	1133.54	24.62	1157.82	1158.15	25.34
1084.31	1084.65	23.19	1108.93	1109.26	23.91	1133.55	1133.88	24.63	1158.16	1158.49	25.35
1084.66	1084.99	23.20	1109.27	1109.60	23.92	1133.89	1134.22	24.64	1158.50	1158.83	25.36
1085.00	1085.33	23.21	1109.61	1109.95	23.93	1134.23	1134.56	24.65	1158.84	1159.18	25.37
1085.34	1085.67	23.22	1109.96	1110.29	23.94	1134.57	1134.90	24.66	1159.19	1159.52	25.38
1085.68	1086.01	23.23	1110.30	1110.63	23.95	1134.91	1135.24	24.67	1159.53	1159.86	25.39
1086.02	1086.36	23.24	1110.64	1110.97	23.96	1135.25	1135.59	24.68	1159.87	1160.20	25.40
1086.37	1086.70	23.25	1110.98	1111.31	23.97	1135.60	1135.93	24.69	1160.21	1160.54	25.41
1086.71	1087.04	23.26	1111.32	1111.65	23.98	1135.94	1136.27	24.70	1160.55	1160.89	25.42
1087.05	1087.38	23.27	1111.66	1112.00	23.99	1136.28	1136.61	24.71	1160.90	1161.23	25.43
1087.39	1087.72	23.28	1112.01	1112.34	24.00	1136.62	1136.95	24.72	1161.24	1161.57	25.44
1087.73	1088.07	23.29	1112.35	1112.68	24.01	1136.96	1137.30	24.73	1161.58	1161.91	25.45
1088.08	1088.41	23.30	1112.69	1113.02	24.02	1137.31	1137.64	24.74	1161.92	1162.25	25.46
1088.42	1088.75	23.31	1113.03	1113.36	24.03	1137.65	1137.98	24.75	1162.26	1162.60	25.47
1088.76	1089.09	23.32	1113.37	1113.71	24.04	1137.99	1138.32	24.76	1162.61	1162.94	25.48
1089.10	1089.43	23.33	1113.72	1114.05	24.05	1138.33	1138.66	24.77	1162.95	1163.28	25.49
1089.44	1089.77	23.34	1114.06	1114.39	24.06	1138.67	1139.01	24.78	1163.29	1163.62	25.50
1089.78	1090.12	23.35	1114.40	1114.73	24.07	1139.02	1139.35	24.79	1163.63	1163.96	25.51
1090.13	1090.46	23.36	1114.74	1115.07	24.08	1139.36	1139.69	24.80	1163.97	1164.30	25.52
1090.47	1090.80	23.37	1115.08	1115.42	24.09	1139.70	1140.03	24.81	1164.31	1164.65	25.53
1090.81	1091.14	23.38	1115.43	1115.76	24.10	1140.04	1140.37	24.82	1164.66	1164.99	25.54
1091.15	1091.48	23.39	1115.77	1116.10	24.11	1140.38	1140.71	24.83	1165.00	1165.33	25.55
1091.49	1091.83	23.40	1116.11	1116.44	24.12	1140.72	1141.06	24.84	1165.34	1165.67	25.56
1091.84	1092.17	23.41	1116.45	1116.78	24.13	1141.07	1141.40	24.85	1165.68	1166.01	25.57
1092.18	1092.51	23.42	1116.79	1117.13	24.14	1141.41	1141.74	24.86	1166.02	1166.36	25.58
1092.52	1092.85	23.43	1117.14	1117.47	24.15	1141.75	1142.08	24.87	1166.37	1166.70	25.59
1092.86	1093.19	23.44	1117.48	1117.81	24.16	1142.09	1142.42	24.88	1166.71	1167.04	25.60
1093.20	1093.54	23.45	1117.82	1118.15	24.17	1142.43	1142.77	24.89	1167.05	1167.38	25.61
1093.55	1093.88	23.46	1118.16	1118.49	24.18	1142.78	1143.11	24.90	1167.39	1167.72	25.62
1093.89	1094.22	23.47	1118.50	1118.83	24.19	1143.12	1143.45	24.91	1167.73	1168.07	25.63
1094.23	1094.56	23.48	1118.84	1119.18	24.20	1143.46	1143.79	24.92	1168.08	1168.41	25.64
1094.57	1094.90	23.49	1119.19	1119.52	24.21	1143.80	1144.13	24.93	1168.42	1168.75	25.65
1094.91	1095.24	23.50	1119.53	1119.86	24.22	1144.14	1144.48	24.94	1168.76	1169.09	25.66
1095.25	1095.59	23.51	1119.87	1120.20	24.23	1144.49	1144.82	24.95	1169.10	1169.43	25.67
1095.60	1095.93	23.52	1120.21	1120.54	24.24	1144.83	1145.16	24.96	1169.44	1169.77	25.68
1095.94	1096.27	23.53	1120.55	1120.89	24.25	1145.17	1145.50	24.97	1169.78	1170.12	25.69
1096.28	1096.61	23.54	1120.90	1121.23	24.26	1145.51	1145.84	24.98	1170.13	1170.46	25.70
1096.62	1096.95	23.55	1121.24	1121.57	24.27	1145.85	1146.18	24.99	1170.47	1170.80	25.71
1096.96	1097.30	23.56	1121.58	1121.91	24.28	1146.19	1146.53	25.00	1170.81	1171.14	25.72
1097.31	1097.64	23.57	1121.92	1122.25	24.29	1146.54	1146.87	25.01	1171.15	1171.48	25.73
1097.65	1097.98	23.58	1122.26	1122.60	24.30	1146.88	1147.21	25.02	1171.49	1171.83	25.74
1097.99	1098.32	23.59	1122.61	1122.94	24.31	1147.22	1147.55	25.03	1171.84	1172.17	25.75
1098.33	1098.66	23.60	1122.95	1123.28	24.32	1147.56	1147.89	25.04	1172.18	1172.51	25.76
1098.67	1099.01	23.61	1123.29	1123.62	24.33	1147.90	1148.24	25.05	1172.52	1172.85	25.77
1099.02	1099.35	23.62	1123.63	1123.96	24.34	1148.25	1148.58	25.06	1172.86	1173.19	25.78
1099.36	1099.69	23.63	1123.97	1124.30	24.35	1148.59	1148.92	25.07	1173.20	1173.54	25.79
1099.70	1100.03	23.64	1124.31	1124.65	24.36	1148.93	1149.26	25.08	1173.55	1173.88	25.80
1100.04	1100.37	23.65	1124.66	1124.99	24.37	1149.27	1149.60	25.09	1173.89	1174.22	25.81
1100.38	1100.71	23.66	1125.00	1125.33	24.38	1149.61	1149.95	25.10	1174.23	1174.56	25.82
1100.72	1101.06	23.67	1125.34	1125.67	24.39	1149.96	1150.29	25.11	1174.57	1174.90	25.83
1101.07	1101.40	23.68	1125.68	1126.01	24.40	1150.30	1150.63	25.12	1174.91	1175.24	25.84
1101.41	1101.74	23.69	1126.02	1126.36	24.41	1150.64	1150.97	25.13	1175.25	1175.59	25.85
1101.75	1102.08	23.70	1126.37	1126.70	24.42	1150.98	1151.31	25.14	1175.60	1175.93	25.86
1102.09	1102.42	23.71	1126.71	1127.04	24.43	1151.32	1151.65	25.15	1175.94	1176.27	25.87
1102.43	1102.77	23.72	1127.05	1127.38	24.44	1151.66	1152.00	25.16	1176.28	1176.61	25.88
1102.78	1103.11	23.73	1127.39	1127.72	24.45	1152.01	1152.34	25.17	1176.62	1176.95	25.89
1103.12	1103.45	23.74	1127.73	1128.07	24.46	1152.35	1152.68	25.18	1176.96	1177.30	25.90
1103.46	1103.79	23.75	1128.08	1128.41	24.47	1152.69	1153.02	25.19	1177.31	1177.64	25.91

Employee's maximum CPP contribution for the year 1997 is **$944.78**

La cotisation maximale de l'employé au RPC pour l'année 1997 est de **$944.78**

B-46

Employment Insurance Premiums   Cotisations d'assurance-emploi

Insurable Earnings Rémunération assurable		EI premium Cotisation d'AE	Insurable Earnings Rémunération assurable		EI premium Cotisation d'AE	Insurable Earnings Rémunération assurable		EI premium Cotisation d'AE	Insurable Earnings Rémunération assurable		EI premium Cotisation d'AE
From - De	To - À		From - De	To - À		From - De	To - À		From - De	To - À	
1092.59	1092.93	31.69	1117.42	1117.75	32.41	1142.25	1142.58	33.13	1167.07	1167.41	33.85
1092.94	1093.27	31.70	1117.76	1118.10	32.42	1142.59	1142.93	33.14	1167.42	1167.75	33.86
1093.28	1093.62	31.71	1118.11	1118.44	32.43	1142.94	1143.27	33.15	1167.76	1168.10	33.87
1093.63	1093.96	31.72	1118.45	1118.79	32.44	1143.28	1143.62	33.16	1168.11	1168.44	33.88
1093.97	1094.31	31.73	1118.80	1119.13	32.45	1143.63	1143.96	33.17	1168.45	1168.79	33.89
1094.32	1094.65	31.74	1119.14	1119.48	32.46	1143.97	1144.31	33.18	1168.80	1169.13	33.90
1094.66	1094.99	31.75	1119.49	1119.82	32.47	1144.32	1144.65	33.19	1169.14	1169.48	33.91
1095.00	1095.34	31.76	1119.83	1120.17	32.48	1144.66	1144.99	33.20	1169.49	1169.82	33.92
1095.35	1095.68	31.77	1120.18	1120.51	32.49	1145.00	1145.34	33.21	1169.83	1170.17	33.93
1095.69	1096.03	31.78	1120.52	1120.86	32.50	1145.35	1145.68	33.22	1170.18	1170.51	33.94
1096.04	1096.37	31.79	1120.87	1121.20	32.51	1145.69	1146.03	33.23	1170.52	1170.86	33.95
1096.38	1096.72	31.80	1121.21	1121.55	32.52	1146.04	1146.37	33.24	1170.87	1171.20	33.96
1096.73	1097.06	31.81	1121.56	1121.89	32.53	1146.38	1146.72	33.25	1171.21	1171.55	33.97
1097.07	1097.41	31.82	1121.90	1122.24	32.54	1146.73	1147.06	33.26	1171.56	1171.89	33.98
1097.42	1097.75	31.83	1122.25	1122.58	32.55	1147.07	1147.41	33.27	1171.90	1172.24	33.99
1097.76	1098.10	31.84	1122.59	1122.93	32.56	1147.42	1147.75	33.28	1172.25	1172.58	34.00
1098.11	1098.44	31.85	1122.94	1123.27	32.57	1147.76	1148.10	33.29	1172.59	1172.93	34.01
1098.45	1098.79	31.86	1123.28	1123.62	32.58	1148.11	1148.44	33.30	1172.94	1173.27	34.02
1098.80	1099.13	31.87	1123.63	1123.96	32.59	1148.45	1148.79	33.31	1173.28	1173.62	34.03
1099.14	1099.48	31.88	1123.97	1124.31	32.60	1148.80	1149.13	33.32	1173.63	1173.96	34.04
1099.49	1099.82	31.89	1124.32	1124.65	32.61	1149.14	1149.48	33.33	1173.97	1174.31	34.05
1099.83	1100.17	31.90	1124.66	1124.99	32.62	1149.49	1149.82	33.34	1174.32	1174.65	34.06
1100.18	1100.51	31.91	1125.00	1125.34	32.63	1149.83	1150.17	33.35	1174.66	1174.99	34.07
1100.52	1100.86	31.92	1125.35	1125.68	32.64	1150.18	1150.51	33.36	1175.00	1175.34	34.08
1100.87	1101.20	31.93	1125.69	1126.03	32.65	1150.52	1150.86	33.37	1175.35	1175.68	34.09
1101.21	1101.55	31.94	1126.04	1126.37	32.66	1150.87	1151.20	33.38	1175.69	1176.03	34.10
1101.56	1101.89	31.95	1126.38	1126.72	32.67	1151.21	1151.55	33.39	1176.04	1176.37	34.11
1101.90	1102.24	31.96	1126.73	1127.06	32.68	1151.56	1151.89	33.40	1176.38	1176.72	34.12
1102.25	1102.58	31.97	1127.07	1127.41	32.69	1151.90	1152.24	33.41	1176.73	1177.06	34.13
1102.59	1102.93	31.98	1127.42	1127.75	32.70	1152.25	1152.58	33.42	1177.07	1177.41	34.14
1102.94	1103.27	31.99	1127.76	1128.10	32.71	1152.59	1152.93	33.43	1177.42	1177.75	34.15
1103.28	1103.62	32.00	1128.11	1128.44	32.72	1152.94	1153.27	33.44	1177.76	1178.10	34.16
1103.63	1103.96	32.01	1128.45	1128.79	32.73	1153.28	1153.62	33.45	1178.11	1178.44	34.17
1103.97	1104.31	32.02	1128.80	1129.13	32.74	1153.63	1153.96	33.46	1178.45	1178.79	34.18
1104.32	1104.65	32.03	1129.14	1129.48	32.75	1153.97	1154.31	33.47	1178.80	1179.13	34.19
1104.66	1104.99	32.04	1129.49	1129.82	32.76	1154.32	1154.65	33.48	1179.14	1179.48	34.20
1105.00	1105.34	32.05	1129.83	1130.17	32.77	1154.66	1154.99	33.49	1179.49	1179.82	34.21
1105.35	1105.68	32.06	1130.18	1130.51	32.78	1155.00	1155.34	33.50	1179.83	1180.17	34.22
1105.69	1106.03	32.07	1130.52	1130.86	32.79	1155.35	1155.68	33.51	1180.18	1180.51	34.23
1106.04	1106.37	32.08	1130.87	1131.20	32.80	1155.69	1156.03	33.52	1180.52	1180.86	34.24
1106.38	1106.72	32.09	1131.21	1131.55	32.81	1156.04	1156.37	33.53	1180.87	1181.20	34.25
1106.73	1107.06	32.10	1131.56	1131.89	32.82	1156.38	1156.72	33.54	1181.21	1181.55	34.26
1107.07	1107.41	32.11	1131.90	1132.24	32.83	1156.73	1157.06	33.55	1181.56	1181.89	34.27
1107.42	1107.75	32.12	1132.25	1132.58	32.84	1157.07	1157.41	33.56	1181.90	1182.24	34.28
1107.76	1108.10	32.13	1132.59	1132.93	32.85	1157.42	1157.75	33.57	1182.25	1182.58	34.29
1108.11	1108.44	32.14	1132.94	1133.27	32.86	1157.76	1158.10	33.58	1182.59	1182.93	34.30
1108.45	1108.79	32.15	1133.28	1133.62	32.87	1158.11	1158.44	33.59	1182.94	1183.27	34.31
1108.80	1109.13	32.16	1133.63	1133.96	32.88	1158.45	1158.79	33.60	1183.28	1183.62	34.32
1109.14	1109.48	32.17	1133.97	1134.31	32.89	1158.80	1159.13	33.61	1183.63	1183.96	34.33
1109.49	1109.82	32.18	1134.32	1134.65	32.90	1159.14	1159.48	33.62	1183.97	1184.31	34.34
1109.83	1110.17	32.19	1134.66	1134.99	32.91	1159.49	1159.82	33.63	1184.32	1184.65	34.35
1110.18	1110.51	32.20	1135.00	1135.34	32.92	1159.83	1160.17	33.64	1184.66	1184.99	34.36
1110.52	1110.86	32.21	1135.35	1135.68	32.93	1160.18	1160.51	33.65	1185.00	1185.34	34.37
1110.87	1111.20	32.22	1135.69	1136.03	32.94	1160.52	1160.86	33.66	1185.35	1185.68	34.38
1111.21	1111.55	32.23	1136.04	1136.37	32.95	1160.87	1161.20	33.67	1185.69	1186.03	34.39
1111.56	1111.89	32.24	1136.38	1136.72	32.96	1161.21	1161.55	33.68	1186.04	1186.37	34.40
1111.90	1112.24	32.25	1136.73	1137.06	32.97	1161.56	1161.89	33.69	1186.38	1186.72	34.41
1112.25	1112.58	32.26	1137.07	1137.41	32.98	1161.90	1162.24	33.70	1186.73	1187.06	34.42
1112.59	1112.93	32.27	1137.42	1137.75	32.99	1162.25	1162.58	33.71	1187.07	1187.41	34.43
1112.94	1113.27	32.28	1137.76	1138.10	33.00	1162.59	1162.93	33.72	1187.42	1187.75	34.44
1113.28	1113.62	32.29	1138.11	1138.44	33.01	1162.94	1163.27	33.73	1187.76	1188.10	34.45
1113.63	1113.96	32.30	1138.45	1138.79	33.02	1163.28	1163.62	33.74	1188.11	1188.44	34.46
1113.97	1114.31	32.31	1138.80	1139.13	33.03	1163.63	1163.96	33.75	1188.45	1188.79	34.47
1114.32	1114.65	32.32	1139.14	1139.48	33.04	1163.97	1164.31	33.76	1188.80	1189.13	34.48
1114.66	1114.99	32.33	1139.49	1139.82	33.05	1164.32	1164.65	33.77	1189.14	1189.48	34.49
1115.00	1115.34	32.34	1139.83	1140.17	33.06	1164.66	1164.99	33.78	1189.49	1189.82	34.50
1115.35	1115.68	32.35	1140.18	1140.51	33.07	1165.00	1165.34	33.79	1189.83	1190.17	34.51
1115.69	1116.03	32.36	1140.52	1140.86	33.08	1165.35	1165.68	33.80	1190.18	1190.51	34.52
1116.04	1116.37	32.37	1140.87	1141.20	33.09	1165.69	1166.03	33.81	1190.52	1190.86	34.53
1116.38	1116.72	32.38	1141.21	1141.55	33.10	1166.04	1166.37	33.82	1190.87	1191.20	34.54
1116.73	1117.06	32.39	1141.56	1141.89	33.11	1166.38	1166.72	33.83	1191.21	1191.55	34.55
1117.07	1117.41	32.40	1141.90	1142.24	33.12	1166.73	1167.06	33.84	1191.56	1191.89	34.56

Yearly maximum insurable earnings are $39,000   Le maximum annuel de la rémunération assurable est de 39 000 $

Yearly maximum employee premiums are $1,131   La cotisation maximale annuelle de l'employé est de 1 131 $

C-12

Saskatchewan
Federal and Provincial Tax Deductions
Biweekly (26 pay periods a year)

Saskatchewan
Retenues d'impôt fédéral et provincial
Aux deux semaines (26 périodes de paie par année)

Pay Rémunération		\multicolumn{12}{c}{If the employee's claim code from the TD1(E) form is / Si le code de demande de l'employé selon le formulaire TD1(F) est}

Pay Rémunération		0	1	2	3	4	5	6	7	8	9	10
From De	Less than Moins de	\multicolumn{11}{c}{Deduct from each pay / Retenez sur chaque paie}
915.-	931.	249.05	182.40	174.20	157.85	141.50	125.60	107.85	92.05	76.20	60.40	44.55
931.-	947.	253.45	186.75	178.60	162.25	145.90	129.85	112.90	97.10	81.25	65.45	49.60
947.-	963.	257.85	191.15	183.00	166.65	150.30	134.10	117.95	102.10	86.30	70.45	54.65
963.-	979.	262.25	195.55	187.40	171.05	154.70	138.40	122.50	106.70	90.85	75.05	59.20
979.-	995.	266.65	199.95	191.80	175.45	159.10	142.80	126.75	110.90	95.10	79.30	63.45
995.-	1011.	271.05	204.35	196.20	179.85	163.50	147.20	131.00	115.15	99.35	83.50	67.70
1011.-	1027.	275.45	208.75	200.60	184.25	167.90	151.55	135.25	119.40	103.60	87.75	71.90
1027.-	1043.	279.85	213.15	205.00	188.65	172.30	155.95	139.65	123.65	107.80	92.00	76.15
1043.-	1059.	284.25	217.55	209.40	193.05	176.70	160.35	144.05	127.90	112.05	96.25	80.40
1059.-	1075.	288.65	221.95	213.80	197.45	181.10	164.75	148.45	132.10	116.30	100.45	84.65
1075.-	1091.	293.05	226.35	218.20	201.85	185.50	169.15	152.85	136.50	120.55	104.70	88.90
1091.-	1107.	297.45	230.75	222.60	206.25	189.90	173.55	157.25	140.90	124.75	108.95	93.10
1107.-	1123.	301.85	235.15	227.00	210.65	194.30	177.95	161.65	145.30	129.00	113.20	97.35
1123.-	1139.	306.25	239.55	231.40	215.05	198.70	182.35	166.05	149.70	133.35	117.45	101.60
1139.-	1155.	311.95	245.25	237.05	220.75	204.40	188.05	171.70	155.35	139.05	122.90	107.05
1155.-	1171.	318.60	251.90	243.75	227.40	211.05	194.70	178.40	162.05	145.70	129.35	113.50
1171.-	1187.	325.30	258.60	250.40	234.10	217.75	201.40	185.05	168.70	152.40	136.05	119.95
1187.-	1203.	331.95	265.25	257.10	240.75	224.40	208.05	191.75	175.40	159.05	142.70	126.40
1203.-	1219.	338.65	271.95	263.75	247.45	231.10	214.75	198.40	182.05	165.75	149.40	133.05
1219.-	1235.	345.30	278.60	270.45	254.10	237.75	221.40	205.10	188.75	172.40	156.05	139.70
1235.-	1251.	352.00	285.30	277.10	260.80	244.45	228.10	211.75	195.40	179.10	162.75	146.40
1251.-	1267.	358.65	291.95	283.80	267.45	251.10	234.75	218.45	202.10	185.75	169.40	153.05
1267.-	1283.	365.35	298.65	290.45	274.15	257.80	241.45	225.10	208.75	192.45	176.10	159.75
1283.-	1299.	372.00	305.30	297.15	280.80	264.45	248.10	231.80	215.45	199.10	182.75	166.40
1299.-	1315.	378.70	312.00	303.80	287.50	271.15	254.80	238.45	222.10	205.80	189.45	173.10
1315.-	1331.	385.35	318.65	310.50	294.15	277.80	261.45	245.15	228.80	212.45	196.10	179.75
1331.-	1347.	392.05	325.35	317.15	300.85	284.50	268.15	251.80	235.45	219.15	202.80	186.45
1347.-	1363.	398.70	332.00	323.85	307.50	291.15	274.80	258.50	242.15	225.80	209.45	193.10
1363.-	1379.	405.40	338.70	330.50	314.20	297.85	281.50	265.15	248.80	232.50	216.15	199.80
1379.-	1395.	412.15	345.45	337.25	320.95	304.60	288.25	271.90	255.55	239.25	222.90	206.55
1395.-	1411.	418.95	352.25	344.05	327.75	311.40	295.05	278.70	262.35	246.05	229.70	213.35
1411.-	1427.	425.90	359.05	350.85	334.55	318.20	301.85	285.50	269.15	252.85	236.50	220.15
1427.-	1443.	433.05	365.85	357.65	341.35	325.00	308.65	292.30	275.95	259.65	243.30	226.95
1443.-	1459.	440.20	372.65	364.45	348.15	331.80	315.45	299.10	282.75	266.45	250.10	233.75
1459.-	1475.	447.40	379.45	371.25	354.95	338.60	322.25	305.90	289.55	273.25	256.90	240.55
1475.-	1491.	454.55	386.25	378.05	361.75	345.40	329.05	312.70	296.35	280.05	263.70	247.35
1491.-	1507.	461.70	393.05	384.85	368.55	352.20	335.85	319.50	303.15	286.85	270.50	254.15
1507.-	1523.	468.95	399.95	391.80	375.45	359.10	342.75	326.40	310.10	293.75	277.40	261.05
1523.-	1539.	476.25	406.90	398.70	382.35	366.05	349.70	333.35	317.00	300.65	284.35	268.00
1539.-	1555.	483.55	413.80	405.65	389.30	372.95	356.60	340.25	323.95	307.60	291.25	274.90
1555.-	1571.	490.80	420.95	412.55	396.20	379.90	363.55	347.20	330.85	314.50	298.20	281.85
1571.-	1587.	498.10	428.25	419.70	403.15	386.80	370.45	354.10	337.80	321.45	305.10	288.75
1587.-	1603.	505.40	435.50	426.95	410.05	393.75	377.40	361.05	344.70	328.35	312.05	295.70
1603.-	1619.	512.65	442.80	434.25	417.15	400.65	384.30	367.95	351.65	335.30	318.95	302.60
1619.-	1635.	519.95	450.10	441.55	424.45	407.60	391.25	374.90	358.55	342.20	325.90	309.55
1635.-	1651.	527.25	457.40	448.80	431.70	414.60	398.15	381.80	365.50	349.15	332.80	316.45
1651.-	1667.	534.50	464.65	456.10	439.00	421.85	405.10	388.75	372.40	356.05	339.75	323.40
1667.-	1683.	541.80	471.95	463.40	446.30	429.15	412.05	395.65	379.35	363.00	346.65	330.30
1683.-	1699.	549.10	479.25	470.70	453.60	436.45	419.30	402.60	386.25	369.90	353.60	337.25
1699.-	1715.	556.40	486.50	477.95	460.85	443.70	426.60	409.50	393.20	376.85	360.50	344.15
1715.-	1731.	563.65	493.80	485.25	468.15	451.00	433.90	416.80	400.10	383.75	367.45	351.10
1731.-	1747.	570.95	501.10	492.55	475.40	458.30	441.15	424.05	407.05	390.70	374.35	358.00
1747.-	1763.	578.25	508.35	499.80	482.70	465.60	448.45	431.35	414.25	397.60	381.30	364.95
1763.-	1779.	585.50	515.65	507.10	490.00	472.85	455.75	438.65	421.50	404.55	388.20	371.85
1779.-	1795.	592.80	522.95	514.40	497.25	480.15	463.05	445.90	428.80	411.70	395.15	378.80

This table is available on diskette (TOD).   Vous pouvez obtenir cette table sur disquette (TSD).

D-9

Modelling Golf Ball Trajectories

CONSIDER THIS SITUATION

In the 1997 Masters Golf Tournament, Tiger Woods consistently hit drives and iron shots much farther than his opponents. On par 5 holes that normally require 3 shots to reach the green, he usually needed only 2. Tiger won the tournament by a record margin of 12 shots.

- What influences how far a golf ball travels when a golfer hits it?
- What determines how high the ball goes?
- When might a golfer be interested in the height of a shot?
- How could we describe the trajectory of a golf shot?

On pages 142–145, you will develop a mathematical model to describe the distances and heights of golf shots with different clubs. You will use the same model to describe the trajectory of a shot.

FYI Visit www.awl.com/canada/school/connections

For information related to the above problem, click on <u>MATHLINKS</u> followed by *Mathematics 11*. Then select a topic under Modelling Golf Ball Trajectories.

MASTERS® TOURNAMENT 1997 OFFICIAL SCORECARD

Hole	1	2	3	4	5	6	7	8	9	Out
Yardage	400	555	360	205	435	180	360	535	435	3465
Par	4	5	4	3	4	3	4	5	4	36

	10	11	12	13	14	15	16	17	18	In	Totals
	485	455	155	485	405	500	170	400	405	3460	6925
	4	4	3	5	4	5	3	4	4	36	

I HAVE CHECKED MY SCORE HOLE BY HOLE.

Graphing a Function

Suppose you are standing on a cliff, 110 m above a beach. You throw a stone straight up at 17 m/s. After reaching its maximum height the stone falls to the beach, just missing the cliff. From physics, its height, h metres, is a function of the elapsed time, t seconds.

$$h(t) = -4.9t^2 + 17t + 110$$

A graph and a table of values for this function are shown.

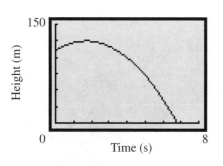

Graphing a function

Follow these steps for the TI-83 calculator. The keystrokes depend on your calculator.

- *Input the function.*

  Enter this equation: $y = -4.9x^2 + 17x + 110$

- *Define the viewing window.*

  On the graph above, x runs from 0 to 8 with tick marks at every unit; and y runs from 0 to 150 with tick marks at every 25 units. Enter these numbers in your calculator.

- *Graph the equation.*

  Press GRAPH. The result should be similar to the graph above.

Displaying a table of values

Your calculator may have a tables feature. Follow these steps.

- *Set up the table.*

  In the table above, x starts at 0 and increases in steps of 1. Press TBLSET, then enter these numbers in your calculator.

- *Display the table.*

Press TABLE. The result should be similar to the one on page 84.

- *Display the graph and the table together.*

Your calculator may have a split screen feature. Consult your manual.

Displaying the coordinates of a point and adjusting the window

- *Use the trace feature.*

Press TRACE, then use the arrow keys to move the cursor along the graph.

The coordinates of points on the graph are shown (below left). We would prefer they contain fewer decimal places, and the *x*-coordinates increase by regular increments.

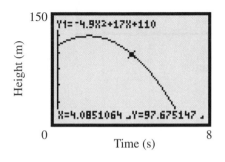

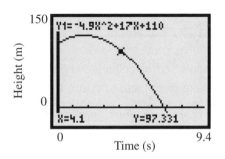

On the TI-83, the display is 95 pixels wide. If we adjust the window so that Xmax $= 9.4$, we obtain values of x in tenths of a second. We use a negative number for Ymin so the horizontal axis appears above the coordinates at the bottom of the screen (above right).

Complete these exercises.

1. Determine the height of the stone after each time.

   a) 1.0 s   **b)** 2.4 s   **c)** 4.5 s   **d)** 6.1 s

2. At what time(s) is the stone at each height?

   a) 110 m   **b)** 90 m   **c)** 60 m   **d)** 30 m

3. What is the maximum height? How long does it take the stone to reach this height?

4. How long does it take the stone to hit the beach?

5. What is the domain of the function?

Zooming to determine the maximum height and time to hit the beach

- *Trace close to the maximum point.*

Press TRACE, then move the cursor close to the maximum point.

- *Set the zoom factors to 10.*

  Zooming will be more efficient if the zoom factors are set to 10.

  Press [ZOOM] [▶] 4 to get the zoom factors menu. Set XFact and YFact to 10.

- *Zoom in to enlarge the graph.*

  Press [ZOOM] 2 [ENTER] to zoom in. Press [TRACE], then use the arrow keys. The screen (below left) shows the approximate coordinates of the maximum point.

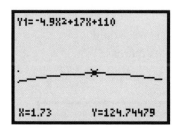
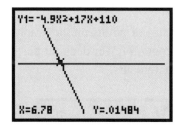

The window settings change when you use the zoom feature. You need to reset them to return to the original graph.

- *Trace close to the horizontal intercept.*

  Press [TRACE], then move the cursor close to the point where $y = 0$.

- *Zoom in to enlarge the graph.*

  Press [ZOOM] 2 [ENTER] to zoom in. Press [TRACE], then use the arrow keys. Repeat for more accuracy, if desired. The screen (above right) shows a value of x for which $y \doteq 0$.

These screens show that the stone takes about 1.73 s to reach a maximum height of about 124.74 m, and it hits the ground after about 6.78 s.

Calculating to determine the maximum height and time to hit the beach

- *Graph the function.*

  To calculate the coordinates of the maximum point, begin by graphing the function as it was originally.

- *Activate the "maximum" feature.*

  Press [CALC] 4 to select "maximum." Since a function can have more than one maximum point, you must indicate the one to be determined. The words "Left Bound?" will appear. Move the cursor to the left of the maximum point, and press [ENTER]. Do the same for "Right Bound?" The word "Guess?" will appear. Move the cursor between the left and right bounds, and press [ENTER]. The screen (below left on the next page) shows the coordinates of the maximum point, to several decimal places.

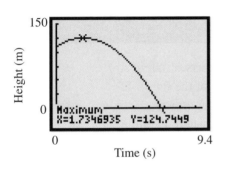

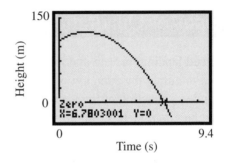

• *Activate the "zero" feature.*

Press CALC 2 to select "zero." The method is the same as the method for finding the maximum. The screen (above right) shows the value of x when $y = 0$.

These screens show the results to many decimal places. Although this much accuracy is not needed in this example, there may be occasions when the results obtained by calculation are preferable to the results obtained by tracing and zooming.

If you have completed the above steps successfully, you are ready to do these exercises.

6. ***Varying the initial velocity of the stone*** Compare the equation $h(t) = -4.9t^2 + 17t + 110$ with the information on page 84.

   a) Which coefficient in the equation is the initial velocity of the stone?

   b) Predict how the graph would change for each initial velocity. Use a graphing calculator to check your predictions.

   i) 22 m/s   **ii)** 27 m/s   **iii)** 12 m/s   **iv)** 7 m/s

7. Suppose a stone is thrown upward with an initial velocity of 25 m/s. Use a graphing calculator.

   a) Determine the maximum height, and the time to reach this height.

   b) Determine the time it takes to hit the beach.

8. What initial velocity is required for each situation?

   a) the stone to reach a maximum height of 150 m

   b) the stone to hit the beach in 10 s

9. ***Varying the height of the cliff*** Compare the equation $h(t) = -4.9t^2 + 17t + 110$ with the information on page 84.

   a) Which coefficient in the equation is the height of the cliff?

   b) Predict how the graph would change for each cliff height. Use a graphing calculator to check your predictions.

   i) 130 m   **ii)** 150 m   **iii)** 90 m   **iv)** 70 m

2.1 Graphs of Quadratic Functions

In case of a forced landing, private and military aircraft often carry a flare pistol that can be fired to attract the attention of rescuers. The height of the flare above the ground is a function of the elapsed time since firing. A typical equation for the height, h metres, might be $h(t) = -5t^2 + 100t$, where the time, t, is in seconds.

The table of values and the graph show how the height varies with time. The heights have been rounded to the nearest metre.

Time, t (s)	Height, h (m)
0	0
2	180
5	375
10	500
15	375
18	180
20	0

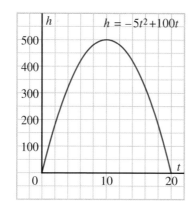

The function $h(t) = -5t^2 + 100t$ is an example of a quadratic function. The function on page 84, $h(t) = -4.9t^2 + 17t + 110$, is also a quadratic function.

> A *quadratic function* has a defining equation that can be written in this form:
>
> $y = ax^2 + bx + c$ or $f(x) = ax^2 + bx + c$
>
> where a, b, and c are constants and a cannot be zero.
>
> The graph of every quadratic function is a curve called a *parabola*.

These are quadratic functions because they can be written in the form $y = ax^2 + bx + c$ or $f(x) = ax^2 + bx + c$.

$y = x^2 - 6$
$f(x) = (2x + 5)(x - 6)$
$h(t) = 2(t - 1)^2 + 5$

These are not quadratic functions because they cannot be written in the form $f(x) = ax^2 + bx + c$.

$r = \dfrac{1}{s^2 - 4}$
$f(z) = \sqrt{z}$

All five equations above represent functions. We can use any letters for the variables. We can use function notation on the left side, or we can use a variable. If the expression on the right side can be written in the form $ax^2 + bx + c$, where $a \neq 0$, the function is a quadratic function.

Every parabola has a line of symmetry that intersects the parabola at a point called the *vertex*. The line of symmetry is called the *axis of symmetry*.

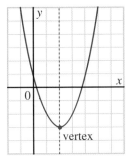

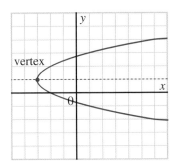

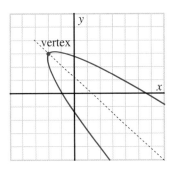

Example 1

a) Graph the function defined by the equation $y = x^2 - 7x + 10$.

b) Identify.

   i) the y-intercept

   ii) the x-intercepts

   iii) the equation of the axis of symmetry

   iv) the coordinates of the vertex

c) What are the domain and range of the function?

Solution

a) To make a table of values, evaluate the function for values of x such as 0, 1, 2, 3, … There are pairs of equal y-values in the table. This pattern is caused by the symmetry of the graph, and it can be used to complete the table for other x-values. This table also includes values of y for $x = -1$ and $y = 18$ to show more of the parabola.

x	y
−1	18
0	10
1	4
2	0
3	−2
4	−2
5	0
6	4
7	10
8	18

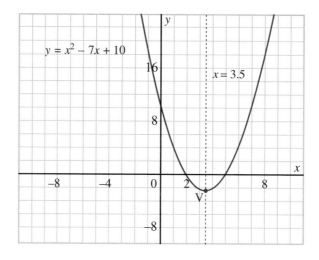

b) **i)** From the graph or the table, the y-intercept is the value of y when $x = 0$. This is 10.

ii) The x-intercepts are the values of x when $y = 0$. These are 2 and 5.

iii) The axis of symmetry is the vertical line halfway between $x = 2$ and $x = 5$. Its equation is $x = 3.5$.

iv) The x-coordinate of the vertex is 3.5.

Substitute 3.5 for x in the equation $y = x^2 − 7x + 10$.
$$y = (3.5)^2 − 7(3.5) + 10$$
$$= −2.25$$

The coordinates of the vertex are V$(3.5, −2.25)$.

c) The domain of this function is the set of all possible values of x. This is the set of all real numbers. The range is the set of all possible values of y. This is the set of real numbers greater than or equal to $−2.25$.

Example 2

a) Use technology to graph the function $f(x) = x^2 - 7x + 4$.

b) Identify.

    i) the coordinates of the vertex   **ii)** the equation of the axis of symmetry

    iii) the y-intercept   **iv)** the x-intercepts

c) What are the domain and range of the function?

Solution

a) Input the function, set the window, then graph the function to obtain a
result similar to the graph (below left).

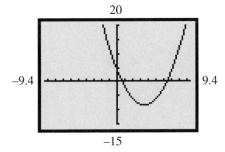 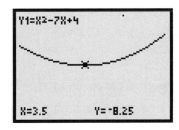

b) **i)** Trace and zoom to find the point with the least possible y-coordinate
(above right). The coordinates of the vertex are $(3.5, -8.25)$.

   ii) The axis of symmetry is the vertical line through the vertex. Its
equation is $x = 3.5$.

   iii) The y-intercept is the value of y when $x = 0$. This is 4.

   iv) The graph (above left) in part a shows there are two x-intercepts.
To find each x-intercept, re-graph the function in its original
window. By tracing and zooming, find the points on the graph with
y-coordinate 0, or very close to 0. The graphs below show that the
x-intercepts are approximately 0.628 and 6.372.

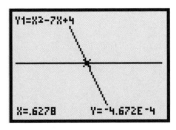 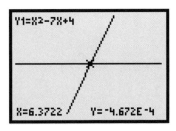

c) The domain is the set of all real numbers. The range is the set of real
numbers greater than or equal to -8.25.

In *Example 2b,* part iv, we will probably never get the *y*-coordinate equal to zero no matter how often we zoom in. Unless otherwise stated, we will zoom in enough times until we are sure that the intercepts are accurate to at least 3 decimal places. This was done in *Example 2.*

An equation that can be written in the form $ax^2 + bx + c = 0$ is called a *quadratic equation.* In *Example 1,* we determined the *x*-intercepts of the function $y = x^2 - 7x + 10$ from the table or the graph.

We could have obtained the same result by substituting $y = 0$ in $y = x^2 - 7x + 10$, then solving the resulting quadratic equation $0 = x^2 - 7x + 10$.

$$x^2 - 7x + 10 = 0$$
$$(x - 2)(x - 5) = 0$$

Either $x - 2 = 0$ or $x - 5 = 0$
$\qquad\qquad\quad x = 2 \qquad\qquad\qquad x = 5$

We say that:

- 2 and 5 are the *roots* of the equation $x^2 - 7x + 10 = 0$.

- 2 and 5 are the *zeros* of the function $y = x^2 - 7x + 10$.

Example 3

Solve the equation $3x^2 - 11x - 4 = 0$.

a) Use algebra.

b) Graph the corresponding function $f(x) = 3x^2 - 11x - 4$.

Solution

a) $3x^2 - 11x - 4 = 0$
$\quad (3x + 1)(x - 4) = 0$

Either $\quad 3x + 1 = 0 \qquad$ or $\qquad x - 4 = 0$
$\qquad\qquad\quad x = -\dfrac{1}{3} \qquad\qquad\qquad x = 4$

b) Use a graphing calculator to graph the function $f(x) = 3x^2 - 11x - 4$.
The graph is shown (below right). Use trace and zoom to determine the
x-intercepts. The screens below show that the x-intercepts are
approximately -0.333 and exactly 4.

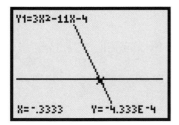

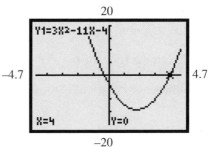

In *Example 3*, the roots of the quadratic equation are the same as the x-intercepts
(or zeros) of the graph.

Example 4

Write the equation of a quadratic function that has zeros $\frac{2}{3}$ and $-\frac{1}{2}$.

Solution

The corresponding quadratic equation has roots $\frac{2}{3}$ and $-\frac{1}{2}$.

> **Think ...**
>
> Just reverse the steps in the solution of *Example 3a*.

If $x = \frac{2}{3}$, then $3x = 2$, or $3x - 2 = 0$

If $x = -\frac{1}{2}$, then $2x = -1$, or $2x + 1 = 0$

A quadratic equation with roots $\frac{2}{3}$ and $-\frac{1}{2}$ is:

$$(3x - 2)(2x + 1) = 0$$
or $\qquad 6x^2 - x - 2 = 0$

Therefore, a quadratic function with zeros $\frac{2}{3}$ and $-\frac{1}{2}$ is $f(x) = 6x^2 - x - 2$.

1. Look at the parabola in the graph on page 88. How can you determine the location of the axis of symmetry and the vertex?

2. In *Example 1,* how can you determine the *x*-intercepts without making a table of values or graphing the function?

3. In *Example 2,* can the roots of $x^2 - 7x + 4 = 0$ be found by factoring? Explain.

4. What is the difference between a "root" and a "zero"? Use an example to illustrate your explanation.

5. In *Example 4,* are there other quadratic functions with the same zeros? Explain.

2.1 EXERCISES

A 1. a) Is each function quadratic?

    i) $y = 3x^2 + 7x - 2$ **ii)** $f(x) = x^2 + \sqrt{x}$ **iii)** $f(x) = 25 - 9x^2$

    iv) $y = 7 - 5x^2$ **v)** $y = 2x^2 + 11 - 4x$ **vi)** $f(x) = \dfrac{1}{4x^2 - 9x + 12}$

    vii) $y = 2x^3 + 6x - 1$ **viii)** $y = (x + 2)^2 - 7$ **ix)** $y = x^3 - 2x^2 + 3$

  b) From part a, choose one quadratic function and one function that is not quadratic. Write to explain how you identified each function.

2. Identify each component of each quadratic function.

    i) the equation of the axis of symmetry

    ii) the coordinates of the vertex

    iii) the *x*- and *y*-intercepts

    iv) the domain and range

a)
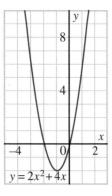
$y = 2x^2 + 4x$

b)
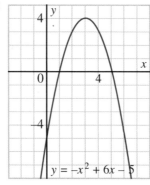
$y = -x^2 + 6x - 5$

c)
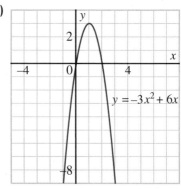
$y = -3x^2 + 6x$

d)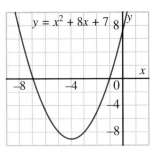
$y = x^2 + 8x + 7$

e)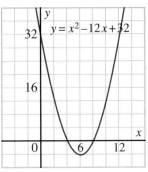
$y = x^2 - 12x + 32$

f)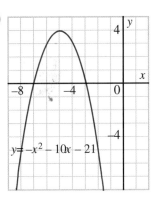
$y = -x^2 - 10x - 21$

3. Write to explain how the vertex of a parabola is related to its axis of symmetry.

4. a) How many different *x*-intercepts could the graph of a quadratic function have? Explain.

   b) How many different *y*-intercepts could it have? Explain.

   c) Is it possible for the graph of a quadratic function to have equal *x*- and *y*-intercepts? Explain.

5. Write to explain how the range of a quadratic function is related to the vertex of the parabola.

B 6. Graph each quadratic function, then identify each component.

   i) the equation of the axis of symmetry
   ii) the coordinates of the vertex
   iii) the horizontal and vertical intercepts
   iv) the domain and range

   a) $y = 2x^2 - 8$ b) $f(x) = x^2 - 6x + 8$ c) $y = 6 + x - x^2$

   d) $f(t) = 2t^2 - 6t + 5$ e) $y = 3 + 2x - 0.5x^2$ f) $f(a) = 6 + 3a - 3a^2$

7. A pebble is dropped from a bridge into a river. Its height, *h* metres, above the river *t* seconds after it is released is modelled by the quadratic function $h(t) = 82 - 4.9t^2$.

   a) Graph the function for reasonable values of *t*.

   b) State the domain and range of the function you graphed.

   c) How high is the pebble after 2.5 s?

8. When a flare is fired vertically, its height, h metres, after t seconds is modelled by the quadratic function $h(t) = -4.9t^2 + 153.2t$.

 a) Graph the function. Adjust the window settings to get a reasonable graph. State the window setting you used. Write to explain why you used it.

 b) How high is the flare after 5 s?

 c) For how many seconds is the flare higher than 1 km?

 d) What are the coordinates of the vertex? Write to explain what these coordinates represent.

 e) Estimate the domain and the range of the function. Write to explain what these represent.

MODELLING the Height of a Projectile

In the example on page 88, the height of a flare was modelled using the function $h(t) = -5t^2 + 100t$. In exercise 8, the height of another flare was modelled using a slightly different function, $h(t) = -4.9t^2 + 153.2t$. Both functions are based on a physics formula. When an object is projected vertically upward, its height, h metres, after t seconds is given by $h(t) = -\frac{1}{2}gt^2 + vt + s$ ①

where g is the acceleration due to gravity ($g \doteq 9.8$ m/s^2), v is the initial speed, and s is the initial height (if any) above the ground.

- Verify that the formula in exercise 8 can be obtained from formula ① by substituting appropriate values for g, v, and s.

- In the example on page 88, a value of 10 was used for g instead of 9.8. How would the table of values and the graph change if the more accurate value of 9.8 had been used?

- Suggest some reasons why the height of a projectile might differ from the height predicted by a model based on formula ①.

9. a) Factor each trinomial.

$$5x^2 + 9x + 4 \qquad 5x^2 - 9x + 4$$
$$5x^2 + 12x + 4 \qquad 5x^2 - 12x + 4$$
$$5x^2 + 21x + 4 \qquad 5x^2 - 21x + 4$$

 b) For which integral values of k can the trinomial $5x^2 + kx + 4$ be factored? Explain how you know there are no others.

 c) In part a, all the trinomials begin with $5x^2$ and end with $+4$. Make up other examples like these, in which several trinomials with two coefficients in common can be factored.

10. If each trinomial can be factored, what values of m are possible? In each case, write to explain how you know that no other values of m are possible.

a) $2x^2 + mx + 4$ b) $2x^2 + mx + 5$ c) $2x^2 + mx + 6$

d) $3x^2 - mx + 4$ e) $3x^2 - mx + 5$ f) $3x^2 - mx + 6$

11. If each trinomial can be factored, what values of t are possible?

a) $x^2 + tx + 1$ b) $2x^2 + tx + 2$ c) $4x^2 + tx + 4$ d) $3x^2 - tx + 2$

12. Solve each quadratic equation.

a) $a^2 + 4a - 21 = 0$ b) $-2t^2 - 5t + 12 = 0$ c) $-9 + 9t + 4t^2 = 0$

d) $24 = 10c^2 - 22c$ e) $11x + 6 = 2x^2$ f) $-3y^2 + 13y - 4 = 0$

13. Write a quadratic equation that has each pair of roots.

a) $3, -5$ b) $6, \frac{2}{3}$ c) $-\frac{4}{5}, 0$ d) $-\frac{1}{2}, -\frac{1}{2}$ e) $-4, -\frac{3}{4}$ f) $-\frac{7}{2}, -\frac{3}{8}$

14. Determine the zeros of each quadratic function.

a) $f(t) = t^2 + 4t - 12$ b) $f(x) = -2x^2 + 3x + 5$ c) $y = 8 - 31x - 4x^2$

d) $h(t) = 5 - 11t + 2t^2$ e) $y = -6 + 11x - 3x^2$ f) $f(x) = 6x^2 - 6x - 12$

15. Write the equation of a quadratic function that has each pair of zeros.

a) $0, 2$ b) $3, -\frac{3}{2}$ c) $-4, -8$ d) $-\frac{7}{4}, -\frac{3}{2}$ e) $10, -18$ f) $\frac{5}{6}, \frac{5}{6}$

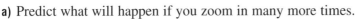

16. In the solution of *Example 2*, we zoomed in a few times to obtain the zeros of the function $f(x) = x^2 - 7x + 4$ to three decimal places.

a) Predict what will happen if you zoom in many more times.

b) Use a calculator to check your prediction. Explain the result.

17. In the example on page 84, the quadratic function that models the height of the stone at any time, t seconds, has one positive zero, which is approximately 6.78. The graph suggests that this function also has a negative zero.

a) Determine the negative zero to two decimal places. Do this in two different ways.

b) How could the negative zero be interpreted in terms of the model?

18. If each trinomial can be factored, what values of k are possible?

a) $x^2 - x + k$ b) $x^2 - 2x + k$ c) $x^2 - 3x + k$

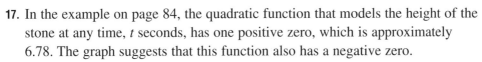

COMMUNICATING THE IDEAS

Compare the algebraic and graphical methods to solve a quadratic equation. What are the advantages and disadvantages of each method? Use examples to illustrate your explanation.

The Parabola of Best Fit

A hockey arena seats 1600 people. The cost of a ticket is $10. At this price, every ticket is sold. To obtain more revenue, the arena management plans to increase the ticket price. It is concerned that if it increases the price too much, attendance will drop and revenue could be less than it is now. A survey was conducted to estimate the revenue generated for different ticket prices. These data were obtained.

Ticket price ($)	Projected revenue ($)
10	16 000
15	19 500
20	20 300
25	14 750

The projected revenue increases, then decreases. This suggests that the projected revenue could be modelled by a quadratic function of the ticket price. We can use a graphing calculator to draw a scatterplot, to calculate the equation of the parabola of best fit, and to graph the parabola along with the data. The screens and keystrokes below are from the TI-83 calculator. Other graphing calculators will produce similar results in similar ways.

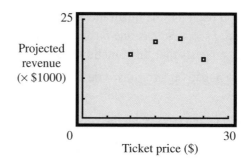

Displaying a scatterplot

- *Clear the graphing screen.*

   Make sure that no equations are selected in the Y= list.

- *Define the viewing window.*

   Choose values so the data will appear on the screen.

- *Enter the data in the list editor.*

  Press [STAT] 1. Enter the prices in the first list and the projected revenues in the second list.

- *Define a statistical plot.*

  Press [STAT PLOT] 1 to select the first plot. In the menu that appears, make sure that "On" is selected in the first line, the first graph type is selected in the second line, and lists L1 and L2 are selected in the third and fourth lines.

- *Graph the data.*

  Press [GRAPH].

Graphing the parabola of best fit

You can use the statistical features of a calculator to determine the equation of the parabola that comes closest to passing through the points. This is called the *parabola of best fit.*

- *Calculate the equation of the parabola.*

  Press [STAT] [▶] 5 [ENTER] to select "QuadReg."

This calculation is called a *quadratic regression*. The next screen shows the result. The values of a, b, and c are shown below the equation $y = ax^2 + bx + c$. The equation of the parabola of best fit is $y = -90.5x^2 + 3108.5x - 6217.5$.

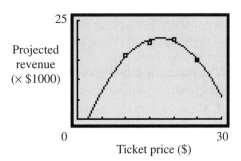

- *Graph the parabola.*

  You can manually enter the equation $y = -90.5x^2 + 3108.5x - 6217.5$ in the Y= list. You can also use the calculator to enter the equation directly. To do this, press [Y=] [CLEAR] [VARS] 5 [▶] [▶] 1.

MODELLING Ticket Price Increases

The parabola of best fit shown above models the projected revenue, R dollars, as a quadratic function of the ticket price, p dollars. The equation of the function is

$$R(p) = -90.5p^2 + 3108.5p - 6217.5$$

- Use the model to determine the ticket price that generates the maximum revenue.
- What ticket prices generate revenue greater than $20 000?
- Do the intercepts have any significance? Explain.
- Suggest a reasonable domain for the function.
- Why is a quadratic function whose graph opens down a reasonable model for this situation?

1. A newspaper article contained these data about Canadians.

Year	1961	1971	1981	1991
Percent of total population over 65	7.64%	8.10%	9.71%	11.6%

a) Create a scatterplot. Just by looking at the graph, state whether a linear or a quadratic function would best describe the data.

b) Carry out the appropriate regression. Determine the equation of the line or parabola of best fit.

c) Use your equation. Predict the percent of the Canadian population over 65 in 2011 and in 2036.

2. In a science experiment, a ball was dropped from a height of 3 m. Stroboscopic camera equipment recorded its position every tenth of a second. Measurements from the photograph were used to calculate its height above the floor every tenth of a second.

Time (s)	0	0.10	0.20	0.30	0.40	0.50	0.60	0.70
Height (m)	3.00	2.95	2.80	2.56	2.22	1.78	1.24	0.60

a) Repeat parts a and b of exercise 1.

b) Determine how long it took the ball to hit the floor, to the nearest hundredth of a second.

c) Determine how long it took the ball to go halfway to the floor. How does the time it took the ball to hit the floor compare with the time it took to go halfway to the floor?

2.2 Modelling Real Situations Using Quadratic Functions

One application of quadratic functions is projectile motion. The height of an object falling under the influence of Earth's gravity is a quadratic function of time. The equation of the quadratic function is based on a physics formula.

Quadratic functions also occur in situations where a certain quantity is the product of two other quantities, one increasing and the other decreasing. The equation of the function is developed from information in the problem.

For example, suppose you have 40 m of fencing to enclose a rectangular pen for your dogs. You might decide to make the pen 15 m long and 5 m wide. Since $15 \times 5 = 75$, the area enclosed by the pen is 75 m². Since the total length of fencing is fixed at 40 m, if one side is lengthened, the other side is shortened by the same amount. Visualize how the shape and the area of the pen change as this happens.

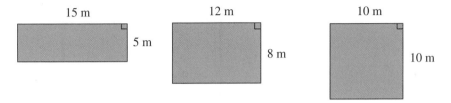

Example 1

There are 40 m of fencing to enclose a rectangular pen.

a) Represent the area of the pen as a function of the length of one side of the pen.

b) Graph the function.

c) What dimensions provide an area greater than 90 m²?

Solution

a) The perimeter is 40 m, so one length and one width are 20 m.
Let x metres represent the length of one side of the pen.

Then, the length of the other side is $20 - x$.

The area of the pen, A square metres, is the product of these dimensions.

$$A(x) = x(20 - x)$$
or $\quad A(x) = -x^2 + 20x$

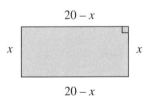

Think …

This is a quadratic function whose graph is a parabola with x-intercepts 0 and 20.
The parabola opens down.
The axis of symmetry is the vertical line halfway between the x-intercepts.
It has equation $x = 10$. When $x = 10$, $A = 100$, the vertex is $(10, 100)$.

b) Use grid paper. Some points on the graph are $(10, 100)$, $(12, 96)$, $(8, 96)$, $(5, 75)$, $(15, 75)$, $(0, 0)$, and $(20, 0)$. Sketch a smooth curve through these points.

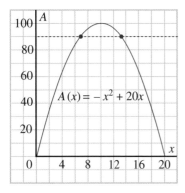

b) Use a graphing calculator. On this graph, the x-values run from 0 to 23.5 to provide x-coordinates increasing by 0.25 when you trace. The y-values start at -25 to provide room for the coordinates under the horizontal axis when tracing and zooming.

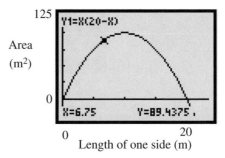

Area (m^2)

Length of one side (m)

c) Visualize starting at 90 on the A-axis and moving across to the parabola. The x-coordinates of the points on the parabola are approximately 7 and 13.

The area of the pen is greater than 90 m^2 when both dimensions are between approximately 7 m and 13 m.

c) By tracing and zooming, find points on the parabola whose y-coordinates are approximately 90. These occur when $x \doteq 6.838$ and $x \doteq 13.162$.

The area of the pen is greater than 90 m^2 when both dimensions are between approximately 6.8 m and 13.2 m.

Here is another situation involving a product in which one quantity is increasing and another is decreasing.

Example 2

A company makes canoes, then sells them for $500 each. At this price, it can sell 60 canoes in a season, generating revenue of $30 000. To increase revenue, management is planning to increase the selling price. It estimates that for every $50 increase in price, the number of canoes sold will drop by 4.

a) Represent the number of canoes sold as a function of the selling price.

b) Represent the revenue as a function of the selling price.

c) Sketch the function. What selling price will provide the maximum revenue? What is the maximum revenue?

d) What range of selling prices will provide revenue greater than $30 000?

Solution

a) Let s dollars represent the selling price. Let n represent the number of canoes sold.

$n = 60 - 4 \times$ (number of $50 increases in the price)

> **Think ...**
>
> The increase in price is represented by $s - 500$.
> The number of $50 increases in price is represented by $\frac{s - 500}{50}$

$$n = 60 - 4 \times \frac{s - 500}{50}$$
$$= 60 - 0.08(s - 500)$$
$$= 100 - 0.08s$$

b) Let R dollars represent the revenue from the sales.

$$R = \text{(selling price)(number of canoes sold)}$$
$$= s(100 - 0.08s)$$
$$= -0.08s^2 + 100s$$

> **Think ...**
>
> The revenue function $R = s(100 - 0.08s)$ is a product of two quantities. The first represents the selling price, which is increasing. The second represents the number of canoes sold, which is decreasing. We can use a graph to solve problems involving this function.

c) Sketch a graph of the function using the s-intercepts. The s-intercepts are the roots of the equation $s(100 - 0.08s) = 0$.

Solve this equation.

Either $s = 0$ or $100 - 0.08s = 0$

$$s = \frac{100}{0.08}$$
$$= 1250$$

The revenue function is a quadratic function whose graph opens down and has s-intercepts 0 and 1250. Hence, the equation of the axis of symmetry is $s = 625$.

To determine the coordinates of the vertex, substitute 625 for s in the equation for R.

$R = s(100 - 0.08s)$
$R = 625(100 - 0.08 \times 625)$
$\quad = 625(50)$
$\quad = 31\ 250$

The coordinates of the vertex are (625, 31 250).

A selling price of $625 provides the maximum revenue, which is $31 250.

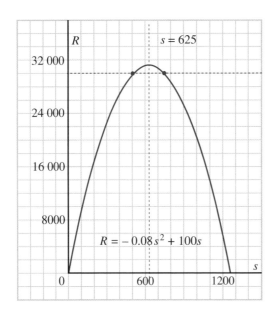

d) Since a selling price of $500 provides revenue of $30 000, the point (500, 30 000) is on the graph. This point is 125 units left of the axis of symmetry. Another point on the graph is 125 units right of the axis of symmetry. The coordinates of this point are (750, 30 000). Selling prices between $500 and $750 will provide revenue greater than $30 000.

1. In *Example 1*, what do the *x*-intercepts represent?

2. **a)** What are the advantages and disadvantages of each method used to solve *Example 1c*?

   b) How could you check the answer to *Example 1c*?

3. In *Example 2a*, why does the expression $\frac{s-500}{50}$ represent the number of $50 increases in price?

4. In *Example 2c*, the *R*-coordinate of the maximum point was obtained by multiplying 625 by 50. What does each of these numbers represent?

5. Why was a graphing calculator not needed to solve *Example 2*?

6. In *Example 2*, *s* represents the selling price, and the revenue is represented as a quadratic function of the selling price.

   a) Suppose *c* represents the number of canoes sold. What expression represents the selling price? How could you represent the revenue as a function of the number of canoes sold?

   b) Suppose *x* dollars represents the increase in selling price. What expression represents the selling price? What expression represents the number of canoes sold? How could you represent the revenue as a function of the increase in selling price?

2.2 EXERCISES

A 1. In *Example 1*, what dimensions of the pen provide an area greater than each given area?

   a) 75 m^2　　　　　　**b)** 50 m^2　　　　　　**c)** 25 m^2

2. Repeat *Example 2*, using each revised estimate.

   a) For every $50 increase in the selling price, the number of canoes sold will drop by 5.

   b) For every $50 increase in the selling price, the number of canoes sold will drop by 6.

B 3. Sixteen metres of fencing are available to enclose a rectangular garden.

   a) Represent the area of the garden as a function of the length of one side.

   b) Graph the function.

   c) What dimensions provide an area greater than 12 m^2?

4. A company that charters a boat for tours around the Gulf Islands can sell 200 tickets at $50 each. For every $10 increase in the ticket price, 5 fewer tickets will be sold.

   a) Represent the number of tickets sold as a function of the selling price.

   b) Represent the revenue as a function of the selling price.

   c) Sketch the function. What selling price will provide the maximum revenue? What is the maximum revenue?

   d) What range of prices will provide a revenue greater than $20 000?

5. When bicycles are sold for $300 each, a cycle store can sell 70 in a season. For every $25 increase in the price, the number sold drops by 10.

   a) Represent the sales revenue as a function of the price. Sketch the function.

   b) The total sales revenue is $17 500. How many bicycles were sold? What is the price of one bicycle?

   c) What range of prices will give a sales revenue that exceeds $18 000?

6. Computer software programs are sold to students for $25 each. Two hundred students are willing to buy them at that price. For every $5 increase in price, there are 20 fewer students willing to buy the software.

   a) Represent the sales revenue as a function of the price. Sketch the function.

   b) What is the maximum revenue?

   c) What range of prices will give a sales revenue that exceeds $5400?

7. Open the *Aircraft* database. Find the record for A300B4 planes in Continental's fleet. Suppose the average ticket price for the flights on these planes was $750. Suppose also that for every $50 decrease in price, an additional 12 passengers would purchase tickets.

   a) One field in the database states the average daily revenue passenger miles (RPMs). Every mile that a paying passenger travels is one revenue passenger mile. To determine the typical number of paying passengers on each flight, divide the RPMs by the average daily miles flown. Round your answer to the nearest whole number.

   b) Let t dollars represent the ticket price. Let p represent the number of paying passengers. Represent the number of paying passengers as a function of the ticket price.

   c) Let R dollars represent the revenue from ticket sales. Represent the revenue as a function of the ticket price.

   d) Sketch the revenue function. What ticket price will provide the maximum revenue? What is the maximum revenue?

   e) For the ticket price in part d, how many paying passengers will there be? Is this within the seating capacity of the aircraft?

8. A company manufactures and sells designer T-shirts. The profit, P dollars, for selling a certain style of T-shirt is projected to be $P = -20x^2 + 1000x - 6720$, where x dollars is the selling price of one T-shirt.

a) What selling price gives the maximum profit? What is the maximum profit?

b) The company hopes to earn a profit in excess of $6000 on this style of T-shirt. Based on its projections, is this possible?

c) Sketch a graph of this function.

MODELLING T-Shirt Price Increases

In exercise 8, the equation $P = -20x^2 + 1000x - 6720$ models the projected profit, P dollars, as a quadratic function of the selling price, x dollars.

• Suggest how the company might have determined this equation, or one similar.
• What is a reasonable domain for the function?
• What do the x-intercepts of the graph represent? What does the y-intercept represent?
• Why is a parabolic graph that opens down a reasonable model for this situation?

9. On a forward somersault dive, a diver's height, h metres, above the water is given by $h(t) = -4.9t^2 + 6t + 3$, where t is the time in seconds after the diver leaves the board.

a) Graph the function.

b) Determine the diver's maximum height above the water.

c) How long does it take the diver to reach the maximum height?

d) For how long is the diver higher than 3 m above the water?

C 10. In *Example 1*, suppose a graph of length against width were drawn, where the length is greater than or equal to the width. How would the graph differ from the one in the solution of *Example 1*?

11. When $f(x) = ax^2 + bx + c$ has a maximum value of 0, what conditions must be satisfied by a, b, and c?

COMMUNICATING THE IDEAS

The vertex of a parabola is an important point. Write a brief paragraph to explain the importance of this point. Assume the person reading your work has no knowledge of quadratic functions.

The Rising Fastball

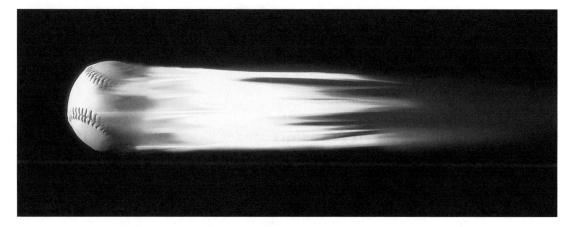

Some baseball fans believe that pitchers can throw a baseball that rises as it approaches the batter. This seems to contradict common sense, that the ball should fall because of gravity. However, a ball pitched with backspin is subject to a lifting force that acts against gravity. This lifting force is about 20% of the force of gravity. This means that a ball with backspin falls about 80% of the distance that a ball without backspin falls.

A rising fastball is an illusion. A ball thrown with backspin may appear to rise simply because it does not fall as much as the batter expects it to. You can confirm this by completing these exercises.

1. It is about 18.4 m from the pitcher's mound to home plate. A reasonably fast pitch travels at about 150 km/h. Calculate how long it takes the ball to reach the batter at this speed, to the nearest hundredth of a second.

2. Assume that a pitcher throws a ball horizontally without backspin. After the ball leaves the pitcher's hand, it begins to fall. From physics, the distance, s metres, that it falls is given by the formula $s = 4.9t^2$, where t is the elapsed time in seconds. Use your answer to exercise 1. Determine how far the ball has fallen by the time it reaches the batter. Give your answer to the nearest centimetre.

3. Suppose the ball is thrown with backspin. About how far would it have fallen by the time it reaches the batter?

2.3 Graphing $y = a(x - p)^2 + q$

Transforming Quadratic Graphs

Use a graphing calculator, or make tables of values; then draw the graphs on grid paper.

In these exercises, you will be asked to decide if two parabolas are congruent. Two parabolas are congruent if a tracing of one parabola can be made to coincide with the other.

Comparing the graphs of $y = x^2$ and $y = (x - p)^2$

1. Graph these functions on the same grid.

$y = x^2$

$y = (x - 2)^2 \qquad y = (x + 2)^2$

$y = (x - 4)^2 \qquad y = (x + 4)^2$

2. For each graph you drew in exercise 1, copy and complete this table.

Function	Value of p	Direction of opening	Vertex	Axis of symmetry	Congruent to $y = x^2$?
$y = x^2$	0	up	$(0, 0)$	$x = 0$	yes
$y = (x - 2)^2$	2				

3. a) What information about the graph does the value of p provide?

   b) What happens to the graph of the function when the value of p is changed? Consider both positive and negative values of p.

   c) Explain why the changes in part b occur.

Comparing the graphs of $y = x^2$ and $y = x^2 + q$

4. a) Graph these functions on the same grid.

$y = x^2 \qquad\qquad y = x^2 + 2$

$y = x^2 - 1 \qquad\quad y = x^2 + 4$

$y = x^2 - 3$

   b) Repeat exercises 2 and 3 for these functions, replacing p with q.

Comparing the graphs of $y = x^2$ and $y = ax^2$

5. a) Graph these functions on the same grid.

$$y = x^2 \qquad y = 2x^2 \qquad y = 0.5x^2 \qquad y = -x^2 \qquad y = -2x^2 \qquad y = -0.5x^2$$

b) Repeat exercises 2 and 3 for these functions, replacing p with a.

6. Use what you discovered in exercises 3, 4, and 5.

a) Without graphing, provide the following information for the function $y = -(x - 2)^2 + 3$.

    i) the values of the constants a, p, and q

    ii) the coordinates of the vertex, the equation of the axis of symmetry, and the direction of opening of the graph of this function

b) Graph this function to check your predictions in part a.

The simplest quadratic function is $y = x^2$. Its graph is a parabola, opening up, with vertex (0, 0) and axis of symmetry the y-axis. In *Investigate*, you made certain changes to the equation $y = x^2$ and discovered what happened to the graph. The results of *Investigate* are summarized below. You can use these results to sketch the graph of any quadratic function without making a table of values or using a graphing calculator.

x	y
−3	9
−2	4
−1	1
0	0
1	1
2	4
3	9

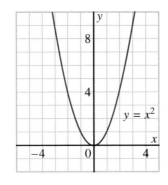

Changing the value of p in $y = (x - p)^2$

To investigate the effect of changing p in $y = (x - p)^2$, you substituted different values for p, then graphed the resulting parabolas. You should have found that the parabola moved horizontally. To see why, consider the function $y = (x - 4)^2$. If we were to graph this function using a table of values, we would start with values of x, subtract 4, then square the results. To give the same y-coordinates as for $y = x^2$, the values of x must be 4 *greater* than they were for $y = x^2$. This means that the x-coordinates of all points on the graph of $y = (x - 4)^2$ are 4 units greater than those on the graph of $y = x^2$. Therefore, the graph of $y = (x - 4)^2$ is translated 4 units to the *right* of the graph of $y = x^2$.

The graph of $y = (x - p)^2$ is the image of the graph of $y = x^2$ under a horizontal translation of p units.

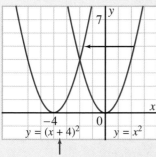 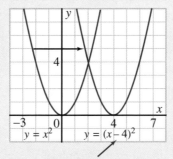

Positive sign, the graph moves to the left.

Negative sign, the graph moves to the right.

Changing the value of q in y = x² + q

To investigate the effect of changing q in $y = x^2 + q$, you substituted different values for q, then graphed the resulting parabolas. You should have found that the parabola moved vertically. To see why, consider the function $y = x^2 + 2$. The y-coordinates of all points on the graph of $y = x^2 + 2$ are 2 greater than those on the graph of $y = x^2$. Therefore, the graph of $y = x^2 + 2$ is 2 units above the graph of $y = x^2$.

The graph of $y = x^2 + q$ is the image of the graph of $y = x^2$ under a vertical translation of q units.

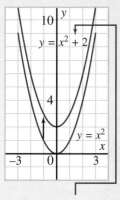 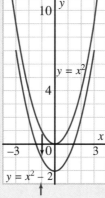

Positive sign, the graph moves up.

Negative sign, the graph moves down.

Changing the value of a in $y = ax^2$

To investigate the effect of changing a in $y = ax^2$, you substituted different values for a, then graphed the resulting parabolas. You should have found that the parabola was stretched or compressed vertically. To see why, consider the function $y = 2x^2$. The y-coordinates of all points on the graph of $y = 2x^2$ are two times those on the graph of $y = x^2$. Therefore, the graph of $y = 2x^2$ is expanded vertically relative to the graph of $y = x^2$. Similarly, the graph of $y = \frac{1}{2}x^2$ is compressed vertically. If a is negative, as in $y = -2x^2$ or $y = -\frac{1}{2}x^2$, there is also a reflection in the x-axis.

> The graph of $y = ax^2$ is the image of the graph of $y = x^2$ under a vertical expansion or compression. If $a < 0$, there is also a reflection in the x-axis.

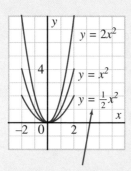

Positive signs, the graphs are expanded or compressed vertically.

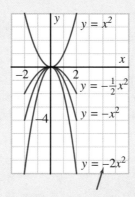

Negative signs, the graphs are expanded or compressed vertically and reflected in the x-axis.

The combination of all these results suggests that, in the equation $y = a(x - p)^2 + q$, the constants a, p, and q have the following geometric meanings.

$x - p = 0$ is the axis of symmetry.

$$y = a(x - p)^2 + q$$

Congruent to the parabola $y = ax^2$

Coordinates of the vertex are (p, q).

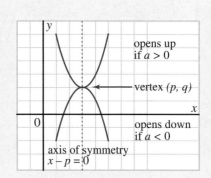

opens up if $a > 0$

vertex (p, q)

opens down if $a < 0$

axis of symmetry $x - p = 0$

Example 1

Graph the function $y = (x + 3)^2 - 2$.

Solution

The equation $y = (x + 3)^2 - 2$ has the form $y = a(x - p)^2 + q$ with $a = 1$, $p = -3$, and $q = -2$. Since a is positive, the parabola opens up and is congruent to $y = x^2$.

The vertex is $(-3, -2)$.

The axis of symmetry has equation $x + 3 = 0$, or $x = -3$.

To sketch the graph, plot the vertex $(-3, -2)$; then draw a vertical broken line to represent the axis of symmetry.

Substitute $x = 0$ to get the y-intercept.

$$y = (0 + 3)^2 - 2$$
$$= 9 - 2$$
$$= 7$$

Plot the point $(0, 7)$, and its image $(-6, 7)$ in the axis of symmetry.
Choose another x-coordinate, such as $x = -2$. Substitute to get $y = -1$.
Plot the points $(-2, -1)$ and its reflection image $(-4, -1)$.
Sketch the parabola through the plotted points.

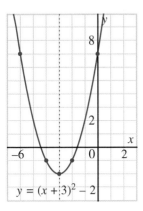

In *Example 1*, at least five points were used to sketch the parabola. A more efficient method to plot the points is based on a pattern in the table of values for $y = x^2$.

The Method of Differences

x	y	Difference
0	0	
		1
1	1	
		3
2	4	
		5
3	9	

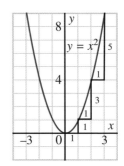

The differences in the y-coordinates are consecutive odd numbers. Starting at the vertex, points on the parabola can be located by moving 1 right and 1 up, then 1 right and 3 up, then 1 right and 5 up, and so on. Other points are obtained using the line of symmetry, or by repeating the same steps but moving to the left each time. This method can be used for any parabola congruent to the parabola $y = x^2$, and it can be modified to apply to other parabolas.

Example 2

Graph the function $f(x) = 2(x - 4)^2 - 3$.

Solution

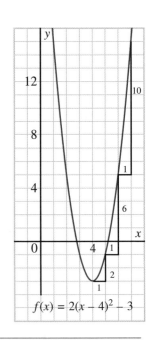

$f(x) = 2(x - 4)^2 - 3$

From the equation, the coordinates of the vertex are $(4, -3)$. Plot this point. Use the method of differences.

For this parabola, $a = 2$; hence, this parabola is congruent to $y = 2x^2$.

> ### Think …
>
> For $y = x^2$, the differences in the y-coordinates are: 1, 3, 5, …
> For $y = 2x^2$, the differences are twice as great: 2, 6, 10, …

Starting at the vertex, move 1 right and 2 up, then 1 right and 6 up, then 1 right and 10 up, and so on. Repeat these steps, moving left each time. Sketch the parabola through the plotted points.

Example 3

The graph of a quadratic function is shown.
What is the equation of the function?

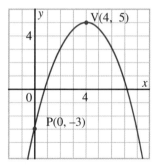

Solution

The coordinates of the vertex are $(4, 5)$.
Hence, the equation of the parabola has
the form

$$y = a(x - 4)^2 + 5 \qquad \text{①}$$

Since the parabola passes through $(0, -3)$,
the coordinates of this point satisfy the equation.

Substitute 0 for x and -3 for y in ①.

$$-3 = a(-4)^2 + 5$$
$$-8 = 16a$$
$$a = -\frac{1}{2}$$

The equation of the function is $y = -\frac{1}{2}(x - 4)^2 + 5$.

DISCUSSING THE IDEAS

1. In *Example 1*, the equation contains the numbers $+3$ and -2. Explain why both p and q are negative.

2. How many points do you need to plot when graphing a parabola? Explain.

3. What other ways are there to solve *Example 3*? Explain.

2.3 EXERCISES

1. The parabola $y = x^2$ is transformed as described below. Its image has the form $y = a(x - p)^2 + q$. Determine each value of a, p, and q for each transformation.

   a) Translate the parabola 3 units left.

   b) Expand the parabola vertically by a factor of 5.

   c) Translate the parabola 4 units up.

   d) Translate the parabola 2 units right and reflect it in the x-axis.

   e) Translate the parabola 5 units down and 3 units right.

   f) Translate the parabola 2 units up and compress it vertically by a factor of $\frac{1}{2}$.

2. Write to explain the difference between translating a parabola vertically up and expanding a parabola vertically.

3. Consider a quadratic function in the form $y = a(x - p)^2 + q$. Explain how its graph will change for each change described.

   a) The sign of a is changed.

   b) The value of q is decreased by 3.

   c) The value of a is decreased, but its sign is not changed.

   d) The value of p is increased by 2.

   e) The value of p is decreased by 4 and the value of q is increased by 3.

   f) The value of p is increased by 5, the value of q is decreased by 3, and the value of a is changed to its opposite.

4. State which graph is represented by each equation.

   a) $y = (x + 3)^2 + 1$
   b) $y = -2(x + 4)^2 + 3$
   c) $y = \frac{1}{2}(x - 2)^2 - 5$
   d) $y = -(x - 3)^2 + 2$

   i)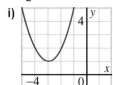

   ii)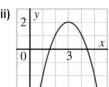

   iii)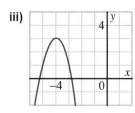

   iv)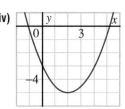

5. For each list of functions, write to explain how their graphs are similar and how they are different.

   a) $y = (x - 5)^2 + 4$
   $y = (x - 5)^2 + 2$
   $y = (x - 5)^2$
   $y = (x - 5)^2 - 2$
   $y = (x - 5)^2 - 4$

   b) $y = (x - 5)^2 + 4$
   $y = (x - 3)^2 + 4$
   $y = (x - 1)^2 + 4$
   $y = (x + 1)^2 + 4$
   $y = (x + 3)^2 + 4$

   c) $y = 3(x - 5)^2 + 4$
   $y = (x - 5)^2 + 4$
   $y = \frac{1}{2}(x - 5)^2 + 4$
   $y = -\frac{1}{2}(x - 5)^2 + 4$
   $y = -(x - 5)^2 + 4$
   $y = -3(x - 5)^2 + 4$

6. Describe what happens to the graph of each function.

   a) $y = a(x - 4)^2 + 3$ as a varies

   b) $y = 2(x - p)^2 + 3$ as p varies

   c) $y = 2(x - 4)^2 + q$ as q varies

B 7. For each parabola:

  i) State the coordinates of the vertex.

  ii) State the equation of the axis of symmetry.

  iii) State the y-intercept.

  a) $y = (x - 5)^2 + 2$ **b)** $y = 2(x + 3)^2 - 8$

  c) $y = -4(x + 1)^2 + 4$ **d)** $y = \frac{1}{2}(x - 2)^2 - 8$

8. Sketch the graphs of the functions in exercise 7.

9. On a sketch of the graph of each parabola:

  i) Label the vertex with its coordinates.

  ii) Label the axis of symmetry with its equation.

  iii) Label two points on the graph with their coordinates.

  a) $y = (x + 2)^2 - 5$ **b)** $y = -(x - 3)^2 + 2$

  c) $y = -\frac{1}{2}(x - 4)^2 - 1$ **d)** $y = 2(x + 1)^2 + 4$

  e) $y = -2(x - 1)^2 + 3$ **f)** $y = 4(x - 5)^2 - 10$

10. Sketch a graph of each parabola.

  a) $k = 2(l - 3)^2 - 1$ **b)** $r = -2(t + 3)^2 + 5$

  c) $m = \frac{1}{2}(n - 4)^2 - 3$ **d)** $p = 3(q - 5)^2 + 1$

  e) $f = -(g + 2.5)^2 + 3$ **f)** $u = -0.2(v + 2)^2 - 1.5$

11. **a)** Visualize what the graph of the equation $y = ax^2$ looks like as a becomes larger and larger.

  b) Predict what the graphs of these equations would look like.

  $y = 10x^2$ $y = 0.1x^2$

  $y = 100x^2$ $y = 0.01x^2$

  $y = 1000x^2$ $y = 0.001x^2$

  $\vdots$ $\vdots$

  $y = 1\ 000\ 000x^2$ $y = 0.000\ 001x^2$

  c) Use a graphing calculator to check your predictions.

12. Write the equation of each parabola.

  a) with vertex $(4, -1)$, that opens up, and is congruent to $y = 2x^2$

  b) with vertex $(-2, 3)$, that opens down, and is congruent to $y = \frac{1}{3}x^2$

  c) with vertex $(-3, 2)$, that opens down, and is congruent to $y = \frac{1}{2}x^2$

  d) with vertex $(3, -4)$, x-intercepts 1 and 5

13. The graphs of three quadratic functions are shown in the screens below. The scales are not shown, and they are not necessarily the same. Which of these statements are true?

  1) In the first graph, the value of *a* is larger than it is in the third graph.

  2) All three graphs could be graphs of the same function.

  3) Any of these graphs could be the graph of any of the functions in exercise 11b.

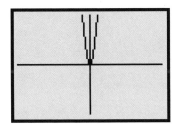 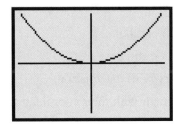 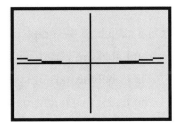

14. Determine the equation of each parabola.

  a) with vertex (0, 2), and passing through (−3, 11)

  b) with vertex (0, 5), and passing through (2, 9)

  c) with vertex (4, 0), and *y*-intercept 16

  d) with vertex (−3, 0), and *y*-intercept 9

15. Determine the equation of each quadratic function.

a)

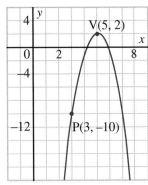

b)

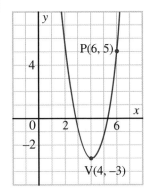

c)

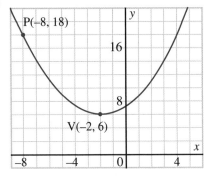

d)

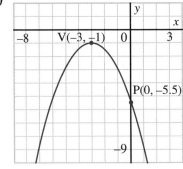

16. a) Write the equation of each parabola.

    i) with vertex $(3, -1)$, x-intercepts 2 and 4

    ii) with vertex $(-1, 4)$, y-intercept 2

    iii) with vertex $(2, -27)$, y-intercept -15

  b) Choose one parabola from part a. Write to explain how you determined its equation.

C 17. Write the equations of the parabolas in each pattern.

  a)

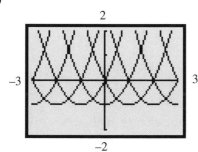

  b)

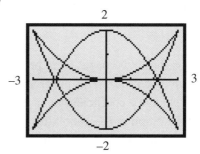

18. a) Use a graphing calculator to make the patterns in exercise 17.

  b) Make a similar pattern of your own.

19. Determine the equation of the parabola, with axis of symmetry the y-axis, that passes through each pair of points.

  a) $(2, 9)$ and $(3, 14)$ b) $(-2, 1)$ and $(4, -5)$

20. A small rocket is launched. It reaches a maximum height of 120 m and lands 10 m from the launching pad. Assume that the rocket follows a parabolic path. Write an equation that describes its height, h metres, as a function of its horizontal distance, x metres, from the launching pad.

21. Suppose the coordinates of two points are given. Is it always possible to find the equation of a parabola that passes through one of these points and has the other point as its vertex? Explain.

COMMUNICATING THE IDEAS

Write a paragraph to explain why changing the value of p in the quadratic function $y = a(x - p)^2 + q$ moves the parabola to the right or to the left, and how you can tell which direction it moves. Include graphs as part of your explanation.

Dynamic Graphs　Part I

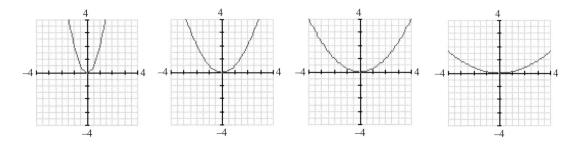

The graphs on these two pages were produced using software called *Graphs! Graphs! Graphs!* This program contains a feature called a "slider" that lets you change a coefficient in an equation gradually from one value to another. The graph moves smoothly in response to the change. This was done to produce the sequence of eight graphs above, and at the top of the next page. They show how $y = ax^2$ changes as a decreases from 2 to -2.

These graphs were obtained from the graph of $y = a(x - b)^2 + c$, which is built into the program. The variables a, b, and c are called "slider variables." The slider at the right is set with $a = 2$, $b = 0$, $c = 0$, and corresponds to the first graph above. The dot beside a indicates that a is active. The slider box is at the top, where $a = 2$. When the box is dragged down, the values of a change gradually from 2 to -2, producing graphs like those above.

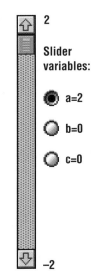

Slider
variables:

$a=2$

$b=0$

$c=0$

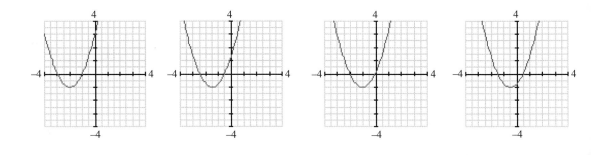

LINKING IDEAS

Mathematics & Technology

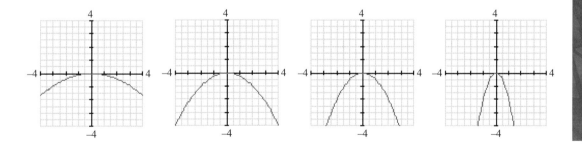

1. **a)** Scan the graphs above and at the top of the preceding page, from left to right. Visualize the parabola gradually changing as *a* decreases from 2 to −2. Then scan them from right to left, and visualize the parabola gradually changing back as *a* increases from −2 to 2.

   b) Estimate the value of *a* for each graph, from right to left.

2. The slider at the right corresponds to one of the graphs in the sequence of eight graphs below and at the bottom of the preceding page.

   a) Use the equation $y = a(x - b)^2 + c$ to identify this graph.

   b) Which slider variable is active: *a*, *b*, or *c*? Scan the graphs below and at the bottom of page 120 from left to right, and from right to left. Visualize the parabola gradually changing as this slider variable changes.

   c) Write the equation of each graph, from right to left.

3. Suppose the programmers had used the equation $y = a(x + b)^2 + c$ instead of $y = a(x - b)^2 + c$. Which graphs, if any, in the sequences above and below would change for the same slider variables? Explain.

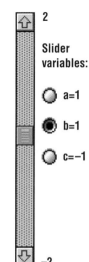

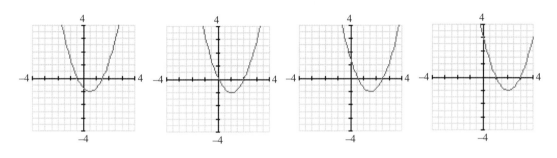

Mathematics & Technology

In Section 2.3, we developed a method to sketch the graph of an equation such as $y = 2(x - 3)^2 - 7$. We used the geometric meaning of the three constants in the equation. When an equation has the form $y = 2x^2 - 12x + 11$, the constants used to sketch the graph are not obvious. To obtain these constants, we use the method of *completing the square.*

Example 1

Write $y = 2x^2 - 12x + 11$ in the form $y = a(x - p)^2 + q$, then sketch the graph.

Solution

Step 1. Remove the coefficient of x^2 as a common factor from the first two terms.

$$y = 2(x^2 - 6x) + 11$$

Step 2. Determine the square of one-half the coefficient of the x-term in *Step 1*; that is, $\left[\frac{1}{2}(-6)\right]^2 = 9$. Add and subtract this number inside the brackets.

$$y = 2(x^2 - 6x + 9 - 9) + 11$$

Step 3. Remove the last term from the brackets and combine with the constant term.

$$y = 2(x^2 - 6x + 9) - 18 + 11$$
$$= 2(x^2 - 6x + 9) - 7$$

Step 4. Factor the expression in the brackets as a complete square.

$$y = 2(x - 3)^2 - 7$$

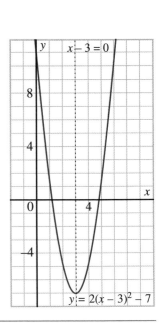

Use the method of differences.

The parabola $y = 2(x - 3)^2 - 7$ is congruent to $y = 2x^2$.

The differences are 2, 6, 10, …

The vertex is $(3, -7)$. From the vertex, move 1 right and 2 up, then 1 right and 6 up, then 1 right and 10 up. Repeat these steps, moving left each time. Sketch the parabola through the plotted points.

The vertex of the graph of a quadratic function is important because it represents the maximum or minimum value of the function. You can determine this value from the equation in the form $y = a(x - p)^2 + q$. You do not need to draw the graph.

Example 2

Consider the function $y = -4x^2 - 12x + 5$.

a) Determine the maximum or minimum value of y and state which it is.

b) For what value of x does the maximum or minimum occur?

Solution

a) Complete the square.
$$y = -4x^2 - 12x + 5$$
$$= -4(x^2 + 3x) + 5$$
$$= -4\left(x^2 + 3x + \tfrac{9}{4} - \tfrac{9}{4}\right) + 5$$
$$= -4\left(x^2 + 3x + \tfrac{9}{4}\right) + 9 + 5$$
$$= -4\left(x + \tfrac{3}{2}\right)^2 + 14$$

Since the coefficient of x^2 is negative, the graph opens down and there is a maximum point. Since the vertex has coordinates $\left(-\tfrac{3}{2}, 14\right)$, the maximum value of y is 14.

b) Using the coordinates of the vertex, the maximum value occurs when $x = -\tfrac{3}{2}$.

VISUALIZING　　Compare the equations in *Examples 1* and *2*.

Positive – curve opens up.
There is a minimum value.

$$y = 2(x - 3)^2 - 7$$

The minimum value occurs when $x - 3 = 0$, or $x = 3$

The minimum value is –7.

Negative – curve opens down.
There is a maximum value.

$$y = -4\left(x + \tfrac{3}{2}\right)^2 + 14$$

The maximum value occurs when $x + \tfrac{3}{2} = 0$, or $x = -\tfrac{3}{2}$

The maximum value is 14.

1. In the solution of *Example 1*, how could you check the result?

2. a) Explain why the method of completing the square works.

   b) Why is the name "completing the square" appropriate?

3. Suppose the graph of a quadratic function has a maximum point. Which coordinate of this point represents the maximum value of the function: the *x*-coordinate or the *y*-coordinate? What does the other coordinate represent?

4. Why it is useful to be able to determine the characteristics of a quadratic function from its equation without drawing a graph?

2.4 EXERCISES

 1. Write each equation in the form $f(x) = a(x - p)^2 + q$. Check by expanding the result.

   a) $f(x) = x^2 - 6x + 8$ b) $f(x) = x^2 + 10x + 14$

   c) $f(x) = 2x^2 + 4x + 7$ d) $f(x) = -2x^2 + 4x + 5$

   e) $f(x) = 3x^2 - 24x + 40$ f) $f(x) = -5x^2 - 20x - 30$

2. Sketch the graph of each parabola in exercise 1.

3. Use the word "maximum" or "minimum," and data from each graph.
   Copy and complete this sentence for each graph:
   "The … value of *y* is … when *x* = …"

a)

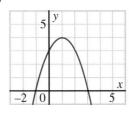

b)

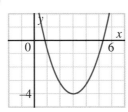

c)

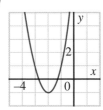

d)

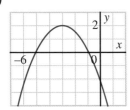

e)

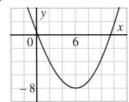

f)

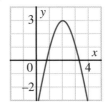

4. Does each function have a maximum value? If it has, for what value of x does it occur?

a) $y = -2(x + 5)^2 - 8$ **b)** $f(x) = \frac{1}{4}(x - 2)^2 - 9$ **c)** $y = -0.5(x - 3)^2 + 7.5$

d) $y = 5 - 3x^2$ **e)** $f(x) = 3\left(x - \frac{5}{2}\right)^2 + \frac{17}{2}$ **f)** $f(x) = -(x + 4)^2 - 19$

5. a) Write each equation in the form $y = a(x - p)^2 + q$.

   i) $y = 2x^2 - 8x + 15$ **ii)** $y = 3x^2 + 12x - 7$ **iii)** $y = x^2 - 6x + 7$

   iv) $y = -2x^2 + 6x + 11$ **v)** $y = -x^2 - 3x - 3$ **vi)** $y = 1.5x^2 - 9x + 10$

b) For each function in part a:

   i) State its maximum or minimum value.

   ii) State the value of x for which the maximum or minimum occurs.

6. How is the constant a in the equation $y = ax^2 + bx + c$ related to the constant a in the equation $y = a(x - p)^2 + q$? Explain.

B **7.** Sketch each parabola.

   i) Label the vertex with its coordinates.

   ii) Label the axis of symmetry with its equation.

   iii) Label two points on the graph with their coordinates.

a) $f(x) = x^2 - 6x + 10$ **b)** $y = 2x^2 + 8x + 7$ **c)** $y = -x^2 + 10x - 13$

d) $f(t) = 3t^2 - 6t + 8$ **e)** $f(n) = -4n^2 - 24n - 20$ **f)** $u = -2v^2 - 16v - 35$

8. Sketch each parabola.

   i) Label the vertex with its coordinates.

   ii) Label the axis of symmetry with its equation.

   iii) Label two other points on the graph with their coordinates.

a) $y = \frac{1}{2}x^2 - 2x + 7$ **b)** $f(t) = 4t^2 + 12t - 5$ **c)** $f(j) = -2j^2 + 14j - 12$

d) $y = 3x^2 - 4x - 6$ **e)** $u = -4v^2 + 10v - 7$ **f)** $f(x) = -2x^2 - 12x - 14$

9. Choose one parabola from exercise 7 or 8. Write to explain how you sketched it.

10. For each parabola, state:

   i) the maximum or minimum value of y

   ii) whether it is a maximum or minimum

   iii) the value of x when it occurs

   iv) the domain and range of the function

a) $y = (x - 3)^2 + 5$ **b)** $y = 2(x + 1)^2 - 3$ **c)** $y = -2(x - 1)^2 + 4$

d) $y = -(x + 2)^2 - 6$ **e)** $y = 0.5x^2 - 9$ **f)** $y = 7 - 2x^2$

11. a) Graph the parabolas represented by these equations on the same grid.

$y = x^2 + 6x + 3$ $y = x^2 - 6x + 3$

$y = x^2 + 4x + 3$ $y = x^2 - 4x + 3$

$y = x^2 + 2x + 3$ $y = x^2 - 2x + 3$

$y = x^2 + 0x + 3$

b) Write to describe how the graph of $y = x^2 + bx + 3$ changes as b changes.

c) What special case occurs when $b = 0$?

12. a) Graph the parabolas represented by these equations on the same grid.

$y = 2x^2 + 4x + 3$ $y = -2x^2 + 4x + 3$

$y = x^2 + 4x + 3$ $y = -x^2 + 4x + 3$

$y = 0.75x^2 + 4x + 3$ $y = -0.75x^2 + 4x + 3$

$y = 0.5x^2 + 4x + 3$ $y = -0.5x^2 + 4x + 3$

$y = 0x^2 + 4x + 3$

b) Write to describe how the graph of $y = ax^2 + 4x + 3$ changes as a changes.

c) What special case occurs when $a = 0$?

13. Write the equation of a quadratic function that satisfies each set of conditions.

a) The function has a minimum value of 4 at $x = -2$.

b) The function has a minimum value of -7 at $x = 3$.

c) The function has a maximum value of -1 at $x = 5$.

d) The parabola is congruent to $y = 2x^2$, and has a maximum value of 14.

e) The range of the function is all values of y that are greater than or equal to -6, and the parabola is congruent to $y = 3x^2$.

C 14. What conditions must be satisfied by a and c for the parabola $y = ax^2 + c$ to have x-intercepts?

15. a) Write $f(x) = ax^2 + bx + c$ in the form $f(x) = a(x - p)^2 + q$.

b) Determine expressions in terms of a, b, and c for the equation of the axis of symmetry, the coordinates of the vertex, and the y-intercept.

c) Determine expressions in terms of a, b, and c for the x-intercepts.

COMMUNICATING THE IDEAS

Given a quadratic function in the form $f(x) = ax^2 + bx + c$, explain how to write this function in the form $f(x) = a(x - p)^2 + q$, and why it works. Use an example to illustrate your explanation.

2.5 Maximum and Minimum Problems

In many problems involving the maximum or minimum of a function, the function is not given; it has to be found from the data. Note the steps involved in the solutions of these examples.

Example 1

Two numbers have a difference of 10. Their product is a minimum. What are the numbers?

Solution

Step 1. Identify the quantity to be maximized or minimized.

The quantity to be minimized is the product, P, of two numbers.

Step 2. Write an algebraic expression for this quantity.

Let the two numbers be x and y. Then, $P = xy$

Step 3. The expression must contain only one variable. If it contains more, use other information to write it in terms of one variable.

Since the numbers have a difference of 10, write $x - y = 10$, where x is the greater number.

Solve for y. $y = x - 10$

Substitute $x - 10$ for y in $P = xy$.
$$P = x(x - 10)$$
$$= x^2 - 10x$$

Step 4. Identify whether the quadratic function has a maximum or minimum value. Then complete the square to determine this value and where it occurs.

Since the coefficient of x^2 is positive, the graph opens up and it has a minimum point.
$$P = x^2 - 10x$$
$$= x^2 - 10x + 25 - 25$$
$$= (x - 5)^2 - 25$$

The minimum value of P is -25, and it occurs when $x = 5$.

Step 5. Answer the question in the problem.

Since $x = 5$ and $x - y = 10$, then $y = -5$

The two numbers are 5 and -5.

Example 2

A rectangular lot is bounded on one side by a river and on the other three sides by a total of 80 m of fencing. Determine the dimensions of the largest possible lot.

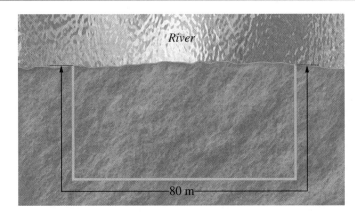

Solution

Step 1. The quantity to be maximized is the area of the lot.
Let A square metres represent the area.

Step 2. Let x metres represent the width of the lot, and y metres the length.
Then, $A = xy$

Step 3. Since the total length of fencing is 80 m,

$$2x + y = 80$$
$$y = 80 - 2x$$

Substitute $80 - 2x$ for y in $A = xy$.

$$A = x(80 - 2x)$$
$$= 80x - 2x^2$$

Step 4. Since the coefficient of x^2 is negative, the graph opens down and it has a maximum point. Determine the coordinates of the maximum point either by graphing or by completing the square.

$$A = -2x^2 + 80x$$
$$= -2(x^2 - 40x)$$
$$= -2(x^2 - 40x + 400 - 400)$$
$$= -2(x^2 - 40x + 400) + 800$$
$$= -2(x - 20)^2 + 800$$

The maximum value of A is 800, and it occurs when $x = 20$.

Step 5. Since $x = 20$ and $2x + y = 80$, then $y = 40$

The dimensions of the largest possible rectangular lot are 40 m by 20 m.

Example 3

Computer software programs are sold to students for $20 each. Three hundred students are willing to buy them at that price. For every $5 increase in price, there are 30 fewer students willing to buy the software. What is the maximum revenue?

Solution

Step 1. The quantity to be maximized is the total revenue, R dollars, from the software sold.

Step 2. Let x dollars be the increase in selling price. Then,

$$R = \text{(cost per program)(number of programs sold)}$$
$$= (20 + x)\left(300 - 30\left(\tfrac{x}{5}\right)\right)$$
$$= (20 + x)(300 - 6x)$$
$$= -6x^2 + 180x + 6000$$

Step 3. The expression contains only one variable.

Step 4. Determine the coordinates of the maximum point either by graphing or by completing the square.

$$R = -6(x^2 - 30x) + 6000$$
$$= -6(x^2 - 30x + 225 - 225) + 6000$$
$$= -6(x^2 - 30x + 225) + 1350 + 6000$$
$$= -6(x - 15)^2 + 7350$$

The maximum value of R is 7350, and it occurs when $x = 15$.

Step 5. The maximum revenue is $7350 when the price increase is $15, or when the selling price is $35.

DISCUSSING THE IDEAS

1. By looking at the equation of a quadratic function, how can you tell whether the function represents a quantity that can be maximized or minimized?

2. Explain how solving a problem involving a quantity that has to be maximized or minimized is different from solving a word problem that can be represented by an equation.

3. In *Example 3*, x represented the increase in selling price. How could this problem be solved in each case?

   a) s represents the selling price.   b) p represents the number of programs sold.

B 1. Two numbers have a difference of 8. Their product is a minimum. Determine the numbers.

2. The sum of two natural numbers is 12. Their product is a maximum. Determine the numbers.

3. The sum of two numbers is 60. Their product is a maximum. Determine the numbers.

4. Two numbers have a constant sum. Prove that their product is a maximum when they are equal.

5. Two numbers have a difference of 20. The sum of their squares is a minimum. Determine the numbers.

6. The sum of two numbers is 16. The sum of their squares is a minimum. Determine the numbers.

7. Two numbers have a difference of 16. The result of adding their sum and their product is a minimum. Determine the numbers.

8. A rectangular lot is bordered on one side by a stream and on the other three sides by 600 m of fencing. The area of the lot is a maximum. Determine the area.

9. Eighty metres of fencing are available to enclose a rectangular play area.

   a) What is the maximum area that can be enclosed?

   b) What dimensions produce the maximum area?

10. A rectangular area is enclosed by a fence and divided by another section of fence parallel to two of its sides. The 600 m of fence used enclose a maximum area. What are the dimensions of the enclosure?

11. A student who wishes to use a computer lab at a local library must buy a membership. The library charges $20 for membership. Four hundred students purchase the membership. The library estimates that for every $4 increase in the membership fee, 40 fewer students will become members. What membership fee will provide the maximum revenue to the library?

12. A bus company carries about 20 000 riders per day for a fare of 90¢. A survey indicates that if the fare is decreased, the number of riders will increase by 2000 for every 5¢ decrease. What ticket price would result in the greatest revenue?

13. A trough is made from a rectangular strip of metal 50 cm wide. The metal is bent at right angles so it is x centimetres high along two sides. For what value of x is the cross-sectional area a maximum?

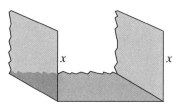

14. The sum of the base and the height of a triangle is 15 cm. What is the maximum area of the triangle?

15. A straight section of railroad track crosses two highways at points that are 400 m and 600 m, respectively, from an intersection. Determine the dimensions of the largest rectangular lot that can be laid out in the triangle formed by the railroad and highways.

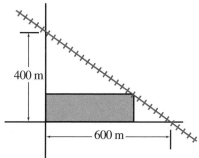

16. A 30-cm piece of wire is cut in two. One piece is bent into the shape of a square. The other piece is bent into the shape of a rectangle with a length-to-width ratio of 2 : 1. The sum of the areas of the square and rectangle is a minimum. What are the lengths of the two pieces?

C 17. Determine the number that exceeds its square by the greatest possible amount.

18. Determine the maximum possible area of a rectangle with a given perimeter.

19. In $\triangle ABC$, $\angle B = 90°$ and AC has a constant length. Prove that the area of $\triangle ABC$ is a maximum when AB = BC.

20. Determine the minimum distance from O(0, 0) to the line $3x + 2y - 12 = 0$.

COMMUNICATING THE IDEAS

Write a summary to explain how to solve a problem that involves a quantity to be maximized or minimized. Illustrate your method by making up a problem that is different from those in the exercises.

Elegance in Mathematics

In exercise 4 of 2.5 Exercises, you proved this general result.

Maximum Principle

If two numbers have a constant sum, their product is a maximum when they are equal.

To illustrate the Maximum Principle, suppose two numbers have a sum of 10. The numbers could be (5, 5), (4, 6), (6, 4), (3, 7), (7, 3), and so on. You can quickly check that the product is a maximum when the numbers are equal. If these ordered pairs are plotted on a grid, they lie on a line with equation $x + y = 10$. If x and y are positive, the product, xy, represents the area of a rectangle. Visualize point P moving along the line. The rectangle has a maximum area when P is (5, 5).

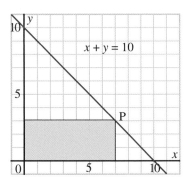

You can use the Maximum Principle to solve many problems. For example, in *Example 2* on page 128, we found the dimensions of the largest lot that could be bounded on one side by a river and on the other three sides by a total of 80 m of fencing. In the solution of that example, we maximized the product xy, and used the fact that $2x + y = 80$.

Think ...

$$2x \ + \ y \ = \ 80$$

Two numbers with...   ... a constant sum of 80

The product $2xy$ is a maximum when the two numbers are equal.

Since the product xy is half the product $2xy$, it will also be a maximum when the two numbers are equal.

Hence, each number is 40. Therefore, $y = 40$ and $2x = 40$, or $x = 20$

The dimensions of the largest possible lot are 40 m by 20 m.

1. Find problems in 2.5 Exercises that can be solved using the Maximum Principle. Use the Maximum Principle to solve each problem.

2. Suggest why "Elegance in Mathematics" was chosen as the title of this feature.

2.6 The Inverse of a Linear Function

In previous sections, we saw that when certain changes were made in the equation of a function, there was a corresponding change in its graph. We shall now investigate the effect of interchanging x and y in the equation of a function.

Consider the linear function $y = 3x + 2$.
If x and y are interchanged, we get $x = 3y + 2$.
Solve for y.

$$x = 3y + 2$$
$$3y = x - 2$$
$$y = \frac{x - 2}{3}$$
$$y = \frac{1}{3}x - \frac{2}{3}$$

This is also a linear function.

Graphical comparison of $y = 3x + 2$ and $y = \dfrac{x - 2}{3}$

To compare the graphs of $y = 3x + 2$ and $y = \frac{x-2}{3}$, consider their tables of values.

$y = 3x + 2$

x	y
0	2
1	5
2	8

$y = \frac{x-2}{3}$

x	y
2	0
5	1
8	2

The tables show that when x and y are interchanged in the equation $y = 3x + 2$, the coordinates of the points that satisfy the equation are interchanged as well.

The graph suggests that the line $y = \frac{x-2}{3}$ is the reflection of the line $y = 3x + 2$ in the line $y = x$. To verify this, consider the line segment AB joining corresponding points A(1, 5) and B(5, 1).

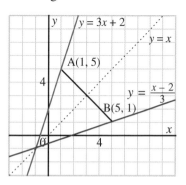

The midpoint of AB is $\left(\frac{1+5}{2}, \frac{5+1}{2} \right)$, or (3, 3).

The slope of AB is $\frac{1-5}{5-1} = \frac{-4}{4}$, or -1.

The midpoint of AB, the point (3, 3), is on the line $y = x$; and the slope of AB, -1, is the negative reciprocal of the slope of the line $y = x$. Therefore, the line $y = x$ is the perpendicular bisector of AB. Other pairs of corresponding points give similar results.

When x and y are interchanged in the equation of a function:

- The coordinates of the points that satisfy the equation of the function are interchanged.
- The graph of the function is reflected in the line $y = x$.

Algebraic comparison of $y = 3x + 2$ and $y = \dfrac{x - 2}{3}$

Let x be any number; for example, $x = 4$.

When $x = 4$, $y = 3x + 2$ becomes:

$y = 3(4) + 2$, or 14 $\longrightarrow$ When $x = 14$, $y = \dfrac{x - 2}{3}$ becomes:

$y = (14 - 2) \div 3$, or 4

Multiply by 3. Add 2. Subtract 2. Divide by 3.

Inverse operations

The function $y = \dfrac{x - 2}{3}$ is the *inverse* of the function $y = 3x + 2$.

To determine the inverse of a function:

- Interchange x and y in the equation of the function.
- Solve the resulting equation for y.

Example 1

Determine the inverse of $y = 3 - 7x$.

Solution

$$y = 3 - 7x$$

Interchange x and y. $\qquad x = 3 - 7y$

Solve for y. $\qquad 7y = 3 - x$

$$y = \frac{3 - x}{7}$$

The equation of the inverse function is $y = \dfrac{3 - x}{7}$.

To express the inverse of a function $f(x)$ in function notation, we use the symbol $f^{-1}(x)$. We say, "f inverse of x." In *Example 1,* the result could be written $f^{-1}(x) = \frac{3-x}{7}$.

When the graph of a function is given, its inverse can be graphed as a reflection in the line $y = x$.

Example 2

Given the graph of $y = f(x)$,
graph $y = f^{-1}(x)$ on the same grid.

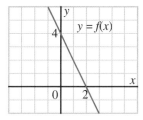

Solution

Since $y = f(x)$ is linear, only two points are needed to graph $y = f^{-1}(x)$.

The simplest points to use are the intercepts.

Reflect these in the line $y = x$.

$(0, 4) \rightarrow (4, 0)$ and $(2, 0) \rightarrow (0, 2)$

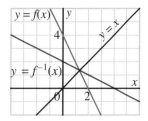

The graph of the inverse function is the straight line through $(4, 0)$ and $(0, 2)$.

This is the reflection of the given line in the line $y = x$.

VISUALIZING

Graph a function on a piece of paper. Make it dark enough so you can see it from the other side.

Hold the paper by the top right and bottom left corners.

Flip the paper.

The graph you see through the paper is the graph of the inverse function.

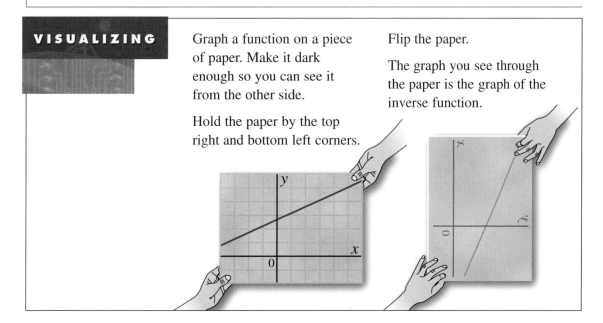

1. Suppose you are given the equations of two functions. How can you tell if one function is the inverse of the other?

2. Suppose you are given the graphs of two functions. How can you tell if one function is the inverse of the other?

2.6 EXERCISES

A 1. On each grid, is one function the inverse of the other function?

a)
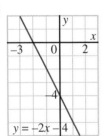

b)
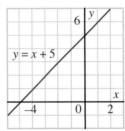

c)
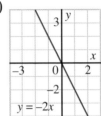

2. Determine the inverse of each function.

a) $y = x + 3$

b) $y = 4x - 1$

c) $y = 2x$

d) $y = 3x - 4$

e) $y = \frac{1}{2}x + 6$

f) $y = \frac{2}{3}x - 1$

3. Graph each function, its inverse, and $y = x$ on the same grid.

a)
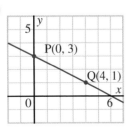

b)
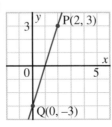

c)
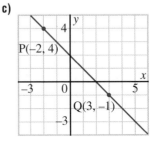

d)
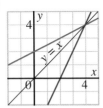

e)

f)
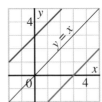

4. Two linear functions are described in words. Is each function the inverse of the other?

   a) i) Multiply by 2, then add 5.   ii) Subtract 5, then divide by 2.

   b) i) Multiply by 2, then add 5.   ii) Divide by 2, then subtract 5.

   c) i) Add 1, then multiply by 6.   ii) Subtract 1, then divide by 6.

   d) i) Add 1, then multiply by 6.   ii) Divide by 6, then subtract 1.

B 5. Determine the inverse of each function.

   a) $f(x) = x + 6$ b) $f(x) = 2x$ c) $f(x) = 3 - x$

   d) $f(x) = \frac{1}{2}x - 3$ e) $f(x) = 5x + 1$ f) $f(x) = 2(1 + x)$

6. Determine if one function is the inverse of the other function.

   a) $y = 2x + 3$, $y = 3x + 2$ b) $y = \frac{1}{2}x - 4$, $y = 2x - \frac{1}{4}$

   c) $y = 4x - 1$, $y = \frac{x + 1}{4}$ d) $y = 3x - 6$, $y = \frac{1}{3}x + 2$

7. a) $y = \frac{1}{2}x - 5$ b) $y = 2x + 5$ c) $y = 5(x - 2)$

   Which of the five functions below is the inverse of each function above?

   i) $y = \frac{1}{2}x + 5$ ii) $y = -\frac{1}{5}x + 2$ iii) $y = 2x + 10$

   iv) $y = \frac{1}{5}x + 2$ v) $y = \frac{1}{2}(x - 5)$

C 8. Show that the inverse of the linear function $y = mx + b$ is a function, provided that $m \neq 0$.

9. Since the inverse of a linear function is a linear function, it also has an inverse. Determine the inverse of the inverse of $f(x) = 2x + 5$. Explain the result both algebraically and geometrically.

10. Find a linear function that is its own inverse. How many such functions are there? How can you be certain that you have found them all?

COMMUNICATING THE IDEAS

The notation for the inverse of a function contains the symbol "–." The meaning of this symbol is different from the meanings you have encountered previously. List as many different ways as you can in which this symbol is used in mathematics. Illustrate each meaning with an example. For each example, explain how you can tell which meaning is intended.

2.7 The Inverse of a Quadratic Function

The inverse of a quadratic function can be found using the same steps as for a linear function.

Example 1

Consider the function $f(x) = x^2 + 4$.

a) Determine the inverse of $f(x)$. Graph it and $y = f(x)$ on the same grid.

b) Is the inverse of $f(x)$ a function? Explain.

Solution

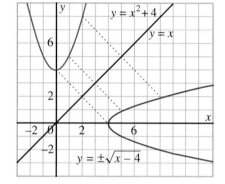

a) To determine the inverse of $f(x)$, interchange x and y in $y = x^2 + 4$.

$$x = y^2 + 4$$

Solve for y.
$$y^2 = x - 4$$
$$y = \pm\sqrt{x - 4}$$

$f(x) = x^2 + 4$ is a parabola with vertex $(0, 4)$, opening up, with axis of symmetry the y-axis.

The inverse, $y = \pm\sqrt{x - 4}$, can be graphed by reflecting $y = f(x)$ in $y = x$.

b) The inverse is not a function because it fails the vertical line test.

As *Example 1* indicates, the graph of the inverse of a quadratic function is a parabola with a horizontal axis of symmetry. *Example 1* also shows that the inverse of a quadratic function is not necessarily a function. For this reason, we do not use the notation $f^{-1}(x)$ to represent the inverse of a quadratic function.

It is sometimes convenient to restrict the domain of a quadratic function so that its inverse is a function.

Example 2

Show two ways to restrict the domain of $f(x) = x^2 + 4$ so that its inverse is a function. Illustrate each way with a graph.

Solution

The domain of $y = x^2 + 4$ is restricted to non-negative real numbers.

That is, $y = x^2 + 4$, $x \geq 0$

Then the inverse is $y = \sqrt{x - 4}$.

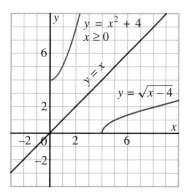

The domain of $y = x^2 + 4$ is restricted to non-positive real numbers.

That is, $y = x^2 + 4$, $x \leq 0$

Then the inverse is $y = -\sqrt{x - 4}$.

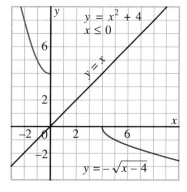

In each case, the inverse is a function because it passes the vertical line test.

DISCUSSING THE IDEAS

1. Is the inverse of a linear function a linear function? Is the inverse of a quadratic function a quadratic function? Use examples to justify your answers.

2. The solution of *Example 2* shows two ways to restrict the domain of $f(x) = x^2 + 4$ so that the inverse is a function. Could this be done in any other ways? Explain.

3. a) In *Example 1*, why is it incorrect to write $f^{-1}(x) = \pm\sqrt{x - 4}$? Explain.
   b) In *Example 2*, could we write $f^{-1}(x) = \sqrt{x - 4}$ or $f^{-1}(x) = -\sqrt{x - 4}$? Explain.

1. Determine the inverse of each function.

a) $y = x^2$ **b)** $y = x^2 - 1$ **c)** $y = x^2 + 3$

d) $y = 2x^2 + 5$ **e)** $y = \frac{1}{4}x^2 - 2$ **f)** $y = \frac{x^2 - 2}{4}$

2. Graph each function, its inverse, and $y = x$ on the same grid.

a)

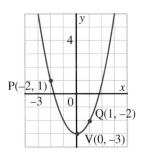

b)

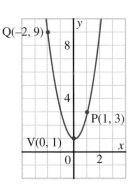

c)

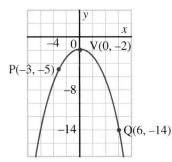

d)

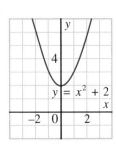

e)

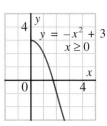

f)

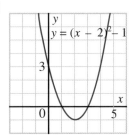

B **3.** Determine the inverse of each function.

a) $f(x) = 4x^2$ **b)** $f(x) = 1 - x^2$ **c)** $f(x) = 2 - 3x^2$

d) $f(x) = (x + 3)^2$ **e)** $f(x) = 5(x - 2)^2$ **f)** $f(x) = \frac{1}{3}(x + 1)^2 - 3$

4. Restrict the domain so that the inverse of each function is a function. Graph the function and its inverse.

a) $y = x^2 - 1$ **b)** $y = x^2 + 2$ **c)** $y = (x + 1)^2$

d) $y = (x - 2)^2 + 1$ **e)** $y = -2x^2 + 3$ **f)** $y = \frac{1}{3}(x - 1)^2 - 2$

5. Determine if the second function is the inverse of the first function.

a) $y = x^2 + 6, x \geq 0$ $y = \sqrt{x + 6}$

b) $y = 2x^2 - 3, x \leq 0$ $y = -\sqrt{\dfrac{x + 3}{2}}$

c) $y = 4(x + 1)^2, x \leq -1$ $y = \dfrac{\sqrt{x} - 2}{2}$

d) $y = \frac{1}{3}(x - 2)^2 + 5, x \leq 2$ $y = -\sqrt{3x + 3}$

6. a) $y = x^2 - 12, x \geq 0$ **b)** $y = -\frac{1}{2}x^2 + 3, x \geq 0$ **c)** $y = 3(x-1)^2 - 2, x \geq 1$

Which of the five functions below is the inverse of each function above?

     i) $y = \frac{1}{3}\sqrt{x+3}$ **ii)** $y = \sqrt{x+12}$ **iii)** $y = 2\sqrt{x-3}$

     iv) $y = \sqrt{\frac{x+2}{3}} + 1$ **v)** $y = \sqrt{6-2x}$

C **7.** Determine the inverse of each function, then graph the function, its inverse, and $y = x$.

     a) $y = 1 + \sqrt{x}$ **b)** $y = 3 - 2\sqrt{x}$

COMMUNICATING THE IDEAS

Your friend has missed the last two mathematics lessons, and she asks you about them. How would you explain to her the meaning of the inverse of a function and how it is determined? How would you explain why the domain must sometimes be restricted? Use a couple of examples to illustrate your explanations.

EXPLORING •WITH A GRAPHING CALCULATOR

Graphing a Function and Its Inverse

Follow these steps to graph a function and its inverse on the TI-83 graphing calculator.

- Input a function as Y1 in the Y= list.

- Define the viewing window. To get an accurate representation of the inverse function, the length : width ratio of the screen should be about 3 : 2.

- Press [DRAW] 8 [VARS] [▶] 1 1 .

  "DrawInv Y1" will appear on the screen.

- Press [ENTER].

  The calculator will graph the inverse of the function Y1.

Practise graphing a function and its inverse from any exercises in 2.6 Exercises or 2.7 Exercises.

MATHEMATICAL MODELLING

Modelling Golf Ball Trajectories

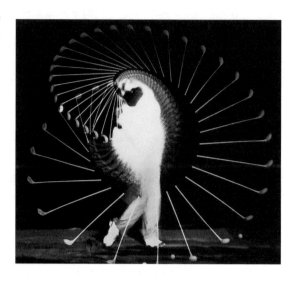

On page 84, we considered the motion of a stone thrown straight upward. Most projectiles are propelled at an angle, and move horizontally as well as vertically. A common example was introduced on page 82. This is the motion of a golf ball when it is struck with different clubs.

The flight of any projectile is determined by several factors. These include the initial speed, the angle of inclination, gravity, and air resistance. A golf ball will be quickly accelerated to a certain initial speed by the club, lofted into the air by the club face, pulled to Earth by gravity, and slowed by air resistance. The model we develop accounts for the first three factors, and we will consider air resistance later.

Consider what happens when a golf ball is hit with a 7 iron. The club face of a 7 iron is set at an angle of about 38° with respect to the shaft. When it hits the ball squarely at the bottom of the swing, the ball is propelled at an initial speed of about 37 m/s perpendicular to the club face, at an angle of inclination of 38° relative to the ground. The initial speed can be determined from stroboscopic photographs similar to the one above.

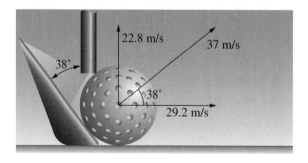

Visualize the motion of the ball. It consists of two separate parts: the vertical motion and the horizontal motion. From trigonometry, the initial vertical speed of the ball is about 22.8 m/s, and the initial horizontal speed is about 29.2 m/s, as shown in the diagram.

 LOOK AT THE IMPLICATIONS

We calculate how long the ball is in the air, then we calculate how far it travels down the fairway.

Consider only the vertical motion of the ball. The situation is similar to the example on page 84. The ball's height in metres is a quadratic function of time, t seconds.

$$h(t) = -4.9t^2 + 22.8t \qquad \text{①}$$

1. Recall that on page 84 you used an equation similar to equation ① to calculate the maximum height of a stone and the time for it to hit the ground. Use equation ① and the same method as before.

   a) Determine the maximum height of the ball that is hit with a 7 iron.

   b) Determine the time it takes the ball to hit the ground.

Next, consider only the horizontal motion of the ball. In the absence of air resistance, the ball moves at a constant speed horizontally. Hence, its distance in metres down the fairway is a linear function of time, t seconds.

$$d(t) = 29.2t \qquad \text{②}$$

2. Use equation ② and your answer from exercise 1b. How far will the ball travel horizontally in that time?

Your answer to exercise 2 should be 136 m. Distances in golf are always measured in yards. One yard is slightly shorter than one metre. To convert metres to yards, add about 10%. Ten percent of 136 is about 14. Hence, the model predicts that the ball will travel about 150 yards down the fairway.

Your answers to exercises 1 and 2 are predictions from the model of a ball hit with a 7 iron: the distance the ball travels down the fairway; how long it takes to get there; and the maximum height it reaches along the way. The distances can be expressed in metres or in yards. You can make similar predictions for other clubs, using the data in the following table.

Club	Club angle	Initial speed (m/s)	Horizontal speed (m/s)	Vertical speed (m/s)
Driver	9.5°	86	84.8	14.2
3 wood	12°	71	69.4	14.8
3 iron	20°	51	47.9	17.4
5 iron	29°	42	36.7	20.4
7 iron	38°	37	29.2	22.8
9 iron	47°	34	23.2	24.9
Pitching wedge	56°	32	17.9	26.5
Sand wedge	65°	31	13.1	28.1

3. Make a table similar to the one above. Include columns for maximum height, time to hit the ground, and horizontal distance. Write your results from exercises 1 and 2 in the table, in either metres or yards.

4. Choose one or more of the other golf clubs. Repeat the calculations from exercises 1 and 2. Write the results in the table.

The curved path of a golf ball is called its *trajectory*. You can use equations ① and ② to develop a model for the trajectories of balls hit with different clubs.

5. The golfer in the diagram is hitting a ball with a 7 iron. Visualize the trajectory of the ball. You can use equations ① and ② to calculate the coordinates of points on the trajectory at any time *t*.

a) Copy and complete the table for the 7 iron. Use either metres or yards.

Time, *t* seconds	*x* (from ②)	*y* (from ①)
0.0		
0.5		
1.0		
1.5		

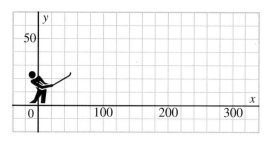

b) Plot the points on a grid, then draw a smooth curve through them. If you use a graphing calculator or a spreadsheet to do this, your graph should look like one of the graphs on page 145.

c) Determine the equation of the trajectory. Try to do this in more than one way.

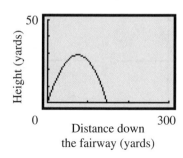

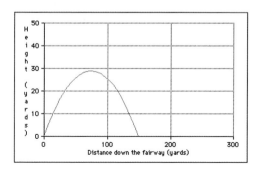

6. Repeat exercise 5 for one or more of the other golf clubs. Use the same horizontal and vertical scale for all graphs.

 **REVISIT THE SITUATION**

Your formulas and graphs form a mathematical model of golf shot trajectories. Results obtained by golfers will differ from those predicted by the model for several reasons.

7. Suppose air resistance is considered.

   a) How would the distances achieved be affected? Explain.

   b) How would the shape of the trajectories be affected? Explain.

8. The proper swing for an iron involves striking the ball on the downswing, before the club reaches its lowest position. How would the results be affected by this type of swing?

9. Suggest some other reasons why the distances achieved by golfers might be different from those predicted by the model.

10. The trajectories you drew in exercises 5 and 6 were based on the assumption that there is no air resistance. What would the trajectories look like if there were no gravity? Explain.

Dynamic Graphs   Part II

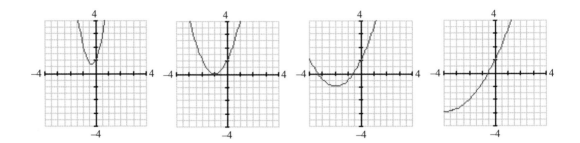

The graphs on these two pages were produced using *Graphs! Graphs! Graphs!* See page 120 for a description of this software.

The sequences of eight graphs above and below were obtained from the graph of $y = ax^2 + bx + c$, which is built into the program.

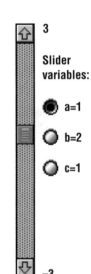

3

Slider variables:

● a=1

○ b=2

○ c=1

−3

1. a) Look at the sequence of eight graphs above and at the top of page 147. Which slider variable is changing: *a*, *b*, or *c*? Explain how you can tell.

   b) Which slider variable is changing in the sequence of eight graphs below and at the bottom of page 147? Explain.

2. The slider at the right corresponds to one of the graphs in the sequence of graphs above and at the top of page 147.

   a) Identify this graph.

   b) Which slider variable is active: *a*, *b*, or *c*? Scan the graphs above from left to right, and from right to left. Visualize the parabola gradually changing as this slider variable changes.

   c) Write the equation of each graph in this sequence, from right to left.

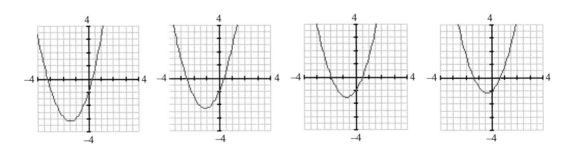

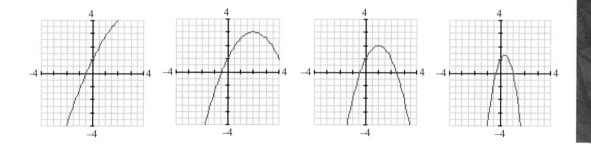

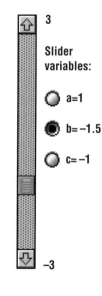

3. The slider at the right corresponds to one graph in the sequence of eight graphs below and at the bottom of page 146. Repeat exercise 2 for this sequence of graphs.

4. One slider variable is not changing in either sequence of graphs shown on these pages.

   a) Which slider variable is this?

   b) Suppose a sequence of graphs is constructed to demonstrate what happens to the graph of $y = ax^2 + bx + c$ as this slider variable changes gradually from one value to another. Describe what this sequence would look like.

5. Compare these sequences of graphs with the corresponding sequences on pages 120 and 121. In what ways are they similar? In what ways are they different? Account for the similarities and the differences.

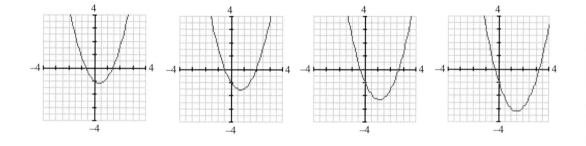

Mathematics & Technology

1. These graphs were produced by a graphing calculator. The equations of the functions are shown. Identify the function that corresponds to each graph. Justify your choices.

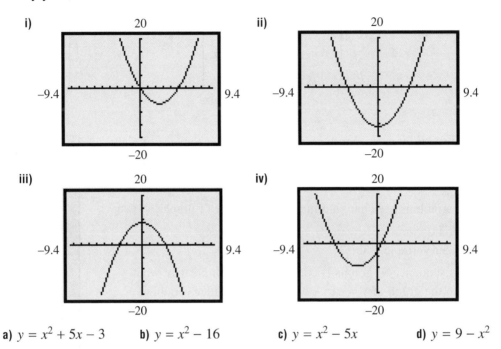

i)
ii)
iii)
iv)

a) $y = x^2 + 5x - 3$ b) $y = x^2 - 16$ c) $y = x^2 - 5x$ d) $y = 9 - x^2$

2. The daily profit, P dollars, of a cotton candy vendor at the fair is described by the function $P = -60x^2 + 240x - 80$, where x dollars is the selling price of a bag of cotton candy.

   a) What should the selling price of a bag of cotton candy be to maximize daily profits?

   b) What is the maximum daily profit?

3. A ball is thrown in the air from the balcony of an apartment building and falls to the ground. The height, h metres, of the ball relative to the ground t seconds after being thrown is given by $h(t) = -4.9t^2 + 10t + 35$.

   a) Graph the function.

   b) Determine the maximum height of the ball above the ground.

   c) How long does it take the ball to reach the maximum height?

   d) What are the domain and range of this function?

4. The sum of two numbers is 28. The sum of their squares is a minimum. Determine the numbers.

5. The equation of the function on page 84 can be written as $h(t) = -\frac{1}{2}gt^2 + 17t + 110$, where g is the acceleration due to gravity. On Earth, g is 9.8 m/s². Suppose you could throw a stone upward at 17 m/s from a 110-m cliff on each heavenly body listed.

Values of g in the Solar System (m/s²)	
Sun	273.0
Moon	1.6
Mercury	3.5
Venus	8.9
Earth	9.8
Mars	3.7
Jupiter	24.9
Saturn	10.6
Uranus	8.9
Neptune	11.7
Pluto	0.59

   a) Predict how the graph on page 84 would change if you use some of these values of g in the equation. Check with a calculator.

   b) On Mars, determine the maximum height, the time to reach that height, and the time to hit the ground.

6. For each parabola:

   i) State the coordinates of the vertex.

   ii) State the equation of the axis of symmetry.

   iii) State the y-intercept.

   iv) Sketch the graph.

   a) $y = 3(x + 7)^2$ b) $y = -2(x - 3)^2 + 4$ c) $y = -\left(x - \frac{1}{2}\right)^2 - \frac{3}{4}$

7. Determine the equation of each parabola.

   a) vertex $(0, 3)$, passing through $(3, 4)$ b) vertex $(4, 0)$, y-intercept 3

8. Sketch each parabola.

   i) Label the vertex with its coordinates.

   ii) Label the axis of symmetry with its equation.

   iii) Label two points on the graph with their coordinates.

   a) $y = x^2 - 6x + 5$ b) $w = 2z^2 - 8z - 5$ c) $p = -3q^2 + 18q - 20$

9. Choose one parabola from exercise 8. Write to explain how you sketched it.

10. The sum of a number and three times another number is 18. Their product is a maximum. Determine the numbers.

11. Determine the inverse of each function.

   a) $f(x) = x + 3$ b) $f(x) = 3x$ c) $f(x) = 3(2 - x)$

12. Choose one function from exercise 11. Write to explain how you found the inverse.

13. Graph each function and its inverse from exercise 11.

14. Determine if one function is the inverse of the other function.

   a) $y = 4x^2$, $y = \frac{\sqrt{x}}{2}$ b) $y = 3x^2$, $y = 3\sqrt{x}$

   c) $y = 2x^2 + 10$, $y = \sqrt{\frac{x - 10}{2}}$ d) $y = 4x^2 + 9$, $y = -\frac{\sqrt{x - 9}}{2}$

1. Determine the missing information.

	Principal ($)	Annual interest rate (%)	Time (days)	Simple interest ($)
a)	2000.00	$5\frac{1}{4}$	50	
b)		4	375	21.58
c)	1610.00		120	39.70
d)	892.00	$3\frac{1}{8}$		55.75

2. A late payment charge of $1.54 was applied to Luis' telephone bill. The telephone company charges a 1.5% late fee. What was Luis' unpaid balance?

3. At an interest rate of 7% compounded annually, how much would you have to invest today to accumulate each amount?

   a) $4250 in 4 years   **b)** $1275 in 2 years

4. Choose one part of exercise 3. Write to explain how you calculated each investment.

5. Determine the interest rate for each investment.

   a) $1000 grows to $1125.51 in 2 years, interest compounded semi-annually

   b) $200 grows to $253.65 in 3 years, interest compounded quarterly

6. Show that these rates are equivalent.
5% compounded semi-annually and 5.0625% compounded annually

7. A family buys a $4000 computer with $800 down and 24 monthly payments of $158.

   a) Determine the finance charge for this payment scheme.

   b) Determine the effective annual rate of interest.

8. Calculate the gross earnings for each person.

   a) A salesperson earns $1200 per month plus 7% commission on sales. Her sales for the month are $10 325.

   b) A server in a restaurant earns $5.75/h plus gratuities. One week, he works 32 h and earns $221 in gratuities.

9. A property has a fair market value of $320 000. The assessed value is 79% of the fair market value. The mill rate is 19 mills.

   a) Determine the property tax.

b) How much should the property owner set aside each month to pay the property taxes?

10. One U.S. dollar is equivalent to $1.38 Can. A computer has a price of $900 U.S or $1300 Can. Which is the better value?

11. Is each function quadratic? Write to explain how you know.

a) $y = 2x^2 + x + 3$ **b)** $f(x) = x + 4$ **c)** $y = x^3 + 2x + 5$

d) $f(x) = 3x^2 + 2\sqrt{x}$ **e)** $y = 4x^2 - 3x + 7$ **f)** $f(x) = \dfrac{x}{x^2 + x + 1}$

12. The power, P watts, supplied to a circuit by a 9-V battery is given by the formula $P(I) = 9I - 0.5I^2$, where I is the current in amperes.

a) Graph P against I.

b) For what value of the current will the power be a maximum?

c) What is the maximum power?

13. A transit company carries about 80 000 riders per day for a fare of $1.25. To obtain more revenue, the management plans to increase the fare. It estimates that for every 5¢ increase in fare, it will lose 1000 riders. What range of fares will provide a revenue greater than $110 000?

14. Determine the zeros of each quadratic function.

a) $f(b) = b^2 + 7b + 12$ **b)** $y = 2x^2 - 5x + 3$ **c)** $f(t) = -3t^2 + 7t + 6$

d) $y = 6x^2 + 5x - 4$ **e)** $f(z) = -3z^2 - 7z + 6$ **f)** $y = x^2 - 2x - 8$

15. Choose one function from exercise 14. Write to explain how you found the zeros.

16. For each parabola:

   i) State the maximum or minimum value of y.

   ii) State whether it is a maximum or minimum.

   iii) State the value of x when it occurs.

   iv) State the domain and range of the function.

a) $y = -2(x + 5)^2 - 8$ **b)** $y = \frac{1}{4}(x - 2)^2 - 9$ **c)** $y = -0.5(x - 3)^2 + 7.5$

17. Choose one parabola from exercise 16. Write to explain how you found the maximum or minimum.

18. A lifeguard marks off a rectangular swimming area at a beach with 200 m of rope. What is the greatest area of water she can enclose?

19. Two linear functions are described in words. Is each function the inverse of the other?

a) i) Multiply by 3, then subtract 4.   **ii)** Divide by 3, then add 4.

b) i) Subtract 2, then divide by 8.   **ii)** Multiply by 8, then add 2.

The Water and the Wine

 CONSIDER THIS SITUATION

This puzzle sometimes appears in puzzle books and magazines.

> A jug holds a quantity of water. Another jug holds an equal quantity of wine. A glass of water is taken from the first jug, poured into the wine, and the contents stirred. A glass of the mixture is then taken and poured into the water.
>
> Is there more wine in the water or is there more water in the wine?

Try to solve this puzzle. It is not particularly difficult, but to solve it you need to think about it in a certain way. You may find that one or more of the following strategies are helpful, or you may want to use another strategy.

- Draw a diagram.
- Solve a simpler problem.
- Make a table.
- Identify a pattern.

On pages 208-209, you will develop mathematical models to solve this puzzle.

 FYI Visit www.awl.com/canada/school/connections

For information related to the above problem, click on <u>MATHLINKS</u> followed by *Mathematics 11*. Then select a topic under The Water and the Wine.

Graphing Polynomial Functions

You have previously studied linear and quadratic functions, similar to these examples.

Linear functions

$y = 3x + 4$ (thin line)
$y = -x - 3$ (thick line)

Quadratic functions

$y = x^2 - 3$ (thin curve)
$y = -0.25x^2 + x + 2$ (thick curve)

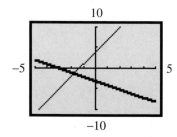

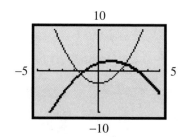

Linear and quadratic functions are examples of *polynomial functions*. In this exploration, you use a graphing calculator to explore the graphs of some polynomial functions.

Many polynomial functions involve higher powers of x such as x^3, x^4, x^5, ... Even for relatively small values of x, such as 3, 4 or 5, these powers can become quite large. Hence, you will often have to adjust the window settings to show the main features of a graph on the screen. You may need a narrower range of values along the x-axis or a wider range of values along the y-axis.

1. Graph each *cubic function* using appropriate window settings. Sketch the results on plain paper. Use the results to form some conclusions about the graphs of cubic functions.

   a) $y = x^3$

   b) $y = 2x^3 - 15x + 7$

   c) $y = -x^3 + 2x^2 - 4x + 8$

   d) $y = -5(x + 2)(x + 1)(x - 1)$

2. Graph each *quartic function* using appropriate window settings. Sketch the results on plain paper. Use the results to form some conclusions about the graphs of quartic functions.

   a) $y = x^4$

   b) $y = x^4 + 4x^3 - 16x - 25$

   c) $y = \frac{1}{2}(x + 4)(x - 4)(x + 2)(x - 2)$

   d) $y = -x^4 + 16x^2 - 3x - 40$

3. Repeat exercise 2 for each fifth-degree function.

   a) $y = x^5$

   c) $y = x^5 + 2x^3 + 9x^2 - 19x + 14$

   e) $y = -x^5 + 3x^4 - 15x^2 + 30x$

   b) $y = x^5 + 5x^4 + 2x^3 - 14x^2 - 19x - 17$

   d) $y = x(x + 3)(x + 1)(x - 2)(x - 3)$

The graphs on page 154 may be described as follows:

$y = 3x + 4$

Straight line extends from the 3rd quadrant to the 1st quadrant.

$y = -x - 3$

Straight line extends from the 2nd quadrant to the 4th quadrant.

$y = x^2 - 3$

Graph extends from the 2nd quadrant to the 1st quadrant, forming a valley.

$y = -0.25x^2 + x + 2$

Graph extends from the 3rd quadrant to the 4th quadrant, forming a hill.

4. This table describes the graphs of polynomial functions. The *degree* of the function is the exponent of the highest power of x in the equation. The *leading coefficient* is the coefficient of this power of x.

Degree	Sign of leading coefficient	Extends from quadrant...	... to quadrant	Greatest number of hills and valleys
1	+	3	1	0
1	−	2	4	0
2	+	2	1	1
2	−	3	4	1

   a) The first four rows in the table describe the graphs of the linear and quadratic functions in the screens on page 154. Verify that these rows are correct.

   b) Copy the table and extend it for 6 more rows.

   c) Complete the remaining rows for the functions in exercises 1, 2, and 3. Use your sketches from those exercises.

5. **a)** Which polynomial functions have graphs that extend from and to the same quadrants as the graphs of linear functions?

   b) Which polynomial functions have graphs that extend from and to the same quadrants as the graphs of quadratic functions?

6. Use the examples from exercise 5. How is the greatest number of hills and valleys related to the degree of the function?

3.1 Polynomial Functions

Linear and quadratic functions are special cases of a more general type of function called a *polynomial function*. Here are some examples of polynomial functions.

Linear function (first degree)	$f(x) = 2x + 3.2$
Quadratic function (second degree)	$f(x) = -3x^2 + 5x + 4$
Cubic function (third degree)	$f(x) = x^3 - 2x^2 + 2x + 3$
Quartic function (fourth degree)	$f(x) = x^4 - 7x^2 + 5$
Quintic function (fifth degree)	$f(x) = 3x^5 + 2x^4 - 5x^2 - 4$

A *polynomial function* of *degree n* is a function whose equation can be written in the form

$$f(x) = a_n x^n + a_{n-1} x^{n-1} + a_{n-2} x^{n-2} + \cdots + a_1 x + a_0$$

where n is a whole number and $a_n, a_{n-1}, a_{n-2}, \cdots, a_1, a_0$, are real numbers. The coefficient of the highest power of x is called the *leading coefficient*.

You can use technology to graph a polynomial function, or make a table of values and plot points on grid paper. The graphs of some polynomial functions are shown below, along with the tables of values used to create them.

Two examples of quadratic functions

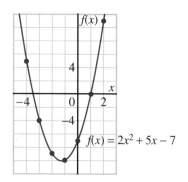

x	-4	-3	-2	-1	0	1	2
$f(x)$	5	-4	-9	-10	-7	0	11

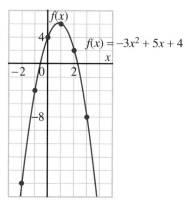

x	-2	-1	0	1	2	3
$f(x)$	-18	-4	4	6	2	-8

Visualize the graphs of quadratic functions extending from the 2nd quadrant to the 1st quadrant (forming a valley), or extending from the 3rd quadrant to the 4th quadrant (forming a hill). All graphs of quadratic functions have an axis of symmetry and either a minimum point or a maximum point.

Two examples of cubic functions

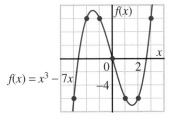

$f(x) = x^3 - 7x$

x	−3	−2	−1	0	1	2	3
$f(x)$	−6	6	6	0	−6	−6	6

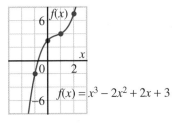

$f(x) = x^3 - 2x^2 + 2x + 3$

x	−1	0	1	2
$f(x)$	−2	3	4	7

The leading coefficient is positive. As x increases, the graph extends from the 3rd quadrant to the 1st quadrant, forming both a hill and a valley. The graph contains both a maximum and a minimum point. There is no axis of symmetry.

The leading coefficient is positive. As x increases, the graph extends from the 3rd quadrant to the 1st quadrant. There are no hills or valleys and no maximum or minimum points. There is no axis of symmetry.

Look at the graph of $f(x) = x^3 - 7x$ (above left). Visualize making a tracing of the curve and rotating the tracing $180°$ about the origin. The tracing will coincide with the original curve. Any line passing through the origin and another point on the curve intersects the curve again at a third point. The origin is the midpoint of the line segment joining the two other points. We say this curve has *point symmetry*.

Two examples of quartic functions

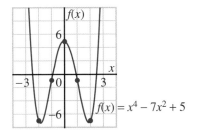

$f(x) = x^4 - 7x^2 + 5$

x	−3	−2	−1	0	1	2	3
$f(x)$	23	−7	−1	5	−1	−7	23

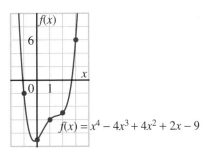

$f(x) = x^4 - 4x^3 + 4x^2 + 2x - 9$

x	−2	−1	0	1	2	3
$f(x)$	51	−2	−9	−6	−5	6

The leading coefficient is positive. As x increases, the graph extends from the 2nd quadrant to the 1st quadrant, forming two valleys and one hill. The graph contains two minimum points and one maximum point. There is an axis of symmetry.

The leading coefficient is positive. As x increases, the graph extends from the 2nd quadrant to the 1st quadrant, forming one valley. The graph contains one minimum point. There is no axis of symmetry.

The equations of polynomial functions are extensions of the equations of linear and quadratic functions. Similarly, the graphs of polynomial functions are extensions of the graphs of linear and quadratic functions.

Properties of the Graphs of Polynomial Functions

Functions with odd degree

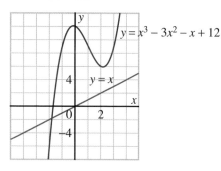

$y = x^3 - 3x^2 - x + 12$

$y = x$

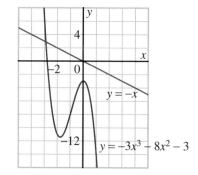

$y = -x$

$y = -3x^3 - 8x^2 - 3$

When the leading coefficient is positive, the graph extends from the 3rd quadrant to the 1st quadrant, as the graph of $y = x$ does.

When the leading coefficient is negative, the graph extends from the 2nd quadrant to the 4th quadrant, as the graph of $y = -x$ does.

Functions with even degree

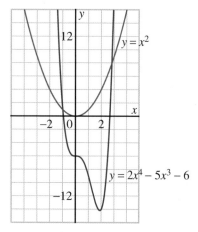

$y = x^2$

$y = 2x^4 - 5x^3 - 6$

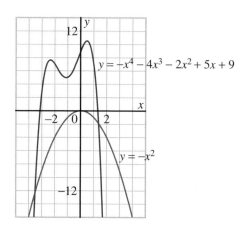

$y = -x^4 - 4x^3 - 2x^2 + 5x + 9$

$y = -x^2$

When the leading coefficient is positive, the graph extends from the 2nd quadrant to the 1st quadrant, as the graph of $y = x^2$ does.

When the leading coefficient is negative, the graph extends from the 3rd quadrant to the 4th quadrant, as the graph of $y = -x^2$ does.

Example

Identify the function that corresponds to each graph. Justify your choices.

$g(x) = x^3 - x^2 \qquad h(x) = x^4 - 3x^3 + x - 1$

$f(x) = -3x^3 + 8x^2 + 7 \qquad g(x) = -x^4 - x^3 + 11x^2 + 9x - 3$

a)

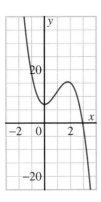

b)

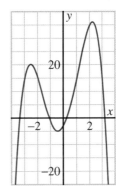

c)

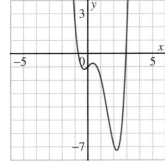

d)

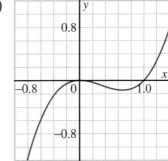

Solution

a) As x increases, this graph extends from the 2nd quadrant to the 4th quadrant, like the line $y = -x$. Hence, its equation is an odd-degree polynomial with a negative leading coefficient. The only possibility among those listed is $f(x) = -3x^3 + 8x^2 + 7$.

b) As x increases, this graph extends from the 3rd quadrant to the 4th quadrant, like the parabola $y = -x^2$. Its equation is an even-degree polynomial with a negative leading coefficient, $g(x) = -x^4 - x^3 + 11x^2 + 9x - 3$.

c) As x increases, this graph extends from the 2nd quadrant to the 1st quadrant, like the parabola $y = x^2$. Its equation is an even-degree polynomial with a positive leading coefficient, $h(x) = x^4 - 3x^3 + x - 1$.

d) As x increases, this graph extends from the 3rd quadrant to the 1st quadrant, like the line $y = x$. Its equation is an odd-degree polynomial with a positive leading coefficient, $g(x) = x^3 - x^2$.

1. Explain why the symmetry of the quadratic functions on page 156 is not evident in their tables of values.

2. Among the graphs on pages 158-159, there are graphs of four cubic functions. Do these graphs appear to have point symmetry? Explain.

3. On page 157, the equations of all four functions have a positive leading coefficient. Suppose the leading coefficient were negative. How would the graphs be affected? Explain.

4. Suppose you are given the graph of a polynomial function.

    a) How can you tell whether the degree of the function is odd or even?

    b) How can you tell whether the leading coefficient is positive or negative?

    c) What else can you tell about the function from its graph?

5. Suppose you are given the graph of a polynomial function. By just looking at the graph can you always tell the degree of the function? Explain.

3.1 EXERCISES

A 1. Determine if each function is a polynomial function or some other type of function. Justify your conclusion.

   a) $f(x) = 2x^3 + x^2 - 5$ b) $f(x) = x^2 + 3x - 2$

   c) $y = 2x + 7$ d) $y = \sqrt{x + 1}$

   e) $y = \dfrac{x^2 + x - 4}{x + 2}$ f) $f(x) = x(x - 1)^2$

   g) $g(x) = 1.2x^2 + \frac{1}{2}x - \pi$ h) $y = \sqrt{2x^3} + 2x - 0.5$

2. Examine these graphs. Which could be graphs of polynomial functions? Explain.

   a)
   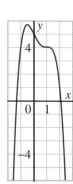

   b)
   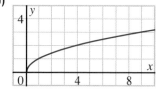

   c)
   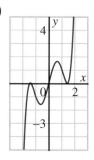

d)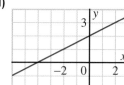

e)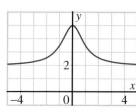

f)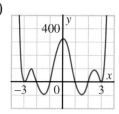

B 3. Look at the graphs of the functions in exercise 2.

   a) Which graphs have line symmetry? Describe the symmetry.

   b) Which graphs have point symmetry? Describe the symmetry.

4. Recall that the graph of every quadratic function has an axis of symmetry.

   a) Which kinds of polynomial functions could have graphs that have an axis of symmetry? Explain.

   b) Do all polynomial functions you identified in part a have graphs with an axis of symmetry? Explain.

5. Identify the function that corresponds to each graph. Justify your choices.

   $f(x) = x^3 - 3x^2 - 5x + 16$

   $g(x) = x^4 - 10x^2 - 5x + 5$

   $k(x) = 5x^4 - 14x^3$

a)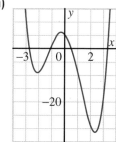

b)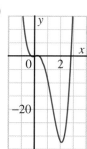

c)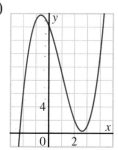

6. a) Use the graphs at the top of page 157 as guides. Sketch the graph of the inverse of each function.

   i) $f(x) = x^3 - 7x$ ii) $f(x) = x^3 - 2x^2 + 2x + 3$

   b) Which, if any, of the functions in part a has an inverse that is a function? Explain.

7. a) Which kinds of polynomial functions could have an inverse that is also a function? Explain.

   b) Do all polynomial functions you identified in part a have an inverse that is also a function? Explain.

8. These screens show the graphs of $y = x^2$, $y = x^4$, and $y = x^6$, respectively. The window settings are $-2 \leq x \leq 2$ and $-1.5 \leq y \leq 1.5$.

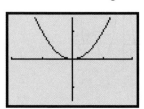 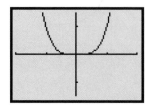 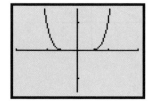

a) In what ways are the graphs similar? In what ways are they different? Explain the similarities and differences.

b) Predict what the graph of $y = x^n$ would look like if n were a much greater even number than those in part a. Use a calculator to check your prediction.

c) Visualize the graphs changing from values of n that are relatively small to values of n that are relatively large. Describe what happens to the graph of $y = x^n$ as n increases through even values of n.

 9. a) Use the same window settings as in exercise 7. Graph the functions $y = x^3$, $y = x^5$, and $y = x^7$.

b) Predict what the graph of $y = x^n$ would look like if n were a much greater odd number than those in part a. Use your calculator to check your prediction.

c) Visualize the graphs changing from values of n that are relatively small to values of n that are relatively large. Describe what happens to the graph of $y = x^n$ as n increases through odd values of n.

COMMUNICATING THE IDEAS

How are the graphs of linear and quadratic functions special cases of the graphs of polynomial functions? Write your ideas in your notebook.

3.2 Properties of the Graphs of Polynomial Functions

Recall *Visualizing* in Section 3.1. We related the graphs of polynomial functions with odd degree to the graphs of $y = x$ or $y = -x$. We related the graphs of polynomial functions with even degree to the graphs of $y = x^2$ or $y = -x^2$. In this section we examine some other properties of the graphs of polynomial functions.

Example 1

a) Graph the function defined by the equation $y = x^3 - 6x^2 + 6x + 3$.

b) Determine the coordinates of any maximum or minimum points.

c) What are the domain and range of the function?

Solution

Use a graphing calculator, or make a table of values and use grid paper.

Using a graphing calculator

a) Input the function, set the window, then graph the function to obtain a result similar to the graph shown.

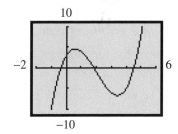

b) Use the "maximum" feature in the calculate menu to determine the coordinates of the maximum point. These are approximately (0.586, 4.657). Similarly, use the "minimum" feature in the calculate menu to determine the coordinates of the minimum point. These are approximately (3.414, −6.657).

c) The domain is the set of all possible values of x. This is the set of all real numbers. The range is the set of all possible values of y. This is also the set of all real numbers.

Using grid paper

a) Use a scientific calculator to make a table of values for $y = x^3 - 6x^2 + 6x + 3$. Plot the ordered pairs on a grid, then draw a smooth curve through them.

b) Use the graph to estimate the coordinates of the maximum and minimum points. These appear to be approximately (0.5, 5) and (3.5, −7), respectively.

c) The domain is the set of all possible values of x. This is the set of all real numbers. The range is the set of all possible values of y. This is also the set of all real numbers.

x	y
−1	−10
0	3
1	4
2	−1
3	−6
4	−5
5	8

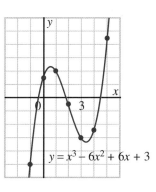

In *Example 1*, the *y*-coordinate of the maximum point is greater than the *y*-coordinates of neighbouring points on either side of the maximum point. However, other points on the graph have greater *y*-coordinates. We say that this point is a *relative maximum* point. Similarly, the minimum point is a *relative minimum* point. It is difficult to determine the coordinates of these points accurately using grid paper. The estimates given in the above solution are only rough estimates.

Example 2

a) Graph the function defined by the equation $y = x^4 - 4x^3 + x^2 + 7x - 3$.

b) Determine the coordinates of any maximum or minimum points.

c) What are the domain and range of the function?

Solution

Use a graphing calculator, or make a table of values and use grid paper.

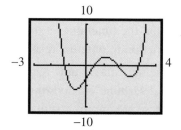

Using a graphing calculator

a) Input the function, set the window, then graph the function to obtain a result similar to the graph shown.

b) Use the "maximum" feature in the calculate menu to determine the coordinates of the maximum point. These are approximately (1.100, 2.050). Similarly, use the "minimum" feature in the calculate menu to determine the coordinates of the two minimum points. These are approximately (−0.629, −5.856) and (2.529, −2.695).

c) The domain is the set of all possible values of *x*. This is the set of all real numbers. The range is the set of all possible values of *y*. This is the set of all real numbers greater than or equal to the *y*-coordinate of the lower minimum point, approximately −5.856.

Using grid paper

a) Use a scientific calculator to make a table of values for $y = x^4 - 4x^3 + x^2 + 7x - 3$. Plot the ordered pairs on a grid, then draw a smooth curve through them.

b) Use the graph to estimate the coordinates of the maximum and minimum points. The maximum point appears to be approximately (1, 2). The minimum points appear to be approximately (−0.5, −5.8) and (2.5, −2.8).

x	y
−2	−35
−1.5	7.3
−1	−4
0	−3
1	2
2	−1
3	0
3.5	12.3

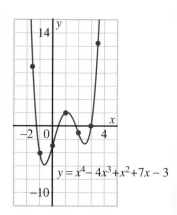

c) The domain is the set of all possible values of x. This is the set of all real numbers. The range is the set of all possible values of y. This is the set of all real numbers greater than or equal to the y-coordinate of the lower minimum point, approximately -5.8.

In *Example 2*, the maximum point is a relative maximum point. The minimum point in the fourth quadrant is a relative minimum point. The minimum point in the third quadrant is called an *absolute minimum point* because there is no other point on the graph that has a lesser y-coordinate.

Example 3

a) Write the equation of the inverse of the function in *Example 2*, then sketch its graph.

b) Determine the domain and the range of the inverse.

c) Is the inverse a function? Explain.

Solution

a) The equation of the inverse is $x = y^4 - 4y^3 + y^2 + 7y - 3$. To graph the inverse, visualize reflecting the graph of the function in the line $y = x$.

b) The domain of the inverse is the range of the function in *Example 2*. This is the set of all real numbers greater than approximately -5.856. The range of the inverse is the domain of the function in *Example 2*. This is the set of all real numbers.

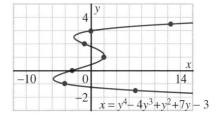

c) The inverse is not a function because it fails the vertical line test.

DISCUSSING THE IDEAS

1. In both *Examples 1* and *2*, the coordinates of the maximum or minimum points obtained using grid paper are not accurate. Suggest how grid paper could be used to obtain these coordinates more accurately.

2. Explain the difference between a relative maximum (or minimum) point and an absolute maximum (or minimum) point.

A 1. Decide if each statement is true or false. If it is true, explain why. If it is false, give a counterexample.

    a) The graph of every cubic function has relative maximum and minimum points.

    b) The domain and range of every cubic function are the set of all real numbers.

    c) The graph of every quartic function has either an absolute maximum point or an absolute minimum point.

    d) The domain and range of every quartic function are the set of all real numbers.

B 2. Consider the function $y = x^3 - 9x^2 + 19x - 3$.

    a) Graph the function, and describe any symmetry.

    b) Determine the coordinates of any maximum or minimum points.

    c) What are the domain and the range of the function?

3. Repeat exercise 2 for these functions.

    a) $y = x^3 + 3x^2 - 9x - 1$ **b)** $f(x) = x^3 - 4x^2 + 6x - 1$

    c) $y = x^4 + x^3 - 4x^2 + 2$ **d)** $f(x) = x^4 - 3x^3 - x + 3$

4. Determine the domain and the range of each function.

    a) $y = x^3 + 3x^2 - 9x - 10$ **b)** $f(x) = -x^3 + x^2 - 7x + 2$

    c) $y = x^4 - 5x^2 + x - 2$ **d)** $f(x) = x^4 - 4x^3 + 7x^2 - 5x + 3$

5. **a)** Use the graph of $y = x^3 - 6x^2 + 6x + 3$ in the solution of *Example 1* as a guide. Sketch the graph of the inverse of $y = x^3 - 6x^2 + 6x + 3$.

    b) Determine the equation, the domain, and the range of the inverse.

    c) Is the inverse a function? Explain.

6. **a)** Sketch the graph of the inverse of each function, and describe any symmetry.

       i) $y = x^3 - 5x^2 + 5$ **ii)** $f(x) = x^4 - 8x^2 + 5$

    b) Determine the domain and range of the inverse of each function in part a.

    c) Which, if any, of the functions in part a has an inverse that is a function? Explain.

C 7. Is it possible for the graph of the inverse of a polynomial function to have a maximum point or a minimum point? Explain.

COMMUNICATING THE IDEAS

Write to explain how to determine the range of a polynomial function. Include some examples to illustrate your explanation.

3.3 Relating Polynomial Functions and Equations

When the graph of a function intersects the x-axis, the x-intercepts are called the *zeros* of the function and the *roots* of the corresponding equation. For example, the quadratic functions below have an equation of the form $y = x^2 - 6x + c$. Visualize what happens to the graph and to the equations as c increases from 8 to 10.

$y = x^2 - 6x + 8$
$\quad = (x - 2)(x - 4)$

$\qquad \uparrow \qquad \uparrow$

$\quad$ zeros 2 and 4

$y = x^2 - 6x + 9$
$\quad = (x - 3)(x - 3)$

$\qquad \uparrow \qquad \uparrow$

$\quad$ zeros 3 and 3

$x^2 - 6x + 8 = 0$
$(x - 2)(x - 4) = 0$

$\qquad \uparrow \qquad \uparrow$

$\quad$ roots 2 and 4

$x^2 - 6x + 9 = 0$
$(x - 3)(x - 3) = 0$

$\qquad \uparrow \qquad \uparrow$

$\quad$ roots 3 and 3

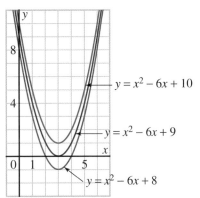

When $c < 9$, the graph intersects the x-axis at two different points. The corresponding quadratic function has two different zeros, and the corresponding quadratic equation has two different roots. When c increases, the two points of intersection with the x-axis come closer together. When $c = 9$, these points coincide at $x = 3$. When $c > 9$, the graph does not intersect the x-axis.

We know the function $y = x^2 - 6x + 8$ has zeros at $x = 2$ and $x = 4$ because the equation of the function can be written as $y = (x - 2)(x - 4)$. We can tell this from the graph because it intersects the x-axis at $(2, 0)$ and $(4, 0)$. We know the quadratic equation $x^2 - 6x + 8 = 0$ has roots $x = 2$ and $x = 4$ because the equation can be written as $(x - 2)(x - 4) = 0$.

We say the function $y = x^2 - 6x + 9$ has a *double zero* at $x = 3$. We know this from the equation of the function, since it can be written as $y = (x - 3)^2$. We can tell this from the graph because the graph intersects the x-axis at $(3, 0)$ but does not cross it. We know the quadratic equation $x^2 - 6x + 9 = 0$ has a *double root*, $x = 3$, because the equation can be written as $(x - 3)^2 = 0$.

The function $y = x^2 - 6x + 10$ has no zeros because its graph does not intersect the x-axis.

Similar results occur with all polynomial functions.

The function $y = x^2 - 6x + 8$ has zeros at $x = 2$ and $x = 4$, but this is not the only quadratic function with these zeros. Any function of the form $y = a(x^2 - 6x + 8)$ has these zeros. Some examples are shown at the right.

The graphs are vertical expansions or compressions of the graph of $y = x^2 - 6x + 8$.

If $a < 0$, there is also a reflection in the x-axis.

Each graph intersects the x-axis at $x = 2$ and at $x = 4$.

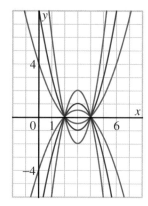

- $y = (x - 2)(x - 4)$
- $y = 2(x - 2)(x - 4)$
- $y = \frac{1}{2}(x - 2)(x - 4)$

- $y = -(x - 2)(x - 4)$
- $y = -2(x - 2)(x - 4)$
- $y = -\frac{1}{2}(x - 2)(x - 4)$

Similarly, $y = x^2 - 6x + 9$ is not the only quadratic function with a double zero at $x = 3$. Any function of the form $y = a(x^2 - 6x + 9)$ has this double zero. The graphs at the right are vertical expansions or compressions of the graph of $y = x^2 - 6x + 9$, combined with a reflection in the x-axis if $a < 0$.

Each graph intersects the x-axis at $(3, 0)$ but does not cross it.

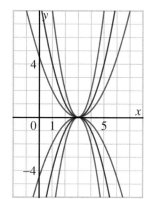

- $y = (x - 3)^2$
- $y = 2(x - 3)^2$
- $y = \frac{1}{2}(x - 3)^2$

- $y = -(x - 3)^2$
- $y = -2(x - 3)^2$
- $y = -\frac{1}{2}(x - 3)^2$

Example 1

Determine the equation of a quadratic function that has zeros -2 and 3, and whose graph passes through the point $(1, 4)$. Graph the function.

Solution

Since the zeros are -2 and 3, let the equation of the function be $y = a(x + 2)(x - 3)$.

Since the graph passes through $(1, 4)$, the coordinates of this point satisfy the equation. Substitute 1 for x and 4 for y.

$$4 = a(3)(-2)$$
$$a = -\frac{2}{3}$$

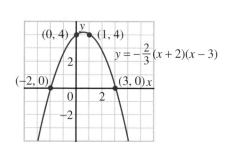

The equation of the function is $y = -\frac{2}{3}(x + 2)(x - 3)$.

The graph opens down. Its axis of symmetry is halfway between $x = -2$ and $x = 3$; that is, $x = 0.5$. By symmetry, the point $(0, 4)$ also lies on the graph. Draw a smooth curve through all known points.

Roots and Factors for Polynomial Equations

VISUALIZING

Each diagram shows two polynomial functions whose graphs intersect the *x*-axis. Factors of the corresponding equations are indicated on the graphs.

Quadratic

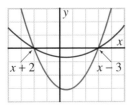

$f(x) = a(x + 2)(x - 3)$

Roots of $f(x) = 0$: $-2, 3$

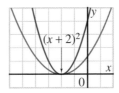

$f(x) = a(x + 2)^2$

Roots of $f(x) = 0$: -2 (double root)

Cubic

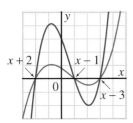

$f(x) = a(x + 2)(x - 1)(x - 3)$

Roots of $f(x) = 0$: $-2, 1, 3$

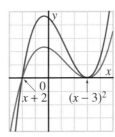

$f(x) = a(x + 2)(x - 3)^2$

Roots of $f(x) = 0$: $-2, 3$ (double root)

Quartic

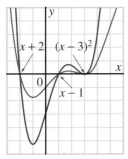

$f(x) = a(x + 2)(x - 1)(x - 3)^2$

Roots of $f(x) = 0$: $-2, 1, 3$ (double root)

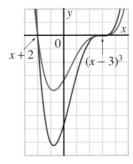

$f(x) = a(x + 2)(x - 3)^3$

Roots of $f(x) = 0$: $-2, 3$ (triple root)

Example 2

Sketch a graph of each function.

a) $y = (x + 2)(x - 5)^2$ **b)** $y = (x + 2)^2(x - 5)^2$

Solution

a) $y = (x + 2)(x - 5)^2$

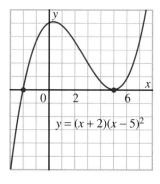

This cubic function has a zero at $x = -2$ and a double zero at $x = 5$. Mark these points on the x-axis. Since the coefficient of x^3 is positive, the graph extends from the 3rd quadrant to the 1st quadrant. Begin in the 3rd quadrant. The curve passes up through $(-2, 0)$, back down to touch the x-axis at $(5, 0)$, then up into the 1st quadrant.

b) $y = (x + 2)^2(x - 5)^2$

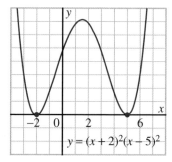

This quartic function has double zeros at $x = -2$ and $x = 5$. Since the coefficient of x^4 is positive, the graph extends from the 2nd quadrant to the 1st quadrant. Begin in the 2nd quadrant. The curve passes down to touch the x-axis at $(-2, 0)$, up into the 2nd quadrant and 1st quadrant, back down to touch the x-axis at $(5, 0)$, then up into the 1st quadrant.

When a polynomial equation is expressed in factored form, its roots can be determined by inspection.

Example 3

Solve for x. $\quad x^3 - 4x^2 - 12x = 0$

Solution

Factor.
$$x(x^2 - 4x - 12) = 0$$
$$x(x + 2)(x - 6) = 0$$

Either $x = 0\quad$ or $\quad x + 2 = 0\quad$ or $\quad x - 6 = 0$
$$x = -2 \qquad\qquad x = 6$$

The roots are 0, -2, and 6.

Example 4

Consider the equation $2x^2 + mx + 9 = 0$. For what value(s) of m is one root double the other root?

Solution

$2x^2 + mx + 9 = 0 \qquad ①$

Let the roots be r and $2r$. Then, the equation can be written in the form

$2(x - r)(x - 2r) = 0$

Expand the left side.

$2(x^2 - 3rx + 2r^2) = 0$
$\quad 2x^2 - 6rx + 4r^2 = 0 \qquad ②$

Think ...

Equation ② must be the same equation as ①.

Compare the coefficients in the equations.

Since the coefficients of x^2 are equal, the other coefficients are equal.

$$2x^2 - 6rx + 4r^2 = 0 \qquad ②$$
$$\downarrow \qquad \downarrow$$
$$2x^2 + mx + 9 = 0 \qquad ①$$

Since ① and ② represent the same equation, the constant terms are equal.

Hence, $4r^2 = 9$

$$r^2 = \frac{9}{4}$$

$$r = \pm\frac{3}{2}$$

Since ① and ② represent the same equation, the coefficients of x are equal.

Hence, $m = -6r$

Substitute $r = \frac{3}{2}$.

$$m = -6\left(\frac{3}{2}\right)$$

$$= -9$$

Substitute $r = -\frac{3}{2}$.

$$m = -6\left(-\frac{3}{2}\right)$$

$$= 9$$

Therefore, when $m = \pm9$, one root of the equation will be double the other root.

DISCUSSING THE IDEAS

1. Explain why the function $y = a(x^2 - 6x + 8)$ has the same zeros as the function $y = x^2 - 6x + 8$.

2. In *Example 1*, why did we not expand the product $(x + 2)(x - 3)$?

3. a) In the solution of *Example 4*, why was a "2" included when we wrote the equation $2(x - r)(x - 2r) = 0$?

   b) Could the solution have been completed using the equation $(x - r)(x - 2r) = 0$? Explain.

4. The method used to solve *Example 4* is sometimes called the *method of equating coefficients*. Why is this name appropriate? Why does the method work?

A **1.** What is the greatest number of roots each equation could have?

a) $x^3 + 5x^2 - 6x - 3 = 0$ b) $7x^2 - 12x + 4 = 0$

c) $2x^3 - x^2 + 8x - 9 = 0$ d) $5x^4 + 3x^2 - x - 12 = 0$

e) $3x^3 + 17x + 15 = 0$ f) $x^3 + 10x = 3x^5 - 8x^2 + 4$

2. Write a cubic function with the given zeros.

a) 1, 2, 3   b) 2, 2, 5   c) −4, 1, 0   d) 2, 2, 2

3. Write a quartic function with the given zeros.

a) 1, 2, 3, 4   b) −2, 1, 2, 2   c) −3, 1, 1, 1   d) 0, 0, −1, −1

4. Choose one function from exercise 2 or 3. Write to explain how you wrote the function.

5. Which statements are true and which are false?

a) Not all quadratic functions have zeros.

b) Not all cubic functions have zeros.

c) Every cubic function has at least one zero.

d) Every quartic function has at least one zero.

B **6.** Sketch an example of a cubic function with the given zeros. Then write an equation of the function. Is the equation unique? Explain.

a) −1, 1, 3   b) 0, 3, 5   c) −2, −2, 2   d) 1, 1, 6

7. Sketch an example of a quartic function with the given zeros. Then write an equation of the function.

a) −2, 1, 3, 6   b) −3, −3, 0, 5   c) −1, −1, 3, 3   d) −1, 3, 3, 3

8. Sketch a graph of each function.

a) $y = (x - 1)(x + 3)$ b) $y = -(x - 3)^2(x + 4)$

c) $y = -x(x - 2)(x - 3)$ d) $y = x^2(x + 4)(x - 5)$

e) $y = -(x - 5)(x + 1)$ f) $y = -(x - 7)^2(x - 1)^2$

g) $y = (x - 2)(x - 6)^2$ h) $y = (x + 2)^2(x - 4)^2$

9. Choose one function from exercise 8. Write to explain how you sketched its graph.

10. A quadratic function has zeros −1 and 4. The graph of the function passes through (5, 9).

a) Determine the equation of the function.

b) Sketch the graph of the function.

11. A cubic function has zeros −3, −1, and 2. The y-intercept of its graph is 12.

    a) Determine the equation of the function.

    b) Sketch the graph of the function.

12. Determine the equation of each function, then sketch its graph.

    a) quadratic function with zeros 2, 2; graph has y-intercept 12

    b) cubic function with zeros −2, 1, 4; graph has y-intercept 24

    c) cubic function with zeros −2, 2, 2; graph has y-intercept −16

    d) cubic function with zeros 0, 2, 4; graph passes through (3, 9)

13. Solve for x.

    a) $x^3 + 10x^2 + 21x = 0$ **b)** $x^3 - 7x^2 + 6x = 0$

    c) $x^3 - x^2 - 20x = 0$ **d)** $x^3 + x^2 - 6x = 0$

    e) $2x^3 - 7x^2 + 3x = 0$ **f)** $3x^3 + 2x^2 - 5x = 0$

    g) $x^3 - x^2 = 0$ **h)** $x^3 - x = 0$

    i) $x^3 - 1 = 0$ **j)** $x^3 + 1 = 0$

14. Determine the zeros of each function.

    a) $f(x) = x^2 - 10x + 16$ **b)** $f(x) = x^3 - 7x^2 + 12x$

    c) $f(x) = x^3 - 5x^2 - 14x$ **d)** $f(x) = x^3 + 2x^2 - 15x$

    e) $g(x) = 2x^3 - x^2 - 3x$ **f)** $h(x) = 3x^3 + 19x^2 - 14x$

    g) $p(x) = x^3 - 25x$ **h)** $q(x) = 4x^2 - 9x$

    i) $f(x) = x^3 - 64x$ **j)** $f(x) = x^3 - 64$

15. Choose one function from exercise 14. Write to explain how you determined its zeros.

16. Determine the value(s) of k in each equation so that one root is double the other root.

    a) $x^2 + kx + 50 = 0$ **b)** $x^2 - kx + 7 = 0$

    c) $4x^2 + kx + 4 = 0$ **d)** $2x^2 - 3x + k = 0$

17. Determine the value(s) of m in each equation so that the two roots are equal.

    a) $x^2 - 6x + m = 0$ **b)** $4x^2 + mx + 25 = 0$

    c) $9x^2 - mx + 1 = 0$ **d)** $9x^2 - 42x + m = 0$

18. Determine the value(s) of n in each equation such that one root is triple the other root.

    a) $3x^2 - 4x + n = 0$ **b)** $4x^2 + nx + 27 = 0$

    c) $x^2 - 16x + n = 0$ **d)** $16x^2 + nx + 27 = 0$

19. Graph $f(x) = x^3 - 2x^2 + k$ for different values of k.

a) Estimate the values of k for which the equation $f(x) = 0$ appears to have a double root.

b) Show that $k = 0$ ensures that $f(x) = 0$ has a double root.

c) Determine another value of k that ensures $f(x) = 0$ has a double root.

C **20.** Consider these three cubic functions.

$$y = x^3 - 9x^2 + 23x - 15$$
$$y = x^3 - 9x^2 + 24x - 16$$
$$y = x^3 - 9x^2 + 25x - 17$$

a) Note the pattern in the coefficients. Assume this pattern continues. Write the next three functions.

b) Graph the three functions on the same screen, using an appropriate viewing window.

c) Describe what happens to the shape of the graph of the function $y = x^3 - 9x^2 + (23 + k)x - (15 + k)$ as k increases through positive values.

d) Suppose the pattern is continued in the opposite direction. Write the preceding three functions.

e) Describe what happens to the shape of the graph of the function $y = x^3 - 9x^2 + (23 + k)x - (15 + k)$ as k decreases through negative values.

21. Both equations $x^3 - 12x + 16 = 0$ and $x^3 - 12x - 16 = 0$ have a double root and one other root that is different from the double root.

a) Use only this information. Determine which of the equations at the right have:

$$x^3 - 12x + 20 = 0$$
$$x^3 - 12x + 10 = 0$$
$$x^3 - 12x - 20 = 0$$

    i) three different roots   **ii)** only one root

b) Determine the values of k for which the equation $x^3 - 12x + k = 0$ has:

    i) 3 different roots

    ii) 2 different roots

    iii) only 1 root

COMMUNICATING THE IDEAS

Suppose you are given the equation or the graph of a polynomial function. Write to explain how you can tell whether the function has a zero and, if so, whether it is a double zero. Use examples to illustrate your explanation.

Visual Proofs

We often use diagrams to prove mathematical results. You may have seen some of these examples, but others may be new.

1. Explain why the diagram (below left) proves that
   $(a + b)^2 = a^2 + 2ab + b^2$.

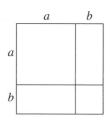 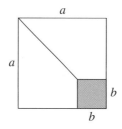 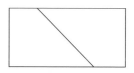

2. Cut a square from a corner of a square piece of paper (above middle). Cut along a diagonal and make a rectangle from the two pieces (above right). Write two different expressions for the area of the rectangle. What algebraic result is proved?

3. This diagram represents the sum of the first five odd numbers, $1 + 3 + 5 + 7 + 9$.

   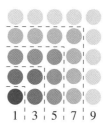

   a) Without adding, use the diagram to determine the sum $1 + 3 + 5 + 7 + 9$.

   b) Visualize a similar diagram representing the sum of the first n odd numbers. Write an expression to represent the sum $1 + 3 + 5 + \cdots + (2n - 1)$.

4. The diagram (below left) shows 15 balls arranged to form a triangle. The number 15 is a *triangular number*. Each triangular number is the sum of consecutive natural numbers.

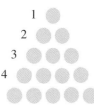 

   a) Two copies of this diagram (above right) form a parallelogram. Without adding, explain why $1 + 2 + 3 + 4 + 5 = 15$.

   b) Visualize a similar diagram representing the sum of the first n natural numbers. Write an expression to represent the sum $1 + 2 + 3 + \cdots + n$.

   c) List the first six triangular numbers, and the 100th triangular number.

5. This diagram shows balls arranged in a square pyramid with 4 layers. The number of balls in the pyramid is a *pyramidal number*. Each pyramidal number is the sum of the squares of consecutive natural numbers.

$1^2 + 2^2 + 3^2 + 4^2$

To determine a formula for the sum $1^2 + 2^2 + 3^2 + \cdots + n^2$, visualize three pyramids similar to this one.

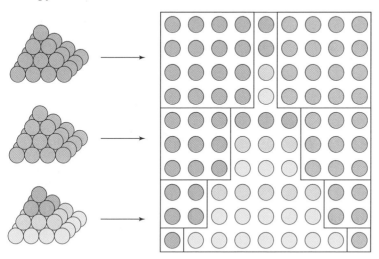

a) Explain how all the balls from the three pyramids fit in the rectangle.

b) What are the dimensions of the rectangle? How many balls are in the rectangle?

c) Use the rectangle to determine the sum $1^2 + 2^2 + 3^2 + 4^2$.

d) Suppose there were n layers in the pyramid. What would the dimensions of the rectangle be? How many balls would be in the rectangle?

e) Write a formula for the sum $1^2 + 2^2 + 3^2 + \cdots + n^2$.

f) List the first six pyramidal numbers and the 100th pyramidal number.

6. a) Write two different expressions for the area of the larger square:

    i) by using the side length $a + b$

    ii) by using the four corner triangles and the inside square

b) Use the result of part a to prove the Pythagorean Theorem: $c^2 = a^2 + b^2$.

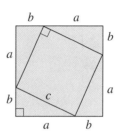

3.4 Solving Polynomial Equations

In Section 3.3, polynomial equations were expressed in factored form. However, many equations are not in factored form. We can also solve the equation graphically using either technology or grid paper.

Example

Solve the equation $x^3 + x = 20$ graphically, using two different methods.

Solution 1

Using a graphing calculator

Enter the equation in the form $y = x^3 + x - 20$. Adjust the window so that all points where the graph intersects the x-axis are visible. On this graph, the x-values run from −4.7 to 4.7 to provide x-coordinates increasing by 0.1 when tracing.

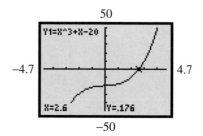

By tracing and zooming, find points on the graph whose y-coordinates are approximately 0, or use the calculate menu.

The equation $x^3 + x = 20$ has one root, $x \doteq 2.59$.

Using grid paper

Let $y = x^3 + x - 20$. Make a table of values, plot the ordered pairs on a grid, then draw a smooth curve through them.

x	y
−2	−30
−1	−22
0	−20
1	−18
2	−10
3	10
4	48

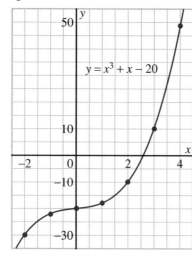

The equation $x^3 + x = 20$ has one root, which is approximately 2.6.

Solution 2

Using a graphing calculator

Enter the equations $y = x^3 + x$ and $y = 20$. By tracing and zooming, or by using the calculate menu, find the coordinates of the point where the line intersects the curve.

The equation $x^3 + x = 20$ has one root, $x \doteq 2.59$.

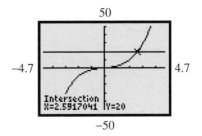

Using grid paper

Graph the function $y = x^3 + x$. Draw the horizontal line $y = 20$. Estimate the x-coordinates of the point where the line intersects the curve.

x	y
−3	−30
−2	−10
−1	−2
0	0
1	2
2	10
3	30

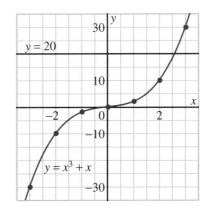

The equation $x^3 + x = 20$ has one root, which is approximately 2.6.

In the *Example*, the root of the equation $x^3 + x = 20$ is the zero of the corresponding function $f(x) = x^3 + x - 20$.

DISCUSSING THE IDEAS

1. What are the advantages and disadvantages of each method used to solve the *Example*?

2. Describe a third method that could have been used to solve the equation in the *Example*. Solve the equation using this method.

3. When you solve a polynomial equation using a graphing calculator, how can you tell when the roots are exact? Use an example to support your answer.

A 1. Use the graph to estimate the zero(s) of each function.

a) $f(x) = x^3 + 2x^2 - 10$

b) $g(x) = -x^3 - 3x^2 + 5x + 16$

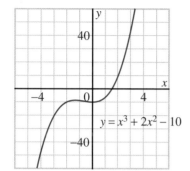

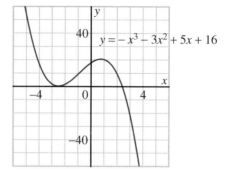

c) $h(x) = x^4 - 10x^2 - 5x + 5$

d) $p(x) = x^5 - 10x^3 + 15x$

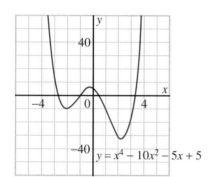

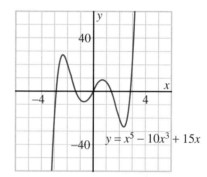

B 2. Solve graphically.

a) $x^3 + 2x = 10$

b) $x^3 + 2x = 5x$

c) $x^3 + 2x = 5x - 10$

d) $x^3 + 2x = 10x - 5$

3. Solve graphically.

a) $x^3 - 10x = 0$

b) $x^3 - 10x + 12 = 0$

c) $x^3 - 10x - 12 = 0$

d) $x^3 - 10x - 24 = 0$

4. Use a graph to approximate the zeros of each function.

a) $p(x) = x^3 - 15x - 10$

b) $q(x) = x^3 + x - 15$

c) $f(x) = x^4 - 15x^2 + 20$

d) $g(x) = x^4 - 5x^2 - 10x - 25$

5. Approximate the zeros of each function.

a) $f(x) = x^3 + 10x - 20$

b) $f(x) = x^3 - 3x^2 + x - 10$

c) $f(x) = x^4 - 10x^2 + 5x + 7$

d) $f(x) = x^4 - 4x^3 + 16x - 25$

6. a) In exercise 1b, can you be sure there are two equal negative zeros? Explain.

    b) If there are not two equal negative zeros, what other possibilities are there for this function?

    c) How could you tell which possibility in part b is correct? Explain.

7. Consider the equation $x^2 - 2x = 20$.

    a) Solve the equation by graphing $y = x^2 - 2x$ and $y = 20$ and using the points of intersection to determine the roots.

    b) Solve the equation by graphing $y = x^2 - 2x - 20$ and using the x-intercepts to determine the roots.

    c) Compare the methods in parts a and b.

      i) Does one method give more accurate results than the other? Explain.

      ii) Is one method more reliable than the other? Explain.

8. Consider the equation $3x^2 + 4 = 9x - 1$.

    a) Solve the equation by graphing $y = 3x^2 + 4$ and $y = 9x - 1$ and using the points of intersection to determine the roots.

    b) Solve the equation by graphing $y = 3x^2 - 9x + 5$ and using the x-intercepts to determine the roots.

    c) Repeat exercise 7c for parts a and b above.

 9. The screen on the right was obtained by graphing the function $y = 0.02x^3 - 5x^2 + 80$ on a graphing calculator. The function appears to have two zeros.

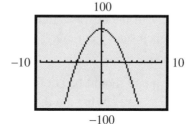

    a) Explain why a third zero must exist for this function.

    b) Is the third zero positive or negative? Explain.

    c) Is the y-axis an axis of symmetry? Explain.

 10. Use a graphing calculator to determine the three zeros of the function in exercise 9.

COMMUNICATING THE IDEAS

Assume you are using either a graphing calculator or grid paper. Write to explain why a polynomial equation can be solved graphically in more than one way. Include some examples to illustrate your explanation.

3.5 Modelling Real Situations Using Cubic Functions

A packaging company has large quantities of cardboard 28 cm long and 21 cm wide. Each piece is used to make a box with no top. Equal squares are cut from the corners and the sides are folded up. When 5-cm squares are removed, the box measures 18 cm by 11 cm by 5 cm.

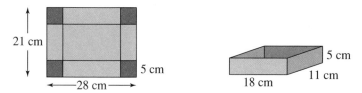

When other sizes of squares are removed, the dimensions of the box change. Visualize how the shape and the volume of the box change as the sides of the squares change.

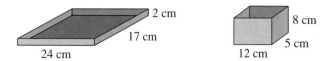

Example

A piece of cardboard 28 cm long and 21 cm wide is used to make a box.

a) Write the volume V of the box as a function of the side length x centimetres of each square cut out.

b) i) Graph V against x.

ii) What size of square should be cut from the corners to have a box with volume 750 cm^3? What are the dimensions of the box?

iii) What size of square should be cut out to have a box with the maximum volume? What are the dimensions of this box?

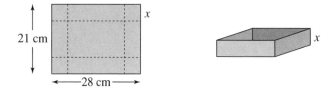

Solution

a) The volume of the box is the product of its height, length, and width.

$$V = x(28 - 2x)(21 - 2x)$$

b) Use a graphing calculator, or make a table of values and use grid paper.

Using a graphing calculator

i) Enter the equation
$y = x(28 - 2x)(21 - 2x)$.

On this graph, the x-values run from 0 to 9.4 to provide x-coordinates increasing by 0.1 when tracing. The y-values start at -150 to provide room for the coordinates under the horizontal axis when zooming.

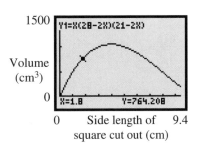

ii) By tracing and zooming, find points on the graph whose y-coordinates are approximately 750. These occur when $x \doteq 1.749$ and $x \doteq 6.665$. The volume of the box is 750 cm^3 when the side length of the squares cut out is approximately 1.75 cm or 6.67 cm.

Substitute each value for x in the three factors of the volume expression.

$$x = 1.749 \qquad 28 - 2x = 28 - 2(1.749) \qquad 21 - 2x = 21 - 2(1.749)$$
$$= 24.502 \qquad\qquad = 17.502$$

The dimensions of a box with volume 750 cm^3 are approximately 24.5 cm by 17.5 cm by 1.7 cm.

Similarly, using $x = 6.665$, the dimensions of another box with volume 750 cm^3 are approximately 14.7 cm by 7.7 cm by 6.7 cm.

iii) By tracing and zooming, find the coordinates of the maximum point. These are approximately (3.960, 1040.1). Substitute 3.960 for x in the three factors of the volume expression.

$$x = 3.960 \qquad 28 - 2x = 28 - 2(3.960) \qquad 21 - 2x = 21 - 2(3.960)$$
$$= 20.080 \qquad\qquad = 13.080$$

The dimensions of a box with the maximum volume are approximately 20.1 cm by 13.1 cm by 4.0 cm. The maximum volume is approximately 1040 cm^3.

Using grid paper

i) Use a scientific calculator to make a table of values for $V = x(28 - 2x)(21 - 2x)$. Include enough values in the table to obtain values of V that are greater than and less than 750. Plot the points and draw a smooth curve through them.

x	V
0	0
1	494
2	816
3	990
4	1040
5	990
6	864
7	686
8	480

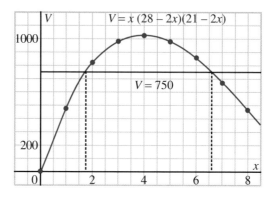

ii) Draw a horizontal line at $V = 750$. It intersects the curve at two points. From the graph, the x-coordinates of those points are approximately 1.8 and 6.7. The volume of the box is 750 cm³ when the side length of the squares cut out is approximately 1.8 cm or 6.7 cm.

Substitute each value for x in the three factors of the volume expression.

$x = 1.8$ $28 - 2x = 28 - 2(1.8)$ $21 - 2x = 21 - 2(1.8)$
                                      $= 24.4$ $= 17.4$

The dimensions of a box with volume 750 cm³ are about 24.4 cm by 17.4 cm by 1.8 cm.

Similarly, using $x = 6.7$, the dimensions of another box with volume 750 cm³ are about 14.6 cm by 7.6 cm by 6.7 cm.

iii) From the graph, the coordinates of the maximum point appear to be (4, 1040). Substitute 4 for x in the three factors of the volume expression.

$x = 4$ $28 - 2x = 28 - 2(4)$ $21 - 2x = 21 - 2(4)$
                                      $= 20$ $= 13$

The dimensions of a box with the maximum volume are about 20 cm by 13 cm by 4 cm. The maximum volume is about 1040 cm³.

The two methods used to solve the *Example* are essentially the same. In each method, a graph of the volume function is used to solve the problem. In principle, it does not matter if the graph is created using technology or on grid paper.

1. What is the domain of the volume function in the *Example*?

2. In the solution of the *Example*, we could have expanded the expression for the volume and written the function in the form $V = 4x^3 - 98x^2 + 588x$.

   a) Why was it not necessary to do this? Explain.

   b) Do you think it is useful to expand the volume expression? Explain.

   c) How can we tell, without expanding, that $x(28 - 2x)(21 - 2x)$ is a cubic function?

3. What are the advantages and disadvantages of each method used to solve the *Example*?

3.5 EXERCISES

A 1. A packaging company makes boxes using cardboard 25.0 cm long and 20.0 cm wide. Determine the size of squares to be cut from the corners for each of these boxes. Determine the dimensions of each box.

   a) a box with volume 500 cm^3

   b) a box with the maximum possible volume

2. Repeat exercise 1 for pieces of cardboard that are 25.0 cm square.

B 3. A packaging company is making another style of box from cardboard 28.0 cm long and 21.0 cm wide. This box has a top that comes from the same piece of cardboard. The diagram (below left) shows how it is made.

   a) Let x centimetres represent the side length of each square cut from the corners. Write the volume of the box as a cubic function of x.

   b) Graph the function in part a. Use a graphing calculator if you have one.

   c) What size of square is cut from the corners to have a box with volume 375 cm^3? What are the dimensions of the box?

   d) What size of square is cut from the corners to have a box with the maximum volume? What are the dimensions of this box?

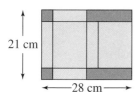
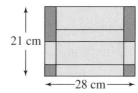

4. The diagram (above right) shows another way to make a box with a top from the same piece of cardboard. Repeat exercise 3 for this box.

5. According to *The Guinness Book of Records*, a square pyramid of empty cans containing 5525 cans was made on two occasions in 1995. In exercise 5 on page 177, you should have found that for a square pyramid with n layers,

$$1^2 + 2^2 + 3^2 + \cdots + n^2 = \frac{n(n+1)(2n+1)}{6}$$

Use this formula to determine the number of layers in the pyramid of cans.

6. The volume, V, of a spherical segment with base radius, r, and height, h, is given by the formula $V = \frac{1}{6}\pi h(3r^2 + h^2)$.

A domed stadium is in the shape of a spherical segment with a base radius of 150 m. The dome must contain a volume of 3 500 000 m^3. Determine the height of the dome at its centre, to the nearest tenth of a metre.

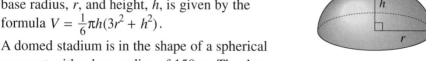

 MODELLING **Domed Stadium Design**

In exercise 6, a domed stadium was modelled by a segment of a sphere.

- Why might it be important for the stadium to have a particular volume?

- Suggest why the height of the dome at the centre might have to be greater than the answer you obtained in exercise 6.

7. A juice box measures 6.4 cm by 4.0 cm by 10.5 cm. It contains 250 mL of juice. The manufacturer wants to design a larger box by increasing each dimension by the same amount. Suppose the larger box holds twice as much juice. What are its dimensions?

C 8. Compare your results from exercises 3 and 4 with the *Example* on page 182.
   a) Describe each relationship.
      i) between the maximum volume for a box with no top and the maximum volume for a box with a top
      ii) among the sizes of squares removed
      iii) among the dimensions of the boxes
   b) Use the equations of the volume functions to explain the relationships you found.

9. A packaging company designs a can with no top. The can is made from 250 cm^2 of metal.

   a) Let r centimetres represent the base radius of the can. Write the volume of the can as a cubic function of r.

   b) Graph the function in part a. Use a graphing calculator if you have one.

   c) What are the dimensions of the can when it has each volume?

      i) 200 mL   ii) the maximum volume

10. The *One Million* book was written to help people visualize one million. There are 200 pages in the book, and each page contains 5000 dots. The dots are arranged in a trapezoidal shape to provide visual appeal. The white space is used to present a few items of interest related to some of the numbers represented by the dots on that page.

 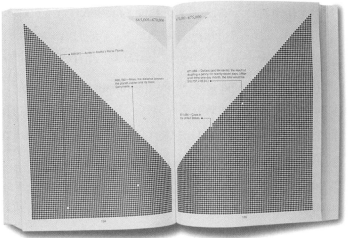

Suppose you are the designer of this book. Visualize the pattern of dots on each page forming a rectangle at the bottom with a right triangle on top. How would you arrange 5000 dots in this way? Determine the numbers of dots along the base and the height of the rectangle, and along the two legs of the right triangle.

COMMUNICATING THE IDEAS

A box is made from a single sheet of cardboard, and contains a given volume. Write to explain why a cubic equation occurs in the solution of this problem, and why there are two different solutions. Use an example, with diagrams, to illustrate your explanation.

Function Operations

Functions that are defined by algebraic expressions can be combined algebraically.

For example, if $f(x) = 2x + 5$ and $g(x) = x^2$, we may write:

$$3f(x) = 3(2x + 5) \qquad f(x) + g(x) = 2x + 5 + x^2 \qquad f(x)g(x) = (2x + 5)(x^2)$$
$$= 6x + 15 \qquad\qquad = x^2 + 2x + 5 \qquad\qquad = 2x^3 + 5x^2$$

1. For each $f(x)$, write the function $3f(x)$.

   a) $f(x) = 6x - 7$
   b) $f(x) = x^2 - 2x + 4$
   c) $f(x) = (x - 5)^2$
   d) $f(x) = 1$

2. For each $f(x)$ and $g(x)$, write the function $f(x) + g(x)$.

   a) $f(x) = 2x + 3$, $g(x) = x - 5$
   b) $f(x) = x^2 - 4$, $g(x) = 2x + 3$
   c) $f(x) = x^2 - x$, $g(x) = x^2 - x + 1$
   d) $f(x) = (x - 2)^2$, $g(x) = (2x - 1)^2$

3. For each $f(x)$ and $g(x)$, write the function $f(x) - g(x)$.

   a) $f(x) = x + 3$, $g(x) = 2 - 3x$
   b) $f(x) = x^2 + 5x$, $g(x) = x^2 - x - 2$
   c) $f(x) = (x + 1)^2$, $g(x) = (x - 1)^2$
   d) $f(x) = (2x - 3)^2$, $g(x) = (3x + 2)^2$

4. For each $f(x)$ and $g(x)$, write the function $f(x)g(x)$.

   a) $f(x) = 3$, $g(x) = 2x + 1$
   b) $f(x) = x - 4$, $g(x) = 7x + 2$
   c) $f(x) = x^2 - 1$, $g(x) = x^2 + 1$
   d) $f(x) = x^2$, $g(x) = 2x - 3$

5. Determine two functions $f(x)$ and $g(x)$ such that each statement is true. Are the functions $f(x)$ and $g(x)$ unique? Explain.

   a) $f(x) + g(x) = x^2 + 3x + 5$
   b) $f(x) + g(x) = x^3 + x^2 - x$
   c) $f(x) - g(x) = 2x^2 + x - 1$
   d) $f(x) - g(x) = x^2 - 4x$
   e) $f(x)g(x) = x^2 - 3x + 4$
   f) $f(x)g(x) = x^2 - 9$

6. Determine two functions $f(x)$ and $g(x)$ such that both statements are true. Are the functions $f(x)$ and $g(x)$ unique? Explain.

   a) $f(x) + g(x) = x^2 + 7x + 10$ $\qquad$ $f(x) - g(x) = x^2 - 7x + 10$

   b) $f(x) + g(x) = x^2 - 1$ $\qquad$ $f(x) - g(x) = x^2 + 1$

   c) $f(x) + g(x) = (x + 1)^2$ $\qquad$ $f(x)g(x) = x(x + 1)^2$

7. Can all functions be combined algebraically? Explain.

Graphing Rational Functions

Functions whose defining equations contain rational expressions are called *rational functions*. Here are some examples.

$$f(x) = \frac{1}{x} \ (x \neq 0) \qquad f(x) = \frac{x}{x^2 - 4} \ (x \neq \pm 2) \qquad f(x) = \frac{x^2 + 1}{x + 1} \ (x \neq -1)$$

A rational function has an equation that can be expressed in the form $f(x) = \frac{p(x)}{q(x)}$, where $p(x)$ and $q(x)$ are polynomial functions, and $q(x) \neq 0$.

Unless suggested otherwise, set the graphing window so that $-4.7 \leq x \leq 4.7$. This provides x-coordinates increasing by 0.1 when tracing. If you use $-3.1 \leq y \leq 3.1$, there will be equal scales on both axes.

Investigating the function $y = \frac{1}{x}$

1. a) Graph the function $y = \frac{1}{x}$.

   b) Use the ⌷TRACE⌷ key to display coordinates of points on the graph. Trace along both parts of the curve. Mentally check that some of these coordinates satisfy the equation of the function.

   c) Use the ⌷TRACE⌷ key to move the cursor to the point where $x = 0$ appears at the bottom of the screen. What happens to the y-coordinate at this point? Explain.

   d) The curve appears to have two parts, called *branches*. Predict whether these two branches are connected. To verify your prediction, turn off the axes. Press ⌷FORMAT⌷, choose AxesOff from the menu that appears, and press ⌷GRAPH⌷. Turn the axes back on before continuing.

2. a) On the same grid, graph both $y = x$ and $y = \frac{1}{x}$.

   b) Press ⌷TRACE⌷, and move the cursor to a point on one of the graphs. Press ⌷▼⌷ to move the cursor vertically to the other graph. Press ⌷▼⌷ again to move it back. Observe the y-coordinates as you do this. What do you notice? Repeat, for other x-coordinates.

3. For the graph of $y = \frac{1}{x}$, state what happens to the value of y as x becomes:

   a) very large and positive   **b)** very small and positive

   c) very large and negative   **d)** very small and negative

Investigating other rational functions

4. Clear the screen, then graph each function. Sketch the results on plain paper. Use the results to form some conclusions about the graphs of rational functions.

a) $y = \dfrac{1}{x + 2}$

b) $y = \dfrac{1}{x^2 - 1}$

c) $y = \dfrac{1}{x^3 + 1}$

d) $y = \dfrac{1}{x^2 - x - 6}$

e) $y = \dfrac{1}{x^2 + 2}$

f) $y = \dfrac{x^2 - 4}{x - 2}$

5. The graphs of six rational functions are shown. Their equations are listed below, but not in the same order as the graphs. Without using a graphing calculator, match each graph with its correct equation. To help you, use your graphs in exercise 4 and the coordinates at the bottom of the screen.

a)

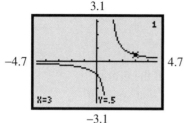

b)

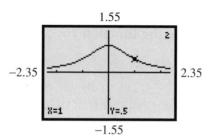

c)

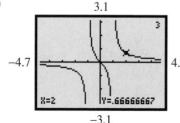

d)

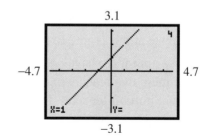

e)

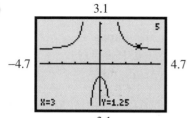

f)

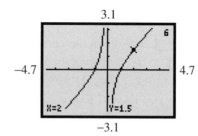

$$y = \frac{x}{x^2 - 1} \qquad y = \frac{x^2 + 1}{x^2 - 1} \qquad y = \frac{1}{x^2 + 1} \qquad y = \frac{x^2 - 1}{x} \qquad y = \frac{1}{x - 1} \qquad y = \frac{x^2 - 1}{x - 1}$$

3.6 Reciprocal Functions

INVESTIGATE Reciprocal Functions

You will graph the functions $y = x$ and $y = \frac{1}{x}$, $(x \neq 0)$, and examine how the graphs are related. Use grid paper, with $-10 \leq x \leq 10$ and $-10 \leq y \leq 10$.

1. a) Use a ruler to graph the line $y = x$. Draw this accurately, passing through the points $(-10, -10)$ and $(10, 10)$. Label this line $f(x)$.

b) Copy and complete this table of values. Use it to graph the equation $y = \frac{1}{x}$. Since there is no y-value when $x = 0$, the table of values is divided into two separate parts. Your graph will also have two separate parts. Draw a smooth curve through the points in each part.

x	−10	−5	−2	−1	−0.5	−0.2	−0.1	0	0.1	0.2	0.5	1	2	5	10
$\frac{1}{x}$								—							

2. Suppose x is a negative number with a relatively large absolute value, such as $x = -100$.

a) How does the value of $\frac{1}{x}$ compare with the value of x? Explain.

b) Visualize the values of x increasing from left to right in the table of values, or from left to right along the graph. What happens to the value of $\frac{1}{x}$ in each case?

   i) when x increases to -1

   ii) when x increases from -1 to 0

   iii) when x increases from 0 to 1

   iv) when x increases to a relatively large value, such as $x = 100$

Recall that the reciprocal of a number x is the number $\frac{1}{x}$, provided that $x \neq 0$.

Similarly, the reciprocal of a function $f(x)$ is the function $\frac{1}{f(x)}$, provided that $f(x) \neq 0$.

For example, for the function $y = x - 2$, the corresponding reciprocal function is $y = \dfrac{1}{x-2}$, where $x \neq 2$. To see how the graphs of these functions are related, we graph them using a graphing calculator or a table of values.

x	x − 2	$\dfrac{1}{x-2}$
−2	−4	−0.25
−1	−3	−0.33
0	−2	−0.5
1	−1	−1
1.5	−0.5	−2
2	0	undefined
2.5	0.5	2
3	1	1
4	2	0.5
5	3	0.33
6	4	0.25

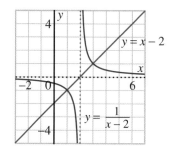

The graph of $y = \dfrac{1}{x-2}$ has two separate parts, or branches, and is called a *hyperbola*.

When $x < 2$, as x increases and comes closer and closer to 2, the hyperbola comes closer and closer to the line $x = 2$, but never reaches this line. For this reason, the line $x = 2$ is called an *asymptote*. When $x \geq 2$, as x decreases and comes closer and closer to 2, the hyperbola also comes closer and closer to the asymptote. The x-axis is another asymptote. The asymptotes are drawn on the grid in a different colour because they are not part of the graph.

The table of values and graph illustrate important properties of reciprocal functions.

Properties of Reciprocal Functions

$f(x)$	$\dfrac{1}{f(x)}$
is less than −1	is between −1 and 0
−1	−1
is between −1 and 0	is less than −1
0	a vertical asymptote may exist
is between 0 and 1	is greater than 1
1	1
is greater than 1	is between 0 and 1

We can use these properties to sketch the graph of the reciprocal of a given function without making a table of values.

Example

On grid paper, sketch a graph of the function $y = \dfrac{1}{x^2 - 4}$, $x \neq \pm 2$.

Solution

Step 1

Graph $y = x^2 - 4$ by translating the graph of $y = x^2$ four units down. Start at $(0, -4)$, then use the method of differences to locate points on the graph. Draw a smooth curve through these points.

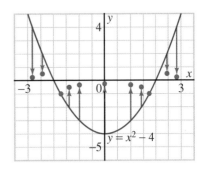

Step 2

On the graph, locate some points that have y-coordinates with absolute values greater than 1. For each point, estimate the location of the corresponding point on the graph of the reciprocal function. These points will be closer to the x-axis than the points on the original graph.

Step 3

On the graph, locate some points that have y-coordinates with absolute values less than 1. For each point, estimate the location of the corresponding point on the graph of the reciprocal function. These points will be farther from the x-axis than the points on the original graph. Points that are very close to the x-axis on the given function will be a great distance from the x-axis on the reciprocal function.

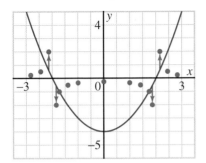
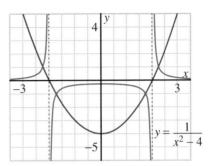

Step 4

Draw a smooth curve through the plotted points. This curve has three distinct branches, separated by the two values of x for which $x^2 - 4$ has a value of 0.

As the examples on pages 192 and 193 suggest, there are different kinds of graphs of reciprocal functions. Here are two more examples.

Other examples of reciprocal functions

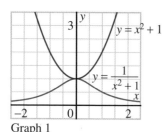

Graph 1

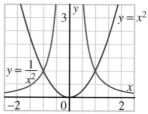

Graph 2

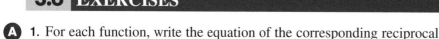

DISCUSSING THE IDEAS

1. How is the graph of $y = \dfrac{1}{x-2}$ related to the graph of $y = \dfrac{1}{x}$?

2. Choose either Graph 1 or Graph 2 above. Choose one function on the graph. Visualize a point moving from left to right along its graph. Explain how the graph of the other function satisfies the properties of reciprocal functions.

3. Look at the graph of $y = \dfrac{1}{x-2}$ on page 192. Describe the symmetry of this graph.

4. The reciprocal of a function $f(x)$ is written as $\dfrac{1}{f(x)}$. Why is it not written as $f^{-1}(x)$?

3.6 EXERCISES

A **1.** For each function, write the equation of the corresponding reciprocal function.

a) $y = 3x - 7$ **b)** $y = 5x^2 - 2x$ **c)** $y = (x - 2)^2 - 1$

d) $y = \sqrt{x + 1}$ **e)** $y = -2x^3 + 3x^2 - 7x$ **f)** $y = 2^x$

2. Copy the graph of each function $f(x)$ on grid paper. Sketch the graph of $\dfrac{1}{f(x)}$ on the same grid, and identify the asymptotes.

a)

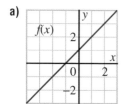

b)

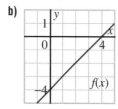

c)

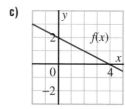

d)

e)

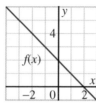

f)

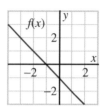

3. Graph each pair of functions on the same grid.

a) $y = x + 2$ and $y = \dfrac{1}{x + 2}$

b) $y = x - 5$ and $y = \dfrac{1}{x - 5}$

c) $y = 2x + 3$ and $y = \dfrac{1}{2x + 3}$

d) $y = \dfrac{1}{2}x - 1$ and $y = \dfrac{1}{\frac{1}{2}x - 1}$

e) $y = -x$ and $y = \dfrac{1}{-x}$

f) $y = -x + 4$ and $y = \dfrac{1}{-x + 4}$

4. Choose one pair of functions from exercise 3. Write to explain how you graphed the functions.

B 5. Graph each function. Graph its reciprocal function on the same grid, then identify the asymptotes.

a) $y = x^2$

b) $y = x^2 - 9$

c) $y = x^2 - 16$

d) $y = x^2 + 5$

e) $y = (x - 2)^2$

f) $y = (x + 3)^2$

6. On plain paper, sketch Graph 1 and Graph 2 from page 194. Use these graphs as guides. Sketch the graph of each pair of functions on plain paper.

a) $y = x^2 + 10$ and $y = \dfrac{1}{x^2 + 10}$

b) $y = x^2 + 2$ and $y = \dfrac{1}{x^2 + 2}$

c) $y = x^2 + 0.5$ and $y = \dfrac{1}{x^2 + 0.5}$

d) $y = x^2 + 0.1$ and $y = \dfrac{1}{x^2 + 0.1}$

7. On plain paper, sketch the graph in the *Solution* on page 193. Use this graph as a guide. Sketch the graph of each pair of functions on plain paper.

a) $y = x^2 - 10$ and $y = \dfrac{1}{x^2 - 10}$

b) $y = x^2 - 2$ and $y = \dfrac{1}{x^2 - 2}$

c) $y = x^2 - 0.5$ and $y = \dfrac{1}{x^2 - 0.5}$

d) $y = x^2 - 0.1$ and $y = \dfrac{1}{x^2 - 0.1}$

8. Look at your graphs in exercises 6 and 7. Visualize how the graphs of $y = x^2 + c$ and $y = \dfrac{1}{x^2 + c}$ change as c varies. Write to explain how the graphs change.

9. a) Use a graphing tool or grid paper to graph each function.

i) $y = \dfrac{1}{2x + 6}$

ii) $y = \dfrac{1}{x + 3}$

iii) $y = \dfrac{1}{3 - x}$

iv) $y = \dfrac{1}{x^2 - 25}$

v) $y = \dfrac{1}{x^2 - 10}$

vi) $y = \dfrac{1}{(x - 5)^2}$

b) Choose one function in part a. Determine the following information about this function, where possible.

i) its domain and range

ii) its zeros

iii) the equations of any asymptotes

iv) a description of any symmetry

v) the coordinates of any maximum or minimum points

10. Write to explain your answer to each question, where $f(x)$ is any given function.

a) Is the graph of $\dfrac{1}{f(x)}$ always curved?

b) Does the graph of $\dfrac{1}{f(x)}$ always have an asymptote?

c) Does $\dfrac{1}{f(x)}$ always have value(s) of x for which it is not defined?

11. Suppose $f(x)$ is a given function.

a) Describe how the domain of the function $\dfrac{1}{f(x)}$ is related to the domain of $f(x)$.

b) Describe how the range of $\dfrac{1}{f(x)}$ is related to the range of $f(x)$.

12. a) Use the graph of $y = \dfrac{1}{x-2}$ on page 192 as a guide. Sketch the graph of the inverse of $y = \dfrac{1}{x-2}$.

b) Determine the equation of the inverse.

c) Is the inverse a function? Explain.

d) Determine the domain and range of the inverse.

13. a) Use the graphs at the top of page 194 as guides. Sketch the graph of the inverse of each function.

i) $f(x) = \dfrac{1}{x^2+1}$ ii) $f(x) = \dfrac{1}{x^2}$

b) Determine the domain and range of the inverse of each function in part a.

c) Which, if any, of the functions in part a has an inverse that is a function? Explain.

 14. These screens show the graphs of $y = x$, $y = x^3$, and $y = x^5$, respectively. The window settings are the same for each screen, $-2 \le x \le 2$ and $-1.5 \le y \le 1.5$.

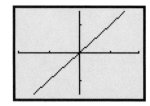 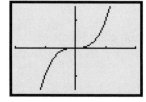 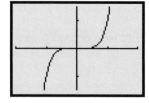

a) Predict what the graph of $y = \dfrac{1}{x^n}$ looks like for $n = 1, 3, 5, 7, \ldots$ (including much larger values, such as 61). Use a calculator to check your prediction.

b) Visualize the graphs of $y = \dfrac{1}{x^n}$ changing from values of n that are relatively small to values of n that are relatively large. Describe what happens to the graph of $y = \dfrac{1}{x^n}$ as n increases through odd values of n.

15. a) Predict what the graph of $y = \dfrac{1}{x^n}$ looks like for $n = 2, 4, 6, \ldots$ (including much larger values, such as 62). Use a calculator to check your prediction.

b) Visualize the graphs of $y = \dfrac{1}{x^n}$ changing from values of n that are relatively small to values of n that are relatively large. Describe what happens to the graph of $y = \dfrac{1}{x^n}$ as n increases through even values of n.

COMMUNICATING THE IDEAS

Suppose you are given the graph of a function $f(x)$. Write to explain how you can use the properties of reciprocal functions to sketch a graph of the function $\dfrac{1}{f(x)}$.

EXPLORING • WITH A GRAPHING CALCULATOR

Graphing Reciprocal Functions

On the TI-83 graphing calculator, follow these steps to graph the reciprocal of a given function, such as $y = x - 2$ (see page 192).

- Define the viewing window.
- Input the given function as Y1 in the Y= list.
- With the cursor beside "Y2=", press 1 [÷] [VARS] [►] 1 1 .
  "1/Y1" will appear beside "Y2=" on the screen.
- Press [GRAPH].

The calculator will graph $f(x)$ and $\dfrac{1}{f(x)}$. You may need to change the viewing window to get a satisfactory result.

Using this method, you can graph a function and its reciprocal by entering the equation of the function. You do not have to enter the equation of the reciprocal.

Use the exercises in this section to practise graphing reciprocal functions.

3.7 Rational Functions

Recall that a rational number is defined as a number that can be written in the form $\frac{m}{n}$, where m and n are integers, and $n \neq 0$. A rational function is defined in a similar way.

A *rational function* is a function whose equation can be written in the form $f(x) = \frac{m(x)}{n(x)}$, where $m(x)$ and $n(x)$ are polynomial functions, and $n(x) \neq 0$. Since a constant is a polynomial, $m(x)$ could be a constant.

A graph of a rational function can be obtained using technology, or by making a table of values and plotting points on grid paper. The reciprocals of the polynomial functions in Section 3.6 are rational functions. The graphs of some other rational functions are shown below, along with their tables of values. Only a few ordered pairs are shown in the tables; many more are needed to produce an accurate graph.

Other examples of rational functions

Graph 1

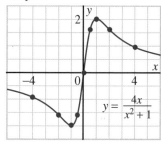

x	−4	−2	−1	−0.5	0	0.5	1	2	4
y	−0.94	−1.6	−2	−1.6	0	1.6	2	1.6	0.94

Graph 1 can be drawn without lifting the pencil. We say that it is *continuous* for all values of x.

Graph 2

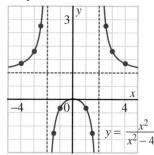

x	±4	±3	±2.5	±2	±1.5	±1	0
y	1.33	1.8	2.78		−1.29	−0.33	0

Graph 2 cannot be drawn without lifting the pencil because there is no value of y when $x^2 = 4$. The graph is *discontinuous* at $x = 2$ and at $x = -2$, where there are vertical asymptotes. There is also a horizontal asymptote at $y = 1$.

Graph 3

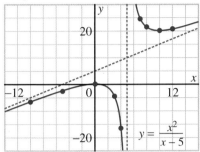

$$y = \frac{x^2}{x-5}$$

x	−10	−5	0	3	4	5	7	8	10	12
y	−6.7	−2.5	0	−4.5	−16		24.5	21.3	20	20.6

This graph is discontinuous at $x = 5$. It has a vertical asymptote there, and an inclined asymptote with equation $y = x + 5$.

Graph 4

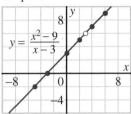

$$y = \frac{x^2 - 9}{x - 3}$$

x	−5	−3	0	2	3	4	6
y	−2	0	3	5		7	9

There is no value of y when $x = 3$. The graph is a straight line with a "hole" at $x = 3$. This is another example of a discontinuous graph.

DISCUSSING THE IDEAS

Explain your answer to each question.

1. a) Look at Graph 1. What is the y-intercept?
   b) Why are all the points on the graph in the 1st quadrant and 3rd quadrant?
   c) Why is the graph continuous?

2. a) Look at Graph 2. What is the y-intercept?
   b) What happens to y when x is very close to 2 or −2?
   c) What happens to y if x is very large?

3. a) Look at Graph 3. What is the y-intercept?
   b) What happens to y when x is very close to 5?
   c) What happens to y if x is very large?

4. a) Look at Graph 4. What is the y-intercept?
   b) Why does the graph look like a straight line?

5. a) Why are reciprocal functions considered rational functions?
   b) Are polynomial functions rational functions?

A 1. Determine if each function is a rational function, a polynomial function, or some other type of function. Justify your conclusion.

a) $y = x^2 - 5x - 9$

b) $y = \dfrac{x}{4x^2 - 12x - 9}$

c) $y = x^3 + x$

d) $y = (x + 2)^{-1}$

e) $y = \dfrac{x^2 - 2x - 8}{4}$

f) $y = 3^x + 1$

g) $y = 2.5x^2 + \dfrac{1}{3}x - \sqrt{5}$

h) $y = \sqrt{5x^3} + x^2 - 1$

2. Examine each graph. Which could be graphs of rational functions, and which could be graphs of polynomial functions? Explain.

a)

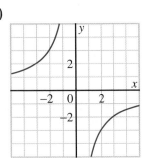

b)

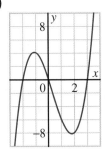

c)

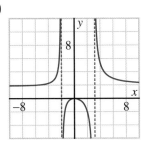

d)

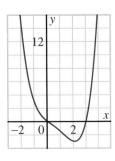

e)

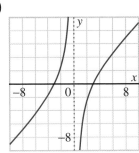

f)

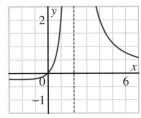

3. a) Is the reciprocal of every polynomial function a rational function? Explain.

b) Is the reciprocal of every rational function a polynomial function? Explain.

B 4. Complete each exercise, where possible, for each graph on pages 198–199.

a) Determine the domain and the range of the function.

b) Determine the zeros of the function.

c) Describe the symmetry of the graph.

d) What are the equations of the vertical asymptote(s)?

e) What are the equations of the horizontal asymptotes(s)?

5. Use Graph 1 on page 198 as a guide. Sketch the graph of each function on plain paper.

a) $y = \dfrac{6x}{x^2 + 1}$

b) $y = \dfrac{2x}{x^2 + 1}$

c) $y = \dfrac{-4x}{x^2 + 1}$

6. Use Graph 2 as a guide. Sketch the graph of each function on plain paper.

a) $y = \dfrac{x^2}{x^2 - 1}$

b) $y = \dfrac{x^2}{x^2 - 9}$

c) $y = \dfrac{x^2}{x^2 - 2}$

7. Use Graph 3 as a guide. Sketch the graph of each function on plain paper.

a) $y = \dfrac{x^2}{x - 3}$

b) $y = \dfrac{x^2}{x - 1}$

c) $y = \dfrac{x^2}{x}$

8. Use Graph 4 as a guide. Sketch the graph of each function on plain paper.

a) $y = \dfrac{x^2 - 9}{x + 3}$

b) $y = \dfrac{x^2 - 1}{x + 1}$

c) $y = \dfrac{x^3 - 3x^2}{x - 3}$

9. Choose one function from exercises 5 to 8. Write to explain how you sketched its graph.

10. a) Use a graphing tool or grid paper. Graph each rational function.

i) $y = \dfrac{x}{x^2 - 4}$

ii) $y = \dfrac{x^2 - 4}{x}$

iii) $y = \dfrac{x}{x^2 + 4}$

iv) $y = \dfrac{x^2 + 4}{x}$

b) Choose one function in part a. Determine the following information about this function, where possible.

   i) its domain and range

   ii) its zeros

   iii) the equations of any asymptotes

   iv) a description of any symmetry

   v) the coordinates of any maximum or minimum points

11. Repeat exercise 10 for each function.

i) $y = \dfrac{x + 3}{x + 1}$

ii) $y = \dfrac{(x + 3)^2}{x + 1}$

iii) $y = \dfrac{x + 3}{(x + 1)^2}$

iv) $y = \dfrac{(x + 3)^2}{(x + 1)^2}$

12. The numbers 3 and 1.5 are unusual because their sum equals their product.

a) Check that $3 + 1.5 = 3 \times 1.5$.

b) What is another example of two numbers whose sum equals their product?

c) Let x and y represent two numbers whose sum equals their product. Write an equation, then solve it for y. The result is a function of x.

d) Substitute some numbers for x to obtain other pairs of numbers whose sum equals their product.

13. a) Use a graphing calculator with $-4.7 \le x \le 4.7$. Graph the function in exercise 12c.

b) Press TRACE , then use the arrow keys to move the cursor along the curve. Pairs of numbers whose sum equals their product will appear on the screen. Check mentally that some of these numbers are correct.

14. a) Use the graphs on page 198 as guides. Sketch the graph of the inverse of each function.

  i) $y = \dfrac{4x}{x^2 + 1}$ **ii)** $y = \dfrac{x^2}{x^2 - 4}$

b) Determine the domain and range of the inverse of each function in part a.

c) Which of the functions in part a has an inverse that is a function? Explain.

15. a) Graph the inverse of each function.

  i) $y = \dfrac{x}{x - 1}$ **ii)** $y = \dfrac{x}{x + 1}$

b) Determine the equation of the inverse of each function in part a.

c) Determine the domain and range of the inverse of each function in part a.

d) Which of the functions in part a has an inverse that is a function? Explain.

C 16. Look at the graphs on pages 198–199 and your graphs in the above exercises. Form some conclusions about how you can tell, just by looking at its equation, when the graph of a rational function:

a) is continuous for all values of x **b)** has a "hole"

c) has a vertical asymptote   **d)** has a horizontal asymptote

e) has an inclined asymptote

17. Suppose $f(x)$ is any given function that has line or point symmetry. Write to explain your answers to these questions.

a) Does the graph of $\dfrac{1}{f(x)}$ have the same kind of symmetry as the graph of $f(x)$?

b) Does the graph of $f^{-1}(x)$ have the same kind of symmetry as the graph of $f(x)$?

COMMUNICATING THE IDEAS

Compare the graphs of rational functions with the graphs of polynomial functions. In what ways, if any, are they similar? In what ways are they different? Write to explain the similarities and differences. Use examples to illustrate your explanations.

3.8 Modelling Real Situations Using Rational Functions

Police sometimes identify speeders by measuring, from the air, the time it takes a car to travel a marked portion of highway. Suppose this distance is 0.5 km, and a car is travelling at v kilometres per hour. Let t seconds represent the time it takes the car to travel 0.5 km.

We can express the speed of a car as a function of the time to travel 0.5 km. Recall that for an object travelling at constant speed,

Distance = Speed × Time

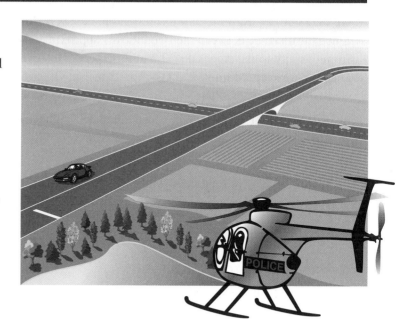

Substitute 0.5 for distance and v for speed. Since speed is in kilometres per hour, the time must be in hours. Substitute $\frac{t}{3600}$ for time.

$$0.5 = v \times \frac{t}{3600}$$
$$v = \frac{1800}{t}$$

The speed of a car is a rational function of the time it takes to travel 0.5 km. We can graph this function using grid paper or a graphing calculator.

Time, t (s)	Speed, v (km/h)
12	150
15	120
18	100
20	90
24	75
30	60

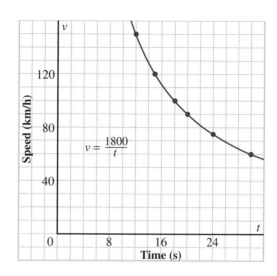

Example

On the highway from Calgary to Edmonton, the speed limit is 110 km/h. Since the distance between the two cities is approximately 330 km, a trip from one city to the other takes 3 h at this speed. However, some cars travel faster than the speed limit, and take less than 3 h. Other cars travel more slowly, and take more than 3 h.

a) Let s represent the change in speed compared with 110 km/h. Let t represent the change in time compared with 3 h. Write t as a function of s.

b) i) Graph t against s.

ii) How much less time does it take at 125 km/h than at 110 km/h?

iii) How much more time does it take at 95 km/h than at 110 km/h?

iv) At what speed does it take 10 min less time?

Solution

a) Use   Distance = Speed × Time

Substitute 330 for distance, $110 + s$ for speed, and $3 + t$ for time.

$$330 = (110 + s)(3 + t)$$

Divide each side by $(110 + s)$.

$$\frac{330}{110 + s} = 3 + t$$
$$t = \frac{330}{110 + s} - 3$$

b) Use a graphing calculator, or make a table of values and use grid paper.

Using a graphing calculator

i) Enter the equation $y = 330/(110 + x) - 3$.

On this graph, the x-values run from −47 to 47 to provide x-coordinates increasing by 1 when tracing.

Change in time (h)

Change in speed (km/h)

ii) Trace to the point where $x = 15$, to determine $y \doteq -0.36$. At 125 km/h, it takes approximately 0.36 h, or 22 min less time than at 110 km/h.

iii) Trace to the point where $x = -15$, to determine $y \doteq 0.4737$. At 95 km/h, it takes approximately 0.4737 h, or 28 min more time than at 110 km/h.

iv) Trace and zoom to the point whose y-coordinate is approximately $-\frac{10}{60}$, or −0.1667. This occurs when $x \doteq 6.471$. At approximately 116.5 km/h, it takes 10 min less time than at 110 km/h.

Using grid paper

i) Use a scientific calculator to make a table of values for $t = \dfrac{330}{110 + s} - 3$.
Plot the points and draw a smooth curve through them.

s	t
−30	1.13
−20	0.67
−10	0.3
0	0.00
10	−0.25
20	−0.46
30	−0.64

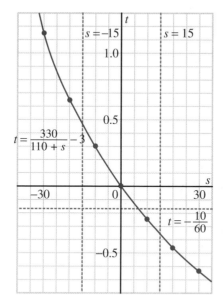

ii) Draw a vertical line at $s = 15$ to obtain $t \doteq -0.35$. It takes about
0.35 h, or 21 min less time at 125 km/h than at 110 km/h.

iii) Draw a vertical line at $s = -15$ to obtain $t \doteq 0.48$. It takes about
0.48 h, or 29 min more time at 95 km/h than at 110 km/h.

iv) Draw a horizontal line at $t = -\dfrac{10}{60}$, or −0.17 to obtain $s \doteq 6$.
At 116 km/h, it takes about 10 min less time than at 110 km/h.

MODELLING How Travelling Time Changes when Speed Changes

In the *Example*, the change in the time, t hours, to travel 330 km (relative to 3 h) is
modelled as a function of the change in speed, s kilometres per hour, (relative to
110 km/h).

- When the change in speed is positive, why is the change in time negative?

- When the change in speed is negative, why is the change in time positive?

- How does the graph of t against s illustrate the answers to these questions?

- Why were positive signs used in both factors on the right side of the equation
  $330 = (110 + s)(3 + t)$? Suppose a negative sign had been used in one of the factors.
  How would the solution be affected?

1. a) For a time t seconds, explain why $\frac{t}{3600}$ represents the time in hours.

   b) In the solution of the *Example*, times in hours were converted to minutes. Explain how this was done.

2. In the solution of the *Example*, we could have simplified the expression $\frac{330}{110 + s} - 3$ and written the function in the form $t = \frac{-3s}{110 + s}$.

   a) Why was it not necessary to do this?

   b) How can we tell, without simplifying, that $\frac{330}{110 + s} - 3$ is a rational function?

3.8 EXERCISES

A 1. Police use aircraft to catch motorists speeding along a highway. The aircraft covers a 0.25-km section of the highway.

   a) Write the speed of a car, v kilometres per hour, as a function of the time, t seconds, it takes to travel this distance.

   b) Graph the function in part a.

   c) What is the domain of the function?

2. Refer to the *Example*. How much less time does it take to travel 330 km:

   a) at 125 km/h than at 95 km/h?

   b) at 130 km/h than at 110 km/h?

   c) at 130 km/h than at 125 km/h?

B 3. The average cost, A dollars, of printing the school yearbook is $A = \frac{2500 + 1.25n}{n}$, where n is the number printed.

   a) Graph the function for $0 < n \le 940$. Use the graph to complete part b.

   b) i) Determine the average cost when 500 yearbooks are printed.

       ii) Determine the number of yearbooks printed when the average cost is $7.50.

4. A candy company sells candies in boxes. Each box has a square base with sides 16.5 cm, and a volume of approximately 1050 cm^3. The company plans to redesign the boxes with a smaller base. The boxes must still have a square base and contain the same volume.

   a) Calculate the height of the box.

   b) Let x centimetres represent the change in the length of the base. Let h centimetres represent the change in height. Write h as a function of x.

   c) Graph the function. Use the graph to complete part d.

d) **i)** Determine the increase in height when the base decreases by 1.5 cm.

ii) Determine the decrease in the base when the height increases by 1.5 cm.

5. The company in exercise 4 also sells candies in a smaller box. The box has a square base with sides 10.4 cm, and a volume approximately 375 cm³. The company plans to redesign these boxes so they have a smaller square base, but contain the same volume.

a) Let *x* centimetres represent the change in the length of the base. Let *h* centimetres represent the change in height. Write *h* as a function of *x*.

b) Graph the function.

c) What is the increase in height for each decrease in length of the base?

    i) 0.5 cm   **ii)** 1.0 cm   **iii)** 1.5 cm

d) What is the decrease in length of the base for each increase in height?

    i) 0.5 cm   **ii)** 1.0 cm   **iii)** 1.5 cm

6. Open the *Aircraft* database. Find the record describing 727-200 planes in TWA's fleet. A typical flight is about 1025 miles. The average speed is about 446 mph.

a) About how long would it take to fly 1025 miles at a speed of 446 mph?

b) Some planes will travel faster or slower than 446 mph. Let *s* represent the change in speed compared with 446 mph. Let *t* represent the change in time compared with your answer to part a. Write *t* as a function of *s*.

c) **i)** Graph *t* against *s*.

ii) How much less time does it take to travel 1025 miles at 470 mph than at 446 mph?

iii) How much more time does it take to travel 1025 miles at 420 mph than at 446 mph?

iv) At what speed does it take 15 min less time?

d) From the database, choose a record about a different airplane. Write a similar exercise for a classmate to solve. Prepare a model solution your classmate could use to check her or his work.

COMMUNICATING THE IDEAS

All the examples and exercises in this section involve two quantities with a constant product. Write to explain why a rational function occurs in each case.

  a) when one quantity is expressed in terms of the other

  b) when a change in one quantity is expressed in terms of the corresponding change in the other

Use an example to illustrate your explanation.

The Water and the Wine

On page 152, you were introduced to this puzzle.

A jug holds a quantity of water. Another jug holds an equal quantity of wine. A glass of water is taken from the first jug, poured into the wine, and the contents stirred. A glass of the mixture is then taken and poured into the water.
Is there more wine in the water or is there more water in the wine?

DEVELOP A MODEL

Use a deck of playing cards. Separate the black and red cards into two piles. Suppose the black cards represent water and the red cards represent wine.

1. Take 10 black cards from the first pile, add them to the pile of red cards, then shuffle thoroughly. Place this pile of mixed cards face down. Take 10 cards from this pile and add them to the pile of black cards.

   a) Do both piles contain the same number of cards? Why?

   b) Compare the number of black cards in the second pile with the number of red cards in the first pile. Are there more black cards in the second pile, or more red cards in the first pile? Explain.

2. Does the playing card model help you solve the puzzle about the water and the wine? What is the solution of the puzzle? Explain the solution.

 ## LOOK AT THE IMPLICATIONS

3. a) Does it matter if the two piles contain the same number of cards at the beginning? Explain.

   b) What does your answer to part a tell you about the quantities of water and wine in the puzzle?

4. a) After the 10 black cards were added to the red cards, does it matter how thoroughly the pile is shuffled?

   b) What does your answer to part a tell you about stirring the contents after the glass of water is added to the jug of wine?

5. a) Does it matter if the numbers of cards transferred in the two steps are different?

b) What does your answer to part a tell you about the volumes of liquid transferred in the puzzle?

 REVISIT THE SITUATION

6. a) A jug holds t litres of water. Another jug holds n litres of wine. From the first jug, x litres of water are removed, poured into the wine, and the contents stirred. Write an expression in terms of t, n, and x for each quantity.

　i) the volume of water in the first jug

　ii) the volumes of water and wine in the second jug

　iii) the fractions of liquid in the second jug that are water and wine.

b) From the second jug, x litres of the mixture are removed and poured back into the water. Write an expression in terms of t, n, and x for each quantity.

　i) the volume of wine transferred back to the first jug

　ii) the volume of water in the second jug

7. In exercise 6, you should have found that the volume of wine in the water is the same as the volume of water in the wine. This amount is a rational function of x: $y = \frac{nx}{x+n}$ 　①

In this equation, n is constant, and represents the volume of wine, in litres, in the second jug at the beginning.

a) Substitute 1, 2, 3, and 4 for n in ① to obtain functions representing the volume of wine in the water (or water in the wine).

b) Why does equation ① depend on n and not on t?

c) What is the domain of each function?

8. In the Y= list, enter the equations at the right. Press MODE, then choose "Dot" in the fifth line. Set the graphing window to show $0 \le x \le 4.7$ and $0 \le y \le 2$. Press TRACE, then use the arrow keys to trace along the curves, and to move from one curve to the other. How much wine is in the water in each situation?

$Y1 = x/(x+1) \{x \le 1\}$
$Y2 = 2x/(x+2) \{x \le 2\}$
$Y3 = 3x/(x+3) \{x \le 3\}$
$Y4 = 4x/(x+4) \{x \le 4\}$

a) $n = 1$, $x = 0.5$ and $x = 0.75$　　　**b)** $n = 2$, $x = 0.75$ and $x = 1.0$

c) $n = 3$, $x = 1.0$ and $x = 1.5$　　　　**d)** $n = 4$, $x = 1.5$ and $x = 2.0$

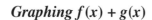 
Graphing f(x) + g(x)

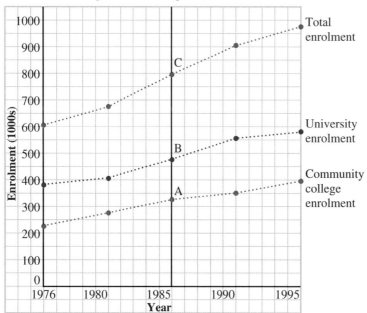

Full time post-secondary enrolment in Canada

In the media, we often see graphs showing two or more functions on the same grid. In the example above, one function is the community college enrolment, the other is the university enrolment. The third function is the total enrolment. The points on the graph of this function are obtained by adding the *y*-coordinates of the corresponding points on the graphs of the other two functions.

For example, in 1986, the community college enrolment is given by point A, and the university enrolment by point B. Since the total enrolment is given by point C, the *y*-coordinate of C is the sum of the *y*-coordinates of A and B.

The TI-82 and TI-83 calculators have a feature that facilitates graphing the sum of two functions. For example, to graph the sum of the functions $f(x) = \dfrac{2}{x^2 + 1}$ and $g(x) = x$, we enter the three equations in the first screen on page 211 in the Y= list. To do this, follow these steps.

Graph the two functions to be added

- Enter the equations in the first two lines of the Y= list. Set the viewing window, then press <kbd>GRAPH</kbd> to obtain the graphs of the two functions to be added.

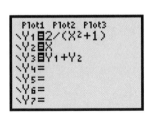

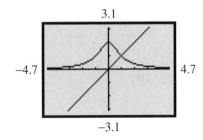

Graph the sum of the functions

- In the Y= list, with the cursor beside "Y3=", press <kbd>VARS</kbd> <kbd>▶</kbd> 1 1.

  "Y1" will appear beside "Y3=" on the screen.

- Press <kbd>+</kbd> <kbd>VARS</kbd> <kbd>▶</kbd> 1 2 <kbd>ENTER</kbd>.

  "Y1 + Y2" will appear beside "Y3=" on the screen.

- Press <kbd>GRAPH</kbd>. The calculator will graph the sum of the functions (below left).

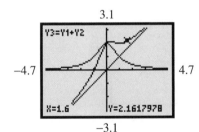

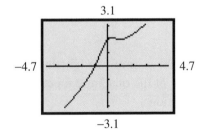

Visualizing the graph of the sum function

- Press <kbd>TRACE</kbd>, then use <kbd>▶</kbd> to move the cursor a short distance along the first graph. Note the y-coordinate of its location.

- Press <kbd>▼</kbd> to move the cursor to the second graph. Note the y-coordinate.

- Press <kbd>▼</kbd> to move the cursor to the third graph. Is the y-coordinate the sum of the other two y-coordinates?

- Repeat for other positions of the cursor on the first graph.

To view the sum of the two functions without the others, turn off the first two equations in the Y= list. The screen (above right) shows the graph of $y = \dfrac{2}{x^2 + 1} + x$. To graph sums of other functions, change the equations in the first two lines of the Y= list, and leave the third line as it is.

Graphing functions related to $y = \frac{2}{x^2+1} + x$

1. Use the graphs on page 211. Predict what the graph of each function looks like. Check your predictions by making appropriate changes to the first two equations in the Y= list.

   a) $y = \frac{2}{x^2+1} + 0.5x$ **b)** $y = \frac{2}{x^2+1} - x$ **c)** $y = \frac{-2}{x^2+1} + x$

Adding functions graphically

2. **a)** Enter these equations. $Y1 = x$ $Y2 = x + 1$

   b) Press [GRAPH]. The calculator will graph these functions and their sum, $y = 2x + 1$

   c) Look at the screen closely. How is the line $y = 2x + 1$ related to the other lines?

3. **a)** Repeat exercise 2, starting with two other linear functions.

   b) Explain why the graphs are related in this way.

Visualizing quadratic functions as sums of functions

This screen shows the graph of the quadratic function
$y = x^2 + 2x + 3$ ①

Think of this function as the sum of two functions. For example, it is the sum of the quadratic function $y = x^2$ and the linear function
$y = 2x + 3$ ②

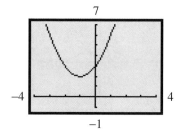

This suggests that the graph of ② might be related to the graph of ①.

4. **a)** Enter these equations. $Y1 = x^2$ $Y2 = 2x + 3$

   b) Press [GRAPH]. The calculator will graph these functions and function ①.

   c) Look at the screen closely. How is the line $y = 2x + 3$ related to the graph of function ①? To see the relationship, turn off the equation of Y1 in the Y= list, then press [GRAPH].

5. **a)** Repeat exercise 4, starting with a quadratic function that is different from ①.

   b) Explain why the graphs are related in this way.

Visualizing cubic functions as sums of functions

This screen shows the graph of the cubic function

$$y = x^3 - 4x^2 + 2x + 3 \qquad ③$$

Think of this function as the sum of two functions. For example, it is the sum of the cubic function $y = x^3$ and the quadratic function

$$y = -4x^2 + 2x + 3 \qquad ④$$

This suggests that the graph of ④ might be related to the graph of ③.

6. a) Enter these equations. $Y1 = x^3$ $Y2 = -4x^2 + 2x + 3$

   b) Press GRAPH. The calculator will graph these functions and function ③.

   c) Look at the screen closely. How is the parabola $y = -4x^2 + 2x + 3$ related to the graph of function ③? To see the relationship, it might help to turn off the equation of Y1 in the Y= list, then press GRAPH.

7. a) Repeat exercise 6, starting with a cubic function that is different from ③.

   b) Explain why the graphs are related in this way.

The cubic function $y = x^3 - 4x^2 + 2x + 3$ is also the sum of the cubic function $y = x^3 - 4x^2$ and the linear function $y = 2x + 3$.

8. a) Enter these equations. $Y1 = x^3 - 4x^2$ $Y2 = 2x + 3$

   b) Press GRAPH. The calculator will graph these functions and function ③.

   c) Look at the screen closely. How is the line $y = 2x + 3$ related to the graph of function ③?

9. a) Repeat exercise 8, starting with a cubic function that is different from ③.

   b) Explain why the graphs are related in this way.

A rational function with a surprising graph

10. a) Enter these equations. $Y1 = \dfrac{4}{x^2 + 1}$ $Y2 = x^2$

    b) Press GRAPH. The calculator will graph these functions and their sum, $y = \dfrac{4}{x^2 + 1} + x^2$.

    c) Explain why the function in part b is a rational function.

    d) What is unusual about the graph of this rational function?

3.9 Composition of Functions

Consider the problem of expressing the cost of fuel, when taking a trip by car, as a function of the distance driven.

The cost of fuel, C cents, is a function of the volume of fuel used. When fuel costs 60¢/L, the cost for x litres of fuel is given by this equation.

$$C = 60x \qquad \text{①}$$

The volume of fuel used is a function of the distance driven. Suppose the car uses fuel at the rate of 8.0 L/100 km, or 0.080 L/km. While travelling d kilometres, the volume of fuel used is given by this equation.

$$x = 0.080d \qquad \text{②}$$

We can express the cost of fuel as a function of the distance driven by substituting $0.080d$ for x in ①.

$$C = 60x$$
$$= 60(0.080d)$$
$$C = 4.8d \qquad \text{③}$$

The cost of fuel to drive a distance of d kilometres is $C = 4.8d$, or about 5 cents per kilometre.

When two functions are applied in succession, the resulting function is called the *composite* of the two given functions. The function described by equation ③ is the composite of the functions described by equations ① and ②.

Function composition can be illustrated with mapping diagrams.

Distance (km)	Fuel (L)	Cost (¢)
100	8.0	480
200	16.0	960
300	24.0	1440
400	32.0	1920

Consider the functions $f(x) = 2x + 3$ and $g(x) = x^2 - 1$. There are two different ways to form the composite of these functions.

Apply f first and g second

Double and square and
add 3, then... subtract 1

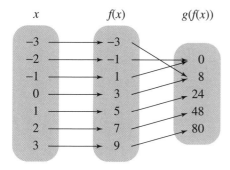

The composite function related the numbers in the first set to those in the third set, and is written as $g(f(x))$. We say, "g of f of x."

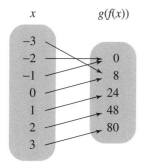

To express $g(f(x))$ as a function of x, substitute $f(x)$ for x in $g(x)$.

$$g(x) = x^2 - 1$$

$$\begin{aligned} g(f(x)) &= (f(x))^2 - 1 \\ &= (2x + 3)^2 - 1 \\ &= 4x^2 + 12x + 9 - 1 \\ &= 4x^2 + 12x + 8 \end{aligned}$$

Apply g first and f second

Square and double and
subtract 1, then ... add 3.

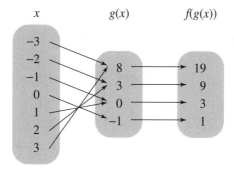

In this case, the composite function is written as $f(g(x))$.

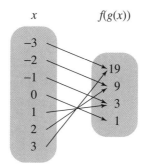

To express $f(g(x))$ as a function of x, substitute $g(x)$ for x in $f(x)$.

$$f(x) = 2x + 3$$

$$\begin{aligned} f(g(x)) &= 2g(x) + 3 \\ &= 2(x^2 - 1) + 3 \\ &= 2x^2 - 2 + 3 \\ &= 2x^2 + 1 \end{aligned}$$

Example 1

Consider the functions $f(x) = 3x - 5$ and $g(x) = x^2 - x$.

a) Determine $f(g(3))$ and $g(f(3))$.

b) Express $f(g(x))$ and $g(f(x))$ as functions of x.

Solution

a)
$$g(x) = x^2 - x$$
$$g(3) = 3^2 - 3$$
$$= 6$$

$$f(x) = 3x - 5$$
$$f(3) = 3(3) - 5$$
$$= 4$$

$$f(g(3)) = f(6)$$
$$= 3(6) - 5$$
$$= 18 - 5$$
$$= 13$$

$$g(f(3)) = g(4)$$
$$= 4^2 - 4$$
$$= 16 - 4$$
$$= 12$$

b) $f(x) = 3x - 5$

$$f(g(x)) = 3(g(x)) - 5$$
$$= 3(x^2 - x) - 5$$
$$= 3x^2 - 3x - 5$$

$g(x) = x^2 - x$

$$g(f(x)) = (f(x))^2 - f(x)$$
$$= (3x - 5)^2 - (3x - 5)$$
$$= 9x^2 - 30x + 25 - 3x + 5$$
$$= 9x^2 - 33x + 30$$

When writing the composite of two functions, it is not necessary for the functions to be different. That is, we can write the composite of a function with itself.

Example 2

Given $f(x) = 3x - 1$, determine $f(f(x))$.

Solution

$$f(x) = 3x - 1$$
$$f(f(x)) = 3(f(x)) - 1$$
$$= 3(3x - 1) - 1$$
$$= 9x - 4$$

Do you think it is possible to have two different functions $f(x)$ and $g(x)$ so that $f(g(x)) = g(f(x))$ for all values of x? If your answer is no, explain why it is not possible. If your answer is yes, give an example.

3.9 EXERCISES

A 1. Given $f(x) = 2x + 1$ and $g(x) = 3x + 1$, determine each value.

   a) $f(3)$ **b)** $g(f(3))$ **c)** $g(3)$ **d)** $f(g(3))$

2. For the functions in exercise 1, determine $g(f(x))$ and $f(g(x))$.

3. Given $f(x) = x^2 + 1$ and $g(x) = 2x$, determine each value.

   a) $f(2)$ **b)** $g(f(2))$ **c)** $g(2)$ **d)** $f(g(2))$

4. For the functions in exercise 3, determine $g(f(x))$ and $f(g(x))$.

5. Determine $f(g(x))$ and $g(f(x))$ for each pair of functions.

   a) $f(x) = 3x + 4$; $g(x) = -2x + 5$

   b) $f(x) = x^2 + 5x$; $g(x) = 2x + 1$

   c) $f(x) = 2x^2 - 3x + 1$; $g(x) = 7 - 4x$

6. Given $f(x) = 3x^2 - 1$, determine $f(g(x))$ and $g(f(x))$ for each function $g(x)$.

   a) $g(x) = x + 2$ **b)** $g(x) = 1 - 2x$ **c)** $g(x) = x^2$

   d) $g(x) = x^2 + x$ **e)** $g(x) = 2x^2 - 3x$

7. The area, A, of a circle is a function of its radius, r, where $A = \pi r^2$. Express the area as a function of the diameter, d.

8. The volume, V, of a sphere is a function of its radius, r, where $V = \frac{4}{3}\pi r^3$. Express the volume as a function of the diameter, d.

B 9. Given $f(x) = 2x - 1$ and $g(x) = 1 - 3x$, determine each value.

   a) $f(g(2))$ **b)** $g(f(2))$ **c)** $f(f(2))$ **d)** $g(g(2))$

10. For the functions in exercise 9, determine each value.

   a) $f(g(x))$ **b)** $g(f(x))$ **c)** $f(f(x))$ **d)** $g(g(x))$

11. For each pair of functions, determine $f(g(x))$, $g(f(x))$, $f(f(x))$, and $g(g(x))$.

   a) $f(x) = \sqrt{x}$; $g(x) = 4 - 2x$ **b)** $f(x) = \sqrt{2x}$; $g(x) = 1 + 3x$

   c) $f(x) = \frac{x}{x+1}$; $g(x) = x^2 - 1$ **d)** $f(x) = 2x$; $g(x) = 3x - 4$

12. The stopping distance, d metres, for a truck can be approximated by the function $d(v) = 0.25v^2 + 4$, where v is the speed of the truck in metres per second. To convert the speed of the truck to kilometres per hour, v must be multiplied by 3.6. Express the stopping distance of the truck, d metres, as a function of s, the speed of the truck in kilometres per hour.

13. The area, A, and perimeter, P, of a square are functions of its side length, s. Express the area as a function of the perimeter.

14. Express the area of a square as a function of the length of its diagonal.

15. Given $f(x) = \frac{1}{x}$ and $g(x) = x^2$, show that $f(g(x)) = g(f(x))$.

16. Consider the functions $f(x) = 1 - x$ and $g(x) = \frac{x}{1-x}$, $x \neq 1$.

    a) Show that $g(f(x)) = \frac{1}{g(x)}$.

    b) Does $f(g(x)) = \frac{1}{f(x)}$? Explain.

17. The temperature of Earth's crust, T degrees Celsius, is given by $T(d) = 0.01d + 20$, where d metres is the distance below the surface. Suppose an elevator travels down a shaft into a mine. The depth of the elevator in metres is given by $d(t) = 5t$, where t is the time of travel in seconds. Express the temperature of the elevator as a function $T(t)$ of time.

18. The velocity, v metres per second, of a ball thrown in the air is given by $v(t) = 49 - 9.8t$. The kinetic energy of the ball, K joules, is a function of its velocity, where $K(v) = 0.4v^2$. Express the ball's kinetic energy as a function $K(t)$ of time.

19. From the functions in the box, find two whose composite function is $h(x)$.

    a) $h(x) = (x + 1)^2$

    b) $h(x) = \sqrt{x - 3}$

    c) $h(x) = x^2 - 6x + 9$

    d) $h(x) = x - 2$

$e(x) = x - 3$	$f(x) = x^2$
$g(x) = \sqrt{x}$	$k(x) = x + 1$

20. Find two functions whose composite function is $k(x)$.

    a) $k(x) = x^6 + 2x^3 + 1$

    b) $k(x) = (x - 4)^2 + 3(x - 4) + 4$

    c) $k(x) = \sqrt{3x - 2}$

    d) $k(x) = \frac{1}{x + 3}$

21. Given $f(x) = x - 3$ and $g(x) = \sqrt{x}$, determine:

    a) $f(g(x))$

    b) the domain of $f(g(x))$

    c) the range of $f(g(x))$

    d) $g(f(x))$

    e) the domain of $g(f(x))$

    f) the range of $g(f(x))$

22. Given $f(x) = x^2 + 1$ and $g(x) = \sqrt{x - 1}$, determine:

    a) $f(g(x))$

    b) the domain of $f(g(x))$

    c) the range of $f(g(x))$

    d) $g(f(x))$

    e) the domain of $g(f(x))$

    f) the range of $g(f(x))$

23. Given $f(x) = ax + b$, $g(x) = cx + d$, and $f(g(x)) = g(f(x))$, how are a, b, c, and d related?

Your friend confuses the composition of two functions with the product of two functions. Write to explain the difference between these concepts, using examples to illustrate your explanation. Include a discussion of whether the order in which the two functions are combined is important.

EXPLORING •WITH A GRAPHING CALCULATOR

Composition of Functions

To graph $f(g(x))$ on the TI-83 graphing calculator, follow these steps.

- Define the viewing window.
- Input a function $f(x)$ as Y1 in the Y= list. Try using $f(x)$ from *Example 1* to start with.
- Input a function $g(x)$ as Y2 in the Y= list. Try using $g(x)$ from *Example 1*.
- With the cursor beside "Y3=", press [VARS] [▶] 1 1.
  "Y1" will appear beside "Y3=" on the screen.
- Press [( ] [VARS] [▶] 1 2 [ )] [ENTER].
  "Y1(Y2)" will appear beside "Y3=" on the screen.
- Press [GRAPH].

The calculator will graph $f(x)$, $g(x)$, and $f(g(x))$. You may need to change the viewing window to get a satisfactory result.

With this method, you can graph two functions and their composite function by entering the equations of the two functions. You do not have to enter the equation of their composite function. You can modify this method to graph $g(f(x))$.

Use the exercises in this section to practise graphing the composition of two functions.

Patterns in Graphs of Functions

This graph was produced by *Graphmatica*. The equations of some of the functions are:

$$y = \frac{-6}{x^2 - 6}, \quad y = \frac{-5}{x^2 - 5}, \quad y = \frac{-4}{x^2 - 4}, \quad \ldots, \quad y = \frac{4}{x^2 + 4}, \quad y = \frac{5}{x^2 + 5}, \quad y = \frac{6}{x^2 + 6}$$

These equations have the form $y = \dfrac{a}{x^2 + a}$, where a ranges from -6 to 6.

The values of a used to create this graph were $-6, -5.5, -5, \ldots, 5, 5.5, 6$.

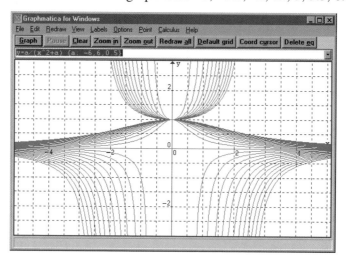

The pattern on the graph has three parts:

Top part	Curves open up.
Middle part	Bell-shaped curves open down, above the x-axis.
Bottom part	Curves with two branches, below the x-axis.

1. a) Graph some of the functions listed above.

b) Use the results to identify the equation of each curve.

   i) the lowest curve in the top part

   ii) the highest curve in the middle part

   iii) the lowest curve in the bottom part

2. One function shown above has a graph that is a straight line. Which function is it? How can you tell?

3. The graphs of all but one function shown above pass through one point.

   a) What are the coordinates of this point?

   b) Which function has a graph that does not pass through this point? Use its equation to explain why it does not pass through this point.

1. Identify the function that corresponds to each graph below.

$h(x) = -x^4 - 4x^3 + 10$ $\qquad\qquad$ $f(x) = x^3 - 3x^2 + 1$

$k(x) = 2x^4 - 3x^2 - 21$ $\qquad\qquad$ $g(x) = -2x^3 + 4x^2 + 7x - 3$

a)

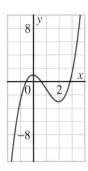

b)

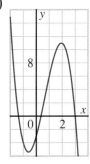

c)

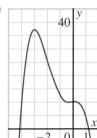

d)

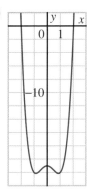

2. Write a cubic function with the given zeros.

a) 3, −2, 1 $\qquad\qquad\qquad$ b) 0, 0, 1

c) −4, 5, −1 $\qquad\qquad\qquad$ d) −3, −3, −3

3. Sketch a graph of each function.

a) $y = (x + 3)(x - 4)$

b) $y = -(x + 3)^2(x - 4)$

c) $y = (x - 1)(x + 2)(x - 3)$

4. Determine the value(s) of k in each equation for which one root is triple the other root.

a) $x^2 - kx + 3 = 0$ $\qquad\qquad$ b) $x^2 + 8x + k = 0$

c) $kx^2 - 8x + 3 = 0$ $\qquad\qquad$ d) $3x^2 + kx + 4 = 0$

5. a) Graph each pair of functions on the same grid.

$\quad$ i) $y = x - 3$ and $y = \dfrac{1}{x - 3}$ $\qquad$ ii) $y = 2x$ and $y = \dfrac{1}{2x}$

b) Determine the domain and range of each function in part a.

c) Choose one pair of functions from part a. Write to explain how you sketched the graphs and determined the domains and ranges.

6. a) Determine if each function is a rational function, a polynomial function, or some other type of function.

i) $y = x^2 - 3x + 2$ **ii)** $y = \dfrac{2x}{3x^3 - 5}$ **iii)** $y = 3x^3 - 2x + 4$

iv) $y = (x^2 - 4)^{-2}$ **v)** $y = \sqrt{x} - 2$ **vi)** $y = x^2 - 3x + \sqrt{x^3}$

b) Write to explain how you identified each type of function in part a.

7. a) Use a calculator to graph the function $y = \dfrac{x^2 - 9}{x - 3}$ (page 199, Graph 4). Set the window to $-4.7 < x < 4.7$ and $-3.7 < y < 7.7$. One pixel on the line is turned off, indicating there is a "hole" at the point (3, 6).

b) Press $\boxed{\text{ZOOM}}$ 2, then use the arrow keys to move the cursor as close as possible to the hole. Press $\boxed{\text{ENTER}}$ to zoom in. Does the pixel turn on?

Press $\boxed{\text{TRACE}}$, then use the arrow keys to move the cursor along the line. What happens at $x = 3$? How close to 3 are the pixels on either side of the hole?

c) Repeat part b several times. Explain the result.

8. A pasta sauce can has a capacity of 725 mL and a base radius of 4.2 cm. The company plans to redesign the can with a larger base. The new can will have the same volume as the original can.

a) Let r centimetres represent the change in base radius. Let h centimetres represent the change in height. Write h as a function of r.

b) Graph the function.

c) What is the decrease in height for each increase in base radius?

     i) 0.5 cm   **ii)** 1.3 cm   **iii)** 1.5 cm

d) What is the increase in base radius for each decrease in height?

     i) 0.5 cm   **ii)** 1.0 cm   **iii)** 1.5 cm

9. A poster has an area of 7200 cm^2.

a) Write its length, l centimetres, as a function of its width, w centimetres.

b) Graph the function in part a. What is the domain of this function?

c) Write the perimeter, p centimetres, of the poster as a function of its width.

d) Graph the function in part c. What is the domain of this function?

10. Determine $f(f(x))$ for each function.

a) $f(x) = \dfrac{1}{1 - x}, x \neq 1$ **b)** $f(x) = \dfrac{x - 1}{x + 1}, x \neq -1$

3 Cumulative Review

1. For each investment, determine the annual simple interest rate.

   a) $300 grows to $360 in 4 years.

   b) $825 grows to $957 in 2 years.

   c) $1250 grows to $1700 in 8 years.

2. a) At an interest rate of 7% compounded annually, how much would you have to invest today to earn each indicated amount?

   i) $4250 in 4 years   ii) $1275 in 2 years

   b) Choose one amount from part a. Write to explain how you calculated the investment.

3. An 8-year Canada Savings Compound Interest Bond of $2000 has an interest rate of 5.5% for the first 2 years, 6.0% for the next 2 years, 6.5% for the next 2 years, and 7.0% for the last 2 years. Determine the value of the savings bond at maturity.

4. The daily profit, P dollars, of an ice cream stand is described by the function $P = -30x^2 + 120x + 625$, where x dollars is the price of one ice cream cone.

   a) What should the selling price of an ice cream cone be for maximum daily profits?

   b) What is the maximum daily profit?

5. a) Sketch each set of graphs on the same grid.

   i) $y = x^2$ $y = x^2 + 4$ $y = x^2 - 2$

   ii) $y = x^2$ $y = 3x^2$ $y = \frac{1}{2}x^2$ $y = -x^2$

   b) Choose one set of graphs from part a. Write to explain how you sketched them.

6. Determine the equation of each parabola.

   a) with vertex $(0, 0)$, and passing through $(2, 7)$

   b) with vertex $(-3, -2)$, opens down, and is congruent to $y = 3x^2$

7. A ball thrown vertically with a velocity of 18 m/s is h metres above the ground after t seconds, where $h = -5t^2 + 18t$. What is the maximum height of the ball, and when does it reach that height?

8. Two numbers have a difference of 24. The result of adding their sum and their product is a minimum. Determine the numbers.

How Can We Model a Spiral?

Spirals occur throughout nature, from spiral galaxies in the heavens, to florets in a daisy, to the beautiful shells of the chambered nautilus.

CONSIDER THIS SITUATION

The diagram at the right is a model of a spiral. Its shape closely resembles the nautilus shell. Find out as much as you can about the squares and rectangles in this diagram by completing these exercises.

- Measure and record the dimensions of all the squares and rectangles.
- Calculate the length : width ratio of each rectangle.
- How could the spiral be extended inward and outward?
- Each rectangle on the diagram has a special property. Try to find out what this property is.
- The sequence of squares has a special property. What is it?

On pages 236-239, you will explore this spiral further. You will learn new skills that you can use to determine some of its properties and applications.

 FYI Visit www.awl.com/canada/school/connections

For information related to the above problem, click on <u>MATHLINKS</u> followed by *Mathematics 11*. Then select a topic under How Can We Model a Spiral?

4.1 Solving Quadratic Equations by Using a Formula

In Chapters 2 and 3 we associated the x-intercept(s) of the graph of a function with the roots of the corresponding equation. We used technology to determine the roots of the equation in decimal form. Although the roots were determined accurately, they were only approximate roots. In this chapter, we will develop strategies to solve certain types of equations without technology, and to express the roots in exact form.

INVESTIGATE

1. Recall on page 122 in Section 2.4, we used the method of completing the square to express the quadratic function $f(x) = 2x^2 - 12x + 11$ in the form $f(x) = 2(x - 3)^2 - 7$. You can use this result to solve the equation $2x^2 - 12x + 11 = 0$.

   Look at the example on page 122. Recall the steps in the method of completing the square. Here is how you would use those steps to begin a solution of $2x^2 - 12x + 11 = 0$:

$$2x^2 - 12x + 11 = 0$$
$$2(x^2 - 6x) + 11 = 0$$
$$2(x^2 - 6x + 9 - 9) + 11 = 0$$
$$2(x^2 - 6x + 9) - 18 + 11 = 0$$
$$2(x - 3)^2 - 7 = 0$$

   a) Continue the solution from this point, and determine the two roots of the equation.

   b) Check the roots you obtained.

2. Solve each equation using the method of completing the square.

   a) $3x^2 - 5x - 2 = 0$ **b)** $x^2 + 2x - 9 = 0$

   c) $2x^2 + 8x + 3 = 0$ **d)** $2x^2 + 3x - 5 = 0$

3. Use the same steps to solve the equation $ax^2 + bx + c = 0$, assuming that $a \neq 0$. The result will be a formula you can use to solve any quadratic equation.

Using the method of completing the square, a formula can be derived for solving any quadratic equation of the form $ax^2 + bx + c = 0$. This is the quadratic formula.

Quadratic Formula

The solution of the quadratic equation $ax^2 + bx + c = 0$, $a \neq 0$, is given by:

$x = \dfrac{-b \pm \sqrt{b^2 - 4ac}}{2a}$, where $b^2 - 4ac \geq 0$.

There are two roots: one obtained by replacing $\pm$ with $+$; the other obtained by replacing $\pm$ with $-$.

Example 1

Solve and check.

a) $3x^2 + x - 2 = 0$ b) $z^2 - 6z + 7 = 0$ c) $t^2 - 5t + 7 = 0$

Solution

a) $3x^2 + x - 2 = 0$

$x = \dfrac{-b \pm \sqrt{b^2 - 4ac}}{2a}$ Substitute $a = 3$, $b = 1$, $c = -2$.

$= \dfrac{-1 \pm \sqrt{1^2 - 4(3)(-2)}}{2(3)}$

$= \dfrac{-1 \pm \sqrt{25}}{6}$

$x = \dfrac{-1 + 5}{6}$ or $x = \dfrac{-1 - 5}{6}$

$= \dfrac{2}{3}$ $= -1$

The roots of the equation are $\dfrac{2}{3}$ and -1.

b) $z^2 - 6z + 7 = 0$

$z = \dfrac{-b \pm \sqrt{b^2 - 4ac}}{2a}$ Substitute $a = 1$, $b = -6$, $c = 7$.

$= \dfrac{-(-6) \pm \sqrt{(-6)^2 - 4(1)(7)}}{2(1)}$

$= \dfrac{6 \pm \sqrt{8}}{2}$

$= \dfrac{6 \pm 2\sqrt{2}}{2}$ Remove 2 as a common factor, then simplify.

$= 3 \pm \sqrt{2}$

The roots of the equation are $3 + \sqrt{2}$ and $3 - \sqrt{2}$.

c) $t^2 - 5t + 7 = 0$

$$t = \frac{-b \pm \sqrt{b^2 - 4ac}}{2a}$$

Substitute $a = 1$, $b = -5$, $c = 7$.

$$= \frac{-(-5) \pm \sqrt{(-5)^2 - 4(1)(7)}}{2(1)}$$

$$= \frac{5 \pm \sqrt{-3}}{2}$$

Since $\sqrt{-3}$ is not defined as a real number, this equation has no real roots.

Check

a) $3x^2 + x - 2 = 0$

When $x = -1$, L.S. $= 3(-1)^2 + (-1) - 2$ R.S. $= 0$
$$= 3 - 1 - 2$$
$$= 0$$

When $x = \frac{2}{3}$, L.S. $= 3\left(\frac{2}{3}\right)^2 + \frac{2}{3} - 2$ R.S. $= 0$
$$= \frac{4}{3} + \frac{2}{3} - 2$$
$$= 0$$

The solution is correct.

b) $z^2 - 6z + 7 = 0$

When $z = 3 + \sqrt{2}$, L.S. $= (3 + \sqrt{2})^2 - 6(3 + \sqrt{2}) + 7$
$$= 9 + 6\sqrt{2} + 2 - 18 - 6\sqrt{2} + 7$$
$$= 0$$
$$= \text{R.S.}$$

When $z = 3 - \sqrt{2}$, L.S. $= (3 - \sqrt{2})^2 - 6(3 - \sqrt{2}) + 7$
$$= 9 - 6\sqrt{2} + 2 - 18 + 6\sqrt{2} + 7$$
$$= 0$$
$$= \text{R.S.}$$

The solution is correct.

We can use a graphing calculator to illustrate the results of *Example 1*. This screen shows the graphs of the three quadratic functions that correspond to the given equations.

The thick curve is the graph of $y = 3x^2 + x - 2$, which intersects the x-axis at $x = -1$ and $x = \frac{2}{3}$.

The thin curve is the graph of $y = x^2 - 6x + 7$, which intersects the x-axis when $x = 3 + \sqrt{2} \doteq 4.4$ and $x = 3 - \sqrt{2} \doteq 1.6$.

The dotted curve is the graph of $y = x^2 - 5x + 7$, which does not intersect the x-axis.

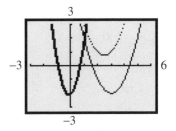

Example 2

Annie Pelletier won a bronze medal in the women's springboard competition at the 1996 Summer Olympics in Atlanta. Pelletier somersaults from a 3-m springboard. Her height h metres above the water t seconds after she leaves the board is given by $h = -4.9t^2 + 8.8t + 3$. How long is it until Pelletier reaches the water?

Solution

$h = -4.9t^2 + 8.8t + 3$

When Pelletier reaches the water, her height $h = 0$.

Substitute 0 for h, then solve the equation.

$0 = -4.9t^2 + 8.8t + 3$

$t = \dfrac{-b \pm \sqrt{b^2 - 4ac}}{2a}$ ⟵ Substitute $a = -4.9$, $b = 8.8$, $c = 3$.

$= \dfrac{-8.8 \pm \sqrt{8.8^2 - 4(-4.9)(3)}}{2(-4.9)}$ ⟵ Divide numerator and denominator by -1.

$= \dfrac{8.8 \pm \sqrt{136.24}}{9.8}$

$\doteq \dfrac{8.8 \pm 11.672}{9.8}$

$t \doteq \dfrac{8.8 + 11.672}{9.8}$ or $t \doteq \dfrac{8.8 - 11.672}{9.8}$

$\doteq 2.09 \qquad\qquad \doteq -0.29$

Since the time is positive, the negative root is rejected. Pelletier reaches the water in about 2.1 s.

In *Example 2*, the quadratic function $h = -4.9t^2 + 8.8t + 3$ is a model of Annie Pelletier's height above the water during her dive. The positive root of the corresponding equation $-4.9t^2 + 8.8t + 3 = 0$ is a model of the time it takes to reach the water.

- Suggest some reasons why the actual time for Annie to reach the water may differ slightly from the time predicted by the model.

- Explain why the equation has two roots, when there is only one solution to the problem.

A simple application of the quadratic formula is to factor trinomials. For example, a trinomial such as $12x^2 - 35x + 18$ can be difficult to factor because the first and last terms have many factors. However, we can factor the trinomial with the help of the quadratic formula.

Example 3

Factor the trinomial $12x^2 - 35x + 18$.

Solution

Write the corresponding quadratic equation, $12x^2 - 35x + 18 = 0$.

Solve the equation using the quadratic formula.

$$x = \frac{-b \pm \sqrt{b^2 - 4ac}}{2a}$$

Substitute $a = 12$, $b = -35$, $c = 18$.

$$= \frac{35 \pm \sqrt{(-35)^2 - 4(12)(18)}}{24}$$

$$= \frac{35 \pm \sqrt{361}}{24}$$

$$= \frac{35 \pm 19}{24}$$

$$x = \frac{35 + 19}{24} \text{ or } x = \frac{35 - 19}{24}$$

$$= \frac{9}{4} \qquad\qquad = \frac{2}{3}$$

Since these are the roots of $12x^2 - 35x + 18 = 0$, this equation may be written as $\left(x - \frac{9}{4}\right)\left(x - \frac{2}{3}\right) = 0$, or $(4x - 9)(3x - 2) = 0$. Therefore, the given trinomial can be factored as: $12x^2 - 35x + 18 = (4x - 9)(3x - 2)$

1. Is it possible for a quadratic equation to have more than two roots? Explain your answer in two different ways.

2. Is it possible for a quadratic equation to have two equal roots? Use the quadratic formula to explain your answer.

3. In *Example 2*, how could you determine the length of time Annie was higher than 3 m?

4. In *Example 3*, how could you check that $12x^2 - 35x + 18 = (4x - 9)(3x - 2)$?

4.1 EXERCISES

A 1. Use the quadratic formula to solve each equation.

a) $x^2 - 6x + 4 = 0$ b) $x^2 + 3x - 1 = 0$ c) $x^2 + 7x + 3 = 0$

d) $x^2 - 5x + 2 = 0$ e) $x^2 + 4x - 1 = 0$ f) $x^2 - 8x - 6 = 0$

2. a) Solve each equation.

    i) $2x^2 - 5x + 2 = 0$ ii) $2x^2 + 7x + 3 = 0$

    iii) $3x^2 - 11x - 14 = 0$ iv) $4x^2 - 9x + 5 = 0$

    v) $5x^2 + 7x = -2$ vi) $6x^2 - 20 = -7x$

b) Choose one equation from part a. Write to explain how you solved it.

3. Solve and check.

a) $6m^2 - 7m + 2 = 0$ b) $2c^2 - 25c + 77 = 0$

c) $6t^2 - t - 1 = 0$ d) $2p^2 - p - 45 = 0$

e) $2x^2 - 5x - 12 = 0$ f) $6x^2 - x - 2 = 0$

4. Determine the roots of each equation to 2 decimal places.

a) $2b^2 - 13b + 10 = 0$ b) $2z^2 - 7z + 4 = 0$

c) $3x^2 - x - 5 = 0$ d) $2a^2 - 9a - 1 = 0$

e) $5t^2 - 3t - 1 = 0$ f) $2y^2 + 5y + 1 = 0$

5. Determine the roots of each equation to 2 decimal places.

a) $5x^2 + 6x - 1 = 0$ b) $2c(c - 3) = 1$

c) $3m^2 + 2m - 7 = 0$ d) $2r(2r + 1) = 3$

e) $3p^2 - 6p + 1 = 0$ f) $2a(a - 3) = -1$

B **6.** An Acapulco diver dives into the sea from a height of 35 m. His height h metres t seconds after leaving the cliff is given by $h = -4.9t^2 + t + 35$.

a) How long is it until he reaches the water?

b) How long does it take him to fall from 35 m to 25 m?

7. Solve each equation.

a) $3x^2 - 4x = 0$

b) $12m^2 - 192 = 0$

c) $25c^2 + 70c + 49 = 0$

d) $\frac{5}{2}y^2 - \frac{3}{2}y - \frac{1}{4} = 0$

e) $0.2s^2 - s - 3.2 = 0$

f) $\sqrt{2}x^2 - 5x - \sqrt{8} = 0$

8. Solve and check.

a) $4m^2 - 17m + 4 = 0$

b) $6a^2 - 11a + 4 = 0$

c) $12x^2 - x - 6 = 0$

d) $3y^2 + 16y - 99 = 0$

e) $5t^2 - 13t - 6 = 0$

f) $15s^2 + 7s - 2 = 0$

9. Solve.

a) $2x^2 + 8x - 5 = 0$

b) $3x^2 + 10x - 8 = 0$

c) $x^2 - 7x + 4 = 0$

d) $(x + 3)(5x + 1) = (2x + 1)(x + 7)$

e) $(2x - 1)(3x + 5) = (x + 2)(2x - 1)$

f) $(5x + 1)(x + 2) = 5x - (2x + 1)(2x + 2)$

10. Use the quadratic formula to factor each trinomial.

a) $8x^2 + 10x + 3$

b) $6x^2 - 5x - 4$

c) $12x^2 - 31x + 20$

d) $18x^2 - 37x - 20$

e) $20x^2 - 12x - 63$

f) $45x^2 + 6x - 16$

11. a) Describe a test you could use to determine whether a given trinomial of the form $ax^2 + bx + c$ can be factored.

b) Use your test to determine which of these trinomials can be factored.

    i) $12e^2 - 13e - 90$

    ii) $55n^2 - 9n - 5$

    iii) $24a^2 - 167a - 66$

    iv) $14d^2 + 143d + 94$

    v) $30m^2 - 27m - 120$

    vi) $28c^2 - 135c + 143$

12. The approximate stopping distance d metres of a car travelling at v kilometres per hour is given by the formula $d = 0.0066v^2 + 0.14v$.

a) What is the stopping distance for a car travelling at each speed?

    i) 80 km/h

    ii) 100 km/h

b) Determine the greatest speed at which a car can be stopped in each distance.

    i) 35 m

    ii) 95 m

13. The surface area A of a certain closed cylinder with radius r is given by the formula $A = 6.28r^2 + 47.7r$.

a) Determine the surface area for each radius.

i) 6 cm **ii)** 25 cm

b) Determine the radius for each surface area.

i) 290 cm^2 **ii)** 2100 cm^2

C 14. Square ABCD has sides with length 6 cm. M and N are points on sides BC and DC so that the areas of $\triangle$ABM, $\triangle$MCN, and $\triangle$ADN are all equal. Determine the lengths of BM and DN to 2 decimal places.

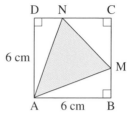

15. a) Solve.

i) $x^2 + x - 1 = 0$ **ii)** $x^2 + x - 2 = 0$

iii) $x^2 + x - 3 = 0$ **iv)** $x^2 + x - 4 = 0$

b) For what values of n does the equation $x^2 + x - n = 0$ have integral roots?

16. a) Determine the sum and product of the roots of each equation.

i) $x^2 + 4x - 45 = 0$ **ii)** $4x^2 + 20x + 21 = 0$

iii) $6x^2 - 29x + 35 = 0$ **iv)** $5x^2 - 6x - 3 = 0$

b) From the results of part a, make a conjecture concerning the sum and product of the roots of $ax^2 + bx + c = 0$.

17. Determine the condition that must be satisfied by a, b, and c when the roots of the equation $ax^2 + bx + c = 0$ are in each ratio.

a) $2 : 3$ **b)** $m : n$

18. The coefficients of the factorable trinomial $x^2 + 3x + 2$ are three consecutive integers, though not in order: 1, 3, 2. Investigate whether there are other factorable trinomials with three consecutive integers as coefficients. Try to find an example in which the consecutive integers are in order.

COMMUNICATING THE IDEAS

Summarize three different methods to solve quadratic equations, and list some advantages and disadvantages of each method. Use examples to illustrate your explanations.

Deriving the Quadratic Formula

Recall that in exercise 3 of *Investigate* on page 226 you solved $ax^2 + bx + c = 0$ using the same steps that were used to solve quadratic equations with numerical coefficients. Compare these steps with the steps on page 226.

$$ax^2 + bx + c = 0 \quad a \neq 0$$

$$a(x^2 + \frac{b}{a}x) + c = 0$$

$$a(x^2 + \frac{b}{a}x + \frac{b^2}{4a^2} - \frac{b^2}{4a^2}) + c = 0$$

$$a(x^2 + \frac{b}{a}x + \frac{b^2}{4a^2}) - \frac{b^2}{4a} + c = 0$$

$$a(x + \frac{b}{2a})^2 = \frac{b^2}{4a} - c$$

$$a(x + \frac{b}{2a})^2 = \frac{b^2 - 4ac}{4a}$$

$$(x + \frac{b}{2a})^2 = \frac{b^2 - 4ac}{4a^2}$$

Dividing each side by a

$$x + \frac{b}{2a} = \frac{\pm\sqrt{b^2 - 4ac}}{2a}$$

Taking the square root of each side, $b^2 - 4ac \geq 0$

$$x = -\frac{b}{2a} \pm \frac{\sqrt{b^2 - 4ac}}{2a}$$

$$x = \frac{-b \pm \sqrt{b^2 - 4ac}}{2a}$$

The development of the quadratic formula about 450 years ago represented an important advance in mathematical thinking. Previously, mathematicians considered only equations with numerical coefficients. This meant repeating the calculation for each equation. Moreover, since they did not recognize negative numbers, a totally different type of calculation was needed for equations such as $x^2 = 8x + 20$ and $x^2 - 8x - 20 = 0$. About 1550, the French mathematician François Vieta introduced the idea of using letters to represent the coefficients of an equation. By that time, negative numbers were becoming accepted. The combination of the two ideas meant that all quadratic equations could be written in the form $ax^2 + bx + c = 0$, and solved using a single formula.

The method above is not the only way to solve $ax^2 + bx + c = 0$. Two other methods are described on page 235. Although the first method uses modern algebraic notation, it was originally developed with more cumbersome notation.

Method 1

About A.D. 1050, the Hindus solved quadratic equations by completing the square in a different way. Their first step was to multiply both sides by 4 times the coefficient of x^2. For example, to solve $3x^2 + x - 2 = 0$:

Step 1. Multiply each side by 12.
$$36x^2 + 12x - 24 = 0$$

Step 2. Add and subtract the square of the original coefficient of x.
$$36x^2 + 12x + 1 - 1 - 24 = 0$$
$$(6x + 1)^2 - 25 = 0$$

1. Complete the solution started above.

2. Solve each equation using this method.

   a) $x^2 + 3x - 9 = 0$ **b)** $2x^2 + 3x - 1 = 0$ **c)** $ax^2 + bx + c = 0, a \neq 0$

Method 2

In the 16th century, François Vieta solved quadratic equations using a novel substitution method. He solved an equation such as $x^2 + 6x + 7 = 0$ this way.

Step 1. Let $x = y + k$, where k is a number to be determined.

Step 2. Substitute $y + k$ for x in the equation. Expand and rearrange the equation as a quadratic equation in y.

Step 3. Determine the value of k so the coefficient of the term containing y is zero. Substitute this into the equation in Step 2, then solve to determine y.

Step 4. Substitute the values for k and y into $x = y + k$ to determine x.

3. Solve each equation using this method.

   a) $x^2 + 6x + 7 = 0$ **b)** $2x^2 + x - 2 = 0$ **c)** $ax^2 + bx + c = 0, a \neq 0$

4. Compare the two methods above with the method on page 234. Which method do you think is simplest? Explain.

5. Recall that the method of completing the square on page 226 was used in Chapter 2 to write the equation of a quadratic function $y = ax^2 + bx + c$ in the form $y = a(x - p)^2 + q$. Determine if either of the two methods above can be used for this purpose.

How Can We Model a Spiral?

See page 224 for an introduction to spirals. When you determined the length : width ratios of the rectangles in the diagram on page 225, you should have found they are all approximately equal. This means that the rectangles are similar and have the same shape.

DEVELOP A MODEL

Compare this rectangle with the diagram on page 225. This could be any one of the rectangles in that diagram.

The rectangle is divided into a square and a smaller rectangle. The smaller rectangle has the same shape as the larger one.

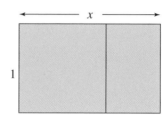

Hence,

$$\frac{\text{length of larger rectangle}}{\text{width of larger rectangle}} = \frac{\text{length of smaller rectangle}}{\text{width of smaller rectangle}} \quad \text{①}$$

1. Let the length of the larger rectangle be x units and its width be 1 unit.

    a) What is the length of the smaller rectangle?

    b) Find an expression for the width of the smaller rectangle.

2. **a)** Use equation ① to write an equation in x.

    b) Solve the equation using the quadratic formula. Express the positive root of the equation in exact form and as a decimal approximation.

In exercises 1 and 2, you should have found that equation ① becomes $x^2 - x - 1 = 0$, and the positive root of this equation is $\frac{1 + \sqrt{5}}{2}$, or

1.618 033 989 .... This number is called the *golden ratio*, and it is often represented by the Greek letter ϕ (phi). A rectangle whose length : width ratio is the golden ratio is called a *golden rectangle*. All the rectangles in the diagram on page 225 are golden rectangles.

Like π, the number ϕ is irrational. Recall that π was also defined as a ratio (the ratio of the circumference of a circle to its diameter), but we write it as a number. Similarly, ϕ is called a ratio, but we write it as a number.

In the diagram on page 225, the spiral passes through several squares. Visualize numbering these squares in order from smallest to largest. This table summarizes their side lengths.

Square	Side length	Side length	Decimal form
1			
2			
3			
4	1	1	
5	ϕ	ϕ	
6			
7			

3. a) Copy the table. Complete column 2 by using the diagram on page 225. Determine an expression involving ϕ for the side length of each square:

   i) square 6   **ii)** square 7   **iii)** square 1   **iv)** square 2   **v)** square 3

  b) What pattern can you find in the sequence of expressions for the side lengths? How does the diagram on page 225 account for this pattern?

4. In exercise 2b, you determined a decimal approximation for ϕ. Substitute this value into the expressions in column 2. Use the results to complete column 4.

On page 224, you may have predicted that the side lengths of the squares form a geometric sequence. However, by using decimal approximations, we cannot be certain that they do. In exercise 5, you will prove that they form a geometric sequence.

5. Observe that ϕ satisfies the equation you wrote in exercise 2a. Keep this in mind as you complete the following exercises.

  a) The side length of square 6 is $\phi + 1$. Show that $\phi + 1 = \phi^2$.

  b) The side length of square 7 is $2\phi + 1$. Show that $2\phi + 1 = \phi^3$.

  c) The side length of square 3 is $\phi - 1$. Show that $\phi - 1 = \phi^{-1}$.

d) Write the results from parts a to c in column 3. Complete the remaining spaces in column 3 in a similar way.

e) Explain why these results prove that the side lengths of the squares form a geometric sequence.

6. In the diagram on page 225, the spiral passes through several rectangles. Visualize numbering these rectangles just like the squares.

a) Copy and complete the table. Summarize the dimensions of the rectangles and their length : width ratios.

b) What do you notice about the rectangles?

Rectangle	Length, *l*	Width, *w*	*l : w* ratio
1			
2			
3	1		
4	ϕ	1	
5		ϕ	
6			
7			

 **REVISIT THE SITUATION**

Look at the rectangles in the diagram on page 225. Visualize how they change as you move outward from the smallest rectangle. The size increases, but the shape remains the same. This models the way the nautilus shell grows. The rectangles represent the stages of growth of the chambers in the shell, which grows at the outer end only. As each new chamber is added, the shape remains the same.

From each stage to the next, the side lengths of the rectangles are multiplied by the golden ratio. This is the only length : width ratio that is possible for all the rectangles to have in a diagram similar to the one on page 225.

Here are some situations involving the golden ratio.

7. The equation you solved in exercise 2b has a negative root.

a) Write an exact expression for this root.

b) How is this root related to the positive root? Explain.

8. In this diagram, a square with sides 2 units long is constructed in a semicircle with radius r.

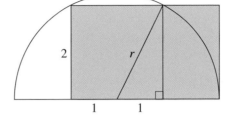

   a) Determine the radius of the semicircle.

   b) Calculate the length of the coloured rectangle.

   c) What is the length : width ratio of this rectangle?

   d) How does this rectangle compare with those on page 225? Explain.

9. There is only one positive number that becomes its own square by adding 1.

   a) Determine this number in exact form. Verify the result.

   b) Determine the number in decimal form. Verify the result.

   c) What negative number becomes its own square by adding 1? Explain.

10. There is only one positive number that becomes its own reciprocal by subtracting 1.

   a) Determine this number in exact form. Verify the result.

   b) Determine the number in decimal form. Verify the result.

   c) What negative number becomes its own reciprocal by subtracting 1? Explain.

Although mathematics is often applied to solve practical problems, it is also worth studying for its own sake. This involves appreciating patterns and interrelationships among mathematical topics.

11. Look over the above exercises and your answers.

   a) Make a list of all the properties of the golden ratio you can find.

   b) Make a list of all the mathematical topics you encountered in these exercises that are related to the golden ratio.

12. Why do you think the ratio you encountered in these exercises is called the golden ratio?

13. Give some reasons why it is important to be able to solve $x^2 - x - 1 = 0$ exactly, so that the roots are expressed as $\frac{1 \pm \sqrt{5}}{2}$ instead of 1.618 … and −0.618 … .

4.2 The Nature of the Roots of a Quadratic Equation

In Chapter 2 you determined the zeros of a quadratic function by graphing. The zeros are the x-intercepts of the graph. Since the graph of a quadratic function is a parabola with a vertical axis of symmetry, it has two different x-intercepts, two equal x-intercepts, or no x-intercepts. Similarly, a quadratic equation has two different real roots, two equal real roots, or no real roots.

We can use the formula $x = \dfrac{-b \pm \sqrt{b^2 - 4ac}}{2a}$ for the roots of the general quadratic equation $ax^2 + bx + c = 0$, $a \neq 0$, to determine the nature of the roots of the equation. Consider how the formula applies to these three equations:

$x^2 - 6x + 5 = 0$, $x^2 - 6x + 9 = 0$, and $x^2 - 6x + 13 = 0$.

$x^2 - 6x + 5 = 0$

$x = \dfrac{6 \pm \sqrt{36 - 20}}{2}$

$= \dfrac{6 \pm \sqrt{16}}{2}$ —————— Positive

The roots are

$\dfrac{6 + 4}{2}$, or 5

and $\dfrac{6 - 4}{2}$, or 1 —————— Two different real roots

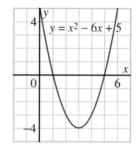

The parabola intersects the x-axis at two points.

$x^2 - 6x + 9 = 0$

$x = \dfrac{6 \pm \sqrt{36 - 36}}{2}$

$= \dfrac{6 \pm \sqrt{0}}{2}$ —————— Zero

The roots are

$\dfrac{6 + 0}{2}$, or 3

and $\dfrac{6 - 0}{2}$, or 3 —————— Two equal roots

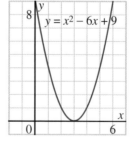

The parabola intersects the x-axis at one point.

$x^2 - 6x + 13 = 0$

$x = \dfrac{6 \pm \sqrt{36 - 52}}{2}$

$= \dfrac{6 \pm \sqrt{-16}}{2}$ —————— Negative

The roots are not defined as real numbers. —————— No real roots

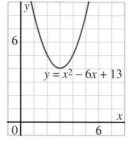

The parabola does not intersect the x-axis.

The expression under the radical sign plays an important role in the calculation of the roots. This expression enables us to determine the *nature of the roots* without solving the equation. By the nature of the roots we mean:

- whether the equation has real roots
- if there are real roots, whether they are different or equal

The above examples suggest the following conclusions about the nature of the roots of the quadratic equation $ax^2 + bx + c = 0$.

If $b^2 - 4ac > 0$, there are two different real roots.

If $b^2 - 4ac = 0$, there are two equal real roots.

If $b^2 - 4ac < 0$, there are no real roots.

The expression $b^2 - 4ac$ is called the *discriminant* of the equation $ax^2 + bx + c = 0$ because it discriminates among the three cases that can occur.

Example 1

Without solving each equation, determine the nature of its roots.

a) $4x^2 - 12x + 9 = 0$ **b)** $2x^2 + 5x - 1 = 0$ **c)** $x^2 - 2x + 3 = 0$

Solution

The nature of the roots is determined by the discriminant, $b^2 - 4ac$.

a) $4x^2 - 12x + 9 = 0$

$b^2 - 4ac = (-12)^2 - 4(4)(9)$ ———— Substituting $a = 4, b = -12, c = 9$

$= 144 - 144$

$= 0$

There are two equal real roots.

b) $2x^2 + 5x - 1 = 0$

$b^2 - 4ac = 5^2 - 4(2)(-1)$ ———— Substituting $a = 2, b = 5, c = -1$

$= 33$

Since $33 > 0$, there are two different real roots.

c) $x^2 - 2x + 3 = 0$

$b^2 - 4ac = (-2)^2 - 4(1)(3)$ ———— Substituting $a = 1, b = -2, c = 3$

$= -8$

Since $-8 < 0$, the equation has no real roots.

Example 2

Determine the values of k for which the equation $x^2 + kx + 4 = 0$ has:

a) equal roots   **b)** two different real roots   **c)** no real roots

Solution

$x^2 + kx + 4 = 0$

$b^2 - 4ac = k^2 - 4(1)(4)$ ———————————— Substituting $a = 1$, $b = k$, $c = 4$

$ = k^2 - 16$

a) For equal roots, $k^2 - 16 = 0$

$$k^2 = 16$$
$$k = \pm 4$$

The equation has equal roots when $k = 4$ or -4.

b) For two different real roots, $k^2 - 16 > 0$

$$k^2 > 16$$

Think ...

Which numbers, when squared, produce a result that is greater than 16?

$$k > 4 \text{ or } k < -4$$

The equation has two different real roots when $k > 4$ or $k < -4$.

This may also be written as $|k| > 4$.

c) For no real roots, $k^2 - 16 < 0$

$$k^2 < 16$$

Think ...

Which numbers, when squared, produce a result that is less than 16?

$$-4 < k < 4$$

The equation has no real roots when $-4 < k < 4$.

This may also be written as $|k| < 4$.

We can use a graphing calculator to illustrate the results of *Example 2*.

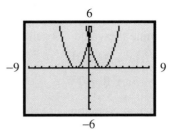

This graph shows the two functions defined by $f(x) = x^2 + kx + 4$, where $k = \pm 4$. The equations of the parabolas on the graph are $y = x^2 + 4x + 4$ and $y = x^2 - 4x + 4$. Each parabola intersects the x-axis at only one point. The x-coordinate of each point is the root of the corresponding equation.

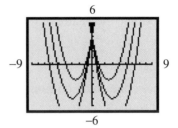

This graph shows examples of functions $f(x) = x^2 + kx + 4$, for values of k that satisfy $k > 4$ or $k < -4$. Each parabola intersects the x-axis at two different points whose x-coordinates are the roots of the corresponding equations.

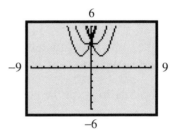

This graph shows examples of functions $f(x) = x^2 + kx + 4$, for values of k that satisfy $-4 < k < 4$. These parabolas do not intersect the x-axis. The corresponding equations have no real roots.

DISCUSSING THE IDEAS

1. Explain how it is possible to determine information about the roots of a quadratic equation without knowing what the roots are.

2. a) In the solution of *Example 2b*, explain why $k^2 > 16$ can be written as $|k| > 4$.

   b) In the solution of *Example 2c*, explain why $k^2 < 16$ can be written as $|k| < 4$.

3. Look at the graphs following the solution to *Example 2*. List as many properties of the parabolas in these graphs as you can. Explain each property in terms of the equation and the graph.

A **1.** Determine the value of each discriminant.

    a) $x^2 + 11x + 24 = 0$ **b)** $x^2 - 4x + 2 = 0$ **c)** $4x^2 - 20x + 25 = 0$

    d) $2x^2 - 5x + 8 = 0$ **e)** $3x^2 + 13x - 10 = 0$ **f)** $7x^2 + 12x + 6 = 0$

2. State which equations in exercise 1 have:

    a) two different real roots   **b)** two equal real roots   **c)** no real roots

3. **a)** Determine the nature of the roots of each equation.

    i) $x^2 - 9x + 7 = 0$ **ii)** $4x^2 + 36x + 81 = 0$ **iii)** $6x^2 + 22x + 20 = 0$

    iv) $2x^2 - 7x - 5 = 0$ **v)** $5x^2 - 8x + 4 = 0$ **vi)** $49x^2 - 70x + 25 = 0$

    b) Choose one equation from part a. Write to explain how you determined the nature of the roots.

B **4.** **a)** Solve each equation.

    i) $x^2 - 2x - 3 = 0$ **ii)** $x^2 - 2x + 1 = 0$ **iii)** $x^2 - 2x + 5 = 0$

    b) On the same grid, graph the quadratic functions corresponding to the equations in part a. Write to explain the results of part a in terms of the graphs.

5. State the condition for the equation $px^2 + qx + r = 0$ to have:

    a) two different real roots   **b)** two equal real roots   **c)** no real roots

6. For what values of k does each equation have two different real roots?

    a) $x^2 + kx + 1 = 0$ **b)** $kx^2 + 4x - 3 = 0$ **c)** $3x^2 + kx + 2 = 0$

7. For what values of m does each equation have two equal real roots?

    a) $x^2 + mx + 7 = 0$ **b)** $(2m + 1)x^2 - 8x + 6 = 0$

8. For what values of p does each equation have no real roots?

    a) $x^2 - px - 4 = 0$ **b)** $px^2 - 8x + 9 = 0$ **c)** $px^2 - 5x + p = 0$

9. Choose one equation from exercises 6 to 8. Write to explain how you determined the values of the variable.

10. When a projectile is fired, the vertical component of its initial velocity is such that its height h metres t seconds after firing is given by $h = 250t - 4.9t^2$.

    a) Determine if the projectile will reach each height.

        i) 2.75 km   **ii)** 4.0 km

    b) Explain the answers in part a in terms of the graphs of the quadratic functions and the discriminants of the corresponding quadratic equations.

11. A cafeteria owner is planning a new cafeteria in a local hospital. She knows that the monthly profit of a cafeteria can be modelled by the quadratic function $y = -0.25x^2 + 17.5x + 1500$, where x is the seating capacity.

a) Suppose this person only owned this cafeteria. Would it be possible for her to earn a monthly profit of $2000? Explain your answer both algebraically and graphically.

b) What range of seating capacity is required to have a monthly profit greater than $170?

C **12.** Determine the values of k for which $x^2 + kx + (8 - k) = 0$ has:

a) equal roots   **b)** real roots   **c)** no real roots

13. These equations are almost the same, differing only in one of the coefficients.

    i) $x^2 + 50x + 624 = 0$ ii) $x^2 + 50x + 625 = 0$ iii) $x^2 + 50x + 626 = 0$

a) Solve each equation using the quadratic formula.

b) What conclusion can you make about what could happen to the roots of an equation if you make a small change in one of its coefficients?

c) Use the graphs of the corresponding functions to explain your conclusion.

14. Determine whether there are any real numbers x and y such that:
$\frac{1}{x+y} = \frac{1}{x} + \frac{1}{y}$. Support your answer with an explanation.

15. Suppose the coefficients of $ax^2 + bx + c = 0$ are integers ($a \neq 0$). Determine which statements are:

    i) always true   ii) never true   iii) sometimes true

a) One root is an integer and the other is rational.

b) One root is rational and the other is irrational.

c) If the roots are equal, then they are real.

d) If $ac > 0$, there are no real roots.

e) If $ac < 0$, the roots are real.

16. a) Show that $x^2 + 4x - 5 = 0$ and $x^2 - 5x + 4 = 0$ have a common real root.

b) What conditions must be satisfied by b and c for these equations to have a common real root?
$$x^2 + bx + c = 0 \qquad x^2 + cx + b = 0$$

COMMUNICATING THE IDEAS

Suppose your friend was absent from today's mathematics class. How would you explain to her how you can tell the nature of the roots of a quadratic equation without solving the equation?

A Short History of Number Systems

From early times, people used the natural numbers to count their possessions. As the centuries passed, the number system has been extended to include integers, rational numbers, and irrational numbers. The result is the system of real numbers that we use today.

It took many centuries to develop the real number system because mathematicians found it difficult to accept negative numbers and irrational numbers. Some regarded negative numbers as impossible or absurd. The ancient Greeks thought that $\sqrt{2}$ is not a number because it cannot be expressed as a fraction.

About 450 years ago, mathematicians knew how to solve a quadratic equation such as $x^2 - 6x + 13 = 0$ to obtain $x = \frac{6 \pm \sqrt{-16}}{2}$. They thought that this equation has no roots, and dismissed it as impossible or absurd. They did not realize that the square root of a negative number can be defined by extending the number system, just as it had been extended in the past to include fractions, irrational numbers, and negative numbers. Eventually, mathematicians extended the real number system to the set of complex numbers, which has many applications in science, engineering, and electronics.

1500 B.C.	1000 B.C.	500 B.C

The Babylonians and the Egyptians used fractions for measurement.

The ancient Greeks knew that the length of the diagonal of a unit square is $\sqrt{2}$.

LINKING IDEAS

Mathematics & History

We define the number i to have the property $i^2 = -1$. Hence, $i = \sqrt{-1}$

We define $\sqrt{-16}$ by writing:

$$\sqrt{-16} = \sqrt{16 \times (-1)}$$
$$= \sqrt{16} \times \sqrt{-1}$$
$$= 4 \times i$$
$$= 4i$$

Using this definition, the roots of $x^2 - 6x + 13 = 0$ are:

$$x = \frac{6 \pm \sqrt{-16}}{2}$$
$$= \frac{6 \pm 4i}{2}$$
$$= 3 \pm 2i$$

An expression of the form $a + bi$, where a and b are real numbers, and $i^2 = -1$, is called a *complex number*. Calculations with complex numbers can be performed by treating i as an algebraic symbol, and replacing i^2 with -1 whenever it occurs.

1. Check that both $3 + 2i$ and $3 - 2i$ satisfy the equation $x^2 - 6x + 13 = 0$.

2. Solve and check.   **a)** $x^2 + 4x + 5 = 0$ **b)** $x^2 - 2x + 3 = 0$

3. Find two numbers with a sum of 2 and a product of 2. Check the result.

A.D. 500	A.D. 1000	A.D. 1500	A.D. 2000

The Hindus used negative numbers to represent the amount of money a person owes.

Mathematicians encountered square roots of negative numbers when they solved certain quadratic equations.

Mathematics & History

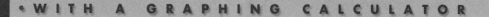
Quadratic Equations and the Butterfly Effect

Here is a novel way to try to solve a quadratic equation such as $x^2 - 2x - 8 = 0$.

$$x^2 = 2x + 8$$

$$x = 2 + \frac{8}{x} \qquad (x \neq 0)$$

> Divide each side by x, $x \neq 0$.

This does not solve the equation because the variable x still occurs on the right side. In the past, this method was usually dismissed as being incorrect. However, with technology we can continue from this point and solve the equation. This method is part of a new branch of mathematics (which has developed during the last 25 years) called *chaos*. You can investigate chaos using a graphing calculator.

To solve the equation, we must find a number x that makes both sides equal. That is, when this number is substituted into the right side, the result simplifies to the number we substituted. We begin by choosing some initial value of x. This can be any number except 0. Suppose we choose $x = 5$.

Substitute 5 for x: R.S. $= 2 + \dfrac{8}{5} = 3.6$

Substitute 3.6 for x: R.S. $= 2 + \dfrac{8}{3.6} = 4.\overline{2}$

Substitute $4.\overline{2}$ for x: R.S. $= 2 + \dfrac{8}{4.\overline{2}}$

$$= 3.894\ 736\ 842$$

Step	x	$2 + \dfrac{8}{x}$
1	5.0000	3.6000
2	3.6000	4.2222
3	4.2222	3.8947
4	3.8947	4.0541
5	4.0541	3.9733

Continue like this, substituting each result back into $2 + \dfrac{8}{x}$.
The first five results are shown in the table, to 4 decimal places.

1. **a)** Ask your teacher for the program called NOVEL1 from the Teacher's Resource Book. Enter the program into your calculator.

   b) To use the program, you must first store the coefficients of the quadratic equation into memory. For the equation above, in which $a = 1$, $b = -2$, and $c = -8$, press:

   1 [STO▸] [ALPHA] A   −2 [STO▸] [ALPHA] B   −8 [STO▸] [ALPHA] C

   c) When you run the program, the calculator will ask for the initial value of x. Enter the number used above, 5.

The calculator will produce the graph below, showing the points whose coordinates are formed by the first two columns in the table. To see the table, press [STAT] [ENTER]. You can use [▼] to scroll down the table to see how the values of x and $2 + \frac{8}{x}$ are getting closer together. Since the values are getting closer and closer to 4, we conclude 4 is a root of the given equation. You can verify this by substitution.

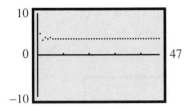

2. **a)** Investigate what happens if you try different initial values of x. Run the program several times, using other initial values. Try very large and very small numbers. Try negative numbers. Do they all lead to 4 as a root? Do any initial values lead to another root?

b) The equation $x^2 - 2x - 8 = 0$ can be expressed as $(x - 4)(x + 2) = 0$. Hence, its roots are 4 and -2. There is only one initial value that leads to the root $x = -2$ (below left). What number is it?

c) What is unusual about the results shown in the screen (below right)? Try to find an initial value that produces a screen like this.

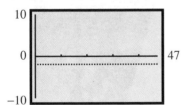
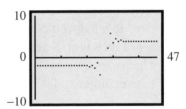

In exercise 2, you should have found that almost all initial values lead to the root $x = 4$. The only initial value that leads to the root $x = -2$ is -2 itself. Does it seem strange that there is this lack of symmetry, when the graph of $y = x^2 - 2x - 8$ is a parabola that has an axis of symmetry?

On page 248, one way to rearrange the equation $x^2 - 2x - 8 = 0$ to get an x on the left side was shown. Here are two other possibilities.

$$x^2 - 2x - 8 = 0$$
$$-2x = -x^2 + 8$$
$$x = \frac{x^2 - 8}{2}$$

$$x^2 - 2x - 8 = 0$$
$$x(x - 2) = 8$$
$$x = \frac{8}{x - 2} \qquad (x \neq 2)$$

3. a) Ask your teacher for the programs NOVEL2 and NOVEL3. These are designed for the equations on the preceding page, respectively.

   b) Choose either program. Run it for several values of x. Are the results similar to those in exercise 2, or are there some differences?

   c) Try using initial values that are very close to the roots of the equation, such as 3.999 999 or 4.000 001, and −1.999 999 or −2.000 001.

   d) What is unusual about the screens below? Try to find initial values that produce results like these.

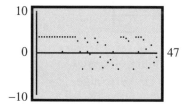 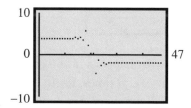

In exercises 2 and 3, you should have discovered that a tiny change in the initial value of x can make a huge difference in what eventually happens. This phenomenon is called the *butterfly effect*. This term originated in the 1960s with a meteorologist who was using computers to model the weather. A butterfly flapping its wings produces a tiny perturbation in the atmosphere. Under the right conditions, this can cause huge changes in the weather across the globe in about a month compared to what would have happened had the butterfly not flapped its wings. Scientists believe that the butterfly effect may help explain why long-range weather forecasting is difficult or even impossible.

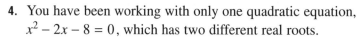

4. You have been working with only one quadratic equation, $x^2 - 2x - 8 = 0$, which has two different real roots.

   a) Investigate what happens for other quadratic equations with two different real roots. Can you predict from the coefficients which root is approached almost all the time and which is approached only once?

   b) Investigate what happens for quadratic equations with:

      i) two equal real roots   ii) no real roots

4.3 The Remainder Theorem

INVESTIGATE

1. Divide the polynomial $x^2 + 5x - 24$ by each binomial, and note the remainder.
   a) $x - 1$ **b)** $x - 2$ **c)** $x + 2$ **d)** $x + 3$

2. Consider the function $f(x) = x^2 + 5x - 24$. Determine each value.
   a) $f(1)$ **b)** $f(2)$ **c)** $f(-2)$ **d)** $f(-3)$

3. Refer to exercises 1 and 2. Assume k is a constant. How is $f(k)$ related to the remainder when $f(x)$ is divided by $x - k$?

4. a) Repeat exercises 1 and 2, replacing $x^2 + 5x - 24$ with a polynomial of your choice.

   b) Is the relation in exercise 3 still true?

In some problems involving division, only the remainder is needed. For example, to find the day of the week 60 days from now, we divide 60 by 7.

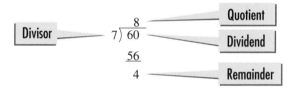

Since the remainder is 4, in 60 days the day of the week will be 4 days after today.

In algebra, we can find remainders without dividing. To understand the method, we have to recognize the relation among the dividend, divisor, quotient, and remainder, in a division problem.

For the division above, we can write:

$$60 \quad = \quad (7)(8) \quad + \quad 4$$

dividend = (divisor)(quotient) + remainder

Division Statement

In any division problem,

dividend = (divisor)(quotient) + remainder

Example 1

Given $f(x) = x^3 + 4x^2 + x - 2$, determine the remainder when $f(x)$ is divided by $x - 1$. Write the division statement.

Solution

$$
\begin{array}{r}
x^2 + 5x + 6 \\
x - 1 \overline{)\; x^3 + 4x^2 + x - 2} \\
\underline{x^3 - x^2} \\
5x^2 + x \\
\underline{5x^2 - 5x} \\
6x - 2 \\
\underline{6x - 6} \\
4
\end{array}
$$

The division statement is:

$x^3 + 4x^2 + x - 2 = (x - 1)(x^2 + 5x + 6) + 4$

In *Example 1*, notice what happens if we substitute 1 for x in each side of the division statement.

On the left side, the result is:

$1^3 + 4(1)^2 + 1 - 2 = 4$

On the right side, the result is:

$(1 - 1)(1^2 + 5(1) + 6) + 4 = 0(12) + 4$
$= 4$

Hence, $f(1) = 4$

This shows that when the polynomial $x^3 + 4x^2 + x - 2$ is divided by $x - 1$, the remainder is $f(1)$. This is an example of a general result called the Remainder Theorem. It is true for any polynomial.

Remainder Theorem

When a polynomial in x is divided by $x - k$, the remainder is equal to the number obtained by substituting k for x in the polynomial.

Suppose a polynomial is divided by a binomial. We can use the Remainder Theorem to determine the remainder without doing the division.

Example 2

Determine the remainder when $x^3 - 4x^2 + 5x + 1$ is divided by each binomial.

a) $x - 2$ **b)** $x + 1$

Solution

a) Let $f(x) = x^3 - 4x^2 + 5x + 1$.

According to the Remainder Theorem, the remainder when $f(x)$ is divided by $x - 2$ is $f(2)$.

$f(2) = 2^3 - 4(2)^2 + 5(2) + 1$
$\qquad = 8 - 16 + 10 + 1$, or 3

The remainder is 3.

b) Since $x + 1 = x - (-1)$, the remainder when $f(x)$ is divided by $x + 1$ is $f(-1)$.

$f(-1) = (-1)^3 - 4(-1)^2 + 5(-1) + 1$
$\qquad = -1 - 4 - 5 + 1$, or -9

The remainder is -9.

Example 3

When $x^3 + 3x^2 + cx + 10$ is divided by $x - 2$, the remainder is 6. Determine the value of c.

Solution

Let $f(x) = x^3 + 3x^2 + cx + 10$.

According to the Remainder Theorem, the remainder when $f(x)$ is divided by $x - 2$ is $f(2)$.

$f(2) = 2^3 + 3(2)^2 + 2c + 10$
$\qquad = 8 + 12 + 2c + 10$
$\qquad = 30 + 2c$

Since the remainder is 6,

$30 + 2c = 6$
$\qquad 2c = -24$
$\qquad\ c = -12$

DISCUSSING THE IDEAS

1. In a division problem, how are the dividend, divisor, quotient, and remainder related?

2. What is the remainder when a polynomial in x is divided by x? Explain.

A 1. Determine the remainder when $x^2 + 2x - 8$ is divided by each binomial.

    a) $x - 1$ **b)** $x - 4$ **c)** $x - 5$ **d)** $x + 1$ **e)** $x + 4$ **f)** $x + 5$

2. Determine the remainder when $x^3 + 3x^2 - 5x + 4$ is divided by each binomial.

    a) $x - 1$ **b)** $x - 2$ **c)** $x - 3$ **d)** $x + 1$ **e)** $x + 2$ **f)** $x + 3$

3. Determine the remainder when each polynomial is divided by $x - 2$.

    a) $x^2 - 5x + 2$ **b)** $x^3 + x^2 - 2x + 3$ **c)** $x^3 + x^2 - 10x + 8$

    d) $3x^3 - 5x^2 + 2x + 8$ **e)** $2x^3 + x^2 + 4x - 7$ **f)** $x^4 - 3x^3 + 2x^2 - 5x - 1$

4. Choose one polynomial from exercise 3. Write to explain how you determined the remainder.

B 5. Determine each value of k.

    a) When $x^3 + kx^2 + 2x - 3$ is divided by $x + 2$, the remainder is 1.

    b) When $x^4 - kx^3 - 2x^2 + x + 4$ is divided by $x - 3$, the remainder is 16.

    c) When $2x^3 - 3x^2 + kx - 1$ is divided by $x - 1$, the remainder is 1.

6. When the polynomial $2x^2 + bx - 5$ is divided by $x - 3$, the remainder is 7.

    a) Determine the value of b.

    b) What is the remainder when the polynomial is divided by $x - 2$?

7. When the polynomial $ax^3 + 2x^2 - x + 3$ is divided by $x + 1$, the remainder is 4.

    a) Determine the value of b.

    b) What is the remainder when the polynomial is divided by $x - 1$?

8. When the polynomial $x^3 + bx^2 + 2x + 9$ is divided by $x - 2$, the remainder is 1.

    a) Determine the value of b.

    b) What is the remainder when the polynomial is divided by $x + 2$?

C 9. Let $f(x)$ represent any polynomial. Suppose $f(x)$ is divided by $x - k$, resulting in a quotient $q(x)$ and a remainder r.

    a) Write the division statement.

    b) Use the division statement to prove the Remainder Theorem.

COMMUNICATING THE IDEAS

Your friend missed today's mathematics lesson and asks you about it. How would you explain the Remainder Theorem to your friend? Write your thoughts in your notebook.

1. a) Use the Remainder Theorem to determine the remainder when $x^2 + 3x - 10$ is divided by each binomial.

    i) $x - 2$ **ii)** $x - 5$ **iii)** $x + 2$ **iv)** $x + 5$

  b) In part a, you should have found that two remainders are 0. What does this tell you about the trinomial $x^2 + 3x - 10$?

2. a) Determine the remainder when $x^3 - 4x^2 + x + 6$ is divided by each binomial.

    i) $x - 1$ **ii)** $x - 2$ **iii)** $x - 3$ **iv)** $x - 4$

  b) What do the zero remainders tell about the polynomial $x^3 - 4x^2 + x + 6$?

3. Here is a table of values for the function $y = x^3 - 5x^2 + 2x + 8$.

x	-2	-1	0	1	2	3	4
y	-24	0	8	6	0	-4	0

  a) Examine the table. Write three factors of $x^3 - 5x^2 + 2x + 8$.

  b) Use your answers to part a to factor the polynomial.

4. Consider the polynomial $x^3 - 2x^2 - 5x + 6$.

  a) Determine three values of x that result in a value of 0 when they are substituted in the polynomial.

  b) Use your answers to part a to factor the polynomial.

5. Explain how you can determine, without dividing, whether $x - k$ is a factor of a given polynomial.

According to the Remainder Theorem, if a number, k, is substituted for x in a polynomial in x, the value obtained is the remainder when the polynomial is divided by $x - k$. If this remainder is 0, then $x - k$ is a factor of the polynomial. This special case of the Remainder Theorem is called the Factor Theorem.

Factor Theorem

If k is substituted for x in a polynomial in x, and the resulting value is 0, then $x - k$ is a factor of the polynomial.

The Factor Theorem provides a simple method to determine whether a binomial of the form $x - k$ is a factor of a given polynomial.

Example 1

Determine which binomials are factors of $x^3 - 6x^2 + 3x + 10$.

a) $x - 2$ **b)** $x - 3$ **c)** $x + 1$ **d)** $x - 5$

Solution

Let $f(x) = x^3 - 6x^2 + 3x + 10$.

If a binomial is a factor, then there is no remainder when the polynomial is divided by the binomial.

a) $f(2) = 2^3 - 6(2)^2 + 3(2) + 10$

$\qquad = 8 - 24 + 6 + 10$

$\qquad = 0$

Since $f(2) = 0$, $x - 2$ is a factor of $x^3 - 6x^2 + 3x + 10$.

b) $f(3) = 3^3 - 6(3)^2 + 3(3) + 10$

$\qquad = 27 - 54 + 9 + 10$

$\qquad = -8$

Since $f(3) \neq 0$, $x - 3$ is not a factor of $x^3 - 6x^2 + 3x + 10$.

c) $f(-1) = (-1)^3 - 6(-1)^2 + 3(-1) + 10$

$\qquad\quad = -1 - 6 - 3 + 10$

$\qquad\quad = 0$

Since $f(-1) = 0$, $x + 1$ is a factor of $x^3 - 6x^2 + 3x + 10$.

d) $f(5) = 5^3 - 6(5)^2 + 3(5) + 10$

$\qquad = 125 - 150 + 15 + 10$

$\qquad = 0$

Since $f(5) = 0$, $x - 5$ is a factor of $x^3 - 6x^2 + 3x + 10$.

In *Example 1*, we found three factors of $x^3 - 6x^2 + 3x + 10$. The product of these three factors is $x^3 - 6x^2 + 3x + 10$. Hence, we can write this polynomial in factored form:

$$x^3 - 6x^2 + 3x + 10 = (x - 5)(x + 1)(x - 2)$$

The product of the constant terms in the factors is $(-5)(+1)(-2) = 10$. This is also the constant term in the polynomial. This suggests the following property of the factors of a polynomial.

Factor Property

If a polynomial in x has any factor of the form $x - k$, then k is a factor of the constant term of the polynomial.

We can use the Factor Theorem and the Factor Property to factor a polynomial. We must find a value of x that results in 0 when it is substituted in the polynomial. The Factor Property helps us decide which values to test.

Example 2

Factor $x^3 + 2x^2 - 5x - 6$.

Solution

Let $f(x) = x^3 + 2x^2 - 5x - 6$.

Find a value of x so that $f(x) = 0$.

According to the Factor Property, the numbers to test are the factors of –6: that is, 1, 2, 3, 6, –1, –2, –3, and –6.

Try $x = 1$. $f(1) = 1^3 + 2(1)^2 - 5(1) - 6$
$$= 1 + 2 - 5 - 6$$
$$\neq 0$$

$x - 1$ is not a factor of $x^3 + 2x^2 - 5x - 6$.

Try $x = -1$. $f(-1) = (-1)^3 + 2(-1)^2 - 5(-1) - 6$
$$= -1 + 2 + 5 - 6$$
$$= 0$$

$x + 1$ is a factor of $x^3 + 2x^2 - 5x - 6$.

Therefore, one factor of $x^3 + 2x^2 - 5x - 6$ is $x + 1$.

Here are two ways to determine the other factors. Other methods can be used as well.

Using long division

$$
\begin{array}{r}
x^2 + x - 6 \\
x + 1 \overline{) x^3 + 2x^2 - 5x - 6} \\
\underline{x^3 + x^2} \\
x^2 - 5x \\
\underline{x^2 + x} \\
-6x - 6 \\
\underline{-6x - 6} \\
0
\end{array}
$$

The other factor is $x^2 + x - 6$. This can be factored as $(x + 3)(x - 2)$.

Therefore, $x^3 + 2x^2 - 5x - 6 = (x + 1)(x^2 + x - 6)$
$$= (x + 1)(x + 3)(x - 2)$$

By substitution

One factor of $x^3 + 2x^2 - 5x - 6$ is $x + 1$.

Let the other factor be $x^2 + bx + c$. Then, $(x + 1)(x^2 + bx + c) = x^3 + 2x^2 - 5x - 6$ ①

> **Think...**
>
> This equation is true for all values of x. It must be true when any particular number is substituted for x.

Substitute 0 for x in ①. $(0 + 1)(0^2 + b(0) + c) = 0^3 + 2(0)^2 - 5(0) - 6$

$$c = -6$$

Hence, equation ① becomes: $(x + 1)(x^2 + bx - 6) = x^3 + 2x^2 - 5x - 6$ ②

> **Think ...**
>
> This equation is also true for all values of x. It must be true when any particular number is substituted for x.

Substitute 1 for x in ②. $2(1 + b - 6) = 1 + 2 - 5 - 6$

$$2(b - 5) = -8$$
$$2b = 2$$
$$b = 1$$

The other factor of $x^3 + 2x^2 - 5x - 6$ is $x^2 + x - 6$.

Therefore, $x^3 + 2x^2 - 5x - 6 = (x + 1)(x^2 + x - 6)$

$$= (x + 1)(x + 3)(x - 2)$$

DISCUSSING THE IDEAS

1. Following *Example 1*, the polynomial $x^3 - 6x^2 + 3x + 10$ was written in factored form. How could you check this is correct?

2. In *Example 2*, would a value of 4 for x be a good choice when you try to find a value of x so that $f(x) = 0$? Explain.

3. In *Example 2*, after we found that $x + 1$ is one factor of $x^3 + 2x^2 - 5x - 6$, two methods to find the other factors were shown.

   a) Which of these two methods do you prefer? Why?

   b) What other method could you use? Does this method have advantages or disadvantages over the two methods in *Example 2*?

4. How could the result of *Example 2* be used to solve the equation $x^3 + 2x^2 - 5x - 6 = 0$?

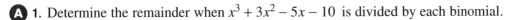

A **1.** Determine the remainder when $x^3 + 3x^2 - 5x - 10$ is divided by each binomial.

  a) $x - 1$ **b)** $x - 2$ **c)** $x - 5$

  d) $x + 1$ **e)** $x + 2$ **f)** $x + 5$

2. Which binomial in exercise 1 is a factor of $x^3 + 3x^2 - 5x - 10$?

3. Suppose $x + 7$ is a factor of $f(x)$. What is the value of $f(-7)$?

4. For each table of values for the function $f(x)$, identify as many binomial factors as possible.

a)

x	f(x)
-4	28
-3	22
-2	0
-1	-20
0	-20
1	28
2	112

b)

x	f(x)
-3	-28
-2	0
-1	6
0	2
1	0
2	12
3	50

c)

x	f(x)
-2	-12
-1	0
0	2
1	0
2	0
3	8
4	30

5. Which polynomials have $x + 3$ as a factor?

  a) $x^3 + 2x^2 - 9x - 18$ **b)** $-x^3 - 2x^2 + 21x - 18$

  c) $x^3 + 6x^2 + 9x$ **d)** $-x^4 - 8x^3 - 14x^2 + 8x + 15$

6. Which polynomial has $x - 2$ as a factor?

  a) $x^3 - 5x^2 - 17x + 21$ **b)** $-x^3 - 5x^2 + 2x + 24$

  c) $x^3 - x^2 - 17x - 15$ **d)** $x^3 + 7x^2 + 7x - 15$

7. Choose one polynomial from exercise 5 or 6. Write to explain how you determined whether the binomial was a factor.

8. For the polynomial $x^3 + 3x^2 - 6x - 8$, which values of x should be chosen to test for factors of this polynomial? Explain your choices.

B **9.** Determine one factor of each polynomial.

  a) $x^3 - 4x + 3$ **b)** $x^3 + x^2 + x + 1$ **c)** $-y^3 - 19y^2 - 19y - 1$

  d) $x^3 - 27$ **e)** $-y^3 + y^2 + y + 2$ **f)** $x^3 + 2x^2 + 5x + 4$

10. a) Show that both $x - 2$ and $x - 3$ are factors of $2x^3 - 11x^2 + 17x - 6$.

  b) Determine another factor of $2x^3 - 11x^2 + 17x - 6$.

11. a) Show that $x + 2$ is a factor of $x^3 - 3x^2 - 6x + 8$.

  b) Determine the remaining factors.

  c) Write to explain how you completed parts a and b.

12. Factor each polynomial completely.

  a) $x^3 + 5x^2 + 2x - 8$ **b)** $x^3 + 9x^2 + 23x + 15$ **c)** $x^3 + 2x^2 - 19x - 20$

  d) $x^3 - 7x - 6$ **e)** $5x^3 - 7x^2 - x + 3$ **f)** $x^3 - 9x^2 + 17x - 6$

  g) $x^3 + 8x^2 + 17x + 10$ **h)** $2x^3 - x^2 - 13x - 6$ **i)** $x^3 - 6x^2 + 10x - 3$

13. Solve by factoring.

  a) $x^3 - 3x^2 - 6x + 8 = 0$ **b)** $x^3 + 4x^2 - 7x - 10 = 0$ **c)** $x^3 - 8x^2 + 20x - 16 = 0$

  d) $x^3 + x^2 - 8x - 12 = 0$ **e)** $x^3 - x^2 - 3x + 2 = 0$ **f)** $x^3 - 4x^2 + 8x - 8 = 0$

14. Solve by factoring.

  a) $2x^3 - 7x^2 + 2x + 3 = 0$ **b)** $3x^3 - 4x^2 - 5x + 2 = 0$ **c)** $6x^3 - 5x^2 - 12x - 4 = 0$

  d) $2x^3 + 7x^2 + x - 10 = 0$ **e)** $x^3 - 5x^2 + 3x + 4 = 0$ **f)** $2x^3 + 5x^2 - 4 = 0$

  g) $2x^3 - 9x^2 + 13x - 12 = 0$ **h)** $3x^3 - 16x^2 - 13x + 6 = 0$ **i)** $x^3 - x - 6 = 0$

15. Determine each value of k.

  a) $x - 2$ is a factor of $x^3 - 6x^2 + kx - 6$.

  b) $x + 4$ is a factor of $3x^3 + 11x^2 - 6x + k$.

  c) $x - 3$ is a factor of $x^3 + kx^2 + kx + 21$.

C 16. Consider the function $f(x) = 2x^3 - 7x^2 - 5x + 4$.

  a) Use division to show that $2x - 1$ is a factor of $f(x)$.

  b) What value of x should be substituted to give a value of 0 for this polynomial? Show that your answer is correct.

17. Without dividing, determine each remainder.

  a) $(6x^2 - 10x + 7) \div (3x + 1)$ **b)** $(-8a^2 - 2a - 3) \div (4a - 1)$

18. Is $2x + 1$ a factor of $2x^3 - x^2 - 13x - 6$? Write to explain how you know.

19. Prove that $x - y$ is a factor of $x^n - y^n$ for all natural number values of n.

20. Prove that $x + a$ is a factor of $(x + a)^5 + (x + c)^5 + (a - c)^5$.

COMMUNICATING THE IDEAS

A friend is confused about the difference between the Remainder Theorem and the Factor Theorem. Explain how these two theorems are related and the difference between them.

4.5 Solving Polynomial Inequalities

The quadratic equation $x^2 - 5x + 4 = 0$ can be solved by factoring to obtain $(x - 1)(x - 4) = 0$, and the roots $x = 1$ and $x = 4$. This equation can also be solved by graphing the function $f(x) = x^2 - 5x + 4$, and noting that the graph intersects the x-axis at $x = 1$ and $x = 4$.

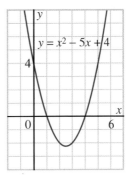

The graph divides the x-axis into three different sets of values for x.

- The values of x that satisfy the equation
  $x^2 - 5x + 4 = 0$

  These are the values of x where the graph intersects the x-axis. The roots of the equation are 1 and 4.

- The values of x that satisfy the inequality
  $x^2 - 5x + 4 > 0$

  These are the values of x where the graph is above the x-axis. The solution of the inequality is $x < 1$ or $x > 4$.

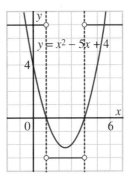

- The values of x that satisfy the inequality
  $x^2 - 5x + 4 < 0$

  These are the values of x where the graph is below the x-axis. The solution of the inequality is $1 < x < 4$.

Any quadratic inequality can be solved by graphing. The roots of the corresponding equation determine the endpoints of the intervals in the solutions of the inequalities. Hence, a more efficient method to solve a quadratic inequality is to solve the corresponding equation, then visualize the graph of the corresponding quadratic function. The solution of the inequality is determined by the intervals on the x-axis defined by the roots of the equation.

Example 1

Solve the inequality $5x^2 + 13x - 6 > 0$.

Solution

Solve the corresponding quadratic equation.

$5x^2 + 13x - 6 = 0$
$(5x - 2)(x + 3) = 0$

Either $\quad 5x - 2 = 0 \qquad$ or $\qquad x + 3 = 0$
$\qquad\qquad x = 0.4 \qquad\qquad\qquad\quad x = -3$

Think ...

Visualize the graph of the corresponding quadratic function, $f(x) = 5x^2 + 13x - 6$.

Since the coefficient of x^2 is positive, the graph opens up. It intersects the x-axis at the points where $x = -3$ and $x = 0.4$.

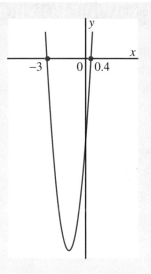

The solution of the inequality $5x^2 + 13x - 6 > 0$ is the values of x for which the graph of $f(x) = 5x^2 + 13x - 6$ lies above the x-axis. Hence, the solution is $x < -3$ or $x > 0.4$.

The solution of the inequality in *Example 1* can be represented on a number line.

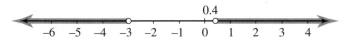

In *Example 1*, the numbers -3 and 0.4 are not part of the solution. If the inequality had been $5x^2 + 13x - 6 \geq 0$, these numbers would satisfy the inequality, and the solution would be $x \leq -3$ or $x \geq 0.4$.

Polynomial inequalities of higher degree can be solved in the same way. The initial step of solving the polynomial equation can be done by factoring (if possible), by using the quadratic formula (if possible), or by using technology. The solution of the inequality can be determined by visualizing the graph of the corresponding polynomial function.

Example 2

Solve $x^3 - 6x^2 + 3x + 10 < 0$.

Solution

Factor $f(x) = x^3 - 6x^2 + 3x + 10$, if possible.

Try $x = 1$. $f(1) = 1^3 - 6(1)^2 + 3(1) + 10$
$$= 1 - 6 + 3 - 18$$
$$\neq 0$$

Try $x = -1$. $f(-1) = (-1)^3 - 6(-1)^2 + 3(-1) + 10$
$$= -1 - 6 - 3 + 10$$
$$= 0$$

Therefore, $x + 1$ is one factor of $x^3 - 6x^2 + 3x + 10$.

Try $x = 2$. $f(2) = 2^3 - 6(2)^2 + 3(2) + 10$
$$= 8 - 24 + 6 + 10$$
$$= 0$$

Therefore, $x - 2$ is another factor of $x^3 - 6x^2 + 3x + 10$.

Since the constant term is $+10$, the third factor must be $x - 5$.

Hence, $x^3 - 6x^2 + 3x + 10 = (x + 1)(x - 2)(x - 5)$, and the given inequality
can be written in the form $(x + 1)(x - 2)(x - 5) < 0$.

Think ...

Visualize the graph of the corresponding
polynomial function, $f(x) = (x + 1)(x - 2)(x - 5)$.

Since the leading coefficient is positive, the graph
extends from the 3rd quadrant to the 1st quadrant.
It intersects the x-axis at the points where $x = -1$,
$x = 2$, and $x = 5$.

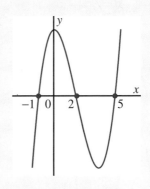

The solution of the inequality $x^3 - 6x^2 + 3x + 10 < 0$ is the values of x
for which the graph of $f(x) = x^3 - 6x^2 + 3x + 10$ lies below the x-axis.
Hence, the solution is $x < -1$ or $2 < x < 5$.

The solution of the inequality in *Example 2* can be represented on a number line.

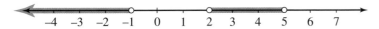

1. The solution of *Example 1* contains a quadratic equation that was solved by factoring.

   a) Do you think you might encounter quadratic inequalities that cannot be solved by factoring in similar examples? Explain.

   b) How would you continue the solution when the inequality cannot be solved by factoring?

2. **a)** In *Example 2*, once we found that $x + 1$ was a factor of $x^3 - 6x^2 + 3x + 10$, what other ways could we have found the other factors?

   b) How would you proceed if you could not determine any factors of $x^3 - 6x^2 + 3x + 10$? Explain.

4.5 EXERCISES

A 1. Use each graph to write the solutions of the inequalities below it.

a)

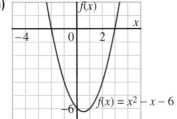

i) $x^2 - x - 6 < 0$

ii) $x^2 - x - 6 \geq 0$

b)

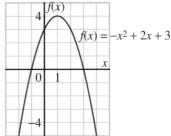

i) $-x^2 + 2x + 3 \leq 0$

ii) $-x^2 + 2x + 3 > 0$

c)

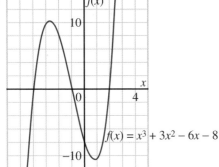

i) $x^3 + 3x^2 - 6x - 8 \leq 0$

ii) $x^3 + 3x^2 - 6x - 8 > 0$

d)

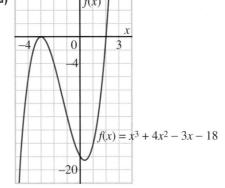

i) $x^3 + 4x^2 - 3x - 18 \geq 0$

ii) $x^3 + 4x^2 - 3x - 18 < 0$

2. Solve each inequality.

a) $(x - 2)(x + 2) > 0$ b) $(x + 1)(x + 2) \leq 0$

c) $x(x - 5) > 0$ d) $x(x - 2) \leq 0$

e) $(a - 2)(a - 3) < 0$ f) $(n - 3)(n + 5) \geq 0$

B   **3. a)** Solve each inequality. Show each solution on a number line.

    i) $x^2 + 5x + 6 > 0$ ii) $m^2 - 2m - 8 > 0$

    iii) $18 - 3y - y^2 < 0$ iv) $3 - x^2 + 2x > 0$

    v) $x^2 - 10x + 9 \leq 0$ vi) $2x^2 - x - 15 < 0$

b) Choose one inequality from part a. Write to explain how you solved it.

4. Solve each inequality. Show each solution on a number line.

a) $x^2 - 4x + 4 > 0$ b) $1 - 4x^2 \leq 0$

c) $3x^2 - 4x - 5 \geq 0$ d) $6x^2 + 7x - 20 < 0$

e) $2x^2 + x + 6 > 0$ f) $2x - 2 - x^2 \geq 0$

g) $12x - 4 - 9x^2 \geq 0$ h) $4x^2 + 10x - 7 \leq 0$

5. a) Sketch the graph of $f(x) = (x - 1)^2 - 4$.

b) Determine the values of x for which $f(x) > 0$.

6. a) Sketch the graph of $g(x) = (x - 2)^2 - 9$.

b) Determine the values of x for which $g(x) \leq 0$.

7. a) Write a quadratic inequality that has each solution.

    i) $x \leq 3$ or $x \geq \frac{1}{2}$ ii) $-3 < x < 4$

    iii) $x < -2$ or $x > 1$ iv) $-\frac{7}{2} \leq x \leq -1$

b) Choose one inequality from part a. Write to explain how you determined it.

8. Determine the values of x for which the graph of $f(x) = x^2 - 4x$:

a) lies above the line $y = 2x - 5$

b) lies below the line $y = x + 14$

9. Solve each inequality.

a) $(x + 3)(x - 1)(x - 3) > 0$ b) $-(x + 3)(x - 1)(x - 3) > 0$

c) $(x + 1)(x - 2)(x - 5) \leq 0$ d) $-x(x - 2)(x + 5) < 0$

e) $(x + 1)(x - 4)^2 > 0$ f) $(x + 3)(x + 1)(x - 2)(x - 5) \leq 0$

10. Solve each inequality.

a) $x^3 - 4x^2 - x + 4 > 0$ **b)** $x^3 - 4x^2 + x + 6 \leq 0$

c) $x^4 + 2x^3 - 7x^2 - 8x + 12 \geq 0$ **d)** $x^4 - 6x^3 + 4x^2 + 6x - 5 < 0$

e) $x^3 + 2x^2 - 8x - 9 < 0$ **f)** $x^3 + 2x^2 + 3x + 6 \geq 0$

11. Write a cubic inequality that has each solution.

a) $-3 \leq x \leq 0$ and $x \geq 2$ **b)** $x < \frac{3}{2}$ and $5 < x < \frac{15}{2}$

c) $3 < x < 7$ and $x > 9$ **d)** $x \leq -5$ and $1 \leq x \leq 4$

12. For what values of x does the graph of $f(x) = x^3 + 2x^2 + 3x + 4$ lie above the graph of $f(x) = x^2 + 2x + 3$?

C **13.** Which statements are true? Explain.

a) Some quadratic inequalities have no solutions.

b) Some quadratic inequalities have only one solution.

c) Every quadratic inequality has infinitely many solutions.

d) Some cubic inequalities have no solutions.

e) Some cubic inequalities have only one solution.

f) Every cubic inequality has infinitely many solutions.

14. Rectangle ABCD is 10 cm long and 5 cm wide. Point P is located x centimetres from A on side AB. Point Q is located y centimetres from A on side AD. For what values of y is it possible to locate a point R on side BC such that $\angle QPR = 90°$?

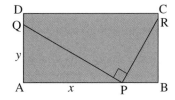

Explain how the zeros of a polynomial function can be used to determine the solutions of polynomial inequalities. Use examples to illustrate your explanation.

4.6 Solving Rational Equations and Inequalities

An equation in which the variable occurs in the denominator of a rational expression is called a *rational equation*.

A quadratic equation can occur in the solution of a rational equation.

Example 1

Solve.

a) $\dfrac{x}{3} - \dfrac{2}{x} = \dfrac{x+6}{3}$ **b)** $\dfrac{4}{x-3} = \dfrac{24}{2x-6}$

Solution

a) $\dfrac{x}{3} - \dfrac{2}{x} = \dfrac{x+6}{3}$

This equation is not defined when $x = 0$.

Hence, this solution is valid only when $x \neq 0$.

Multiply each side by the lowest common denominator, $3x$.

$$\frac{x}{3} \times 3x - \frac{2}{x} \times 3x = \frac{x+6}{3} \times 3x$$
$$x^2 - 6 = x^2 + 6x$$
$$-6 = 6x$$
$$x = -1$$

b) $\dfrac{4}{x-3} = \dfrac{24}{2x-6}$

This equation is not defined when

$$x - 3 = 0 \qquad \text{or} \qquad 2x - 6 = 0$$
$$x = 3 \qquad\qquad\qquad x = 3$$

Hence, this solution is valid only when $x \neq 3$.

Multiply each side by $(x - 3)(2x - 6)$.

$$\frac{4}{x-3} \times (x-3)(2x-6) = \frac{24}{2x-6} \times (x-3)(2x-6)$$
$$4(2x-6) = 24(x-3)$$
$$8x - 24 = 24x - 72$$
$$16x = 48$$
$$x = 3$$

Since the solution is not valid when $x = 3$, there is no solution.

We can use a graphing calculator to illustrate the solutions of the equations in *Example 1*. The first screen below shows the graphs of $y = \frac{x}{3} - \frac{2}{x}$ (the curve with two branches) and $y = \frac{x+6}{3}$ (the straight line). The graphs intersect at $x = -1$.

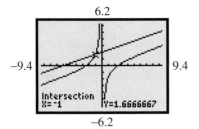

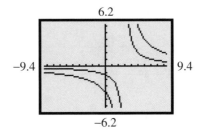

The second screen shows the graphs of $y = \frac{4}{x-3}$ (the two branches closer to the *x*-axis) and $y = \frac{24}{2x-6}$ (the other two branches). The graphs do not intersect.

In *Example 1b*, we could have predicted that there would be no solution because the equation can be written in the form $\frac{4}{x-3} = \frac{12}{x-3}$. If we substitute any number other than 3 for *x*, the denominators will be equal. Since the numerators are not equal, there is no solution. On the screen (right above), this also shows one graph is a vertical expansion of the other.

To solve a rational equation algebraically:

1. Identify the values of the variable for which the rational expressions in the equation are not defined.

2. Multiply each side by the lowest common denominator and solve the equation that results.

3. Reject any roots that are values of the variable for which the rational expressions are not defined.

Example 2

Solve $\frac{2}{x-3} + \frac{3}{x} = 2$.

Solution

$\frac{2}{x-3} + \frac{3}{x} = 2$

This equation is not defined when $x = 3$ or $x = 0$.

Hence, this solution is not valid for these values of x.

Multiply each side by $x(x-3)$.

$$\frac{2}{x-3} \times x(x-3) + \frac{3}{x} \times x(x-3) = 2 \times x(x-3)$$
$$2x + 3(x-3) = 2x(x-3)$$
$$2x + 3x - 9 = 2x^2 - 6x$$
$$2x^2 - 11x + 9 = 0$$
$$(2x-9)(x-1) = 0$$

Either $\quad 2x - 9 = 0 \qquad$ or $\qquad x - 1 = 0$

$\qquad\qquad x = 4.5 \qquad\qquad\qquad\quad x = 1$

Since these are both values of x for which the equation is defined, they are both roots of the equation.

We can use a graphing calculator to illustrate the solution in *Example 2*. The screens below show the graphs of $y = \frac{2}{x-3} + \frac{3}{x}$ and $y = 2$. They intersect when $x = 1$ and when $x = 4.5$.

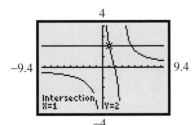
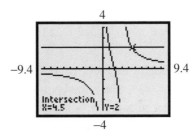

An inequality in which the variable occurs in the denominator of a rational expression is called a *rational inequality*. Recall that when solving an inequality, the direction of the inequality sign must be reversed if each side is multiplied or divided by a negative number.

To solve a rational inequality algebraically:

1. Follow the same steps as for solving a rational equation.

2. When both sides of the inequality are multiplied by an expression containing the variable, consider separately the cases when this expression is positive and negative.

Example 3

Solve the inequality $x + 1 < \frac{6}{x}$.

Solution

$x + 1 < \frac{6}{x}$

This inequality is not defined when $x = 0$.

Hence, this solution is valid only when $x \neq 0$.

Case 1. Let $x > 0$.

This part of the solution is valid only when $x > 0$.

Multiply each side by x.

$$x(x + 1) < 6$$
$$x^2 + x - 6 < 0$$
$$(x + 3)(x - 2) < 0$$

> ### Think ...
> Visualize the graph of the corresponding quadratic function,
> $f(x) = (x + 3)(x - 2)$. Since the coefficient of x^2 is positive, the graph opens up.
> It intersects the x-axis at $x = -3$ and $x = 2$. The solution of $(x + 3)(x - 2) < 0$
> consists of the values of x for which the graph of $f(x) = (x + 3)(x - 2)$ lies
> below the x-axis.

The solution of $(x + 3)(x - 2) < 0$ is $-3 < x < 2$.

However, since this part of the solution is valid only when $x > 0$,
the solution of the given inequality in this case is $0 < x < 2$.

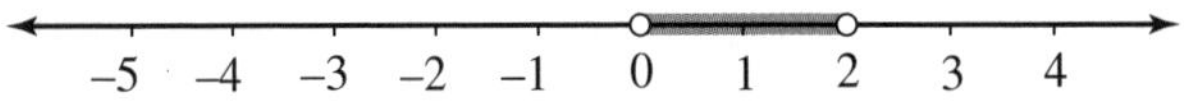

Case 2. Let $x < 0$.

This part of the solution is valid only when $x < 0$.

Multiply each side by x. Since x is negative, reverse the inequality sign.

$$x(x + 1) > 6$$
$$x^2 + x - 6 > 0$$
$$(x + 3)(x - 2) > 0$$

Think ...

The solution of $(x + 3)(x - 2) > 0$ consists of the values of x for which the graph of $f(x) = (x + 3)(x - 2)$ lies above the x-axis.

The solution of $(x + 3)(x - 2) > 0$ is $x < -3$ or $x > 2$.

However, since this part of the solution is valid only when $x < 0$, the solution of the given inequality in this case is $x < -3$.

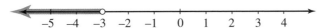

Combining the results of *Case 1* and *Case 2*, the solution of the given inequality is $x < -3$ or $0 < x < 2$.

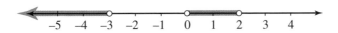

We can use a graphing calculator to illustrate the solution of *Example 3*. This screen shows the graphs of $y = x + 1$ (the line) and $y = \dfrac{6}{x}$ (the curve with two branches).

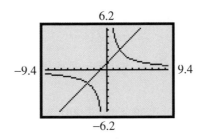

Case 1 corresponds to points to the right of the y-axis. The line is below the curve when $0 < x < 2$.

Case 2 corresponds to points to the left of the y-axis. The line is below the curve when $x < -3$.

Example 4

Solve the inequality $\dfrac{8}{x+3} \leq \dfrac{6}{x+1}$.

Solution

$$\frac{8}{x+3} \leq \frac{6}{x+1}$$

This inequality is not defined when $x = -3$ or when $x = -1$.

Hence, this solution is valid only when $x \neq -3$ or $x \neq -1$.

Case 1. Let $x < -3$.

This part of the solution is valid only when $x < -3$.

$$\frac{8}{x+3} \leq \frac{6}{x+1}$$
$$8(x+1) \leq 6(x+3)$$
$$8x + 8 \leq 6x + 18$$
$$2x \leq 10$$
$$x \leq 5$$

> Multiply each side by $(x+3)(x+1)$. Since $x < -3$, both $x + 3$ and $x + 1$ are negative. So the inequality sign is unchanged.

Since this part of the solution is valid only when $x < -3$, the solution in this case is $x < -3$.

Case 2. Let $x > -3$ and $x < -1$; that is, $-3 < x < -1$.

This part of the solution is valid only when $-3 < x < -1$.

$$\frac{8}{x+3} \leq \frac{6}{x+1}$$
$$8(x+1) \geq 6(x+3)$$
$$8x + 8 \geq 6x + 18$$
$$x \geq 5$$

> Multiply each side by $(x+3)(x+1)$. Since $-3 < x < -1$, then $x + 3$ is positive and $x + 1$ is negative. Reverse the inequality sign.

Since this part of the solution is valid only when $-3 < x < -1$, there is no solution in this case.

Case 3. Let $x > -1$.

This part of the solution is valid only when $x > -1$.

$$\frac{8}{x+3} \leq \frac{6}{x+1}$$
$$x \leq 5$$

> Multiply each side by $(x+3)(x+1)$. Since $x > -1$, both $x + 3$ and $x + 1$ are positive. So the inequality sign is unchanged. The solution from this line is the same as in *Case 1*.

Since this part of the solution is valid only when $x > -1$, the solution in this case is $-1 < x \leq 5$.

Combining the results of *Cases 1* to *3*, the solution of the given inequality is $x < -3$ or $-1 < x \le 5$.

1. In *Example 1b*, we could have multiplied each side by the lowest common denominator, $2(x - 3)$. Do this, then compare the result with the solution on page 267. How do the results compare? Explain.

2. In *Example 2*, the quadratic equation in the solution was solved by factoring. How would you continue the solution if the equation cannot be solved by factoring?

3. Look at *Example 2* and the graphs following its solution. What are the solutions of each inequality? Explain.

a) $\dfrac{2}{x-3} + \dfrac{3}{x} > 2$ **b)** $\dfrac{2}{x-3} + \dfrac{3}{x} \le 2$

4. Why must the inequality sign be reversed if you multiply or divide each side of an inequality by a negative number?

5. In *Example 3*, why can't you simply multiply each term by x to obtain $x^2 + x - 6 < 0$, then solve this inequality?

6. Look at *Example 3* and the graph following its solution. What are the solutions of each equation and inequality? Explain.

a) $x + 1 = \dfrac{6}{x}$ **b)** $x + 1 \ge \dfrac{6}{x}$

4.6 EXERCISES

A **1.** Solve each equation.

a) $\dfrac{10}{x} = -4$ **b)** $\dfrac{16}{x-1} = -4$ **c)** $\dfrac{12}{x-2} = -2$

d) $\dfrac{-5}{x-8} = -1$ **e)** $\dfrac{3x-4}{x+4} = 7$ **f)** $\dfrac{x^2-4}{x+2} = 3x + 2$

2. a) Solve each equation.

     i) $\dfrac{8}{x-1} = x - 3$ **ii)** $\dfrac{4}{x-2} = x + 1$ **iii)** $\dfrac{x+2}{x-2} = x - 4$

     iv) $\dfrac{x-1}{x+1} = x + 5$ **v)** $\dfrac{6-x}{x+4} = x + 6$ **vi)** $\dfrac{4-2x}{x+2} = x - 2$

b) Choose one equation from part a. Write to explain how you solved it.

3. Choose one equation from exercise 2. Use a graphing calculator to illustrate the solution. Sketch what you see on the screen.

B 4. Solve.

a) $\frac{4}{x} + \frac{3}{x+2} = 5$

b) $\frac{-2}{x+3} - \frac{5}{x} = 2$

c) $\frac{4}{x-2} + \frac{6}{x+1} = -7$

d) $\frac{8}{x+2} + \frac{7}{1-x} = -1$

e) $\frac{8}{x+3} - \frac{6}{x-1} = -3$

f) $\frac{5}{3x-1} - \frac{9}{6x-1} = 2$

5. Solve each inequality.

a) $\frac{12}{x} > -3$

b) $\frac{-8}{x} < -2$

c) $\frac{6}{x-9} \le 2$

d) $\frac{4}{2x+5} < -4$

e) $\frac{x+1}{x-5} \ge 4$

f) $\frac{x+4}{2x+3} \le 3$

6. a) Solve.

i) $3 + x > \frac{4}{x}$

ii) $\frac{5}{x} \ge 2x - 3$

iii) $-8 + \frac{3}{x} \ge 3x$

iv) $\frac{5}{2x} + 3x \le \frac{13}{2}$

v) $\frac{7}{x} + 2 < 5x$

vi) $2x - \frac{9}{2x} < -\frac{5}{2}$

b) Choose one inequality from part a. Write to explain how you solved it.

7. Solve.

a) $\frac{6}{x+3} > x + 8$

b) $\frac{-4}{3x-4} \ge x - 4$

c) $\frac{x-5}{x-9} \le x - 8$

d) $\frac{x-4}{2x+1} > x + 4$

e) $\frac{x+6}{x+1} < 6 - x$

f) $\frac{4x+3}{4x-1} \ge 2x + 4$

8. a) Sketch the graph of $f(x) = \frac{6}{x}$.

b) Determine the values of x for which $f(x) > -2$.

9. a) Sketch the graph of $g(x) = \frac{1}{x}$.

b) Determine the values of x for which $g(x) \le 1$.

10. Solve.

a) $\frac{x-1}{x+5} + \frac{x-3}{x+2} = x + 13$

b) $\frac{x-3}{x-4} + \frac{3x-5}{x-3} = x + 2$

c) $\frac{x^2-x-2}{x^2-4} = x + 5$

d) $\frac{x^2+2x-3}{x^2-x-12} = x - 5$

11. Solve.

a) $\frac{x^2+6}{3} - \frac{7}{2} = \frac{x+10}{2}$

b) $\frac{3}{x+2} = \frac{1}{4} - \frac{1}{x-4}$

c) $\frac{4}{x} + \frac{x}{4} = \frac{5}{2}$

d) $\frac{50}{x} - \frac{40}{x+10} = 1$

12. Solve.

a) $\frac{x}{x-3} = \frac{2x}{x+2}$

b) $\frac{3x}{x+1} - \frac{x}{x-1} = \frac{2x+3}{x+1}$

c) $\frac{3x}{x-2} + \frac{x}{x+2} = \frac{2x-1}{x+2}$

d) $\frac{5x+2}{x+3} = \frac{2x}{x+3} - \frac{x}{x-3}$

e) $\frac{2x-1}{x+5} = \frac{x+2}{x+3}$

f) $\frac{2x-3}{x+2} - \frac{x+2}{x-1} = \frac{3x}{x-1}$

13. Solve.

a) $\dfrac{x^2}{x^2-4} = \dfrac{2x}{x+2}$

b) $\dfrac{3x^2}{x^2-1} = \dfrac{x}{x+1} + \dfrac{x}{1-x}$

c) $\dfrac{3x}{x+2} + \dfrac{2x^2}{x^2+5x+6} = \dfrac{4}{x+3}$

d) $\dfrac{2x}{x-1} - \dfrac{4}{3-x} = \dfrac{5x^2-7}{x^2-4x+3}$

e) $\dfrac{2x^2}{x^2-9} - \dfrac{x}{3-x} = \dfrac{-5}{x+3}$

f) $\dfrac{3x+1}{x^2+x-2} = \dfrac{2x-3}{x^2-x-6} - \dfrac{5}{x^2-4x+3}$

C 14. The graph of $f(x) = \dfrac{10}{x^2+1}$ is shown in the graphing calculator display below. The table shows the coordinates of some points on the graph.

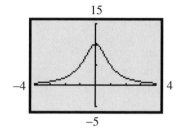

X	Y1
0	10
.5	8
1	5
1.5	3.0769
2	2
2.5	1.3793
3	1

X=0

Use the information in these displays. Determine the values of x for each inequality.

a) $f(x) > 0$ b) $f(x) > 2$ c) $f(x) > 5$ d) $f(x) > 8$

15. Solve each inequality. Show each solution on a number line.

a) $\dfrac{1}{x} < \dfrac{3}{x+4}$ b) $\dfrac{2}{x-3} < \dfrac{1}{2x}$

c) $\dfrac{1}{x-1} < \dfrac{2}{x-2}$ d) $\dfrac{x}{x-1} \geq \dfrac{2}{x-2}$

16. Solve each inequality.

a) $\dfrac{8}{x-3} + \dfrac{8}{x+1} > -3$ b) $\dfrac{12}{x-6} + \dfrac{6}{2x-1} \leq -1$

c) $\dfrac{8}{x+6} + \dfrac{16}{x-8} < x+3$ d) $\dfrac{4}{x+3} + \dfrac{4}{2x-4} \geq x-2$

COMMUNICATING THE IDEAS

Explain why we cannot solve a rational inequality algebraically by just multiplying each side by a common denominator and solving the polynomial inequality that results.

The Most Famous Problem in Mathematics

1. One of the hardest problems in mathematics originated with the great French mathematician, Pierre de Fermat (1601–1665). Fermat usually wrote his results, without the proofs, in the margins of his books. Other mathematicians had to re-create the proofs for themselves. The last of his results to be proved is known as Fermat's Last Theorem. It took more than 350 years to prove it.

Fermat's Last Theorem

If x, y, z, and n are natural numbers, the equation $x^n + y^n = z^n$ has no solution if $n > 2$.

2. Sophie Germain (1776–1831) was one of the most important mathematicians of her time to try to prove Fermat's Last Theorem. She established the following result, which gained her much recognition.

Sophie Germain's Theorem

If x, y, and z are integers, and if $x^5 + y^5 = z^5$, then either x, y, or z is divisible by 5.

Sophie later extended her theorem for exponents up to 100.

3. Many mathematicians spent years trying to prove Fermat's Last Theorem for all values of n. But they were only able to prove it for certain values of n.

Leonhard Euler (1707–1783) $n = 3$
Adrien Legendre (1752–1833) $n = 5$
Ernst Kummer (1810–1893) $n < 100$

4. By the 1970s, it was known that Fermat's Last Theorem is true for all values of n up to at least 125 000. But there are infinitely many values of n larger than this. Hence, no one knew if the theorem was true for all natural numbers n. What was needed was a completely general proof that was true for all values of n.

5. During the 20th century, mathematicians from all over the world developed new ideas that eventually contributed to the proof of Fermat's Last Theorem.

6. In 1993, Andrew Wiles presented a proof of Fermat's Last Theorem at a conference lecture. He had worked on it for seven years. The achievement was reported in the media around the world.

Wiles' proof was 200 pages long, and depended on results established by many mathematicians. Unfortunately, it was soon discovered that he had made an error in reasoning at one point in the proof.

7. It took Wiles two more years to fix the mistake. His corrected proof has been checked by scores of experts, and everyone now accepts that Fermat's Last Theorem has been proved conclusively.

This problem has been so difficult to solve that one wonders if Fermat actually had a proof of his famous theorem. Most mathematicians think not. But we cannot be certain.

Although proving Fermat's Last Theorem was extremely difficult, several related problems are much easier to solve.

1. Find examples of natural numbers that satisfy $x^2 + y^2 = z^2$.

2. A set of three natural numbers that represent the lengths of the sides of a right triangle is called a *Pythagorean triple*. Your answers to exercise 1 are examples of Pythagorean triples. Other Pythagorean triples may be found by substituting natural numbers for m and n $(m > n)$ in:
$$m^2 - n^2 \qquad 2mn \qquad m^2 + n^2$$

   a) Use these expressions to determine more Pythagorean triples.

   b) Verify that $(m^2 - n^2)^2 + (2mn)^2 = (m^2 + n^2)^2$.

3. Find examples of natural numbers that satisfy each equation.

   a) $x^2 + y^2 + z^2 = w^2$ b) $x^3 + y^3 + z^3 = w^3$

   c) $x^2 + y^2 = z^2 + w^2$ d) $x^3 + y^3 = z^3 + w^3$

4. Assume there are no natural numbers that satisfy the equation $x^3 + y^3 = z^3$. Explain why there cannot be any natural numbers that satisfy $x^6 + y^6 = z^6$.

5. Although n must be a natural number in Fermat's Last Theorem, suppose we investigate what happens if we use other values of n in the equation $x^n + y^n = z^n$. For example, when $n = -1$, the equation becomes $x^{-1} + y^{-1} = z^{-1}$, or $\frac{1}{x} + \frac{1}{y} = \frac{1}{z}$.

   a) Find examples of natural numbers x, y, and z that satisfy this equation.

   b) Find another value of n such that you can find examples of natural numbers x, y, and z that satisfy the equation $x^n + y^n = z^n$.

The design of a domed stadium has a roof that is part of a sphere. Suppose the diameter of the base of the stadium is 200 m, and the roof is 75 m above the centre of the playing field. What is the radius of the sphere?

This question can be answered by solving the following equation for r, where h is the height of the roof, and d is the diameter of the base.

$$\sqrt{4h(2r - h)} = d$$
$$\sqrt{300(2r - 75)} = 200$$

> Substitute 200 for d and 75 for h.

This equation is called a *radical equation* because the variable occurs under a radical sign. We can solve it by squaring each side to eliminate the radical.

$$\left(\sqrt{300(2r - 75)}\right)^2 = 200^2$$
$$300(2r - 75) = 40\,000$$
$$6r - 225 = 400$$
$$6r = 625$$
$$r \doteq 104.2$$

> Divide each side by 100.

The radius of the sphere is approximately 104 m.

The steps used to solve the above equation are used to solve other radical equations.

Example 1

Solve each equation, then check.

a) $\sqrt{x - 3} - 3 = 0$ b) $\sqrt{x - 3} + 3 = 0$

Solution

> Isolate the radical.

a) $\sqrt{x - 3} - 3 = 0$
$\sqrt{x - 3} = 3$
$\left(\sqrt{x - 3}\right)^2 = 3^2$
$x - 3 = 9$
$x = 12$

> Square each side.

> Isolate the radical.

b) $\sqrt{x - 3} + 3 = 0$
$\sqrt{x - 3} = -3$
$\left(\sqrt{x - 3}\right)^2 = (-3)^2$
$x - 3 = 9$
$x = 12$

> Square each side.

Check	Check
$\sqrt{x-3} - 3 = 0$	$\sqrt{x-3} + 3 = 0$
When $x = 12$,	When $x = 12$,
L.S. $= \sqrt{12-3} - 3$	L.S. $= \sqrt{12-3} + 3$
$= \sqrt{9} - 3$	$= \sqrt{9} + 3$
$= 0$	$= 6$
R.S. $= 0$	R.S. $= 0$
The solution is correct. That is, 12 is the only root of $\sqrt{x-3} - 3 = 0$.	The solution is not correct. That is, 12 is not a root of $\sqrt{x-3} + 3 = 0$. This equation has no real roots.

In *Example 1b*, we could have predicted that the equation has no real roots. Since the radical sign always denotes the positive square root, it is impossible for the left side of the equation, $\sqrt{x-3} + 3$, to be equal to 0. This example shows that the operation of squaring both sides of an equation may lead to numbers that do not satisfy the original equation. These numbers are called *extraneous roots*. They are roots of the equation that was obtained after squaring, but they are not roots of the original equation.

We can use a graphing calculator to explain the results in *Example 1*. Consider the line in the solution after the radical was isolated, $\sqrt{x-3} = 3$. The first screen below shows $y = \sqrt{x-3}$ and $y = 3$. The two graphs intersect when $x = 12$. The second screen shows $y = \sqrt{x-3}$ and $y = -3$, which do not intersect.

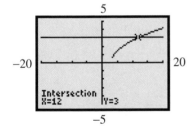

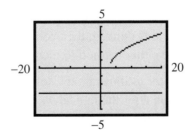

To solve a radical equation algebraically:

1. Isolate the radical on one side of the equation.

2. Square each side, then solve the equation that results.

3. Identify extraneous roots and reject them.

Example 2

Solve the equation $\sqrt{2x + 7} - x = -4$.

Solution

$$\sqrt{2x + 7} - x = -4$$

The radical is defined only when $2x + 7 > 0$, or $x > -3.5$.

Isolate the radical.

$$\sqrt{2x + 7} = x - 4$$

Since the left side of this equation is positive, $x - 4 \geq 0$. That is, $x \geq 4$.

From this point on, the solution is valid only when $x \geq 4$.

Square each side.

$$2x + 7 = (x - 4)^2$$
$$2x + 7 = x^2 - 8x + 16$$
$$x^2 - 10x + 9 = 0$$
$$(x - 9)(x - 1) = 0$$

Either $\quad x - 9 = 0 \qquad$ or $\qquad x - 1 = 0$
$$x = 9 \qquad\qquad\qquad x = 1$$

Reject the root $x = 1$ because it is less than 4.

The equation has only one root, $x = 9$.

In *Example 1* the extraneous roots were identified by checking the roots obtained after squaring. In *Example 2* they were identified by first noting the values of x for which the radical is defined, then comparing the roots obtained with these values. You can use either method to identify extraneous roots.

We can use a graphing calculator to illustrate the solution of *Example 2*. Consider the line in the solution after the radical was isolated, $\sqrt{2x + 7} = x - 4$. The first screen below shows the graphs of $y = \sqrt{2x + 7}$ and $y = x - 4$. There is only one point where the graphs intersect, and $x = 9$ at this point. This is the only root of the given equation.

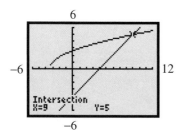

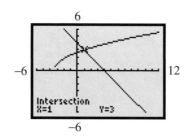

After squaring, the equation became $2x + 7 = (x - 4)^2$, which has two roots, $x = 9$ and $x = 1$. The equation before squaring could have been $\sqrt{2x + 7} = -x + 4$. The second screen on page 280 shows the graphs of $y = \sqrt{2x + 7}$ and $y = -x + 4$, which intersect when $x = 1$. We obtained two roots after squaring because the algebra cannot tell whether we squared each side of $\sqrt{2x + 7} = x - 4$ or whether we squared each side of $\sqrt{2x + 7} = -x + 4$.

An inequality in which the variable occurs under a radical sign is called a *radical inequality*. The operation of squaring each side is not valid for all inequalities. For example:

- Squaring each side of $3 > 2$ leads to $9 > 4$, which is correct.
- Squaring each side of $3 > -2$ leads to $9 > 4$, which is correct.
- Squaring each side of $3 > -5$ leads to $9 > 25$, which is not correct.

This means that we cannot formulate steps that will work for solving all radical inequalities. If we wish to square each side, we must check that this operation is valid. If both sides are positive, we can square each side.

Example 3

Solve for x. $\sqrt{3x + 1} < x - 9$

Solution

The radical is defined only when $3x + 1 \geq 0$, or $x \geq -\frac{1}{3}$.

The expression on the left side is positive, and it is less than the expression on the right side. Hence, the expression on the right side is positive. That is, $x - 9 > 0$, or $x > 9$. From this point on, the solution is valid only when $x > 9$. Since both sides of the inequality are positive, we can square each side.

$$\sqrt{3x + 1} < x - 9 \qquad \text{Square each side.}$$
$$3x + 1 < (x - 9)^2$$
$$3x + 1 < x^2 - 18x + 81$$
$$-x^2 + 21x - 80 < 0 \qquad \text{Multiply each side by } -1.$$
$$x^2 - 21x + 80 > 0$$
$$(x - 5)(x - 16) > 0$$

The left side is a quadratic function, $f(x) = (x - 5)(x - 16)$. Visualize the graph of this function. It opens up and has x-intercepts 5 and 16. Hence, the inequality is satisfied when $x < 5$ or $x > 16$. Since the solution is valid only when $x > 9$, we reject $x < 5$. Hence, the solution of the given inequality is $x > 16$.

We can use a graphing calculator to illustrate the solution of *Example 3*. The first screen below shows the graphs of $y = \sqrt{3x + 1}$ and $y = x - 9$. The graphs intersect when $x = 16$. The graph of $y = \sqrt{3x + 1}$ lies below the graph of $y = x - 9$ when $x > 16$.

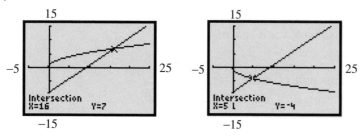

After squaring, the inequality became $3x + 1 < (x - 9)^2$, which led to the results $x < 5$ or $x > 16$. These results would occur if the inequality before squaring were $-\sqrt{3x + 1} < x - 9$. The second screen shows the graphs of $y = -\sqrt{3x + 1}$ and $y = x - 9$, which intersect when $x = 5$.

Example 4

Solve for x. $\sqrt{x + 1} > \dfrac{x}{x + 1}$

Solution

> ***Think ...***
>
> We cannot square each side because it would be difficult to determine when the inequality sign would have to be reversed. Instead, visualize how the values of the expressions are related for different values of x.

$$\sqrt{x + 1} > \frac{x}{x + 1}$$

The radical is defined only when $x \geq -1$. The rational expression is not defined when $x = -1$. Hence, we need only consider values of x greater than -1.

Case 1. Let $-1 < x \leq 0$.

> For these values of x, $\sqrt{x + 1}$ is positive and $\dfrac{x}{x + 1}$ is negative or zero.
>
> Hence, $\sqrt{x + 1} > \dfrac{x}{x + 1}$ for all values of x in this case.

Case 2. Let $x > 0$

For these values of x, $\sqrt{x+1}$ and $\frac{x}{x+1}$ are both positive.

However, $\sqrt{x+1}$ is greater than 1. But $\frac{x}{x+1}$ is less than 1 because the numerator is less than the denominator.

Hence, $\sqrt{x+1} > \frac{x}{x+1}$ for all values of x in this case.

The solution of the inequality is $x > -1$.

We can use a graphing calculator to illustrate the solution of *Example 4.* This screen shows the graphs of $y = \sqrt{x+1}$ (the upper part of a parabola) and $y = \frac{x}{x+1}$ (the hyperbola). *Case 1* corresponds to the points with x-coordinates between -1 and 0. The parabola is above the x-axis and the hyperbola is below it. *Case 2* corresponds to the points to the right of the y-axis. Visualize the line $y = 1$ on this screen. The parabola is above this line and the hyperbola is below it.

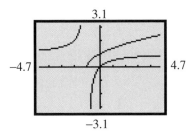

DISCUSSING THE IDEAS

1. Why is isolating the radical the first step in solving a radical equation or inequality?

2. What is the only difference in the steps for solving an inequality, compared with the steps for solving an equation?

3. Look at *Example 2* and the graphs following its solution. What is the solution of each inequality? Explain.

   a) $\sqrt{2x+7} < x - 4$ **b)** $\sqrt{2x+7} \geq x - 4$

4. Look at *Example 3* and the graphs following its solution. What are the solutions of each equation and inequality? Explain.

   a) $\sqrt{3x+1} = x - 9$ **b)** $\sqrt{3x+1} \geq x - 9$

5. Observe how factoring was used in the solutions of *Example 2* and *Example 3.*

   a) In similar examples, do you think you might encounter quadratic equations or inequalities that cannot be solved by factoring? Explain.

   b) How would you continue the solution if the equation or inequality cannot be solved by factoring?

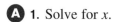

4.7 EXERCISES

A 1. Solve for x.
   a) $\sqrt{x} = 5$
   b) $2\sqrt{x} = 12$
   c) $2\sqrt{x} = 4\sqrt{3}$
   d) $-4\sqrt{x} = -1$
   e) $\sqrt{3x} + 1 = 7$
   f) $\sqrt{6x} - 1 = 5$

2. Solve for x.
   a) $\sqrt{x} > 4$
   b) $3\sqrt{x} \le 15$
   c) $2\sqrt{x} < 3\sqrt{3}$
   d) $-5\sqrt{x} \ge -3$
   e) $\sqrt{2x} + 2 > 4$
   f) $\sqrt{8x} - 3 \le 6$

3. Solve.
   a) $\sqrt{3x + 1} = 7$
   b) $\sqrt{2x + 7} = 5$
   c) $2\sqrt{x} = 8$
   d) $12\sqrt{x} = 30$
   e) $\sqrt{x} + 3 = 4$
   f) $\sqrt{x} - 6 = -3$

4. Solve.
   a) $\sqrt{2x + 4} < 5$
   b) $\sqrt{3x - 2} > 10$
   c) $3\sqrt{x} \ge 21$
   d) $28 > 7\sqrt{x}$
   e) $\sqrt{x} + 9 \le 10$
   f) $\sqrt{x} - 5 > -1$

B 5. a) Solve.
   i) $\sqrt{x - 2} - 5 = 0$
   ii) $\sqrt{5x + 2} - 3 = 1$
   iii) $\sqrt{2x + 1} + 5 = 8$
   iv) $\sqrt{3x - 1} + 7 = 10$
   v) $-2\sqrt{6x + 1} = 14$
   vi) $-3\sqrt{2x + 1} + 5 = -4$

   b) Choose one equation from part a. Write to explain how you solved it.

6. a) Solve.
   i) $\sqrt{x - 4} - 7 \ge 0$
   ii) $\sqrt{2x + 7} - 9 < 0$
   iii) $\sqrt{7x - 3} - 2 > 3$
   iv) $2 < 3\sqrt{2x - 5}$
   v) $5 + \sqrt{4x - 3} \le 9$
   vi) $-7 + 5\sqrt{2x + 3} \ge 8$

   b) Choose one inequality from part a. Write to explain how you solved it.

7. Choose one equation from exercise 3 and one inequality from exercise 4. Use a graphing calculator to illustrate the solutions. Sketch the graphs you see on the screen.

8. Determine, by inspection, which equations have extraneous roots.
   a) $\sqrt{x + 3} + 5 = 0$
   b) $\sqrt{3x + 2} - 2 = 3$
   c) $4 + \sqrt{2x - 7} = 0$
   d) $-4 + \sqrt{3x + 1} = 0$
   e) $7 + 5\sqrt{2x + 3} = 4$
   f) $3\sqrt{x + 1} + 2 = 8$

9. The formula for the length d of the diagonal of a rectangle with sides of length a and b is $d = \sqrt{a^2 + b^2}$. Solve the formula for a.

10. Point P lies on the x-axis. The coordinates of A and B are $(0, 6)$ and $(0, -15)$ respectively. The sum of the distances PA and PB is 27 units. Determine the possible coordinates of P.

11. Point P lies on the *y*-axis. The coordinates of R and S are $(-21, 0)$ and $(15, 0)$ respectively. The sum of the distances PR and PS is 54 units. Determine the coordinates of P.

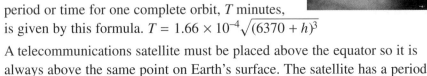

12. Point P lies on the line $y = x$. The coordinates of Q and T are $(-1, 8)$ and $(8, 1)$ respectively. The sum of the distances QP and TP is 18 units. Determine the coordinates of P.

13. When a satellite is *h* kilometres above Earth, the period or time for one complete orbit, *T* minutes, is given by this formula. $T = 1.66 \times 10^{-4}\sqrt{(6370 + h)^3}$

A telecommunications satellite must be placed above the equator so it is always above the same point on Earth's surface. The satellite has a period of 24 h.

a) What is the period of a telecommunications satellite, in minutes?

b) How high must the satellite be above the equator?

 MODELLING a Telecommunications Satellite

To an observer on Earth, a satellite that is always above the same point on the equator would appear to be stationary. This important property permits the use of dish antennas that are mounted permanently facing the satellite. In exercise 13, you modelled the period of a telecommunications satellite and determined its height.

- The figure 6370 in the formula for the period is Earth's radius in kilometres. Use this value and the height you calculated in exercise 13 to determine how far the satellite travels in 24 h.

- Calculate the satellite's speed, in kilometres per hour.

14. a) Solve.

   i) $\sqrt{2x + 4} = \sqrt{x + 7}$ ii) $\sqrt{14x - 3} = \sqrt{7 - 6x}$

   iii) $\sqrt{x + 3} = \sqrt{2x - 1}$ iv) $\sqrt{2x + 25} = \sqrt{x + 14}$

   v) $\sqrt{2x - 3} + \sqrt{x + 1} = 0$ vi) $\sqrt{9x + 10} - \sqrt{3x + 7} = 0$

b) Choose one equation from part a. Write to explain how you solved it.

15. Choose one equation from exercise 14. Use a graphing calculator to illustrate the solution. Sketch the graphs you see on the screen.

16. Solve.

a) $\sqrt{x} + \sqrt{x - 3} = 9$ b) $\sqrt{x - 2} + \sqrt{3x - 3} = 3$

c) $\sqrt{2x - 3} + \sqrt{2x + 1} = 5$ d) $\sqrt{3x + 10} - \sqrt{3x + 7} = 1$

17. In $\triangle ABC$, $\angle B = 90°$, and AB is 1 cm longer than BC. The perimeter of the triangle is 70 cm. Determine the lengths of the three sides.

C **18.** The graph of $f(x) = 5\sqrt{x}$ is shown below. The table shows the coordinates of some points on the graph.

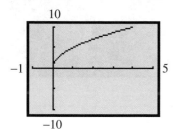

Use the information in these displays. Determine the values of x for which each inequality is true.

a) $f(x) > 0$ **b)** $f(x) > 5$ **c)** $f(x) > 10$

d) $f(x - 1) > 0$ **e)** $f(x - 1) - 4 > 0$ **f)** $f(x + 1) - 1 > 0$

19. Solve.

a) $\sqrt{x + 2} > \dfrac{x}{x + 2}$ **b)** $\sqrt{x + 2} > \dfrac{1}{x + 2}$ **c)** $\sqrt{x + 2} > x$

d) $\sqrt{x - 2} > \dfrac{1}{x - 2}$ **e)** $\sqrt{x + 2} + \dfrac{1}{x + 2} > 0$ **f)** $\sqrt{x + 2} + \dfrac{x}{x + 2} > 0$

20. Use a graphing calculator to illustrate your solutions to exercise 19.

21. a) Solve.

     i) $\sqrt{4x - 12} > \sqrt{2x + 12}$ **ii)** $\sqrt{8x + 5} < \sqrt{2x + 2}$

     iii) $\sqrt{3x} + \sqrt{x - 3} \geq 0$ **iv)** $\sqrt{-3x} - \sqrt{x + 4} \leq 0$

     v) $\sqrt{x - 1} \geq \sqrt{2x - 4}$ **vi)** $\sqrt{x + 12} \leq \sqrt{x - 4}$

b) Choose one inequality from part a. Write to explain how you solved it.

COMMUNICATING THE IDEAS

Write to explain why extraneous roots are sometimes introduced in the solution of a radical equation. Are extraneous roots introduced in the solutions of other kinds of equations? Explain.

On the number line, the numbers –4 and 4 are each located 4 units from 0.
Each number is said to have an absolute value of 4.
We write $|-4| = 4$ and $|4| = 4$.

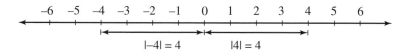

Example 1

Simplify.

a) $|12|$ **b)** $|-7|$

Solution

a) The absolute value of a positive number is the number itself.
$|12| = 12$

b) The absolute value of a negative number is the opposite number.
$|-7| = 7$

As *Example 1* indicates, the definition of the absolute value of a number
depends on whether the number is positive or negative.

Definition of Absolute Value

If a number is positive or zero, its
absolute value is the number itself.
If $x \geq 0$, then $|x| = x$

If a number is negative, its
absolute value is the opposite number.
If $x < 0$, then $|x| = -x$

An equation containing the variable inside an absolute-value sign is called an
absolute-value equation. To solve an absolute value equation, we use the above
definition. Consider the two parts of the definition separately.

Example 2

Solve for x. $|x - 2| = 3$

Solution

Case 1. Let $x - 2 \geq 0$
$$x \geq 2$$

This part of the solution is valid only when $x \geq 2$.

For these values of x, $|x - 2| = x - 2$

The given equation becomes: $x - 2 = 3$
$$x = 5$$

Case 2. Let $x - 2 < 0$
$$x < 2$$

This part of the solution is valid only when $x < 2$.

For these values of x, $|x - 2| = -(x - 2)$
$$= -x + 2$$

The given equation becomes: $-x + 2 = 3$
$$x = -1$$

Therefore, the given equation has two roots, $x = 5$ and $x = -1$.

We can use a graphing calculator to explain the solution in *Example 2*. In each screen below, the V-shaped graph is the graph of $y = |x - 2|$, and the horizontal line is the graph of $y = 3$. The graphs intersect at two points, where $x = 5$ and $x = -1$.

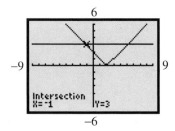

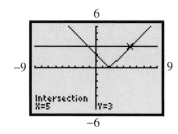

To solve an absolute-value equation algebraically:

1. Use the definition of absolute value. This means that two separate cases must be considered, corresponding to the two parts of the definition.

2. Identify extraneous roots and reject them. You can reject extraneous roots either by checking the roots obtained or by identifying the values of the variable for which the equation is defined.

In some equations, one or both of the roots obtained are not possible values of the variable. These roots must be rejected.

Example 3

Solve and check. $|x - 3| = 2x$

Solution

Case 1. Let $x - 3 \geq 0$

$\qquad x \geq 3$

This part of the solution is valid only when $x \geq 3$.

For these values of x, $|x - 3| = x - 3$

The given equation becomes: $x - 3 = 2x$

$\qquad\qquad -3 = x$

$\qquad\qquad x = -3$

This root is rejected because it is not among the possible values of x.

Case 2. Let $x - 3 < 0$

$\qquad x < 3$

This part of the solution is valid only when $x < 3$.

For these values of x, $|x - 3| = -(x - 3)$

$\qquad\qquad\qquad\qquad = -x + 3$

The given equation becomes: $-x + 3 = 2x$

$$3 = 3x$$
$$x = 1$$

Therefore, the given equation has only one root, $x = 1$.

Check. When $x = 1$, L.S. $= |x - 3|$ R.S. $= 2x$

$$= |1 - 3| \qquad\qquad = 2(1)$$
$$= |-2| \qquad\qquad = 2$$
$$= 2$$

Therefore, $x = 1$ is correct.

We can use a graphing calculator to explain the solution in *Example 3*. In the screen at the right, the V-shaped graph is the graph of $y = |x - 3|$, and the line is the graph of $y = 2x$. The graphs intersect at only one point, where $x = 1$.

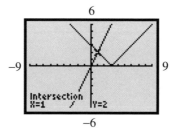

An inequality containing the variable inside an absolute-value sign is called an *absolute-value inequality*.

To solve an absolute-value inequality algebraically:

1. Follow the same steps as for solving an absolute-value equation.

2. When both sides are multiplied or divided by a negative number, the direction of the inequality sign must be reversed.

Example 4

Solve for x. $|2x - 5| \geq x + 1$

Solution

Case 1. Let $2x - 5 \geq 0$

$$x \geq 2.5$$

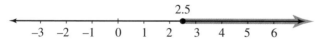

This part of the solution is valid only when $x \geq 2.5$.

For these values of x, $|2x - 5| = 2x - 5$

The given inequality becomes: $\quad 2x - 5 \geq x + 1$

$$x \geq 6$$

Since the only possible values of x are those shown on the number line above, the solution in this case is $x \geq 6$.

Case 2. Let $2x - 5 < 0$

$$x < 2.5$$

This part of the solution is valid only when $x < 2.5$.

For these values of x, $|2x - 5| = -(2x - 5)$

$$= -2x + 5$$

The given inequality becomes: $\quad -2x + 5 \geq x + 1$

$$-3x \geq -4 \quad \boxed{\text{Divide each side by } -3.}$$

$$x \leq \frac{4}{3}$$

Since the only possible values of x are those shown on the number line above, the solution in this case is $x \leq \frac{4}{3}$.

Combining the results of *Case 1* and *Case 2*, the solution of the given inequality is $x \geq 6$ or $x \leq \frac{4}{3}$.

We can use a graphing calculator to illustrate the solution of *Example 4*. This screen shows the graphs of $y = |2x - 5|$ and $y = x + 1$, which intersect when $x = \frac{4}{3}$ and when $x = 6$. The graph of $y = |2x - 5|$ is above the graph of $y = x + 1$ when $x > 6$ and when $x < \frac{4}{3}$.

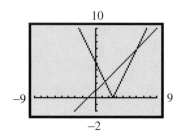

Example 5

Solve for x. $|x - 3| + |x + 2| \geq 7$

Solution

Case 1. Let $x - 3 \geq 0$

$$x \geq 3$$

This part of the solution is valid only when $x \geq 3$.

For these values of x, $|x - 3| = x - 3$ and $|x + 2| = x + 2$

The inequality becomes:

$$x - 3 + x + 2 \geq 7$$
$$2x \geq 8$$
$$x \geq 4$$

Since the only possible values of x are $x \geq 3$, the solution is $x \geq 4$.

Case 2. Let $x - 3 < 0$

$$x < 3$$

This part of the solution is valid only when $x < 3$.

For these values of x, $|x - 3| = -x + 3$

The value of $|x + 2|$ depends on whether $x + 2 \geq 0$ or $x + 2 < 0$.

Case 2a. Let $x + 2 \geq 0$

$$x \geq -2$$

This part of the solution is valid only when $-2 \leq x < 3$.

For these values of x, $|x + 2| = x + 2$ and (from above) $|x - 3| = -x + 3$

The inequality becomes:

$$-x + 3 + x + 2 \geq 7$$
$$5 \geq 7$$

This statement is not true. There is no solution for this case.

Case 2b. Let $x + 2 < 0$
$$x < -2$$

This part of the solution is valid only when $x < -2$.

For these values of x, $|x - 3| = -x + 3$
and $|x + 2| = -x - 2$

The inequality becomes:

$$-x + 3 - x - 2 \geq 7$$
$$-2x \geq 6$$
$$x \leq -3$$

Since the only possible values of x are $x < -2$, the solution is $x \leq -3$.

Combining the results of *Case 1* and *Case 2b*, the solution of the given inequality is $x \geq 4$ or $x \leq -3$.

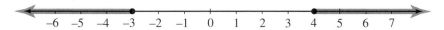

DISCUSSING THE IDEAS

1. a) In *Example 2*, could we have solved the equation by writing $x - 2 = \pm 3$ and solving $x - 2 = 3$ and $x - 2 = -3$ separately? Explain.

   b) Could we have used the same strategy to solve *Example 3*? Explain.

2. a) In *Example 3*, why do we not have to check the root $x = -3$ that was rejected in *Case 1*?

   b) Use the graph to explain what this root represents and why it occurred in the solution.

3. Look at *Example 3* and the graph following its solution. Use the graph to determine the solution of each inequality. Explain your reasoning.

   a) $|x - 3| < 2x$ b) $|x - 3| \geq 2x$

4. Look at *Example 4* and the graph following its solution. Use the graph to determine the solution of each equation and inequality. Explain your reasoning.

   a) $|2x - 5| = x + 1$ b) $|2x - 5| < x + 1$

5. Could the strategy of squaring each side be used to solve absolute-value equations and inequalities? Explain your answer, with examples.

A 1. Solve for x.

a) $|x| = 5$ b) $|x| = 0$ c) $|x - 2| = 7$

d) $|x - 4| = 2$ e) $|x + 1| = 5$ f) $0 = |x - 5|$

2. Solve for x.

a) $|x| < 3$ b) $|x| \geq 4$ c) $|x - 2| < 5$

d) $|x - 1| \leq 2$ e) $|x + 1| \geq 7$ f) $|x + 1| < 9$

B 3. Solve.

a) $|x - 5| = 2$ b) $|x - 2| = 5$ c) $|x + 1| = 2$

d) $|x + 2| = 1$ e) $|x - 3| = 9$ f) $4 = |3 - x|$

4. Solve and graph each solution on a number line.

a) $|x - 3| \leq 1$ b) $|x - 3| > 2$ c) $|x + 2| < 6$

d) $|x - 5| \geq 4$ e) $|x + 1| \geq 10$ f) $|x + 5| \leq 9$

5. a) Sketch the graph of $f(x) = |x| - 4$.

b) Determine the values of x for which $f(x) > 0$.

6. a) Sketch the graph of $g(x) = |x - 1| - 3$.

b) Determine the values of x for which $g(x) \leq 0$.

7. Solve each equation.

a) $|x + 1| = x - 1$ b) $2x = |x + 1|$ c) $|4 + 3x| = 7x$

d) $|2 - 5x| = 2 - 3x$ e) $x = |4x - 1|$ f) $|x + 4| = x - 1$

8. Solve each inequality.

a) $|x| > x - 2$ b) $|3x + 2| < 5x + 1$ c) $|6 - 3x| \leq x - 2$

d) $|3x - 2| \leq x + 1$ e) $|2 - 3x| < 3x - 4$ f) $3x + 1 > |5x + 2|$

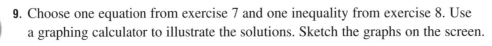

9. Choose one equation from exercise 7 and one inequality from exercise 8. Use a graphing calculator to illustrate the solutions. Sketch the graphs on the screen.

10. Solve each equation. Illustrate each solution with a graph that explains the results.

a) $|x - 3| = x - 2$ b) $|x - 3| = x - 4$ c) $|x - 3| = x - 3$

11. Solve.

a) $|2x + 1| \leq 0$ b) $|3x - 2| < 6$ c) $|4 + 2x| > 3$

d) $|5x + 2| > x$ e) $|\frac{1}{2}x + 1| \leq x + 2$ f) $3x > |2 - 3x|$

12. Three of these equations have no roots. Which equations are they? Explain how you can tell.

a) $|2x + 3| = -5$ b) $3x = |x - 2|$ c) $|x| = 5|x| - 8$

d) $|x - 2| + |2x + 6| = 0$ e) $x = |2x + 1|$ f) $|5x - 1| = |1 - 5x|$

13. Write an absolute-value inequality whose solution is represented by each graph.

a)

b)

c)

d)

14. Write an absolute-value inequality with each solution.

a) $x \geq 3$ b) $0 \leq x \leq 4$ c) $x < 0$ or $x > 4$ d) $x \neq 2$

 15. Solve each equation.

a) $|x - 3| + |x - 8| = 17$ b) $|x| + |x - 1| = 5$

c) $|2x - 1| - |1 - 2x| = 4$ d) $|x - 1| + |x - 3| = 6$

16. Solve each inequality.

a) $|x - 1| + |x - 3| \leq 2$ b) $|x + 2| + |2 - x| < 8$

c) $|x + 1| + |2x - 5| > 5$ d) $|x + 2| - |x| \geq 4$

17. Choose one equation from exercise 15 and one inequality from exercise 16. Use a graphing calculator to illustrate the solutions. Sketch the graphs on the screen.

18. Write an absolute-value inequality that has:

a) no solution   b) every real number as a solution   c) only one solution

19. Solve.

a) $|x + 2| \geq \dfrac{1}{x + 2}$ b) $|x + 2| \geq \dfrac{x}{x + 2}$

COMMUNICATING THE IDEAS

The inequality in *Example 3* has the form $|f(x)| = g(x)$, where $f(x)$ and $g(x)$ are linear functions of x. The graph following the solution shows that the graph of $y = |f(x)|$ is V-shaped and the graph of $y = g(x)$ is a straight line. Summarize the different ways in which a line can intersect a V-shaped graph, and explain how this affects the solutions of the equation $|f(x)| = g(x)$. Use examples to illustrate your explanations.

1. Solve each equation. Give the answers to 2 decimal places.

   a) $(2x + 1)(3x - 2) = (4x - 7)(x - 3)$ b) $(x - 8)(2x - 3) = 7x + 7$

   c) $8x + (2x - 5)(3x - 1) = (6x - 5)(2x - 1)$ d) $2x(3x - 7) = 7x - 1$

2. Select one equation from exercise 1. Write to explain how you solved it.

3. Determine the value of each discriminant.

   a) $2x^2 - 3x + 1 = 0$ b) $3x^2 + 10x - 4 = 0$

   c) $x^2 - 2x + 1 = 0$ d) $7x^2 - 10x + 9 = 0$

4. State which equations in exercise 3 have:

   a) two different real roots   b) two equal real roots   c) no real roots

5. Determine the remainder when each polynomial is divided by $x - 1$.

   a) $x^3 + x^2 - x - 1$ b) $2x^3 - 4x^2 - 7x + 8$ c) $4x^3 - 2x^2 + 7$

   d) $x^4 - 2x^3 + 3x^2 + 4x + 5$ e) $3x^3 - 2x^2 + 7x - 6$ f) $x^3 - 5x^2 + 4x - 8$

6. Choose one polynomial from exercise 5. Write to explain how you determined the remainder.

7. Solve by factoring.

   a) $2x^3 + 3x^2 - 11x - 6 = 0$ b) $x^3 + x^2 + 2x - 4 = 0$

   c) $x^3 - 5x^2 + x + 10 = 0$ d) $2x^3 - 3x^2 - 14x + 15 = 0$

8. a) Sketch a graph of $f(x) = (x - 3)^2 - 9$.

   b) Determine the values of x for which $f(x) < 0$.

9. Write a quadratic inequality that has each solution.

   a) $x < 2$ or $x > 4$ b) $-\frac{1}{2} \le x \le \frac{3}{2}$

   c) $x \le -2.2$ or $x \ge -1.7$ d) $3 < x < 7$

10. Solve each equation. Give each answer to 2 decimal places.

   a) $\dfrac{x}{x - 1} = \dfrac{3x}{x - 4}$ b) $\dfrac{2x}{x - 3} - \dfrac{x}{x + 3} = \dfrac{3x - 2}{x - 3}$

   c) $\dfrac{x + 1}{x - 5} + \dfrac{x - 2}{x} = \dfrac{3x - 1}{x - 5}$ d) $\dfrac{3x - 2}{x + 1} - \dfrac{x}{x - 1} = \dfrac{2x}{x + 1}$

11. Solve.

   a) $\sqrt{2x + 5} = 5$ b) $\sqrt{3x - 2} = 10$

   c) $\sqrt{x + 2} - 4 = 0$ d) $\sqrt{3x - 5} + 5 = 0$

12. Choose one equation from exercise 11. Write to explain how you solved it.

13. Determine, by inspection, which equations have extraneous roots.

    a) $\sqrt{x+1} + 2 = 0$ b) $\sqrt{3x-1} - 7 = 0$

    c) $6 + \sqrt{4x+7} = 0$ d) $-8 + \sqrt{3x-7} = 0$

14. Solve.

    a) $\sqrt{3x+1} = \sqrt{x-2}$ b) $\sqrt{8x-2} = \sqrt{3+7x}$

    c) $\sqrt{x-1} = \sqrt{5x+7}$ d) $\sqrt{x} + \sqrt{x-1} = 7$

15. The diagram shows a sector of a circle with radius r. When h is defined as on the diagram, the chord length c is given by this formula.

    $c = 2\sqrt{h(2r-h)}$

    Solve the formula for h.

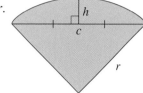

16. To check the results of exercise 15:

    a) Substitute the value of c that would result if the sector were a semicircle.

    b) Determine a condition that must be satisfied by r and c.

17. Solve for x, and graph each solution on a number line.

    a) $|x| > 1$ b) $|x| \leq 4$

    c) $|x| + 5 < 7$ d) $|x + 1| > 2$

18. Solve each inequality.

    a) $|x| \geq 2x + 1$ b) $|4x - 1| < 2x + 1$

    c) $3x - 2 \leq |2x + 1|$ d) $4x > |3x + 2| - 1$

19. Choose one inequality from exercise 18. Write to explain how you solved it.

20. Three of these equations have no roots. Which equations are they? Explain how you can tell.

    a) $|3x + 1| = -2$ b) $2x = |x - 1|$ c) $2|x| = 4|x| + 3$

    d) $2|x| = 4|x| - 3$ e) $|2x - 1| = |1 - 2x|$ f) $|x - 3| + |3x - 2| = 0$

21. Write an absolute-value inequality with each solution.

    a) $x \geq 4$ b) $-2 \leq x \leq 2$

    c) $x < -2$ or $x > 5$ d) $x \neq 5$

22. a) Solve each equation.

    i) $(x - 3)^2 = x - 3$ ii) $\dfrac{1}{x-3} = x - 3$

    iii) $\sqrt{x-3} = x - 3$ iv) $|x - 3| = x - 3$

    b) Compare the roots of the four equations in part a, listing the similarities and differences. Explain the similarities and the differences. Use graphs to illustrate your explanations.

1. The balance in a daily interest savings account on March 4 is $625.25. The annual interest rate is $\frac{3}{4}$%. A deposit of $100 is made on March 20. Determine the accumulated amount on March 31.

2. Determine the accumulated amount when $400 earns 2% annual interest for 1 year.

3. An investor deposits $10 000 at an interest rate of 10% compounded annually. In about how many years will the investment double?

4. For what interest rate, with annual compounding, will $775 accumulate to $920.46 in 5 years?

5. Determine the present value of an investment that yields $15 000 after 2 years at 6% interest compounded semi-annually.

6. Use the payroll deduction tables on pages 78 to 81. Determine the deduction for each income.

   a) CPP, monthly gross income $1170

   b) EI, monthly gross income $1096.05

   c) income tax deduction, claim code 5, biweekly taxable income $1616.11

7. Graph each function.

   a) $y = 2x^2 - 18$ b) $f(x) = 2x^2 - 4x + 7$ c) $y = 7 - x^2 + 2x$

8. Choose one function from exercise 7. Write to explain how you graphed it.

9. Write a quadratic function that has each pair of zeros.

   a) 2, −6 b) 1, 4 c) 3, $\frac{3}{2}$

10. Choose one function from exercise 9. Write to explain how you wrote the quadratic function.

11. School spirit T-shirts are sold for $20 each. Three hundred students are willing to buy them at that price. For every $2 increase in the price, there are 10 fewer students willing to buy the T-shirts.

    a) Represent the sales revenue as a function of the price. Sketch the function.

    b) What is the maximum revenue?

    c) What range of prices will give sales revenue that exceeds $3500?

12. Determine the inverse of each function.

    a) $f(x) = 9x^2$ b) $f(x) = 4 - x^2$ c) $f(x) = 1 - 4x^2$

13. Restrict the domain so that the inverse of each function is a function.

    a) $y = x^2 + 3$ b) $y = 2(x - 1)^2 + 4$ c) $y = 3x^2 + 7$

14. Is each function polynomial?

    a) $f(x) = 3x^4 - x$ **b)** $y = \dfrac{1}{x^2 + 1}$ **c)** $y = x$

    d) $y = 3x^3 + 2x^2 + 4x - \sqrt{x}$ **e)** $f(x) = 7x^3 - 8x + 4$ **f)** $y = \dfrac{5x^4 + 3x^2 - 2x}{x}$

15. Choose one function from exercise 14. Write to explain how you decided whether it was a polynomial function.

16. Sketch an example of each cubic function with the given zeros.

    a) $-2, 1, 3$ **b)** $-1, 0, 1$ **c)** $-4, -4, -4$

17. Determine the equation of each function, then sketch its graph.

    a) a cubic function with zeros $-3, 4, 4$; graph passes through $(5, 12)$

    b) a quartic function with zeros $-2, 0, 0, 1$; graph passes through $(-3, -12)$

18. Solve for x.

    a) $x^3 + x^2 - 5x + 3 = 0$ **b)** $x^3 - 4x^2 = 0$ **c)** $x^3 - 27 = 0$

19. Use a graph to approximate the zeros of each function.

    a) $g(x) = -x^3 + 2x^2 - 3x + 4$ **b)** $p(x) = x^4 - 2x^3 + 4x$

20. For each function, write the equation of the corresponding reciprocal function.

    a) $y = 2x + 1$ **b)** $y = 3x^2 - 5$ **c)** $y = 3^x - 2$

21. a) Use a graphing tool or grid paper. Graph each rational function.

       i) $y = \dfrac{2x}{x^2 - 9}$ **ii)** $y = \dfrac{x^2 - 9}{2x}$

    b) For each function in part a, determine the following information about it, where possible.

       i) its domain and range   **ii)** its zeros

       iii) the equations of any asymptotes   **iv)** a description of any symmetry

       v) the coordinates of any maximum or minimum points

22. Determine two functions whose composite function is $k(x)$.

    a) $k(x) = 4x^4 + 4x^2 + 1$ **b)** $k(x) = x + 2\sqrt{x} + 1$ **c)** $k(x) = \sqrt{2x + 1}$

23. A grappling iron is thrown vertically to catch on a ledge 7.5 m above the thrower. The height of the grappling iron, h metres, at a time, t seconds, after it is thrown is given by $h = -4.9t^2 + 11t + 1.5$. Will the grappling iron reach the ledge? Explain.

24. For what values of k does each equation have

       i) two different real roots?   **ii)** two equal real roots?   **iii)** no real roots?

    a) $kx^2 - 3x + 7 = 0$ **b)** $7x^2 + kx + 12 = 0$ **c)** $4x^2 - 32x + k - 5 = 0$

25. Solve each inequality.

    a) $(x - 2)(x + 3)(x - 1) \geq 0$ **b)** $(x + 1)(x - 4)(x - 2) < 0$

Will There Be Enough Food?

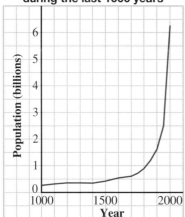

Estimated world population during the last 1000 years

(graph: Population (billions) on vertical axis from 0 to 6, Year on horizontal axis from 1000 to 2000, showing a curve rising sharply near 2000)

 CONSIDER THIS SITUATION

The world population has been increasing steadily for centuries, but it has never grown as rapidly as it has during the last 50 years. Many scientists are asking how many people Earth can support.

• Use the graph. About when did the world population reach each number?

    a) 1 billion   **b)** 2 billion   **c)** 3 billion

    d) 4 billion   **e)** 5 billion

• Do you think the world population will continue to grow indefinitely, or is there some maximum population that can be supported?

• Can the food supply grow at the same rate as the population, so that everyone can be fed?

On pages 328-329 and 360-363, you will use different graphical models to compare world population and food supply.

 FYI Visit www.awl.com/canada/school/connections

For information related to the above problem, click on <u>MATHLINKS</u> followed by *Mathematics 11*. Then select a topic under Will There Be Enough Food?

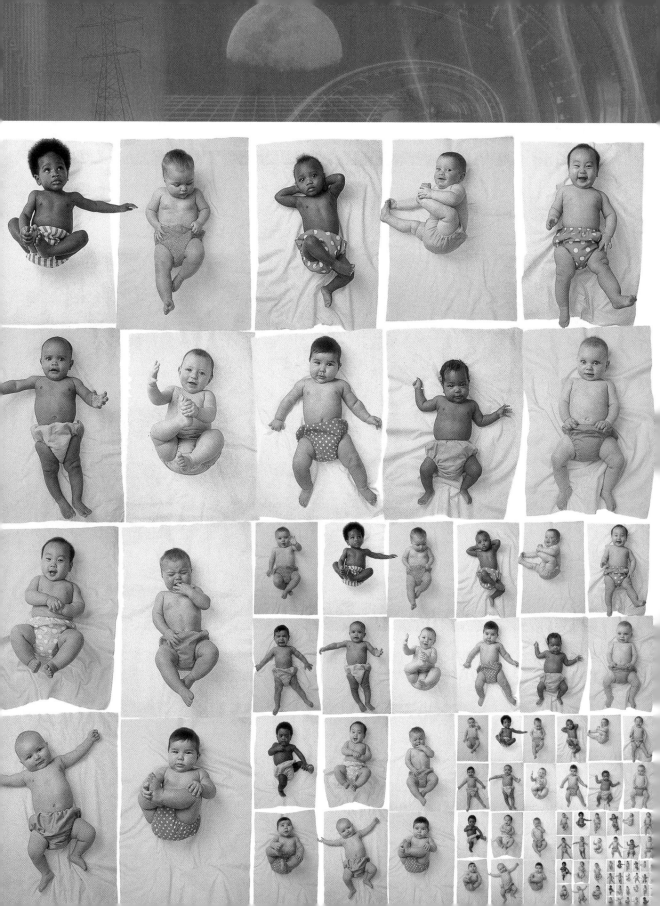

Sales personnel at a sporting goods store have a choice of two methods of remuneration.

Plan A

A straight 5% commission on all sales

Plan B

A monthly salary of $200 plus a 2% commission on all sales

Which is the better plan for an employee? This question can be answered using graphs. Let x dollars represent the total monthly sales.

Plan A

The monthly remuneration is

$y = 0.05x$

This is the equation of a linear function whose graph has slope 0.05 and passes through the origin.

Plan B

The monthly remuneration is

$y = 200 + 0.02x$

This is the equation of a linear function whose graph has slope 0.02 and vertical intercept 200.

To draw the graph, use two points whose coordinates satisfy the equation, such as (0, 0) and (10 000, 500).

To draw the graph, use two points whose coordinates satisfy the equation, such as (0, 200) and (10 000, 400).

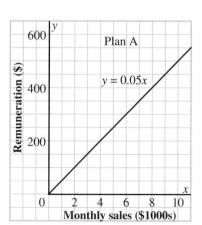

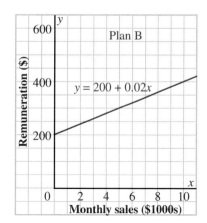

When the two plans are graphed on the same grid, the two lines intersect. The point of intersection indicates that Plan B is the better plan if sales are less than approximately $7000. If sales are greater than $7000, then Plan A is better.

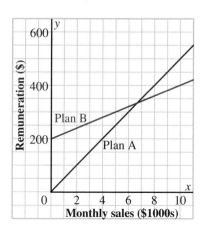

Recall that equations such as $y = 0.05x$ and $y = 200 + 0.02x$ are *linear equations*. A pair of linear equations, considered together, is called a *linear system*. To solve a linear system means to determine all the ordered pairs (x, y) that satisfy *both* equations. These ordered pairs may be determined by graphing both equations on the same grid. If the lines intersect, the coordinates of the point of intersection satisfy both equations. This gives the solution of the linear system.

Recall that the equation of a straight line may be expressed in the form $y = mx + b$, where m represents its slope and b represents its y-intercept.

Example 1

Solve this linear system graphically, then check the result.

$3x + y = 11$ ⓵

$x - 2y = 6$ ⓶

Solution

Solve each equation for y to express it in the form $y = mx + b$.

Equation ⓵ becomes $y = -3x + 11$, which has slope -3 and y-intercept 11.

Equation ⓶ becomes $y = \frac{1}{2}x - 3$, which has slope $\frac{1}{2}$ and y-intercept -3.

Graph each equation on the same grid. The only point common to both lines is the point of intersection, $(4, -1)$. The solution of the linear system is $(4, -1)$.

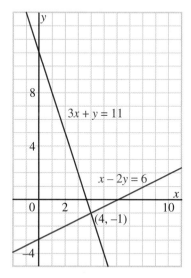

Check

Substitute $x = 4$ and $y = -1$ in *both* equations ⓵ and ⓶.

$3x + y = 11$ $x - 2y = 6$

L.S. $= 3x + y$ R.S. $= 11$ L.S. $= x - 2y$ R.S. $= 6$

$\quad = 3(4) + (-1)$ $\quad = 4 - 2(-1)$

$\quad = 12 - 1$ $\quad = 4 + 2$

$\quad = 11$ $\quad = 6$

The solution $(4, -1)$ is correct.

It is not always possible to obtain the exact solution of a linear system by graphing. If the lines do not intersect on the grid lines, it is necessary to estimate the solution. Exact solutions can only be found algebraically.

If one or both of the equations in a system are not linear equations, the system is a *non-linear* system. Some non-linear systems can be solved graphically.

Example 2

Solve this non-linear system graphically.

$y = x^2$ ①
$y = x + 2$ ②

Solution

Equation ① is a parabola with vertex $(0, 0)$ opening up, and passing through the points $(0, 0)$, $(\pm1, 1)$, $(\pm2, 4)$, $(\pm3, 9)$. Equation ② is a line with slope 1 and y-intercept 2.

Graph each equation on the same grid. The line intersects the parabola at the points $(-1, 1)$ and $(2, 4)$. Hence, the solution of the system is $(-1, 1)$ and $(2, 4)$.

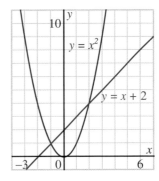

DISCUSSING THE IDEAS

1. In *Example 1*, the equations were graphed by solving for y then using the slope y-intercept form of the equation of a line. What other method could have been used to graph these equations?

2. a) What does the "solution of a linear system" mean?

   b) Why do the coordinates of the point where the two lines intersect correspond to the solution of a linear system?

3. Do you think it would be possible for two lines to appear to intersect on the grid lines when they actually do not? How would this affect a solution obtained graphically?

4. In *Example 2*, the line intersects the parabola at two points. What other possibilities are there for a line to intersect a parabola?

---

 1. Solve by graphing, then check.

  a) $x + y = 5$
    $3x + y = 3$

  b) $x - y = -2$
    $4x + 2y = 16$

  c) $x + y = 7$
    $3x + 4y = 24$

  d) $x - y = 2$
    $3x + y = -14$

  e) $x - y = 4$
    $2x + y = -4$

  f) $5x + 4y = 40$
    $5x + 6y = 50$

  g) $6x - 2y = -20$
    $4x + 2y = -10$

  h) $2x + 8y = 8$
    $-2x + y = 10$

B 2. Solve by graphing.

a) $x - 2y = 10$
$3x - y = 0$

b) $x - y = 1$
$3x + 2y = -12$

c) $5x - y = -23$
$4x + 6y = 2$

d) $5x - 2y = -5$
$3x + y = -14$

3. Choose one linear system from exercise 1 or 2. Write to explain how you solved it.

4. Solve by graphing.

a) $2x - y = 80$
$x + 3y = -30$

b) $3x + 2y = 60$
$3x - 5y = -150$

c) $x + y = -5$
$2x + y = 20$

d) $x + 2y = -6$
$3x + 2y = -34$

5. Solve by graphing.

a) $x + y = 4$
$3x - 2y = 2$

b) $x + 2y = 5$
$x - 2y = 13$

c) $2x - y = 8$
$x + 2y = -21$

d) $2x - 3y = 0$
$2x + 3y = 12$

e) $3x - 6y = 30$
$2x + 3y = 6$

f) $2x + 3y = 2$
$4x - 3y = 4$

g) $5x + 4y = 8$
$2x - 3y = -29$

h) $2x + 3y = 3$
$3x - 5y = 33$

6. For each pair of coordinates, determine a system of linear equations for which that point is the solution.

a) $(4, -6)$

b) $(5, 8)$

c) $(-3, 7)$

d) $(-2, -4)$

7. These graphs were produced by a graphing calculator. The equations of the systems are shown. Estimate the solution of the system on each graph. Then identify the equations that correspond to the graph.

i)
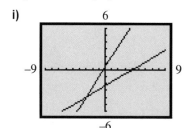

ii)
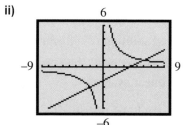

iii)
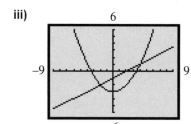

iv)
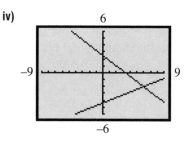

a) $3x + 4y = 10$
$2x - 5y = 22$

b) $3x - 2y = -1$
$4x - 7y = 16$

c) $y = 0.5x - 1$
$y = 0.25x^2 - 3$

d) $y = 0.5x - 2$
$y = \frac{6}{x}$

8. If two lines are drawn on the same grid, do they always intersect at only one point? What other possibilities are there for two lines? Explain.

9. Solve each system by graphing.

a) $y = x^2$
   $y = x$

b) $y = x^2 - 4$
   $y = -x + 6$

c) $y = x^2 - 2$
   $y = 6 - x^2$

d) $y = |x|$
   $y = \frac{1}{2}x + 3$

e) $y = \frac{6}{x}$
   $y = 2x + 4$

f) $y = \sqrt{x}$
   $y = x - 2$

10. Choose one non-linear system from exercise 9. Write to explain how you solved it.

C 11. a) Solve this system by graphing.

$$x + 2y = 8$$
$$3x - y = 3$$

b) Form a new equation by adding the two equations in part a. Graph this equation on the same grid. What do you notice?

c) Form another equation by subtracting the two equations in part a. Graph this equation on the same grid. What do you notice?

12. Solve by inspection.

a) $x + y = 6$
   $2x + y = 8$

b) $x - y = 1$
   $5x + 2y = 5$

c) $x + y = -8$
   $2x + y = -11$

d) $2x + y = 13$
   $x + 2y = 7$

13. A system of two equations is formed as follows. One equation is $y = mx + b$. The other equation is given below. For each equation, how many different intersection possibilities are there? Illustrate your answer with sketches.

a) $y = |x|$

b) $y = \frac{6}{x}$

c) $y = \sqrt{x}$

COMMUNICATING THE IDEAS

Your friend has missed today's mathematics lesson and he asks you about it. How would you explain to him what a system of equations is and how he can solve it graphically? Include any advantages or disadvantages of this method to solve a system that you think are important to mention. Write your ideas in your notebook.

Solving Systems of Equations

On page 302, two plans for remuneration of sales staff were given. Here are two other plans.

Plan A

A monthly salary of $325.50 plus a 1.8% commission on all sales

$y = 325.50 + 0.018x$

Plan B

A monthly salary of $94.25 plus a 5.3% commission on all sales

$y = 94.25 + 0.053x$

The equations form a linear system whose graph is shown below. The screen images and keystrokes are from the TI-83 calculator. Other graphing calculators will produce similar results.

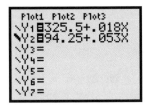

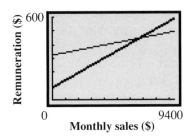

Displaying the graphs

- Enter the equations in the Y= list. Use the graph icons at the left of the screen to select a heavy line for the second equation. This helps to distinguish the graphs on the screen.

- Define the viewing window. Choose values so the graphs intersect on the screen. If you use 0 for Xmin and 9400 for Xmax, you will get simpler results when you trace to the point of intersection.

- Press GRAPH. The line for Plan B is heavier than the line for Plan A.

Solving the system by zooming in on the point of intersection

- Press TRACE, and use the arrow keys to move the flashing cursor close to the point of intersection.

- Press ZOOM 2 ENTER to zoom in. Press TRACE, and use the arrow keys again. The first screen on the next page shows the approximate coordinates of the intersection point.

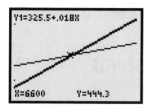 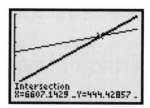

Solving the system by calculation

- To calculate the coordinates of the point of intersection, graph the functions as they were originally.

- Press [CALC] 5 to select "intersect." Since there could be more than two lines on the screen, you must indicate the ones to be used for solving. A flashing cursor will appear on one line. Press [ENTER] to select it. Press [ENTER] again to select the other line. Press [ENTER] a third time to begin the calculation. The screen (above right) shows the coordinates of the intersection point to several decimal places.

If you have successfully completed the above steps, you are ready to do these problems.

1. **a)** In the above example, which is the better plan?

   b) Suppose Plan B is changed to a monthly salary of $117.50 plus a 5.6% commission on all sales. Determine when each plan is the better one.

2. Use your graphing calculator to solve each system. Determine the solutions to at least two decimal places. Make sure you obtain the coordinates of all the points of intersection.

   a) $y = 3x - 2$
   $y = -0.5x + 4$

   b) $y = 0.35x + 3$
   $y = 0.65x + 1$

   c) $y = 0.5x^2 - 3$
   $y = 2x - 1$

   d) $y = x + 1$
   $y = \frac{1}{x} - 1$

   e) $y = 3x + 2$
   $y = 2^x$

   f) $y = x - 3$
   $y = \sqrt{x}$

3. What step is required before you can solve these systems with your calculator? Carry out that step, then determine the solutions.

   a) $2x + y = 7$
   $3x - 2y = 6$

   b) $x + y = 5$
   $x^2 - y = 3$

   c) $2x - y = 2$
   $2^x - y = 4$

4. Make up an example to show that a line and a curve could appear to intersect when they are graphed on the calculator screen, when they actually do not intersect. How could you convince someone that they do not intersect?

INVESTIGATE **Properties of Linear Systems**

Linear systems have two basic properties you should discover in this *Investigate*.

Multiplying an Equation by a Constant

1. a) Solve this linear system graphically.　$3x - y = 11$ ①
$x + 2y = 6$ ②

b) Write this system of equations. Solve it graphically on the same grid as part a.

Multiply each side of ① by 2.　　$6x - 2y = 22$ ③
Multiply each side of ② by 5.　　$5x + 10y = 30$ ④

c) Compare the results in parts a and b. What do you notice? Explain why this occurs.

d) Write another system of equations that has the same solution as these two systems.

2. Exercise 1 illustrates a property of linear systems. Describe this property.

Adding or Subtracting the Equations

3. a) Write another equation by adding equations ① and ②.
You should obtain: $4x + y = 17$ ⑤

b) Graph equation ⑤ on the same grid as exercise 1. What do you notice? Explain.

c) Write another equation by subtracting equation ② from equation ①.
You should obtain: $2x - 3y = 5$ ⑥

d) Graph equation ⑥ on the same grid as exercise 1. What do you notice? Explain.

4. Exercise 3 illustrates another property of linear systems. Describe this property.

The properties you discovered allow us to combine the equations of a linear system without changing the solution.

5. For each linear system, write another linear system that has the same solution.

a) $3x + 2y = 18$
$x - y = 1$

b) $2x + y = 9$
$x - 3y = 8$

c) $5x - y = 4$
$x + 4y = 5$

6. Choose one system in exercise 5. Try to find a way to use the properties you discovered to solve this system algebraically.

In *Investigate*, you should have discovered the following properties of linear systems.

> ### Properties of Linear Systems
> 1. Multiplying both sides of either equation of a linear system by a constant does not change the solution.
>
> 2. Adding or subtracting the equations of a linear system does not change the solution.

Solving linear systems by graphing is time-consuming and does not always give exact solutions. Hence, it is important to develop algebraic methods to solve them. There are different ways to do this, and they all involve eliminating one of the variables in some way. One method depends on the above properties of linear systems.

Example 1

Solve for x and y. Check the solution.

$$3x - 5y = -9 \qquad ①$$
$$4x + 5y = 23 \qquad ②$$

Solution

Since $-5y$ and $5y$ occur in the equations, eliminate y by adding the equations. According to the properties of linear systems, this will not change the solution.

$$3x - 5y = -9$$
$$4x + 5y = 23$$

Add. $\qquad 7x = 14$

$\qquad\qquad x = 2$

Since the value of x is 2, determine the value of y by substituting 2 for x in either equation ① or ②.

Use equation ②.

$$4(2) + 5y = 23$$
$$5y = 23 - 8$$
$$5y = 15$$
$$y = 3$$

The solution of this linear system is $(2, 3)$.

Check

Substitute $x = 2$ and $y = 3$ in both equations ① and ②.

$3x - 5y = -9$ $4x + 5y = 23$

L.S. = $3x - 5y$ R.S. = -9 L.S. = $4x + 5y$ R.S. = 23
  $= 3(2) - 5(3)$ $= 4(2) + 5(3)$
  $= 6 - 15$ $= 8 + 15$
  $= -9$ $= 23$

The solution $(2, 3)$ is correct.

Not all systems can be solved by adding the two equations. We frequently must multiply one, or both, of the equations by a constant before one variable can be eliminated this way. According to the properties of linear systems, this will not change the solution.

Example 2

Solve for x and y. $x - 2y = 7$ ①
                               $3x + 4y = 1$ ②

Solution

Multiply equation ① by 3, so the coefficients of x will be the same in both equations.

Multiply ① by 3. $3x - 6y = 21$
Copy ②. $3x + 4y = 1$
Subtract. $\overline{ - 10y = 20}$
                              $y = -2$

Substitute -2 for y in equation ①.

$x - 2(-2) = 7$
    $x + 4 = 7$
        $x = 3$

The solution of this linear system is $(3, -2)$.

In *Example 2*, after obtaining $y = -2$, we chose to substitute into equation ①. At this point, we have replaced equation ② with the equation $y = -2$. That is, we have replaced the given linear system with this system:

$x - 2y = 7$
    $y = -2$

After obtaining $x = 3$, we have the solution, which replaces the previous linear system with this system:

$x = 3$
$y = -2$

We can interpret the solution of *Example 2* graphically, as shown below. Systems of equations that have the same solution are called *equivalent systems*.

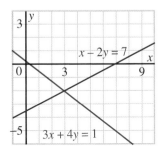

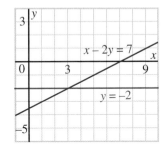

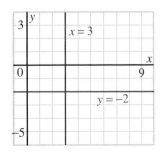

The given linear system:
$x - 2y = 7$
$3x + 4y = 1$

Subtracting the equations gives the equivalent system:
$x - 2y = 7$
$y = -2$

The final equivalent system:
$x = 3$
$y = -2$

Example 3

Solve for x and y. $2x + 3y = 8$ ①
                               $5x - 4y = -6$ ②

Solution

Multiply ① by 5. $10x + 15y = 40$
Multiply ② by 2. $\underline{10x - 8y = -12}$
Subtract. $23y = 52$
                           $y = \frac{52}{23}$

Multiply ① by 4. $8x + 12y = 32$
Multiply ② by 3. $\underline{15x - 12y = -18}$
Add. $23x = 14$
                          $x = \frac{14}{23}$

The solution is $\left(\frac{14}{23}, \frac{52}{23}\right)$.

We can use a linear system to determine the equation of a function when certain information about the function is known.

Example 4

The temperature of Earth's crust, T degrees Celsius, is a linear function of the depth, d metres, below the surface.

$$T = md + b$$

In a mine shaft, the temperature is 23°C at a depth of 240 m. At 620 m, the temperature is 31°C. Determine the values of m and b, then write T as a function of d.

Solution

Use the given equation, $T = md + b$.

Substitute 240 for d and 23 for T. $23 = 240m + b$ ①

Substitute 620 for d and 31 for T. $\underline{31 = 620m + b}$ ②

Subtract. $-8 = -380m$

$$m = \frac{8}{380}$$
$$\doteq 0.021\ 05$$

Substitute 0.021 05 for m in equation ①.

$$23 \doteq 240(0.021\ 05) + b$$
$$b \doteq 23 - 5.052$$
$$\doteq 17.948$$

The temperature function is $T \doteq 0.021d + 17.95$.

MODELLING the Temperature of Earth's Crust

The linear function $T \doteq 0.021d + 17.95$ in *Example 4* models the temperature of Earth's crust at depths of d metres.

- What do the two numbers in the equation represent?
- What would the graph of temperature against depth look like?
- According to *The Guinness Book of Records,* the world's deepest mine is 3581 m deep in South Africa. The deepest penetration into Earth's surface is 15 000 m at a geological drilling in Russia. At these depths, what temperatures are predicted by the model?
- Suggest a reasonable domain for this function. Justify your choice.

1. In *Example 1*, why is it necessary to check the solution in both original equations?

2. What other ways can you think of to solve *Example 2*?

3. In *Example 3*, after we found that $y = \frac{52}{23}$, why didn't we substitute this value in one of the given equations to determine x? What would have happened if we had done this?

4. In *Example 4*, the given equation has four variables. How do you know which ones are the variables in the linear system to be solved?

5. In *Example 1*, how do we know that $(2, 3)$ is the only solution of the system?

5.2 EXERCISES

A 1. For which linear systems is $(-1, 1)$ a solution?

   a) $5x + 6y = 1$ **b)** $3x + 4y = 1$ **c)** $3x - 4y = -6$ **d)** $2x + 3y = 1$
      $6x + 2y = -3$ $5x - 3y = -8$ $3x + 3y = 1$ $4x + 6y = 2$

2. Solve and check.

   a) $2x + 3y = 18$ **b)** $3x + 5y = 12$ **c)** $7x - 4y = 26$ **d)** $3x - 4y = 0$
      $2x - 3y = -6$ $7x + 5y = 8$ $3x + 4y = -6$ $5x - 4y = 8$

   e) $7x + 6y = 4$ **f)** $4x - 3y = 20$ **g)** $3x + 5y = 4$ **h)** $-5x + 2y = -1$
      $5x - 6y = -28$ $6x - 3y = 24$ $3x + 2y = 7$ $5x - 4y = -13$

3. Solve and check.

   a) $3x + y = 3$ **b)** $5x + 2y = 5$ **c)** $4x + 3y = 9$ **d)** $2x + 6y = 26$
      $2x + 3y = -5$ $3x - 4y = -23$ $2x - 7y = 13$ $5x - 3y = 11$

   e) $2x - 5y = -18$ **f)** $4x + y = -11$ **g)** $6x - 5y = -2$ **h)** $3x - 10y = 16$
      $8x - 13y = -58$ $3x - 5y = 9$ $2x + 3y = 18$ $4x + 2y = 6$

4. Choose one linear system from exercises 2 and 3. Write to explain how you solved it.

B 5. Solve.

   a) $8x - 3y = 38$ **b)** $3x + 4y = 29$ **c)** $6a - 5b = \frac{4}{3}$ **d)** $3s + 4t = 18$
      $4x - 5y = 26$ $2x - 5y = -19$ $10a + 3b = 6$ $2s - 3t = -5$

   e) $2m - 5n = 29$ **f)** $5x + 8y = -2$ **g)** $6x - 2y = 21$ **h)** $3c + 7d = 3$
      $7m - 3n = 0$ $4x + 6y = -2$ $3y + 4x = 1$ $4d - 5c = 42$

6. Solve.

a) $7x + 6y = 2$
$x + 8y = -4$

b) $8x - y = 16$
$2x - 3y = 2$

c) $5x = y$
$-x + 3y = 3$

d) $3x + 2y = 8$
$x - 12y = -10$

e) $3x - y = 5$
$2x + 3y = 10$

f) $4x + 3y = 3$
$3x - 2y = -19$

g) $\frac{x}{3} + \frac{y}{2} = \frac{1}{6}$
$x - 6y = 8$

h) $\frac{1}{2}x - \frac{2}{3}y = 6$
$\frac{1}{4}x + \frac{1}{3}y = -1$

7. At points on Earth's surface, the temperature of the atmosphere, T degrees Celsius, is usually a linear function of the altitude, h metres.

$$T = mh + b$$

a) The temperature is 12°C at an altitude of 2500 m and −50°C at an altitude of 12 500 m. Determine the values of m and b.

b) Write T as a function of h.

c) What do m and b represent?

d) Suggest a reasonable domain for the function. Write to explain how you determined the domain.

8. The annual cost of owning and operating a car, C dollars, is a linear function of the distance, d kilometres, it is driven.

$$C = md + b$$

a) The cost is $4600 for 10 000 km and $9100 for 25 000 km. Determine the values of m and b.

b) Write C as a function of d.

c) What do m and b represent?

d) Suggest a reasonable domain for the function.

9. In hockey, the number of points a player scores is equal to the sum of her or his goals and assists.

a) In the 1995-96 NHL season, Mario Lemieux scored 161 points. He scored 23 fewer goals than assists. Write a system of equations to represent this situation. Solve the system to determine his goals and assists that season. Check your answer using the *Hockey* database.

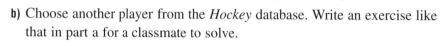

b) Choose another player from the *Hockey* database. Write an exercise like that in part a for a classmate to solve.

10. The number of baskets of apples, n, that can be produced in a small orchard is a quadratic function of the number of trees, t.

$$n = mt^2 + bt$$

a) Why is there no constant term in the quadratic function?

b) It is estimated that 60 trees will produce 1140 baskets and 80 trees will produce 1360 baskets. Substitute these values for t and n in the above equation to obtain a linear system in m and b.

c) Solve the system. Write n as a function of t.

11. The value of a wheat crop per hectare, v dollars, is a quadratic function of the number of days after planting, d.
$$v = md^2 + bd$$

a) Why is there no constant term in the quadratic function?

b) The crop is worth \$225/ha after 50 days and \$240/ha after 60 days. Determine the values of m and b.

12. Solve.

a) $x + 2y = 8$
$y + x = 5$

b) $2x - y = 19$
$3y - 5x = -46$

c) $2x + 3y = 12$
$4y - 3x = -1$

d) $2x + 3y = 32$
$2y + 22 = 3x$

e) $3y - 7x = x$
$3x - 1 = y$

f) $3x - 10 = 4y$
$6y + 2x = 11$

g) $y + 2x = 10 + 4y$
$4(x + y) = 42 - y$

h) $2(x - 2y) = 26 - 5y$
$3(y - x) = -2(y - 7)$

13. In the equation $2x + 5y = 8$, the coefficients form a pattern:

a) Write two equations that form the pattern above.

b) Solve the system in part a.

c) Write a linear equation with a different pattern in its coefficients. Repeat parts a and b for your equation. Write to explain what you notice.

14. The polynomial $p(x) = x^3 + bx^2 + cx - 4$ has a remainder of -8 when divided by $x + 1$ and a remainder of 4 when divided by $x - 1$. Determine the values of b and c.

15. When $2x^3 + bx^2 + cx + 10$ is divided by $x - 1$, the remainder is 15. When this polynomial is divided by $x + 2$, the remainder is 18. Determine the values of b and c.

16. The polynomial $p(x) = 4x^3 + bx^2 + cx + 11$ has a remainder of -7 when divided by $x + 2$ and a remainder of 14 when divided by $x - 1$. Determine the values of b and c.

17. Determine the coordinates of the vertices of the triangle formed by each set of three lines.

a) $2x - y - 7 = 0$
$3x + 4y - 16 = 0$
$x - 6y + 24 = 0$

b) $x - 2y + 10 = 0$
$3x + y - 19 = 0$
$2x + 3y - 8 = 0$

18. a) For each linear system, for what values of m and n is $(5, -3)$ the solution?

i) $mx - y = 23$
$nx + y = 12$

ii) $mx + ny = 12$
$mx - ny = 18$

iii) $mx + ny = -11$
$2mx - 3ny = 8$

iv) $3mx + 4ny = 30$
$2mx - 5ny = 20$

b) Choose one system from part a. Write to explain how you determined the values of m and n.

C **19. a)** Solve this linear system by graphing.

$$7x - 11y = -22$$
$$2x + 3y = 18$$

b) Solve the system in part a by addition or subtraction. Give the solution to two decimal places.

c) Compare the results of part a and part b. Explain any differences in the results.

20. Consider this linear system.

$$2x - 3y = 10$$
$$-4x + 6y = -20$$

a) Check that the ordered pairs $(2, -2)$, $(5, 0)$, and $(8, 2)$ are all solutions.

b) Find two more ordered pairs that are solutions.

c) Explain what would happen if you tried to solve the system graphically.

d) Find out what happens if you try to solve the system algebraically. Explain the result.

21. Consider this linear system.

$$2x - 3y = 10$$
$$-4x + 6y = -30$$

a) Find out what happens if you try to solve the system algebraically. Explain the result.

b) Explain what would happen if you tried to solve the system graphically.

c) Does the system have a solution? Explain.

COMMUNICATING THE IDEAS

In your notebook, list ways in which solving a linear system of two equations is different from solving a single equation. With a partner, decide what you think is the most important skill you used to solve a linear system of two equations.

5.3 Number of Solutions of a Linear System

In Sections 5.1 and 5.2, each linear system had one solution: the coordinates of the point of intersection of the corresponding lines. But not every linear system has one solution. We can see this by solving certain systems graphically.

Example 1

Solve by graphing.　$3x + y = 9$　①
$$6x + 2y = 6 \quad ②$$

Solution

Make a table of values for each equation:

For $3x + y = 9$　　　　For $6x + 2y = 6$

x	y
0	9
3	0

x	y
0	3
1	0

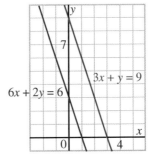

Graph each equation on the same grid. The lines are parallel. Therefore, there are no ordered pairs (x, y) that satisfy both equations. This system has no solution.

In *Example 1*, we can see from the equations that there is no solution. If each side of equation ① is multiplied by 2, the result is $6x + 2y = 18$. No ordered pair can satisfy this equation and also equation ②. We say the system is *inconsistent*.

Example 2

Solve by graphing.　$3x + y = 9$　①
$$6x + 2y = 18 \quad ③$$

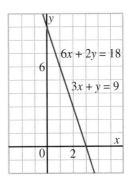

Solution

If each side of equation ① is multiplied by 2, the result is equation ③. Therefore, the graphs of the two equations coincide, and there is only one line. Any point on this line has coordinates that satisfy both equations. The coordinates of all points on the line are solutions of the system.

In *Example 2*, we can see from the equations that there are infinitely many solutions because the second equation is a multiple of the first. Since the system has solutions, we say that it is *consistent*. A system that has only one solution is also consistent.

When we solve a linear system of two equations in two variables, there are three possibilities.

Intersecting lines

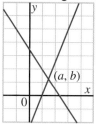

Parallel lines

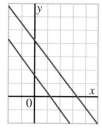

Coincident lines

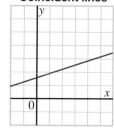

(a, b) is the only solution. The system is consistent.

There is no solution. The system is inconsistent.

There are infinitely many solutions. The system is consistent.

Example 3

Consider the equation $x - 2y = 8$. Write a second equation to form a linear system with:

a) infinitely many solutions   **b)** no solution   **c)** only one solution

Solution

a) Multiply each side of the given equation by any non-zero number, for example, 3.

$$3x - 6y = 24$$

This linear system has infinitely many solutions. $x - 2y = 8$
$$3x - 6y = 24$$

b) Multiply one side of the given equation by any non-zero number, for example, 2.

$$x - 2y = 16$$

This linear system has no solution. $x - 2y = 8$
$$x - 2y = 16$$

c) Write any equation in which the left side is not a multiple of the left side of the given equation, for example, $x + 3y = 12$.

This linear system has only one solution. $x - 2y = 8$
$$x + 3y = 12$$

In *Example 3c*, we can say that the linear system has only one solution, without knowing what the solution is.

DISCUSSING THE IDEAS

1. In the graph in *Example 1*, the lines appear to be parallel. How can we be certain that they are parallel?

2. In *Example 2*, suppose a table of values is made for each equation. Would you be able to tell from the tables of values that the lines are coincident? Explain.

3. In *Example 3a* and *b*, why did we say to multiply by any *non-zero* number?

4. In the solution of *Example 3b*, which side of the given equation was multiplied by 2? Could the other side have been multiplied by 2 instead? Explain.

5. Explain what it means to say that a system is:

   a) consistent   b) inconsistent

5.3 EXERCISES

A 1. Determine whether each linear system has infinitely many solutions, no solution, or one solution.

a) $x + 2y = 6$
$x + 2y = 2$

b) $3x + 5y = 9$
$6x + 10y = 18$

c) $2x - 5y = 30$
$4x - 10y = 15$

d) $3x + 2y = 12$
$6x + 4y = 24$

2. Six equations are listed. Use only equations from this list. Write two different linear systems that have:

   a) no solution

   b) only one solution

   c) infinitely many solutions

$$4x + 2y = 20$$
$$x - 3y = 12$$
$$5x - 15y = -60$$
$$2x + y = 10$$
$$6x + 3y = 5$$
$$2x - 6y = 24$$

B 3. Consider the equation $3x - 4y = 12$. Write a second equation to form a linear system that has:

   a) no solution   b) only one solution   c) infinitely many solutions

4. a) Determine if each system is consistent or inconsistent.

   i) $4x + y = 9$
   $3x + y = 7$

   ii) $x - 2y = -3$
   $2x - 4y = -6$

   iii) $3x - 4y = 12$
   $6x - 8y = 20$

   iv) $2x + 5y = 10$
   $6x + 15y = 30$

   b) Choose one system from part a. Write to explain how you determined whether it was consistent or inconsistent.

5. Four equations are listed. Use only equations from this list. Write two different linear systems that are:

   a) consistent   b) inconsistent

   $$4x + 2y = 20$$
   $$x - 3y = 12$$
   $$5x - 15y = -60$$
   $$2x + y = 20$$

6. For which of these systems is $(2, -3)$ a solution?

   a) $x - y = 5$
      $3x + 4y = -6$

   b) $2x + y = 7$
      $x - 3y = 10$

   c) $4x - y = 11$
      $-12x + 3y = -33$

   d) $5x - 3y = 19$
      $-2x + 4y = -16$

7. For which systems in exercise 6 is $(2, -3)$ the only solution? Explain.

C 8. Refer to the Properties of Linear Systems on page 311. Those properties were developed using linear systems that have only one solution. Determine whether the properties apply to a linear system that has:

   a) infinitely many solutions   b) no solution

9. Recall that the solution of a linear system is an ordered pair that satisfies both equations. When a system has infinitely many solutions, we can write the solution as a single ordered pair that satisfies both equations, as follows.

   a) Refer to *Example 2*. Check that these ordered pairs satisfy both equations:
      $(1, 6), (2, 3), (3, 0), (4, -3), (5, -6), ...$

   b) Suppose the list of ordered pairs in part a is continued. Write an expression for the nth ordered pair. Then check that this ordered pair satisfies both equations.

   c) In your ordered pair, what values of n are possible? Explain.

   d) One system in exercise 6 has infinitely many solutions. Express the solutions of this system as a single ordered pair. Verify that this ordered pair satisfies both equations.

10. Linear systems in two variables can have more than two equations.

    a) Make a list of all the different ways three lines can intersect.

    b) Determine if each system is consistent or inconsistent.

       i) $x + 2y = 5$
          $x - 7y = -1$
          $2x + y = 4$

       ii) $2x - 3y = 1$
           $3x + 4y = -3$
           $x - 10y = 5$

       iii) $3x + y = -2$
            $x - 2y = 7$
            $5x - 3y = 6$

COMMUNICATING THE IDEAS

Suppose a linear system of two equations in two variables is given. Write to explain how you can decide, just by looking at the equations:

a) whether the system is consistent or inconsistent

b) whether there are infinitely many solutions, or only one solution

5.4 Solving Systems by Substitution

In Section 5.2, a linear system was solved by adding or subtracting equations to eliminate one variable. Variables can also be eliminated by solving one equation for one variable, then substituting the result into the other equation.

Example 1

Solve. $3x + y = 3$ ①
           $7x - 2y = 20$ ②

Solution

Solve equation ① for y. $3x + y = 3$
$$y = 3 - 3x \quad ③$$

Substitute this expression for y in equation ②, then solve for x.

$$7x - 2(3 - 3x) = 20$$
$$7x - 6 + 6x = 20$$
$$13x = 26$$
$$x = 2$$

Substitute 2 for x in equation ③, then solve for y.

$$y = 3 - 3(2)$$
$$= -3$$

The solution of the linear system is $(2, -3)$.

The procedure in *Example 1* may be better seen graphically.

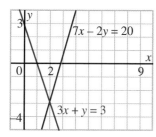

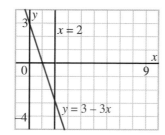

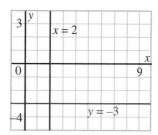

The given linear system:
$3x + y = 3$
$7x - 2y = 20$

Substituting 2 for x into $y = 3 - 3x$ gives the equivalent system:
$x = 2$
$y = 3 - 3x$

The final equivalent system:
$x = 2$
$y = -3$

A special case of the substitution method occurs when the given equations are already solved for y.

Example 2

Determine the coordinates of the points of intersection of the parabola defined by $y = x^2 - 2$ and the line defined by $y = 2x + 1$. Illustrate the solution on a graph.

Solution

This is the system of equations to be solved.

$$y = x^2 - 2 \qquad ①$$
$$y = 2x + 1 \qquad ②$$

Substituting the expression for y from one equation into the other equation results in a single equation formed from the two right sides.

$$x^2 - 2 = 2x + 1$$
$$x^2 - 2x - 3 = 0$$
$$(x - 3)(x + 1) = 0$$

Either $x - 3 = 0$ or $x + 1 = 0$

$\qquad\qquad x = 3 \qquad\qquad\qquad x = -1$

Substitute each value of x in equation ② to determine y.

When $x = 3, y = 2(3) + 1$ When $x = -1, y = 2(-1) + 1$

$\qquad\qquad\quad = 7 \qquad\qquad\qquad\qquad\qquad\qquad = -1$

The system has two solutions: $(3, 7)$ and $(-1, -1)$.

The graph of equation ① is a parabola with vertex $(0, -2)$ that is congruent to the graph of $y = x^2$ and opens up.

The graph of equation ② is a line with slope 2 and y-intercept 1.

The graph shows that the line intersects the parabola at $(3, 7)$ and $(-1, -1)$.

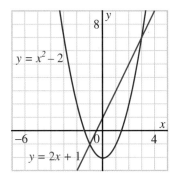

In *Example 1*, we solved the first equation for y, then substituted the result into the second equation.

a) Do you think we could have solved the second equation for y, then substituted the result into the first equation? Try it to see.

b) Do you think we could have solved one equation for x, then substituted the result into the other equation? Try it to see.

5.4 EXERCISES

A **1.** For which linear systems is $(-2, 5)$ a solution?

a) $3x + y = 1$
$2x + 3y = 11$

b) $5x - 3y = -5$
$3x + 2y = 4$

c) $-5x - 3y = -5$
$3x + 2y = 4$

d) $\frac{3}{2}x + \frac{2}{5}y = -1$
$\frac{5}{4}x - \frac{3}{10}y = -4$

2. Solve by substitution.

a) $x + y = 9$
$2x + y = 11$

b) $x + y = 1$
$3x - y = 11$

c) $x - y = 7$
$2x + y = -10$

d) $3x + y = 7$
$5x + 2y = 13$

e) $2x + 3y = 11$
$5x - y = -15$

f) $4x + y = -5$
$2x + 3y = 5$

g) $3x + 2y = 19$
$2x - 3y = -9$

h) $5y + 2x = -2$
$5x - 2y = 24$

3. Choose one linear system from exercise 2. Write to explain how you solved it.

B **4.** Solve by substitution, then check.

a) $3x - 4y = -15$
$5x + y = -2$

b) $2x + y = 2$
$3x - 2y = 10$

c) $3m - n = 5$
$5m - 2n = 8$

d) $4s - 3t = 9$
$2s - t = 5$

5. Solve.

a) $3x + 6y = 4$
$x - 2y = 1$

b) $7x + y = 13$
$3x - 2y = 8$

c) $4x + 6y = 1$
$x + y = 4$

d) $5x + 3y = 5$
$2x + y = 8$

6. In each system, one equation is the same, and the other equations are related. Solve each system. Use a graph to explain why the number of solutions is different in each case.

a) $y = x^2$
$y = x + 6$

b) $y = x^2$
$y = x - \frac{1}{4}$

c) $y = x^2$
$y = x - 2$

7. a) Solve the system formed by $y = x^2$ and each equation in this list.

$y = 2x - 1$
$y = 2x + 1$
$y = 2x + 3$
$y = 2x + 5$
$\vdots$

b) What patterns can you find in the results of part a? Predict the next two equations in the list, and the solutions of the system formed by $y = x^2$ and each of those equations.

c) Write to explain why the patterns occur.

8. Solve each system.

a) $y = x^2$
$y = 8 - x^2$

b) $y = x^2$
$y = -(x - 2)^2 + 2$

c) $y = x^2$
$y = \frac{1}{2}x + 5$

9. Solve each system.

a) $y = \frac{4}{x}$
$y = 3x - 1$

b) $y = \frac{4}{x}$
$y = 2x - 2$

c) $y = \frac{4}{x}$
$y = x - 3$

10. Solve each system.

a) $y = \sqrt{x}$
$y = x - 6$

b) $y = \sqrt{x}$
$y = x + \frac{1}{4}$

c) $y = \sqrt{x}$
$y = x + 2$

11. Solve each system.

a) $y = |x|$
$y = x + 2$

b) $y = |x|$
$y = x - 1$

c) $y = |x|$
$y = \frac{1}{2}x + 4$

12. Choose one system from exercises 8 to 11. Write to explain how you solved it.

13. Point P is on the line $3x + y = 26$, and is 10 units from the origin. Determine the coordinates of P, then illustrate your solution with a graph.

14. The perimeter of a rectangle is 13 cm, and its area is 10 cm^2. Determine its length and width.

15. The length of the diagonal of a rectangle is $\sqrt{20}$ cm, and its area is 10 cm^2. Determine the length and the width of the rectangle.

16. A skydiver jumped from an airplane and fell freely for several seconds before releasing her parachute. Her height, h metres, above the ground at any time is given by these equations.

$h = -4.9t^2 + 5000$ before she released her parachute
$h = -4t + 4000$ after she released her parachute

a) How long after jumping did the skydiver release her parachute?

b) How high was she above the ground at that time?

The equations in exercise 16 form a model for the height of the skydiver after she jumps from the airplane.

- Why does the model consist of two different functions?
- Why is the first function quadratic and the second function linear?
- What would the graph of height against time look like?
- Give some reasons why the actual heights may differ from those predicted by the model.

C 17. **a)** Solve each system.

    i) $y = 3x^2 + 4x + 2$ **ii)** $y = 5x^2 - 2x + 3$
        $y = 0.5x^2 + 4x + 2$ $y = -2x^2 - 2x + 3$

  b) Your results in part a should suggest that the parabolas defined by $y = a_1x^2 + bx + c$ and $y = a_2x^2 + bx + c$ have a special property. What is this property? Why does it occur? Illustrate your answers with diagrams.

18. Determine a value of b so that the line defined by $y = 2x + b$ intersects the parabola defined by $y = (x - 3)^2 + 2$ in only one point.

19. **a)** In each system below, one equation is the same, and the other equations are related. Solve each system. Use a graph to explain why the number of solutions is different in the two cases.

    i) $y = \dfrac{6}{x}$ **ii)** $y = \dfrac{6}{x}$
       $y = 1 - x^2$ $y = 10 - x^2$

  b) Is there a value of k so that the parabola $y = k - x^2$ does not intersect the graph of $y = \dfrac{6}{x}$? Explain algebraically and graphically.

  c) Is there a value of k so that the parabola $y = k - x^2$ intersects the graph of $y = \dfrac{6}{x}$ in exactly two points? Explain algebraically and graphically.

COMMUNICATING THE IDEAS

You can solve a linear system using either the method of addition and subtraction or the method of substitution. Is one method always easier than the other, or does it depend on the particular linear system you are solving? Write to explain your answer by referring to examples from the above exercises.

MATHEMATICAL MODELLING

Will There Be Enough Food? (Part I)

On page 300, we considered world population growth. To determine if there is enough food for each person, we model the growth in population and the growth of the population that can be fed by the food supply.

 **DEVELOP A MODEL**

One model of population growth is similar to compound interest. Like an investment earning interest at a fixed annual rate, this model assumes that the growth rate is constant. In 1995 the world population was about 5.73 billion and was growing at about 1.8% per year. Assuming that this growth rate continues, the population in future years is modelled by the equation below, where n is the number of years after 1995. Since the variable occurs in an exponent, this is called an *exponential* model.

$$P_1 = 5.73(1.018)^n \qquad ①$$

Grain is an important part of the human diet. In 1995, the world grain production was sufficient to feed about 7.16 billion people. Data from previous years indicated that the annual increase in grain production was sufficient to feed an additional 92 million people each year. Assuming that this growth pattern continues, the number of people who could be fed in future years, in billions, can be modelled by the equation below. Since the equation has the form $y = mx + b$, this is a *linear* model.

$$P_2 = 0.092n + 7.16 \qquad ②$$

1. Equations ① and ② model the projected world population, P_1, and the projected population that could be fed, P_2, for years after 1995. Predict what will happen if you graph these equations on the same grid.

2. Graph equations ① and ②. You may have to experiment with the viewing window to get a satisfactory result. Does the result agree with your prediction in exercise 1?

LOOK AT THE IMPLICATIONS

3. a) Use the intersection feature of your calculator. Determine the coordinates of the point of intersection of the graphs of P_1 and P_2.

  b) For what value(s) of n is each statement true?

  i) $P_1 = P_2$ **ii)** $P_1 < P_2$ **iii)** $P_1 > P_2$

4. What are the implications of your answers in exercise 3b?

5. What assumptions are implied by the models in equations ① and ②? Do you think these assumptions are reasonable? Explain.

REVISIT THE SITUATION

In 1798, Thomas R. Malthus published his *Essay on the Principle of Population as It Affects the Future Improvement of Society*. This essay describes his theory that the increase in world population was exceeding the development of food supplies. He reasoned that the population doubles about every 25 years, or increases exponentially. But the food supply increases only linearly at best. Thus, sooner or later, there could be widespread starvation.

Here is an excerpt from his essay.

> Taking the whole earth … and, supposing the present population to equal a thousand millions, the human species would increase as the numbers 1, 2, 4, 8, 16, 32, 64, 128, 256, and subsistence as 1, 2, 3, 4, 5, 6, 7, 8, 9. In two centuries the population would be to the means of subsistence as 256 to 9; in three centuries 4096 to 13; and in two thousand years the difference would be almost incalculable.

6. Describe Malthus' models to represent population growth and food supply.

7. According to the excerpt:

  a) what was the world population in 1798?

  b) what would the world population be today?

8. What changes, if any, would you make to Malthus' model for population growth? Explain.

9. Give some reasons to explain why the population increases exponentially but the food supply increases only linearly.

Solving a Two-Variable Linear System

You can use technology to solve a linear system in two variables. A systematic approach is required because the calculator or computer uses the same method every time. For example, consider the system below. The data are repeated at the right, without x and y, and without the $+$ and $=$ signs. These numbers will appear on the calculator or computer display.

$$3x + 2y = 19 \qquad\qquad 3 \quad 2 \quad 19$$
$$4x - 3y = 14 \qquad\qquad 4 \quad -3 \quad 14$$

Step 1 Replace with an equivalent system in which the first equation is the same and the second equation has x eliminated.

$$3x + 2y = 19 \qquad\qquad 3 \quad 2 \quad 19$$
$$0x + 17y = 34 \qquad\qquad 0 \quad 17 \quad 34$$

Step 2 Replace with an equivalent system in which the second equation is the same and the first equation has y eliminated.

$$-51x + 0y = -255 \qquad\qquad -51 \quad 0 \quad -255$$
$$0x + 17y = 34 \qquad\qquad 0 \quad 17 \quad 34$$

Step 3 Replace with an equivalent system by dividing the first equation by the coefficient of x and the second equation by the coefficient of y.

$$1x + 0y = 5 \qquad\qquad 1 \quad 0 \quad 5$$
$$0x + 1y = 2 \qquad\qquad 0 \quad 1 \quad 2$$

Look at the numbers beside each system above. Numbers in the first row represent the first equation of the system; those in the second row represent the second equation. Numbers in the first two columns represent the coefficients of x and y; those in the third column represent the constant term. The third column of the last system gives the solution of the original system: $(5, 2)$.

If you have a graphing calculator, complete exercises 1 and 2, and exercises 5 to 9.
If you have a computer, complete exercises 3 to 8, and exercise 10.

1. Ask your teacher for the program called LINSYST2 from the Teacher's Resource Book. Enter the program in your calculator. When you run the program, you are prompted to enter the coefficients for two equations. If you enter the numbers from the system above, the first screen on page 331 will appear. To continue, press ENTER, and the second screen will appear. The solution of the system appears in the column at the right in Step 3. Press ENTER again, and the third screen will appear.

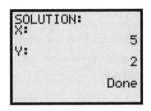

```
GIVEN SYSTEM:      ⋮
      [[3 2  19]
       [4 -3 14]]
STEP 1:
      [[3 2  19]
       [0 17 34]]
```

```
STEP 2:            ⋮
      [[-51 0  -255]
       [0   17 34  ]]
STEP 3:
       [[1 0 5]
        [0 1 2]]
```

```
SOLUTION:
X:
                    5
Y:
                    2
               Done
```

2. Test your program by using it to solve some systems of equations in the examples or exercises of Section 5.4. Do the results agree with the previous results?

3. **a)** Start a new spreadsheet document. Enter the text and formulas below. Pay particular attention to the dollar signs in some formulas. These are necessary to prevent certain cell addresses from changing when the formulas are copied to the right.

	A	B	C	D
1	Equations	3	2	19
2		4	-3	14
3				
4	Step 1	=B1		
5		=B1*$B2-B2*$B1		
6				
7	Step 2	=B5*$C4-B4*$C5		
8		=B5		
9				
10	Step 3	=B7/$B7		
11		=B8/$C8		

b) Copy the formulas in cells B4 to B11 to the right to column D.

c) Compare the numbers in your spreadsheet with those on page 330. They should be the same. The solution of the system appears in cells D10 and D11.

4. Test your spreadsheet by using it to solve some systems of equations in the examples or exercises of Section 5.4. Enter the coefficients for the equations in the appropriate cells in the first two rows. Do the results agree with the previous results?

Use technology to complete exercises 5 to 8.

5. Sue has a part-time job at *Snack Shack*. On Saturday, she sold 76 cones and 49 drinks, for a total revenue of $179.55. On Sunday, she sold 54 cones and 37 drinks, for a total revenue of $129.65. How much do one cone and one drink cost?

6. Glen works at *Music Melodies*. He has to order CDs and tapes from the manufacturer. In September, he ordered 425 CDs and 250 tapes, at a total cost of $1631.25. In October, he ordered 675 CDs and 360 tapes, at a total cost of $2522.25. What is the cost of one CD and one tape?

7. A bridge is designed with expansion joints to allow for thermal expansion. The exact length of a steel girder, L millimetres, is a linear function of the temperature, T degrees Celsius.

$$L = mT + b$$

a) At 5°C, a particular girder is 9982 mm long. At 34°C, it is 10 016 mm long. Determine the values of m and b for this girder.

b) What do m and b represent?

8. On a particular road surface, the stopping distance of a car, d metres, is a quadratic function of its speed, v kilometres per hour.

$$d = mv^2 + bv$$

A car travelling at 50 km/h takes 48 m to stop. A car travelling at 100 km/h takes 170 m to stop. Determine the values of m and b.

9. When you use the LINSYST2 program, you may obtain one of these screens.

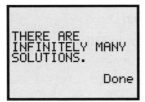  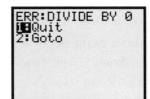

a) By experimenting or otherwise, find an example of a linear system that leads to each screen.

b) Explain why the results in part a occurred.

10. When you use the spreadsheet to solve linear systems in two variables, you may obtain division by zero errors in Step 3.

a) Find examples of linear systems that lead to division by zero errors.

b) Explain why these systems lead to division by zero errors.

LINKING IDEAS

Mathematics & Technology

5.5 Problems Involving Linear Systems

Many problems involve two unknown quantities. The quantities can often be determined if two equations relating them can be formed.

Example 1

A play-off football game drew 36 500 fans. Depending on seat location, the ticket prices were $35 and $20. The total revenue from the ticket sales was $940 000. How many $35 tickets and how many $20 tickets were sold?

Solution

Let x represent the number of $35 tickets sold.
Let y represent the number of $20 tickets sold.

The total number of tickets is 36 500. $x + y = 36\ 500$ ①
The total revenue was $940 000. $35x + 20y = 940\ 000$ ②

Multiply ① by 20. $20x + 20y = 730\ 000$
Copy ②. $35x + 20y = 940\ 000$
Subtract. $-15x \qquad\ \ = -210\ 000$
                     $x = 14\ 000$

Substitute 14 000 for x in ①. $14\ 000 + y = 36\ 500$
                                   $y = 22\ 500$

There were 14 000 tickets sold at $35 and 22 500 tickets sold at $20.

Example 2

A person invested $2000, part at 8% per annum and the rest at 10% per annum. After one year, the total interest earned was $190. How much was invested at each rate? Check the result.

Solution

Let x dollars represent the amount invested at 8%.

The interest earned on this amount in one year is $0.08x$ dollars.

Let y dollars represent the amount invested at 10%.

The interest earned on this amount in one year is $0.10y$ dollars.

Total money invested is $2000. $x + y = 2000$ ①
Total interest earned is $190. $0.08x + 0.10y = 190$ ②

Multiply ② by 100. $8x + 10y = 19\ 000$
Multiply ① by 8. $8x + \ 8y = 16\ 000$
Subtract. $2y = 3000$
  $y = 1500$
Substitute 1500 for y in ①. $x = 500$

$500 were invested at 8% and $1500 at 10%.

Check

Interest on $500 at 8%: $0.08 \times \$500 = \ \$40$
Interest on $1500 at 10%: $0.10 \times \$1500 = \underline{\$150}$
Total interest earned: $190

The solution is correct.

DISCUSSING THE IDEAS

1. How would you check the answers in *Example 1*?

2. Could you solve *Example 1* using only one variable? How would the solution compare with the solution given for *Example 1*?

3. Do you think it would be easy to solve *Example 2* using only one variable? Explain.

5.5 EXERCISES

A 1. Two shirts and one sweater cost $60. Three shirts and two sweaters cost $104. Determine the prices of one shirt and one sweater.

2. At a sale, all CDs are one price and all tapes are another. Three CDs and two tapes cost $72. One CD and three tapes cost $52. What are the prices of one CD and one tape?

B 3. A sports club charges an initiation fee and a monthly fee. At the end of 5 months, a member had paid a total of $170. At the end of 10 months, she had paid a total of $295. What is the initiation fee?

4. A tennis club charges an annual fee and an hourly fee for court time. One year, one member played for 39 h and paid $384. Another member played for 51 h and paid $456. Determine the annual fee and the hourly fee.

5. Three footballs and one soccer ball cost $155. Two footballs and three soccer balls cost $220. Determine the cost of one football and the cost of one soccer ball.

6. Tickets for a school play cost $4.00/adult and $2.50/student. Nine hundred tickets were sold for $2820. How many of each kind were sold?

7. For the athletic banquet, one adult ticket cost $15.00 and one student ticket cost $10.00. One hundred forty tickets were sold. The total receipts were $1600. How many student tickets were sold?

8. For the school play, one adult ticket cost $5.00 and one student ticket cost $3.00. Twice as many student tickets as adult tickets were sold. The total receipts were $1650. How many of each kind of ticket were sold?

9. A football stadium has 20 000 seats between the goal lines and 5000 in the end zones. An end-zone seat is $5 cheaper than one between the goal lines. For one game, the revenue when all seats were sold was $350 000. What is the cost of each type of seat?

10. When 20 bolts are placed in a box, the total mass is 340 g. When there are 48 bolts in the box, the total mass is 760 g. Determine the mass of the box and the mass of each bolt.

11. A crate of 36 grapefruit has a total mass of 4 kg. When 12 grapefruit are removed, the total mass is 3 kg. Determine the mass of the crate and the mass of one grapefruit.

12. Yasmin invested $2100 in the stock market. She purchased shares of World Oil at $7.50/share, and of Zinco Mines at $3.25/share. The total number of shares was 450. How many shares of each stock did Yasmin buy?

13. Jennifer invested $500, part at 7% per annum and the rest at 10% per annum. After one year, the total interest earned was $44. How much did Jennifer invest at each rate?

14. Vien invested $800, part at 9% per annum and the rest at 12% per annum. After one year, the total interest earned was $79.50. How much did Vien invest at each rate?

15. Mee Ha invested $2500, part at 8% per annum and the rest at 12% per annum. In one year, the two parts earned equal amounts of interest. How much did Mee Ha invest at each rate?

16. Southwest Airlines fleet includes B737-300s and B737-500s. Each B737-300 can carry 137 passengers. Each B737-500 can carry 122 passengers. Altogether, Southwest has 122 of these planes. The planes can carry a total of 16 339 passengers. How many of each type of plane does Southwest have? Check your answer using the *Aircraft* database.

17. Vito invested $500, part at 9% per annum and the rest at 11% per annum. After one year, the interest earned on the 9% investment was $20 less than the interest earned on the 11% investment. How much did Vito invest at each rate?

18. The cost of renting a car depends on the number of days for which it is rented and the distance it is driven. The cost for one day and 240 km is $39. The cost for three days and 800 km is $125. What is the cost per day and the cost per kilometre?

C 19. A school play ran for two nights, with audiences totalling 1390 adults and students. They paid $4285 for admission. One adult ticket cost $4.00 and one student ticket cost $2.50. The ratio of adults to students was 3 : 5 on the first night and 2 : 3 on the second night. How many students attended each night?

20. A patrol plane can carry fuel for 8 h flying time, and can fly at 300 km/h in still air. Its outbound patrol is against a 30 km/h head wind. It returns with a 30 km/h tailwind. How far can it fly against the head wind and return safely?

21. An aircraft travels 5432 km from Montreal to Paris in 7 h and returns in 8 h. The wind speed is constant. Determine the wind speed and the speed of the aircraft in still air.

22. Five kilograms of tea and 8 kg of coffee cost $58. The price of tea increases by 15% and that of coffee by 10%. The new total is $65.30. What are the new prices for 1 kg of tea and 1 kg of coffee?

23. A lifeguard earns an hourly rate for 20 h work in one week and an increased rate for overtime. One week, Theresa worked 24 h and received $166.40. The next week, she worked 27.5 h and received $200.00. Determine her hourly rate and her overtime rate of pay.

COMMUNICATING THE IDEAS

Some problems are solved using one variable and one equation. Others are solved using two variables and two equations. Choose one of the above problems. Write to explain how it can be easily solved using either one unknown or two unknowns. Choose another problem that would be harder to solve using one unknown than two unknowns. Explain why.

5.6 Solving Linear Systems in Three Variables

INVESTIGATE

You can use the same methods to solve three equations in three variables as you have been using to solve two equations in two variables.

1. Given these three equations, try to determine the values of x, y, and z. You can use the method of addition or subtraction, the method of substitution, or any algebraic manipulations that work.

$$2x + y - z = 1 \qquad ①$$
$$x + 3y + z = 10 \qquad ②$$
$$x + 2y - 2z = -1 \qquad ③$$

2. Check that your solution is correct.

3. Do you think you could use a similar method to solve any linear system similar to this one? Explain.

A linear system of two equations in two variables represents two lines in a plane. When the lines intersect at a single point, the coordinates of this point represent the solution of the system.

VISUALIZING

A linear equation in three variables, such as $x + 3y + z = 10$, represents a plane in space. In space, a point has three coordinates, (x, y, z). To visualize planes in space, look at the walls and the floor of a room. A system of three equations represents three planes. If all the planes intersect at a single point (like two adjacent walls and the floor), the coordinates of this point represent the solution of the system. The planes representing some linear systems intersect in other ways.

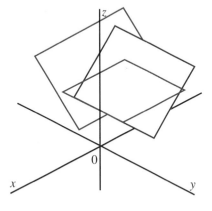

To solve a linear system of three equations in three variables, we use the same principles as for two equations in two variables.

- Multiply any equation by a constant without changing the solution.
- Add or subtract any two equations without changing the solution.

Example 1

Solve this system.

$$x + 4y + 3z = 5 \quad ①$$
$$x + 3y + 2z = 4 \quad ②$$
$$x + y - z = -1 \quad ③$$

Solution

> **Think ...**
>
> If we subtract equation ② from equation ①, we will obtain an equation in y and z.
> If we subtract equation ③ from equation ②, we will obtain another equation in y and z.
> The result will be a system of two equations in y and z, which we can solve.

Consider equations ① and ②.   Consider equations ② and ③.

$$\begin{array}{c} x + 4y + 3z = 5 \\ x + 3y + 2z = 4 \\ \hline y + z = 1 \end{array} \qquad \begin{array}{c} x + 3y + 2z = 4 \\ x + y - z = -1 \\ \hline 2y + 3z = 5 \end{array}$$

Subtract. (left)   Subtract. (right)

The given system has been reduced to a system of two equations in two variables, which can be solved as before.

$$y + z = 1 \quad ④$$
$$2y + 3z = 5 \quad ⑤$$

Multiply ④ by 2. $2y + 2z = 2$
Copy ⑤. $\underline{2y + 3z = 5}$
Subtract. $-z = -3$
                          $z = 3$

Substitute 3 for z in equation ④.

$$y + 3 = 1$$
$$y = -2$$

Substitute 3 for z and -2 for y in equation ③.

$$x - 2 - 3 = -1$$
$$x = 4$$

The solution of this linear system is $(4, -2, 3)$.

Although a three-variable linear system has three equations, we can combine only two equations at a time. Each time we combine two equations, we eliminate one variable. We must do this twice, eliminating the same variable each time.

Example 2

Solve this system.

$$3x - 3y - 2z = 14 \quad ①$$
$$5x + y - 6z = 10 \quad ②$$
$$x - 2y + 4z = 9 \quad ③$$

Solution

Think ...

We must combine pairs of equations in two different ways to eliminate the same variable. We could multiply ② by 3 and add to ① to obtain an equation in x and z. To get another equation in x and z, we could multiply ② by 2 and add to ③.

Copy ①.	$3x - 3y - 2z = 14$		② × 2	$10x + 2y - 12z = 20$
② × 3	$15x + 3y - 18z = 30$		Copy ③.	$x - 2y + 4z = 9$
Add.	$18x - 20z = 44$		Add.	$11x - 8z = 29$
Divide by 2.	$9x - 10z = 22$			

The given system has been reduced to this system in two variables.

$$9x - 10z = 22 \quad ④$$
$$11x - 8z = 29 \quad ⑤$$

④ × 4	$36x - 40z = 88$
⑤ × 5	$55x - 40z = 145$
Subtract.	$-19x = -57$
	$x = 3$

Substitute 3 for x in equation ④.

$$27 - 10z = 22$$
$$10z = 5$$
$$z = 0.5$$

Substitute 3 for x and 0.5 for z in equation ②.

$$15 + y - 3 = 10$$
$$y = -2$$

The solution of this linear system is $(3, -2, 0.5)$.

Example 3

The temperature, T, in an oven is a quadratic function of the number of minutes, n, since it was turned on. $T = an^2 + bn + c$

The temperature is 82°C after 1 min, 134°C after 2 min, and 174°C after 3 min. Determine the values of a, b, and c.

Solution

Use the given equation $T = an^2 + bn + c$.

Substitute 1 for n and 82 for T.	$82 = a + b + c$	①
Substitute 2 for n and 134 for T.	$134 = 4a + 2b + c$	②
Substitute 3 for n and 174 for T.	$174 = 9a + 3b + c$	③

Copy ③. $9a + 3b + c = 174$ Copy ②. $4a + 2b + c = 134$
Copy ②. $4a + 2b + c = 134$ Copy ①. $\underline{a + b + c = 82}$
Subtract. $5a + b = 40$ ④   Subtract. $3a + b = 52$ ⑤

Copy ④. $5a + b = 40$
Copy ⑤. $\underline{3a + b = 52}$
Subtract. $2a = -12$
                   $a = -6$

Substitute -6 for a in equation ④.

$$5(-6) + b = 40$$
$$b = 70$$

Substitute -6 for a and 70 for b in equation ①.

$$82 = -6 + 70 + c$$
$$c = 18$$

The temperature function is $T = -6n^2 + 70n + 18$.

MODELLING the Temperature of an Oven

The function $T = -6n^2 + 70n + 18$ in *Example 3* models the temperature of an oven.

- The graph of this function is a parabola opening down. What part of this parabola would be a reasonable one to model the increasing temperature in the oven?
- Determine the coordinates of the vertex of the parabola.
- Suggest a reasonable domain for this function. Justify your choice.
- Sketch a graph showing how the temperature of the oven depends on time.

1. a) In *Example 1*, after finding $z = 3$ we used equation ④ to determine y. What other equation could we have used?

   b) After finding $y = -2$, we used equation ③ to determine x. What other equations could we have used?

   c) What other ways could we have solved *Example 1*?

2. a) In *Example 2*, after the first addition we divided each side of the equation $18x - 20z = 44$ by 2 to obtain $9x - 10z = 22$. Is this necessary? Explain.

   b) What other ways could we have solved *Example 2*?

3. Why are three equations needed when there are three variables?

4. Why do you think systems of equations like the ones in the examples are called "linear," when the equations represent planes rather than lines?

5. Given a system of three linear equations in three variables representing planes in space, do you think every system has a unique solution? Explain.

5.6 EXERCISES

A 1. Which is the solution of this linear system? Explain.

$2x + y + z = -4$
$x - y + 3z = -19$
$3x + 2y + 2z = -7$

a) $x = 1$
   $y = 2$
   $z = -2$

b) $x = 3$
   $y = -1$
   $z = 0$

c) $x = -1$
   $y = 3$
   $z = -5$

2. Which linear systems have a solution of $(-1, 2, -5)$? Explain.

a) $x + y + z = -4$
   $x - y + 2z = -13$
   $2x + y - 3z = 15$

b) $3x + 2y - z = 6$
   $x + y + z = 5$
   $2x - 3y + 2z = -10$

c) $x + y + 2z = -8$
   $3x - y - z = 0$
   $2x + 2y - z = -2$

B 3. Solve.

a) $x + 3y + 4z = 19$
   $x + 2y + z = 12$
   $x + y + z = 8$

b) $x + 5y + 3z = 4$
   $2x + y + 4z = 1$
   $2x - y + 2z = 1$

c) $4x + 2y - 3z = 7$
   $x + 3y + z = 2$
   $x + 4y - 2z = -9$

4. Choose one system of equations from exercise 3. Write to explain how you solved it.

5. Solve and check.

a) $5x + y + 2z = 13$
  $x + y - z = 0$
  $2x - 3y + z = -1$

b) $2x + 8y + 3z = 13$
  $4x + 6y - 5z = 7$
  $x + 4y - 2z = 10$

c) $3x + y + 2z = 5$
  $2x - y + z = -1$
  $4x + 2y - z = -3$

6. Solve each system.

a) $3x - 2y = 8$
  $x + y - 3z = -11$
  $2x - y + z = 9$

b) $2x - y + 2z = 0$
  $5x + 2y - z = 3$
  $x + y + z = 3$

c) $a - b - c = -1$
  $a + c = 2$
  $a - 2b = -7$

7. Solve each system.

a) $p - r = 5$
  $2p - q + 2r = -3$
  $5p - 4q - 3r = 15$

b) $3d + e - 2f = -14$
  $2d - 3e + 4f = -23$
  $5d + 4e - 10f = -13$

c) $3a + b - 5c = -4$
  $6a - 2b + 5c = 9$
  $12a + 5b - 10c = -12$

8. Two sets of values of x, y, and z in exercise 1 were not solutions of the given linear system. For each set of values, determine a system for which the values are a solution.

9. The sum of three numbers is 15. The sum of the first, the second, and twice the third is 20. The sum of the first, twice the second, and three times the third is 27. What are the numbers?

10. The sum of two numbers is equal to a third number. The sum of twice the first, three times the second, and twice the third is 7. The third number is 3 more than the first. What are the numbers?

11. In a sale of art books, all books are sold for the same price. The total revenue, r dollars, from the sale of books is a quadratic function of the price, p dollars, of one book.

$r = ap^2 + bp + c$

a) The revenue is $6000 at a price of $30, $6000 at a price of $40, and $5000 at a price of $50. Determine the values of a, b, and c.

b) Write r as a function of p.

c) Suggest a reasonable domain for the function. Write to explain how you determined the domain.

12. If a hockey arena increases its ticket prices by x dollars, the predicted revenue, r dollars, is a quadratic function of x.

$$r = ax^2 + bx + c$$

a) The predicted revenue is $35 000 when the price increase is $1, $34 000 when the price increase is $2, and $32 000 when the price increase is $3. Determine the values of a, b, and c.

b) What is the predicted revenue when the price increase is 50¢?

c) Suggest a reasonable domain for this function.

13. a) Write a linear system in three variables whose solution is $x = -1, y = 1, z = 3$.

b) Is there more than one linear system with this solution? Explain.

14. This linear system has the solution $(a, b, 2)$.

$$2x - y - z = -3$$
$$x + 2y - 2z = -2$$
$$x + ky + z = 5$$

a) Determine the values of a, b, and k.

b) Is it possible to have more than one value of k? Explain.

15. In each example and exercise above, the linear system has a unique solution. The planes represented by the equations intersect at a single point, like the walls and floor in a corner of a room.

a) What are some other ways in which three planes can be situated in space?

b) Give an example of a system of three equations that has no solution. How might the planes be situated in this case?

c) Give an example of a system of three equations that has infinitely many solutions. How might the planes be situated in this case?

COMMUNICATING THE IDEAS

Write to describe the similarities and the differences between a linear system in two variables and its solution, and a linear system in three variables and its solution. Restrict your description to systems that have unique solutions.

LINKING IDEAS

Solving a Three-Variable Linear System

The method to solve a two-variable system on page 330 can be extended to a three-variable system. For example, consider the system below.

$$\begin{aligned}2x + 3y + z &= 4 \\ 5x + 6y + 3z &= 5 \\ 4x - y + 3z &= -1\end{aligned}$$

2	3	1	4
5	6	3	5
4	−1	3	−1

Step 1 Replace with an equivalent system in which the first equation is the same, and the second and third equations have x eliminated.

$$\begin{aligned}2x + 3y + z &= 4 \\ 0x + 3y - z &= 10 \\ 0x + 14y - 2z &= 18\end{aligned}$$

2	3	1	4
0	3	−1	10
0	14	−2	18

Step 2 Replace with an equivalent system in which the first and second equations are the same, and the third equation has y eliminated.

$$\begin{aligned}2x + 3y + z &= 4 \\ 0x + 3y - z &= 10 \\ 0x + 0y - 8z &= 86\end{aligned}$$

2	3	1	4
0	3	−1	10
0	0	−8	86

Step 3 Replace with an equivalent system in which the third equation is the same, and the first and second equations have z eliminated.

$$\begin{aligned}16x + 24y + 0z &= 118 \\ 0x + 24y + 0z &= -6 \\ 0x + 0y - 8z &= 86\end{aligned}$$

16	24	0	118
0	24	0	−6
0	0	−8	86

Step 4 Replace with an equivalent system in which the second and third equations are the same, and the first equation has y eliminated.

$$\begin{aligned}-384x + 0y + 0z &= -2976 \\ 0x + 24y + 0z &= -6 \\ 0x + 0y - 8z &= 86\end{aligned}$$

−384	0	0	−2976
0	24	0	−6
0	0	−8	86

Step 5 Replace with an equivalent system by dividing the first equation by the coefficient of x, the second by the coefficient of y, and the third by the coefficient of z.

$$\begin{aligned}1x + 0y + 0z &= 7.75 \\ 0x + 1y + 0z &= -0.25 \\ 0x + 0y + 1z &= -10.75\end{aligned}$$

1	0	0	7.75
0	1	0	−0.25
0	0	1	−10.75

The third column of the last system gives the solution of the original system: $(7.75, -0.25, -10.75)$.

Mathematics & Technology

If you have a graphing calculator, complete exercises 1 and 2, and exercises 5 to 9.

If you have a computer, complete exercises 3 to 8, and 10.

1. Ask your teacher for the program called LINSYST3 from the Teacher's Resource Book. Enter the program in your calculator. When you run the program, you are prompted to enter the coefficients for three equations. If you enter the numbers from the system on page 344, the first screen below will appear. To continue, press [ENTER] for each step. Compare the numbers appearing in each step with those on page 344. The solution of the system appears in the column at the right of the screen in Step 5, and also in the last screen.

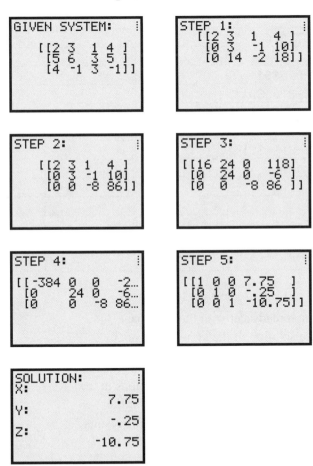

2. Test your program. Use it to solve some systems of equations in the examples or exercises of Section 5.6. Do the results agree with the previous results?

Mathematics & Technology

3. a) Start a new spreadsheet document. Enter the text and formulas below.

	A	B	C	D	E
1	Equations	2	3	1	4
2		5	6	3	5
3		4	-1	3	-1
4					
5	Step 1	=B1			
6		=B1*$B2-B2*$B1			
7		=B1*$B3-B3*$B1			
8					
9	Step 2	=B5			
10		=B6			
11		=B6*$C7-B7*$C6			
12					
13	Step 3	=B11*$D9-B9*$D11			
14		=B11*$D10-B10*$D11			
15		=B11			
16					
17	Step 4	=B14*$C13-B13*$C14			
18		=B14			
19		=B15			
20					
21	Step 5	=B17/$B17			
22		=B18/$C18			
23		=B19/$D19			

b) Copy the formulas in cells B5 to B23 to the right to column E.

c) Compare the numbers in your spreadsheet with those on page 344. They should be the same. The solution of the system appears in cells E21 to E23.

4. Test your spreadsheet. Use it to solve some systems of equations in the examples or exercises of Section 5.6. Enter the coefficients for the equations in the appropriate cells in the first three rows. Do the results agree with the previous results?

Use technology to complete exercises 5 to 8.

5. Tom has a part-time job at the ballpark. On Friday, he sold 12 posters, 18 pennants, and 7 caps, for a total of $168.65. On Saturday, he sold 37 posters, 29 pennants, and 18 caps, for a total of $380.15. On Sunday, he sold 22 posters, 19 pennants, and 9 caps, for a total of $224.15. How much do one poster, one pennant, and one cap cost?

6. Kathy works at *The Clothing Store*. She orders clothing from the manufacturer. In August, she ordered 54 shirts, 33 sweaters, and 25 coats, at a total cost of $3245.60. In September, she ordered 92 shirts, 56 sweaters, and 37 coats, at a total cost of $5255.35. In October, she ordered 77 shirts, 45 sweaters, and 28 coats, at a total cost of $4196.70. What is the cost of one shirt, one sweater, and one coat?

7. The number of accidents in one month, n, is a quadratic function of the age of the driver, x years. $n = ax^2 + bx + c$

   Eighteen-year old drivers had 2478 accidents. Thirty-five-year old drivers had 1875 accidents. Sixty-year old drivers had 2765 accidents. Determine the values of $a, b,$ and c.

8. The total number of oranges, N, in a square pyramid of oranges is a cubic function of the number of layers, x.

   $N = ax^3 + bx^2 + cx$

   a) How many oranges are there in a pyramid with each number of layers?

   i) one layer   ii) two layers   iii) three layers

   b) Use the results of part a. Determine the values of $a, b,$ and c.

9. When you use the LINSYST3 program, you will obtain one of the screens on page 332.

   a) By experimenting or otherwise, find an example of a linear system that leads to each screen.

   b) Explain why the results in part a occurred.

10. When you use the spreadsheet to solve linear systems in three variables, you may obtain division by zero errors in Step 5.

   a) Find some examples of linear systems that lead to division by zero errors.

   b) Explain why these systems lead to division by zero errors.

5.7 | Graphing Linear Inequalities in Two Variables

The graph shows the line defined by the equation $y = x$. The y-coordinate of every point on this line is equal to the x-coordinate.

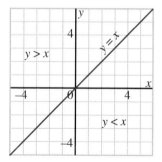

In the region *above* the line, the y-coordinate of every point is *greater* than the x-coordinate. This region is the graph of the inequality $y > x$.

In the region *below* the line, the y-coordinate of every point is *less* than the x-coordinate. This region is the graph of the inequality $y < x$.

In general, the graph of any linear equation is a straight line that divides the plane into two *half-planes*. The half-planes are the graphs of the corresponding inequalities.

To graph an inequality, follow these steps.

Step 1. Graph the corresponding equation.

Step 2. Determine the coordinates of any point that satisfies the inequality.

Step 3. Plot the point on the graph. The half-plane in which the point is located is the graph of the inequality.

Example 1

Graph the inequality $4x - 5y < 20$.

Solution

Step 1

The corresponding equation is $4x - 5y = 20$. Graph this equation using any method. For example, when $y = 0, x = 5$, and when $x = 0, y = -4$

The graph is a line with y-intercept -4 and x-intercept 5.

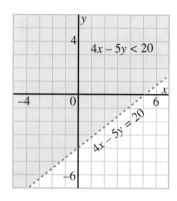

Step 2

A point that satisfies the inequality is (0, 0), since $4(0) - 5(0) = 0$, which is less than 20.

Step 3

The point (0, 0) is the origin. Since it lies in the region above the line, this region is the graph of the inequality $4x - 5y < 20$.

Example 2

A farmer uses up to 80 ha of land to plant two crops, corn and wheat. Draw a graph to show the area of each crop that could be planted.

Solution

Let *c* and *w* represent the area in hectares of corn and wheat, respectively, the farmer could plant. Since *up to* 80 ha of land can be planted, $c + w \leq 80$.

Step 1

The corresponding equation is $c + w = 80$. The graph of this equation is shown.

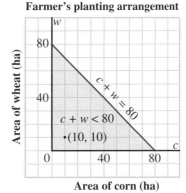

Farmer's planting arrangement

Area of corn (ha)

Steps 2 and 3

A point that satisfies the inequality is (10, 10). Since (10, 10) lies in the region below the line, this region and the line is the graph of the inequality $c + w \leq 80$.

MODELLING a Farmer's Planting Arrangement

The graph in *Example 2* is a model for the areas of two different crops that could be planted on 80 ha of land.

- Why does the model include points in the shaded region, and not just points on the line?
- Suppose you were the farmer. Give some reasons why you might decide to rule out some points in the shaded region when deciding how much of each crop to plant.
- The graph is symmetrical about the line defined by $w = c$. Explain why.

1. The inequality signs in *Examples 1* and *2* involve "less than." Why is the shaded region above the line in *Example 1* and below the line in *Example 2*?

2. In *Example 1*, the line defined by $4x - 5y = 20$ is a broken line but, in *Example 2,* the line defined by $c + w = 80$ is a solid line. Explain why one line is broken and the other is solid.

3. In *Example 2,* why are no points shown in the second, third, or fourth quadrants?

5.7 EXERCISES

A 1. State the coordinates of any point that satisfies each inequality.

a) $2x + y < 7$ b) $3x - 2y > 12$ c) $x - 4y \leq 8$

d) $5x + 3y > 9$ e) $2x + 3y \geq 15$ f) $3x + 4y \leq 18$

2. State the coordinates of any point on the graph that satisfies each inequality, and whether it lies in the region above or below the line.

e) $x - 2y > -4$ b) $x + 2y < 4$ c) $x - y < -3$

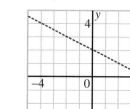

B 3. These graphs were produced by a graphing calculator. The equation of the line at the edge of the shaded region is shown. Use either > or < to write the equation of the inequality for each shaded region.

a) $x + 3y = -7$

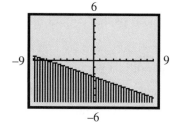

b) $7x - 2y = 10$

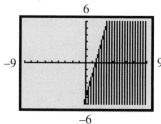

c) $3x + y = -4$

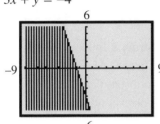

d) $x - 3y = -6$

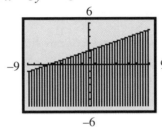

4. Graph each inequality.

a) $x + y < 5$ b) $x + y \geq 2$ c) $x - y \leq -3$

d) $x + 2y < 4$ e) $3x - 2y \geq -6$ f) $y \geq x + 8$

g) $y \leq -2x + 8$ h) $5x + 2y > -10$ i) $2x - 7y \geq 14$

5. Choose one inequality from exercise 4. Write to explain how you graphed it.

6. A company makes motorcycles and bicycles. In any given week, a total of up to 400 vehicles can be made. Draw a graph to show the number of motorcycles and bicycles that could be made in 1 week.

7. Teri plans to spend up to 12 h reviewing science and French in preparation for examinations. Draw a graph to show how much time she could spend studying each subject.

8. Graph each inequality.

a) $x - 2y \geq 4$ b) $3x - 2y \leq 6$ c) $4x - y < -4$

d) $2x + y > -4$ e) $y \geq 3x + 6$ f) $3x - 4y < 12$

g) $5x + 2y > 10$ h) $4x - 5y > 20$ i) $4x - 6y \geq 12$

C **9.** Fiona plans to start a physical fitness program that requires that she jog and do aerobics up to a maximum of 10 h each week. She must spend twice as much time jogging as she does doing aerobics. Draw a graph to show the amount of time she can give to each activity.

COMMUNICATING THE IDEAS

When you graph a linear inequality, you graph the corresponding equation, then shade on one side of the line. How can you tell which side of the line to shade? Use examples to illustrate your explanation. Write your ideas in your notebook.

5.8 Graphing Systems of Linear Inequalities

From Section 5.7, recall that the graph of any linear equation is a straight line that divides the plane into two half-planes. The half-planes are the graphs of the corresponding inequalities. A system of linear inequalities can consist of two or more inequalities, and the solution is the region where the corresponding half-planes overlap.

Example 1

A region is defined by these inequalities.

$$x \geq 0$$
$$y \geq 2$$
$$2x + y \leq 8$$

a) Graph this region.

b) Describe the shape of the region, then determine its area.

Solution

a) Graph each inequality on the same grid.

The graph of $x \geq 0$ is the points on the y-axis and in the half-plane to the right of it.

The line $y = 2$ is horizontal. The graph of $y \geq 2$ is this line together with the half-plane above it.

The equation $2x + y = 8$ can be written as $y = -2x + 8$. Its graph is a line with slope -2 and y-intercept 8. The graph of $2x + y \leq 8$ is this line together with the half-plane below it. This can be checked with a test point such as $(0, 0)$.

Shade the region covered by all three half-planes.

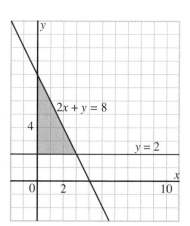

b) The region defined by the given inequalities is a right triangle with legs 3 units and 6 units.

$$\text{Area} = \tfrac{1}{2}(3)(6)$$
$$= 9$$

The area of the region is 9 square units.

Example 2

A sporting goods manufacturer makes footballs and soccer balls. The length of time each ball requires on the cutting and stitching machines is shown.

Type of ball	Time on cutting machine (min)	Time on stitching machine (min)
Football	2	1
Soccer ball	2	4

Draw a graph to show the number of each type of ball that could be made in one hour or less.

Solution

Let x represent the number of footballs made in one hour or less. Let y represent the number of soccer balls made in one hour or less.
x and y are subject to these restrictions.

- The number of balls cannot be negative.

  $x \geq 0$ ①
  $y \geq 0$ ②

- The total time on the cutting machine is less than or equal to 60 min.

  $2x + 2y \leq 60$, which simplifies to
  $x + y \leq 30$ ③

- The total time on the stitching machine is less than or equal to 60 min.

  $x + 4y \leq 60$ ④

The shaded region shows the ordered pairs (x, y) that satisfy all four inequalities. Since only whole numbers of balls can be made, only the ordered pairs with integral coordinates are solutions of the problem. Any of these ordered pairs represents a combination of the numbers of footballs and soccer balls that could be made in one hour.

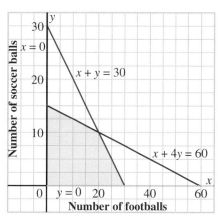

MODELLING the Manufacture of Footballs and Soccer Balls

The graph in *Example 2* is a model for the number of footballs and soccer balls that could be made in one hour.

• Why does the model include points in the shaded region, and not just points on the line?

• Explain why the graph is not symmetrical about the line defined by $y = x$.

• How would the graph change if there were:

   a) two cutting machines and two stitching machines?

   b) three cutting machines and two stitching machines?

DISCUSSING THE IDEAS

1. In *Example 1*, suppose the $\leq$ sign was replaced with $<$, and the two $\geq$ signs were replaced with $>$. What change, if any, would there be in:

   a) the graph?

   b) the area of the region defined by the inequalities?

2. In *Example 2*, suppose both the cutting machine and the stitching machine were in use for the entire hour. What part of the graph would show the numbers of footballs and soccer balls that could be made?

5.8 EXERCISES

 1. Write the system of inequalities that represents the shaded region on each graph.

a)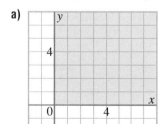

b)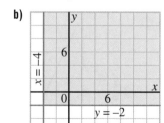

c)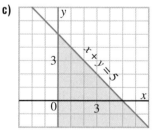

d)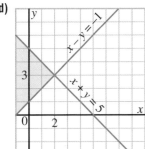

e)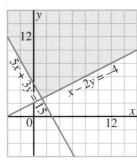

f)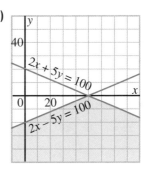

2. These graphs were produced by a graphing calculator. The equations of the lines at the edges of the shaded region are shown. Use either > or < to write the inequalities corresponding to each overlapping region.

a) $y = 2x + 3$
$y = 0.5x + 1$

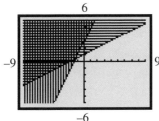

b) $3x + 2y = -5$
$2x - 3y = 3$

B 3. Graph the region defined by each system of inequalities. Describe the shape of the region, then determine its area.

a) $x \le 0$
$x \ge -3$
$y \le 0$
$y \ge -3$

b) $x \le 0$
$x \ge -4$
$y \ge 0$
$y \le 7$

c) $x \ge 0$
$x \le 4$
$y \ge -2$
$y \le 0$
$x - y \ge 2$

4. Determine the area of the region defined by each system.

a) $y \le -2x + 16$
$y \le x + 4$
$y \ge 0$

b) $y \ge -4x - 16$
$y \le 4$
$3y \ge 2x - 6$

c) $7y \le -2x + 49$
$x \le 7$
$2y \ge x - 9$
$x \ge -7$

5. Choose one region from exercise 4. Write to explain how you determined its area.

6. Write the system of inequalities that represents each shaded region.

a)

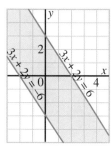

b)

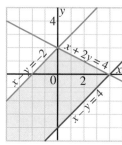

c)

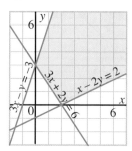

7. Graph each system.

a) $x + y \leq 6$
$x - 2y \geq 4$

b) $2x + y \leq 6$
$3x - 2y \leq 6$

c) $x + 3y < -3$
$4x - y < -4$

8. Graph each system.

a) $2x - y > 2$
$2x - y < 8$

b) $x + 3y \leq 6$
$x + 3y \geq -3$

c) $3x - 4y \geq -12$
$3x - 4y \leq 24$

d) $x - y \geq -2$
$y \leq 2$
$x + y \leq 6$

e) $5x + 2y > 10$
$x + 3y > 6$
$x - 2y < 4$

f) $x + y \leq 7$
$x + 3y \leq 12$
$x \geq 0$
$y \geq 0$

9. A company makes motorcycles and bicycles. The physical dimensions of the work area limit the number of both kinds that can be made in one day.

No more than 20 motorcycles can be made.
No more than 30 bicycles can be made.
No more than 40 vehicles in all can be made.

Draw a graph of the ordered pairs that show the numbers of motorcycles and bicycles that can be made in one day.

10. An 80-ha farm is to be planted with corn and wheat. Planting and harvesting costs, for which no more than $12 000 are available, are $300/ha for corn and $100/ha for wheat. Draw a graph to show the area of each crop that can be planted.

COMMUNICATING THE IDEAS

Your friend has missed today's mathematics lesson and she asks you about it. How would you explain to her how graphing a system of inequalities is similar to, yet different from, graphing a system of equations? Write your ideas in your notebook.

Solving Linear Systems

In the technology features on pages 330 and 344, you used a graphing calculator or a spreadsheet to solve linear systems. Other calculators and computer software have similar capabilities. The screens below were obtained using the TI-92 calculator, using the "rref" command. This command instructs the calculator to carry out the steps on page 330 (for a two-variable system) or page 344 (for a three-variable system).

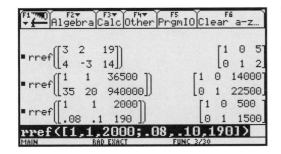
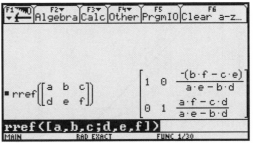

1. On the first screen, the calculator has solved the systems from pages 330, 333, and 334. Compare the results on the screen with the results on those pages. Where are the solutions of the systems shown on the screen?

2. In this chapter, no formula was given for solving a linear system. Since the TI-92 calculator can do algebraic computation, it can solve the system below. The result is shown on the second screen.

$$ax + by = c$$
$$dx + ey = f$$

   a) Use the result on the screen. Write a formula for the solution of this system.

   b) Compare the expressions in the solution with the coefficients in the equation. What patterns can you find? How do these patterns arise from the steps on page 330?

   c) Would using a formula be an efficient method to solve a two-variable linear system? Explain.

   d) According to the formula, what condition must be satisfied by the coefficients if the system is to have a unique solution? Why does this condition occur?

3. The TI-92 can solve a three-variable system similar to the one in exercise 2, using the steps on page 344. Refer to those steps and try to visualize how complicated the resulting formulas would be. Do you think formulas are a practical way to solve linear systems? Explain.

Mathematics & Technology

Maximum-Minimum Problems

Refer to *Example 2* on page 353. In that example, a sporting goods manufacturer makes footballs and soccer balls. The graph in the solution of that example is repeated here. The coloured region shows all the combinations of footballs and soccer balls that can be made in one hour.

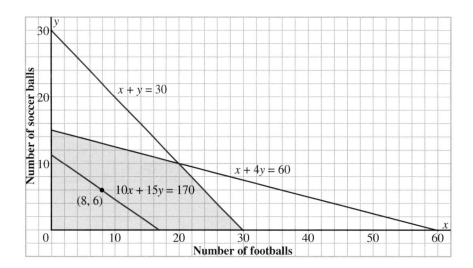

Suppose the manufacturer makes a profit of $10 per football and $15 per soccer ball. This problem is an example of a *maximum-minimum* problem.

How many of each kind of ball should be made per hour to yield the maximum profit?

Suppose we select any point in the coloured region, such as (8, 6). The profit on 8 footballs and 6 soccer balls is $10 \times 8 + \$15 \times 6 = \$170$. However, other combinations of footballs and soccer balls would also yield a profit of $170.

For x footballs and y soccer balls, and a profit of $170: $10x + 15y = 170$

The line represented by this equation is also shown on the graph. All points with integral coordinates that are on this line and in the coloured region represent combinations of footballs and soccer balls that yield a profit of $170.

For a greater profit, we would replace 170 in the above equation with a greater number, but the coefficients of x and y would remain the same. This means that the line corresponding to a greater profit is parallel to the line shown, and farther from (0, 0). For a maximum profit, we require a parallel line that is as far as possible from (0, 0).

Visualize the line above moving away from (0, 0) and remaining parallel to the one shown. Can you see that the line that is as far as possible from (0, 0) must pass through one vertex of the coloured region? This is the key to solving the problem above. We need only determine the profit for the balls represented by the coordinates of each vertex of the region. The one that yields the maximum profit is the solution to the problem.

1. Solve the problem on the preceding page. How many of each kind of ball should be made per hour to yield the maximum profit?

2. In the example on page 358, suppose the profit on each ball is as shown. How many of each ball should be made per hour to yield the maximum profit?

   a) profit on a football: $20
      profit on a soccer ball: $4

   b) profit on a football: $4
      profit on a soccer ball: $20

3. Refer to exercise 9 on page 356. Suppose the profit on each vehicle is as shown. How many of each vehicle should be produced per day to maximize the profit?

   a) profit on a motorcycle: $50
      profit on a bicycle: $25

   b) profit on a motorcycle: $25
      profit on a bicycle: $50

4. Refer to exercise 10 on page 356. Suppose the profit for each crop is as shown. What areas of corn and wheat should be sown to maximize the profit?

   a) corn: $200/ha
      wheat: $300/ha

   b) corn: $300/ha
      wheat: $200/ha

   c) corn: $400/ha
      wheat: $100/ha

5. A farmer expects a profit of at least $8000 by planting 40 ha with wheat and barley. The ground requires fertilizer: 400 kg/ha for the wheat and 240 kg/ha for the barley. The fertilizer used is kept to a minimum. Suppose the profit for each crop is as given. How much land should the farmer sow with each grain?

   a) $240/ha for the wheat and $160/ha for the barley

   b) $140/ha for the wheat and $240/ha for the barley

Will There Be Enough Food? (Part II)

On page 328, we used an exponential model $P_1 = 5.73(1.018)^n$ for world population, which has the graph (below left). This is not a realistic model because the growth cannot continue indefinitely. A better model has a graph like the one (below right), showing that Earth's population eventually reaches a maximum. According to United Nations projections, this could be between 8 and 28 billion, depending on the success of population control efforts. In the model below, we will use 20 billion for the maximum population.

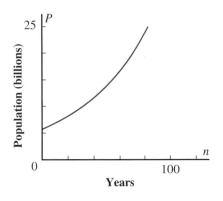

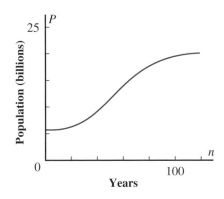

DEVELOP A MODEL

In the exponential model, the population increases by 1.8% every year. If the population in any year is x, the addition to the population that year is $0.018x$. The annual addition varies directly as the population. A graph of the annual addition is a straight line with equation $y = 0.018x$ (upper left on next page).

In the improved model, the maximum population is 20 billion. When the population is small, the annual addition is close to 0 and increasing. When the population reaches 10 billion, the annual addition is a maximum. As the population continues to rise, the annual addition declines. The annual addition eventually becomes 0 when the maximum population of 20 billion is reached. In this model, the annual addition is a quadratic function of the population. Its graph is a parabola with x-intercepts 0 and 20. Hence, the equation of the graph has the form $y = ax(20 - x)$, where a is a constant.

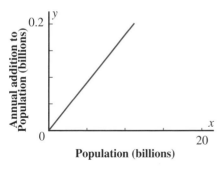

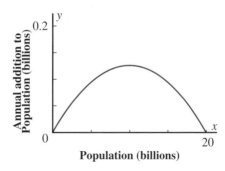

The constant a is a measure of how rapidly the population increases toward the maximum. We can determine its value as follows. In 1995, the population was 5.73 billion and the growth rate was 1.8% per annum. The increase in population that year was 0.018×5.73 billion $\doteq 0.103\,14$ billion. Hence, the expression $ax(20 - x)$ should equal $0.103\,14$ when $x = 5.73$. Substitute 5.73 for x in this expression.

$$ax(20 - x) = 0.103\,14$$
$$a(5.73)(20 - 5.73) = 0.103\,14$$
$$a = \frac{0.103\,14}{5.73(20 - 5.73)}$$
$$\doteq 0.001\,26$$

Hence, the annual addition to the population is modelled by the equation below, where x represents the population.

$$y = 0.001\,26x(20 - x) \quad \text{③}$$

This is the equation of the parabola in the graph above.

1. According to equation ③, what is the annual addition for each population?

    a) 8 billion   **b)** 12 billion   **c)** 16 billion   **d)** 20 billion

2. According to equation ③, what is the population for each annual addition?

    a) maximum   **b)** zero

 LOOK AT THE IMPLICATIONS

Equation ③ does not give projected populations in future years. However, we can use it to calculate populations. In 1995 the population was approximately 5.73 billion. Substitute 5.73 for x to calculate the increase in population that year.

$$y = 0.001\,26(5.73)(20 - 5.73)$$
$$\doteq 0.103\,03$$

The projected population in 1996, in billions, is $5.73 + 0.103\,03 = 5.833\,03$

Substitute 5.833 03 for x to calculate the increase in population in 1996.

$$y = 0.001\ 26(5.833\ 03)(20 - 5.833\ 03)$$
$$\doteq 0.104\ 12$$

The projected population in 1997, in billions, is $5.833\ 03 + 0.104\ 12 = 5.937\ 15$

We can repeat this to calculate the projected population in any future year. A graphing calculator can be programmed to do these calculations.

Ask your teacher for the program called EARTHPOP from the Teacher's Resource Book. Enter it in your calculator. When you run the program, it will prompt you to enter the initial population, the maximum sustainable population, and the constant a. Use the figures above (5.73, 20, and 0.001 26, respectively). When the word "DONE" appears, press [STAT] [1]. The screen (below left) will appear. The calculator has produced a table of values showing the population for the next 200 years in increments of 10 years.

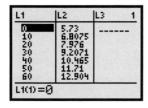
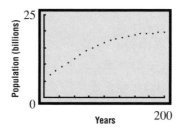

To graph the data, press [STAT PLOT] [1] to select the first plot. In the menu that appears, make sure that "On" is selected, the first graph type is selected, and lists L1 and L2 are selected. Choose the dot as the plotting mark. Then press [GRAPH]. The graph above will appear (the EARTHPOP program set the viewing window to the one shown, but you can change it). You can trace along the graph to see the same data that are in the table.

3. Approximately how many years does it take for the population to reach each number?

a) 15 billion   b) 19 billion   c) 19.8 billion   d) 19.99 billion

4. Predict how the graph would change in each situation. Use the program to verify your predictions.

a) The maximum sustainable population is greater than 20 billion or less than 20 billion.

b) The constant a is greater than 0.001 26 or less than 0.001 26.

c) The initial population is greater than 5.73 billion or less than 5.73 billion. Try the most recent population figure you can find.

5. According to the United Nations, the most likely maximum population is 12 billion, which could be reached around the year 2150. Use the program to try to confirm this prediction. Explain the result.

REVISIT THE SITUATION

If you have a TI-83 calculator, you can determine the equation of the curve formed by the points generated by the EARTHPOP program. Run the program again, using 5.73 for the initial population, 20 for the maximum sustainable population, and 0.001 26 for the constant a. Then obtain the graph as before. Press STAT ▶. In the menu that appears, scroll down until the cursor is beside "Logistic" (this is a type of curve shaped like the second one on page 360). Press ENTER VARS ▶ 1 1 ENTER. The calculator will determine the equation of the logistic curve of best fit, then copy it into the Y= list beside Y1 (it takes between one and two minutes to do this). The next screen shows the result. To view the graph of the equation, press GRAPH.

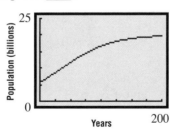

The equation of the function is approximately $y = \dfrac{20}{1 + 2.5e^{-0.0252x}}$. The letter e represents the irrational number 2.71828…, which often occurs in problems involving growth, just as π occurs in problems involving circles. Using this function, we can model Earth's population using the equation below, where n is the number of years after 1995. This is called the *logistic* model.

$$P_3 = \dfrac{20}{1 + 2.5e^{-0.0252n}}$$

6. a) Graph P_3 by entering this equation in the Y= list:

   Y1 = 20 / (1 + 2.5 e^(–0.0252x))

   b) Recall that on page 328 we used a linear model $P_2 = 0.092n + 7.16$ for the population that can be fed by the available food supply. Use your calculator to graph the equation Y2 = 0.092x + 7.16 on the same screen as part a.

   c) For what value(s) of n is each statement true?

   i) $P_3 = P_2$ ii) $P_3 < P_2$ iii) $P_3 > P_2$

7. What are the implications of your answers in exercise 6c?

1. Solve by graphing.

   a) $x + y = -8$
   $x - 2y = 7$

   b) $2x + y = 8$
   $4x - 9y = 5$

   c) $x + 2y = -2$
   $-2x + y = 6$

   d) $x - 2y = -5$
   $-3x + y = 4$

   e) $x + 2y = 6$
   $2x - 3y = 5$

   f) $3x + y = 6$
   $2x + 4y = -1$

2. Choose one linear system from exercise 1. Write to explain how you solved it.

3. Determine two other linear systems that have the same solution as this system:
   $2x + 5y = 8$
   $x - 3y = -7$

4. For each pair of coordinates, determine a system of linear equations for which that point is the solution.

   a) $(-2, 5)$ **b)** $(4, -1)$ **c)** $(7, 2)$ **d)** $(-5, -6)$

5. Solve by addition or subtraction.

   a) $3x - 4y = 1$
   $3x - 2y = -1$

   b) $3a + 2b = 5$
   $9a - 2b = 15$

   c) $3x - 4y = -2$
   $4x - 3y = -5$

   d) $2s + 3t = 6$
   $5s + 10t = 20$

   e) $3a + 2b = 5$
   $2a + 3b = 0$

   f) $2x + y = -5$
   $3x + 5y = 3$

6. Solve by substitution.

   a) $5v + u = -17$
   $3u - 4v = 6$

   b) $x + 5y = -11$
   $4x - 3y = 25$

   c) $\frac{x}{2} + \frac{y}{2} = 7$
   $3x + 2y = 48$

   d) $\frac{a}{2} + \frac{b}{3} = 1$
   $\frac{a}{4} + \frac{2b}{3} = -1$

   e) $9x + 2y = 2$
   $1 - y = 4x$

   f) $8x + 4y = 1$
   $7x = -2y$

7. Choose one system from exercises 5 and 6. Write to explain how you solved it.

8. Solve each non-linear system.

   a) $y = \frac{2}{x}$
   $y = 4x - 2$

   b) $y = x^2$
   $y = 18 - x$

   c) $y = |x|$
   $y = -x + 3$

   d) $y = \sqrt{x}$
   $y = x - 4$

9. The cost of 4 L of oil and 50 L of gasoline is $42.50. The cost of 3 L of oil and 35 L of gasoline is $30.30. Determine the cost of 1 L of oil and 1 L of gasoline.

10. When Chana rented a car for 3 days and drove 160 km, the charge was $124. When she rented the same car for 5 days and drove 400 km, the charge was $240. What was the charge per day and the charge per kilometre?

11. Yasser invests \$4500, part at 7%, the balance at $8\frac{1}{2}\%$. After one year, the interest earned on the 7% investment was \$150 less than the interest earned on the $8\frac{1}{2}\%$ investment. How much was invested at each rate?

12. Solve.

a) $2x - y + 3z = 2$
   $3x + 2y - 4z = 5$
   $x - 4y + z = -10$

b) $4x + 3y - z = 0$
   $x - 2y + 3z = 5$
   $6x + y + 4z = 5$

c) $x + 5y - 3z = -25$
   $-2x + 3y - z = -3$
   $4x - y + 5z = -3$

13. Write the inequality that represents:

a) each shaded region

b) each unshaded region

i)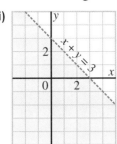

ii)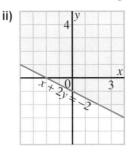

iii)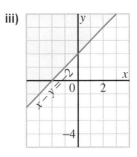

14. Emily plans to spend up to 11 h each week exercising. She will do weight-training and swimming. Draw a graph to show the amount of time she could spend on each activity.

15. Use the information in the news item. Draw a graph to show the ages at which Chinese men and women may marry.

> ### In China, Women Must Be 20 to Marry
>
> *Beijing.* A Chinese law sets minimum legal ages for marrying. The minimum legal age is 22 for men and 20 for women. In addition, couples are urged not to marry until the ages of the bride and groom total more than 52.

16. Write the system of inequalities that represents the shaded region on each graph.

a)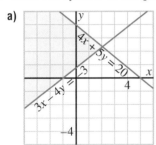

b)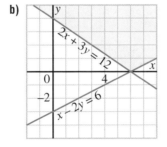

c)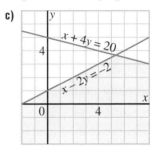

5 Cumulative Review

1. Determine the principal that should be deposited today to accumulate to $2000 in 5 years at 5% compounded:

   a) semi-annually   b) monthly

2. One bank offers an annual interest rate of $3\frac{1}{2}\%$ compounded monthly. Another bank offers an annual interest rate of $3\frac{1}{2}\%$ compounded quarterly. Suppose you deposit $1000 for 5 years. How much more interest would you earn at the first bank?

3. The water company offers an equalization plan that charges a fixed amount every month. The company uses the previous year's usage to calculate the fixed amount. The LeBlanc family's water bills for last year are summarized in the table.

January	$11.97	July	$22.54
February	$12.02	August	$23.21
March	$10.99	September	$20.24
April	$13.58	October	$17.74
May	$15.95	November	$14.55
June	$17.89	December	$12.81

   a) Explain why the water usage is so high in July and August.

   b) What monthly amount should the water company charge the LeBlancs?

4. Sketch each set of graphs on the same grid.

   a) $y = x^2$ $y = (x - 1)^2$ $y = (x + 3)^2$

   b) $y = x^2$ $y = 2(x + 1)^2 + 3$ $y = \frac{1}{2}(x - 2)^2 - 4$ $y = -(x - 3)^2 + 5$

5. A bus company carries about 40 000 riders per day for a fare of $1.00. A survey indicates that if the fare is decreased, the number of riders will increase by 2500 for every 5¢ decrease. What fare will result in the greatest revenue?

6. Write a quartic function with the given zeros.

   a) –4, 3, –2, 1   b) 1, 1, 3, 5   c) –2, 6, –5, 4   d) –2, –2, 3, 3

7. A can of soup has a volume of 540 mL and a base radius of 4.3 cm. The company plans to redesign the can with a larger base. The new can will have the same volume as the original can.

   a) Let r centimetres represent the change in base radius. Let h centimetres represent the change in height. Write h as a function of r.

b) Graph the function.

c) What is the decrease in height for each increase in base radius?

    i) 0.5 cm   **ii)** 1 cm   **iii)** 1.5 cm

d) What is the increase in base radius for each decrease in height?

    i) 0.5 cm   **ii)** 1 cm   **iii)** 1.5 cm

8. When $ax^3 + bx^2 - x + 3$ is divided by $x - 1$, the remainder is 4. When this polynomial is divided by $x - 2$, the remainder is 21. Determine the values of a and b.

9. At the scene of a traffic accident, police can estimate the minimum speed a car had been travelling by the length of the skid marks. One formula used for this purpose is $v = \sqrt{48.3df}$, where v is the speed in kilometres per hour, d is the length of the skid marks in metres, and f is a friction coefficient.

a) An officer determines that the friction coefficient for the level, dry, normal asphalt surface on which an accident occurred is 4.2. Substitute this value in the formula and solve for d.

b) Determine the length of the skid marks for a car braking from each speed.

    i) 60 km/h   **ii)** 90 km/h   **iii)** 120 km/h

c) Determine the speed of the car for each length of its skid marks.

    i) 10 m   **ii)** 25 m   **iii)** 50 m

10. Suppose a hockey team offers players two salary packages. Package A has a base salary of x dollars and a $1000 bonus for each goal scored. Package B has a base salary $10 000 dollars less than that in Package A, but pays a bonus of $1500 for each goal scored.

a) Let y represent the number of goals a player scores in a season. Write a system of equations to represent the two salary packages.

b) Solve the system to determine the number of goals for which the salaries are equal.

c) Open the *Hockey* database. Find the 66 records for the 1995-96 season. Sort them according to the number of goals scored. Suppose you are a hockey agent. Players are signing contracts for the 1996-97 season. For how many of the players in the database would you recommend Package A? For how many would you recommend Package B?

11. Naomi invested $1000, part at 8% per annum and the rest at 10% per annum. In one year, the two parts earned equal amounts of interest. How much did Naomi invest at each rate?

Incredible Journeys

CONSIDER THIS SITUATION

Suppose you participate in an experiment requiring you to travel through space at close to the speed of light. As you board the spaceship, you say goodbye to your friends and family.

One year has passed and you step out of the spaceship, greeting the people near you. You don't recognize them at first, yet they are all strangely familiar. Your best friend steps out from the crowd and you are stunned by her appearance. She is over 50 years old. She has grandchildren! What has happened? You are still only 18!

Albert Einstein used mathematics and geometry to predict that time is not constant in how it changes. The closer one travels to the speed of light, the slower time moves relative to someone at rest. Recent experiments have confirmed that Einstein's prediction is correct.

Geometry has helped scientists to predict changing time, black holes, and curved space. The basics of two-dimensional geometry are points, lines, and curves.

• Would you want to make a trip such as the one described above?

• Does this mean that it is possible to travel back in time?

• What do we mean by a "point" in geometry?

On pages 378 and 379, you will visualize another incredible journey to help you understand the meaning of a "point."

FYI Visit www.awl.com/canada/school/connections

For information related to the above problem, click on MATHLINKS followed by *Mathematics 11*. Then select a topic under Incredible Journeys.

If we examined many different snowflakes, we would likely find that no two are exactly alike. We might then conclude that no two snowflakes are ever exactly alike. But we can never be certain because we would have to examine every snowflake that ever existed.

If the same result occurs over and over again, we may conclude that it will always occur. This kind of reasoning is called *inductive reasoning*. We use inductive reasoning to conclude that two snowflakes are never exactly alike.

Example 1

Suppose you multiply an odd number by an even number.

$23 \times 14 = 322$ $\qquad$ $17 \times 24 = 408$ $\qquad$ $57 \times 32 = 1824$

The products are all even numbers. What conclusion might you make?

Solution

The product of an odd number and an even number is even.

Example 2

Suppose you draw some intersecting lines and measure the angles formed.

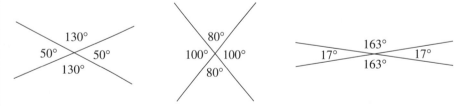

The angles opposite each other are equal. What conclusion might you make?

Solution

When two lines intersect, the opposite angles are equal.

When we use inductive reasoning, we can never be certain the conclusion is *always* true. Just by looking at the three products in *Example 1*, we cannot be certain that *every* time we multiply an odd number by an even number the product is even. Just by looking at the three diagrams in *Example 2*, we cannot be certain that the opposite angles are equal for *every* pair of intersecting lines.

DISCUSSING THE IDEAS

1. Make up another arithmetic example similar to *Example 1*. How does it illustrate the meaning of inductive reasoning?

2. Describe another geometry example similar to *Example 2*. How does it illustrate the meaning of inductive reasoning?

6.1 EXERCISES

A 1. Suppose you multiply two 2-digit numbers.

$$28 \times 16 = 448 \qquad 35 \times 29 = 1015 \qquad 68 \times 75 = 5100$$

All the products have either 3 digits or 4 digits.

a) Check this by multiplying other pairs of 2-digit numbers.

b) What conclusion might you make?

2. Suppose you add consecutive odd numbers starting at 1.
The results are all perfect squares.

a) Check this by continuing the pattern.

b) What conclusion might you make?

$$1 + 3 = 4$$
$$1 + 3 + 5 = 9$$
$$1 + 3 + 5 + 7 = 16$$

B 3. a) Draw any △ABC.

b) Use a protractor to measure its angles.

c) Do this for several triangles.

d) What conclusion might you make?

4. For exercise 3, write to explain another way to demonstrate the conclusion is true.

5. For exercise 3, what can you conclude about the angles in △ABC in each case?

a) Triangle ABC is equilateral.   b) Triangle ABC is a right triangle.

6. Two lines in the same plane that never meet are *parallel lines*.

  a) Use both sides of a ruler to draw two parallel lines. Draw a third line to intersect these lines. This line is a *transversal*.

  b) Measure the angles formed by the transversal and the parallel lines.

  c) Repeat for other transversals.

  d) What conclusions might you make?

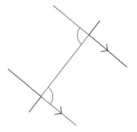

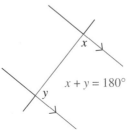

$x + y = 180°$

Angles forming a
Z-pattern are
alternate angles.

Angles forming an
F-pattern are
corresponding angles.

Angles forming a
C-pattern are
interior angles.

7. In each triangle, M is the midpoint of AB, and N is the midpoint of AC.

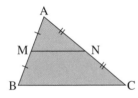
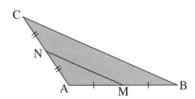
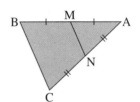

  a) How does MN compare with BC in each case?

  b) What conclusions might you make?

C **8. a)** Draw any quadrilateral ABCD. By measuring, locate the midpoints of its sides.

  b) Join the midpoints to form another quadrilateral. What kind of quadrilateral have you drawn?

  c) Repeat parts a and b for at least one more quadrilateral. What conclusion might you make?

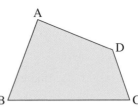

COMMUNICATING THE IDEAS

This section deals with some geometric properties you learned in a previous grade. Make a list of these properties.

6.2 Deductive Reasoning

To promote sales between 12 noon and 2 p.m., a restaurant offers a pizza free of charge if it takes more than five minutes to come to your table after you order it. When Raji was there, the following statements were true:

- It was between 12 noon and 2 p.m.

- It took more than five minutes to take the pizza to Raji's table.

Raji concluded that she would receive the pizza free of charge, and she did.

When we make a conclusion based on statements that we accept as true, we are using *deductive reasoning*.

In *Example 2* on page 370, there are three diagrams. We used inductive reasoning to conclude that when two lines intersect, the opposite angles are equal. But we cannot draw all possible pairs of intersecting lines to examine every pair. How can we be sure that the opposite angles are equal, even for diagrams we have not drawn? We use deductive reasoning, as follows.

This diagram represents any two intersecting lines.

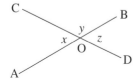

Since ∠AOB is a straight angle: $x + y = 180°$ ①
Since ∠COD is a straight angle: $y + z = 180°$ ②
Comparing ① and ②: $x + y = y + z$
                      $x = z$

Therefore, ∠AOC = ∠DOB

In the same way, we can prove that ∠AOD = ∠COB .

Opposite Angles Theorem
When two lines intersect, the opposite angles are equal.

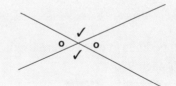

When we use deductive reasoning in geometry, we arrive at certain conclusions. Any conclusions we obtain can be used to arrive at other conclusions. The conclusions that are most useful for this purpose are called *theorems*.

Here are some other examples of deductive reasoning.

Example 1

Suppose you add two odd numbers.

$$3 + 7 = 10 \qquad\qquad 25 + 13 = 38 \qquad\qquad 437 + 129 = 566$$

Explain why the sums are always even.

Solution

Let $2m + 1$ and $2n + 1$ represent the odd numbers, where m and n are natural numbers. Then their sum is:

$$2m + 1 + 2n + 1 = 2m + 2n + 2$$
$$= 2(m + n + 1)$$

Since 2 is a factor of the sum, the sum is always even.

In *Example 1*, we *proved* that the sum of two odd numbers is always an even number, including numbers we have not added. This illustrates the power of deductive reasoning. When we use deductive reasoning, we are certain that the conclusion is true.

Example 2

ABCD is a square and $\triangle$ABE is an equilateral triangle. Explain why $\triangle$BCE is isosceles.

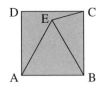

Solution

Since ABCD is a square:	AB = CB ①
Since $\triangle$ABE is an equilateral triangle:	AB = EB ②
Comparing ① and ②, we see that:	CB = EB
Therefore, $\triangle$BCE is isosceles.	

In *Example 2*, certain geometric words were used. The definitions of these words are on the next page.

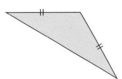

A *square* is a rectangle with four equal sides.

An *equilateral triangle* has all three sides equal.

An *isosceles triangle* has at least two equal sides.

DISCUSSING THE IDEAS

1. What is the difference between inductive reasoning and deductive reasoning?

2. How are inductive reasoning and deductive reasoning similar? Explain.

6.2 EXERCISES

A 1. Use deductive reasoning. Copy and complete each conclusion.

a) All students are teenagers. Lynn is a student.
   Conclusion: Lynn is a _____.

b) All musicians have long hair. Sharon is a musician.
   Conclusion: Sharon has _____.

c) If it is raining, the ground is wet. It is raining.
   Conclusion: The ground is _____.

d) If the temperature goes above 5°C, the rink will melt.
   The temperature is 8˚C.
   Conclusion: The rink will _____.

e) A square is a rectangle. A rectangle has four right angles.
   Conclusion: A square has _____.

f) Parallel lines never meet. Lines l_1 and l_2 are parallel.
   Conclusion: Lines l_1 and l_2 _____.

2. Use deductive reasoning. Copy and complete each conclusion.

a) Anyone who jogs regularly will be fit. Manuel jogs regularly. Therefore, …

b) If a person takes English, her or his writing will improve. Jennifer takes English. Therefore, …

c) The sum of the angles in any triangle is 180°. In $\triangle ABC$, $\angle A = 90°$; therefore, $\angle B + \angle C = $ …

d) In $\triangle PQR$, $\angle P = 90°$ and $\angle Q = \angle R$; therefore, the measure of $\angle Q$ is …

e) A square has four equal sides. Some rectangles are squares. Therefore, …

B **3.** Decide if each conclusion follows logically from the given statement(s).

a) All mathematics students can compute. Lisa is a mathematics student. Therefore, Lisa can compute.

b) Some students like pizza. Sophie is a student. Therefore, Sophie likes pizza.

c) Squares ABCD and EFGH have the same area. Therefore, their sides have the same length.

d) Rectangles STUV and WXYZ have the same area. Therefore, they have the same length and width.

e) An isosceles triangle has at least two equal sides. An equilateral triangle has three equal sides. Therefore, an equilateral triangle is isosceles.

f) A square has four equal sides. ABCD has four equal sides. Therefore, ABCD is a square.

4. a) Bahamas, Barbados, Jamaica, and Trinidad and Tobago are four countries in the Caribbean. All the following statements about their land areas are true. List the countries in order of increasing size. Use the *International Transportation* database to check your answer.

• Barbados is smaller than Trinidad and Tobago.
• Bahamas is neither the largest nor the smallest.
• At least two countries are larger than Trinidad and Tobago.

b) Use the database to make up a similar exercise for a classmate to complete.

5. In a previous grade you probably drew a triangle, measured its angles, and found that their sum is 180°. You may have repeated this for several triangles. But you cannot do this for all triangles. Here is a way to use deductive reasoning to prove that the sum of the angles in any triangle is 180°, without drawing and measuring.

a) Draw any △ABC.

b) Draw a line DE through A parallel to BC.

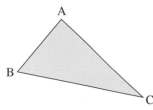

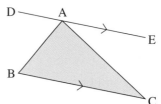

c) What can you conclude about ∠DAB and ∠B? Explain.

d) What can you conclude about ∠EAC and ∠C? Explain.

e) Explain how you can use the results of parts c and d to conclude that the sum of the angles in △ABC is 180°. This result is called the *Angles in a Triangle Theorem.*

6. For the diagram (below left), write to explain why $x + y < 180°$.

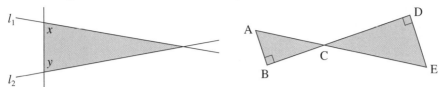

7. In the diagram (above right), explain why $\angle A = \angle E$.

8. Triangle ABC is any triangle. The angle formed by extending one side, as shown, is an *exterior angle* of the triangle. Explain why $\angle ACD = \angle A + \angle B$. This result is called the ***Exterior Angle Theorem.***

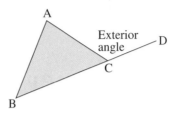

9. a) Prove that the sum of two even numbers is always even.

   b) Prove that the product of two odd numbers is always odd.

10. a) Prove that the difference of the squares of two even numbers is always divisible by 4.

   b) Prove that the difference of the squares of two odd numbers is always divisible by 4.

11. Alice, Brittany, Carol, Dini, and Elise are on the school basketball team. All the following statements are true. List the girls in order of increasing height.

   • There are at least two girls shorter than Alice.

   • Dini is shorter than Carol.

   • Brittany is not the shortest girl.

   • Dini is taller than Alice.

C **12.** JKLM is a square. Triangle JXK and △KYL are both equilateral. Prove that △KXY is isosceles.

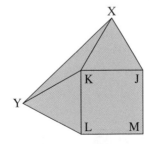

COMMUNICATING THE IDEAS

How would you explain the difference between inductive reasoning and deductive reasoning? Write your ideas in your notebook.

0.1 m

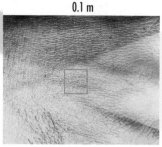

Back of hand

0.01 m

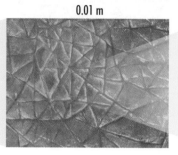

Skin creases appear to meet at a point

0.001 m

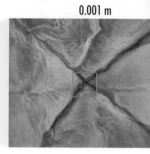

Close-up of the point

Incredible Journeys

On page 368, you considered an incredible journey that would affect your age relative to those who stayed behind. The study of geometry provides an opportunity to engage in another incredible journey illustrated by the images on these pages. These images are taken from an extraordinary film called *Powers of Ten,* made by the noted designers/filmmakers Charles and Ray Eames.

DEVELOP A MODEL

Look closely at the skin creases on your hand. Use a magnifying glass if necessary. Find creases that appear to meet. The place where the creases meet is a model of a point.

LOOK AT THE IMPLICATIONS

Visualize taking an imaginary journey along a straight line toward the point. In this sequence of diagrams, each illustration is one-tenth as wide as the one before it.

10^{-14} m

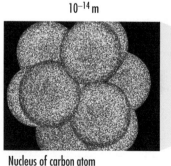

Nucleus of carbon atom

10^{-13} m

Nucleus of carbon atom appears

10^{-12} m

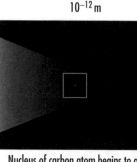

Nucleus of carbon atom begins to a

Images: Eames Office © 1982, 1998 www.eamesoffice.com

10^{-4} m

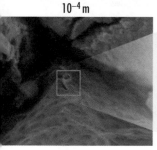

Unexpected detail appears

10^{-5} m

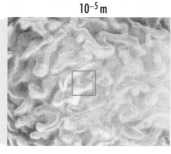

White blood cell

10^{-6} m

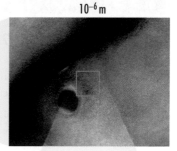

Cell nucleus

Physicists can continue this incredible journey for about 3 or 4 more steps, to subatomic particles called quarks. In mathematics we can continue the journey as long as we like, but we will never get to the object called a "point." We say that a point is infinitely small.

10^{-7} m

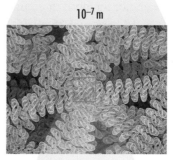

DNA molecules appear

 REVISIT THE SITUATION

1. a) Visualize what we mean by a line. How would you model a line?

   b) Suppose we took the imaginary journey toward a line. How would the sequence of diagrams be different from those on these pages?

2. Repeat exercise 1 for a plane.

Points infinitely small, lines and planes infinitely thin, and measurements that are absolutely precise are the basics of geometry.

10^{-8} m

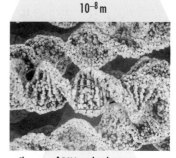

Close-up of DNA molecule

10^{-11} m

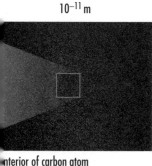

Interior of carbon atom

10^{-10} m

Close-up of carbon atom

10^{-9} m

Carbon and hydrogen atoms appear

Images: Eames Office © 1982, 1998 www.eamesoffice.com

A Famous Example of Intuitive Reasoning

For centuries people believed that Earth was flat, and the stars revolved around it. They rejected the ideas that Earth was spherical and it moves around the sun, because these ideas are contrary to intuition. By *intuition* we mean an instinctive knowledge or feeling without attention to reasoning.

When we use our intuition, we can never be certain that our conclusions are correct.

For example: Suppose three lines l_1, l_2, and l_3 are such that $l_1 \parallel l_2$ and $l_2 \parallel l_3$. Our intuition might tell us that l_1 should be parallel to l_3. In this case the conclusion is correct, as we can see from a diagram.

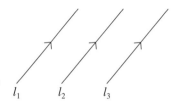

$l_1 \parallel l_2$
$l_2 \parallel l_3$
$l_1 \parallel l_3$

But suppose three lines l_1, l_2, and l_3 are such that $l_1 \perp l_2$ and $l_2 \perp l_3$. Again, our intuition might suggest that l_1 is perpendicular to l_3. But in this case our intuition is incorrect, since l_1 is *not* perpendicular to l_3. The diagram shows that l_1 and l_3 are parallel.

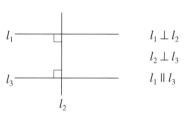

$l_1 \perp l_2$
$l_2 \perp l_3$
$l_1 \parallel l_3$

Use your intuition to answer exercises 1, 2, and 3. Then reason out the answers.

1. A bottle and a cork together cost $2.10. The bottle costs $2 more than the cork. How much does the cork cost?

2. Naomi's average in English, history, and science is 80%. She calculates that if she receives a mark of 60% in French, her average for the four subjects will be 70%. Is this correct?

3. A car travels 240 km at an average speed of 80 km/h, then returns at an average speed of 60 km/h. What is the average speed for the entire trip?

4. a) Suggest how people first became convinced Earth was spherical.

   b) List as many different ways as you can to prove, to someone living today, that Earth is spherical.

6.3 Conjectures and Counterexamples

Many natural numbers can be written as the sum of consecutive numbers. Here are some examples.

$9 = 4 + 5$ $6 = 1 + 2 + 3$

$12 = 3 + 4 + 5$ $10 = 1 + 2 + 3 + 4$

$22 = 4 + 5 + 6 + 7$ $29 = 14 + 15$

Based on examples like these, you might think that every natural number can be written as the sum of consecutive numbers.

1. Check that the examples above are correct.

2. Try writing some other natural numbers as the sum of consecutive numbers.

3. Find a natural number that cannot be written as the sum of consecutive numbers. How can you be certain that your number is correct?

At the beginning of *Investigate*, you may have concluded that every natural number is the sum of consecutive natural numbers. Since this conclusion was based on examples, we call it a *conjecture*.

Conjecture: Every natural number is the sum of consecutive natural numbers.

We cannot prove this conjecture using examples because there are infinitely many natural numbers, and we would have to check every one. However, if we could find even *one* natural number that cannot be expressed in this way, we will know that every natural number cannot be written as the sum of consecutive numbers.

An example that shows a conjecture is false is a *counterexample*. The number you found in exercise 3 in *Investigate* is a counterexample that shows not every natural number can be written as the sum of consecutive numbers.

DISCUSSING THE IDEAS

1. What is the difference between a conjecture and a theorem?

2. Why do you need only one counterexample to show that a conjecture is false?

3. Why can't inductive reasoning be used to explain why a conjecture is true?

A 1. Each example below suggests a conjecture. Find a counterexample to show each conjecture is false.

a) $3 = 1^2 + 1^2 + 1^2$ $59 = 1^2 + 3^2 + 7^2$

    $14 = 1^2 + 2^2 + 3^2$ $61 = 3^2 + 4^2 + 6^2$

    $24 = 2^2 + 2^2 + 4^2$ $89 = 2^2 + 2^2 + 9^2$

    Conjecture: Every natural number can be written as the sum of three perfect squares.

b)

 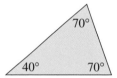 

    Conjecture: In every triangle, all three angles are less than 90°.

c)

    Conjecture: Every square has two horizontal sides and two vertical sides.

2. Three conjectures are given. For which conjectures is this diagram a counterexample?

a) If a quadrilateral has four equal angles, it has four equal sides.

b) The length of a rectangle is always double its width.

c) The opposite sides of a parallelogram are equal in length.

B 3. Several conjectures are given. Decide whether each conjecture is true or false. If it is true, write to explain why. If it is false, give a counterexample.

a) A number that is not positive is negative.

b) If 1 is added to an odd number, the result is always an even number.

c) The square of a number is always greater than the number.

d) If two angles are acute, their sum is less than 180°.

e) The altitude of a triangle always lies inside the triangle.

f) Every rectangle is a square.

g) Every square is a rectangle.

4. Two students were discussing the Pythagorean Theorem. Their discussion went as follows:

Frank: "In these problems we had to find the value of *x*. In the first triangle I got $\sqrt{7}$, using the Pythagorean Theorem. Then I noticed that $4 + 3 = 7$. So, a faster way to find the value of *x* is just to add the two given lengths and take the square root. This method gives the correct answer for the other three triangles, too."

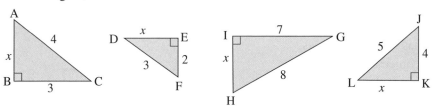

Susan: "Your method doesn't work for all triangles. I have found a counterexample."

a) Find a counterexample to support Susan's claim.

b) Explain why Frank's method works for some right triangles but not for others.

5. a) In the *Hockey* database, find the record for each of these players for the 1995-1996 season. Record the number of shots taken and goals scored by each player. Based on these data, copy and complete this conjecture: "The more shots a player takes..."

   i) Jamie Baker   **ii)** Dave Reid
  iii) Peter Forsberg   **iv)** Teemu Selanne
   v) Eric Lindros   **vi)** Jaromir Jagr

b) Find a record in the database that is a counterexample to the conjecture you made in part a.

6. In 1742, the German mathematician C. Goldbach conjectured that every even number greater than 2 is the sum of two prime numbers (for example, $18 = 13 + 5$). No one has been able to prove this is true for all even numbers, and no one has ever found a counterexample. This problem is known as *Goldbach's Conjecture*. It is a famous unsolved problem in mathematics.

a) Choose three other even numbers. Verify that each number can be written as the sum of two primes.

b) To obtain a counterexample, what would you have to find?

7. a) Is every odd number the sum of two prime numbers? If so, explain why. If not, give a counterexample.

b) Is every prime number the sum of two prime numbers? If so, explain why. If not, give a counterexample.

C **8.** A science textbook states that water boils at 100°C. Find a counterexample.

9. Two students used their graphing calculators to graph the function $y = x^x$. The results and their discussion are shown.

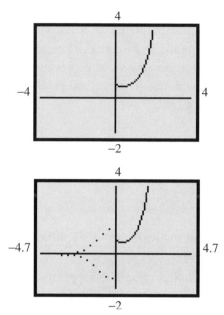

Bruce: "The screen is blank for $x < 0$. Hence, I conjecture that x^x is undefined when $x < 0$.

Lisa: "My screen has counterexamples to show that your conjecture is false. I conjecture that x^x is undefined when $x < 0$, except for the 15 values of x where the dots appear on my screen."

a) Find a counterexample to show that Bruce's conjecture is false.

b) Find a counterexample to show that Lisa's conjecture is false.

10. Use a graphing calculator to obtain Lisa's graph in exercise 9. Then find out as much as you can about the function $y = x^x$ as follows.

a) Use the $\boxed{\text{TRACE}}$ key. What are the values of x corresponding to the dots on the screen?

b) Turn off the axes in the FORMAT menu.

c) Use the tables feature. In the TABLE SETUP menu use, for example, −3 for TblStart and 0.01 for △Tbl. Try other values for TblStart.

d) Explain some of the results you obtained in part c.

COMMUNICATING THE IDEAS

Explain what the word "conjecture" means. Look up this word in a dictionary. Is the meaning consistent with your explanation? Explain.

6.4 Statements Involving "not," "and," "or"

Simple sentences can be combined to form more complex sentences in different ways. For example, consider these two sentences:

The Flames won.   ...①

The Canucks won.   ...②

We can create sentences using ① and ②, and the words "not," "and," "or".
Here are some examples:

The Flames did **not** win.   ...③

The Flames won **and** the Canucks won.   ...④

The Flames won **or** the Canucks won.   ...⑤

Suppose ① and ② are both true. Then ③ is false, and ④ and ⑤ are both true.

Using "not"

Sentence ③ is called the *negation* of ①. Sentence ① is also the negation of ③. When a sentence is true, its negation is false, and vice versa.

In mathematics, we form negations using symbols such as ≠, ≮, and ≯.

Example 1

On a number line, indicate the location of the numbers corresponding to each statement.

a) $x \neq 2.5$ **b)** $x \not< 2.5$

Solution

a) $x \neq 2.5$

These are all possible values of x except 2.5. Draw a solid line with an open circle at 2.5 to indicate that this number is not included.

b) $x \not< 2.5$

These are the values of x that are not less than 2.5. That is, they are greater than or equal to 2.5. Draw a solid circle at 2.5 to indicate that this number is included.

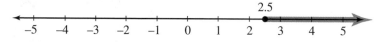

Using "and"

Sentence ④ is formed using the word "and." For the combined sentence to be true, both parts of the sentence must be true. Otherwise, the combined sentence is false.

Example 2

What numbers x satisfy this statement?

x is a multiple of 6 and x is a multiple of 4.

Solution

The multiples of 6 are: 6, **12**, 18, **24**, 30, **36**, 42, …

The multiples of 4 are: 4, 8, **12**, 16, 20, **24**, 28, 32, **36**, 40, …

The numbers satisfying the given statement are in both lists.

These are the numbers: 12, 24, 36, …

We can illustrate *Example 2* with two overlapping loops. One loop contains multiples of 6. The other loop contains multiples of 4. The numbers satisfying the given statement are those in both loops. This diagram is a *Venn diagram*.

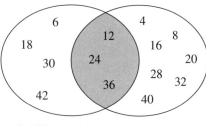

Multiples of 6   Multiples of 4

Example 3

On a number line, indicate the location of the numbers described by each statement.

a) $x > 3$ and $x < 8$ **b)** $x < 3$ and $x > 8$

Solution

a) $x > 3$ and $x < 8$

The first inequality describes all the numbers greater than 3. These are represented by the arrow in the first diagram. The second inequality describes all the numbers less than 8. These are represented by the arrow in the second diagram.

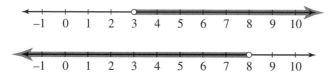

The combined statement describes all the numbers represented by *both* arrows. These are the numbers between 3 and 8 (exclusive).

b) $x < 3$ and $x > 8$

The first inequality describes all the numbers less than 3. The second inequality describes all the numbers greater than 8.

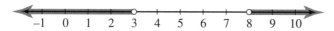

The combined statement describes all the numbers represented by *both* arrows. Since there are no such numbers, the combined statement does not describe any numbers on the number line.

Using "or"

On page 385, sentence ⑤ is formed using the word "or." For the combined sentence to be true, at least one of the two parts (or both parts) must be true.

Example 4

What numbers *x* satisfy this statement?

x is a factor of 12 or *x* is a factor of 20.

Solution

The factors of 12 are: 1, 2, 3, 4, 6, and 12

The factors of 20 are: 1, 2, 4, 5, 10, and 20

All these numbers satisfy the given statement.

These are the numbers: 1, 2, 3, 4, 5, 6, 10, 12, and 20

We can illustrate *Example 4* in a Venn diagram.

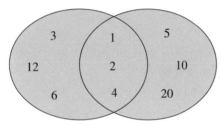

Factors of 12   Factors of 20

Example 5

On a number line, indicate the location of the numbers described by each statement.

a) $x > 3$ or $x < 8$ **b)** $x < 3$ or $x > 8$

Solution

a) $x > 3$ or $x < 8$

The first inequality describes all the numbers greater than 3.
The second inequality describes all the numbers less than 8.

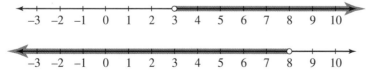

The combined statement describes all the numbers represented by at least one of the arrows. These are all real numbers.

b) $x < 3$ or $x > 8$

The first inequality describes all the numbers less than 3. The second inequality describes all the numbers greater than 8.

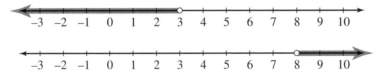

The combined statement describes all the numbers represented by at least one of the arrows. These are the numbers less than 3 and the numbers greater than 8.

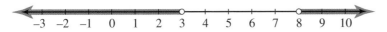

Example 6

A student survey gives these data.

60% of those surveyed have seen the movie *Return of the Jedi.*

40% have seen *Jurassic Park.*

30% have seen *Twister!*

30% have seen both *Return of the Jedi* and *Jurassic Park.*

20% have seen both *Return of the Jedi* and *Twister!*

15% have seen both *Jurassic Park* and *Twister!*

10% have seen all three movies.

What percent of the students have seen at least one of the three movies?

Solution

Draw a loop to represent each movie. Inside the loops, write the percent of students who have seen each movie. Since 10% of the students have seen all three movies, mark 10 in the central region.

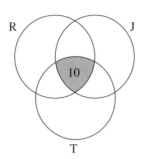

Fill in the data for students who have seen two movies. This includes those who have seen all three. For example, since 30% have seen both *Return of the Jedi* and *Jurassic Park*, and 10% have seen all three, mark 20% in the overlapping region above the central region. Similarly, mark 10% and 5% in the other overlapping regions.

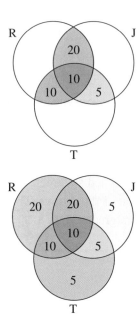

Fill in the data for students who have seen one movie. This includes those who have seen two and three movies. For example, since 60% saw *Return of the Jedi*, there must be a total of 60% in loop R. Mark 20% in the outer loop marked R. Similarly, mark 5% and 5% in the other two outer loops.

The total of the percents in the loops is the percent of students who have seen at least one of the three movies, which is 75%.

DISCUSSING THE IDEAS

1. Why isn't "The Flames lost last night" the negation of "The Flames won last night"?

2. In *Example 2*, why is the term *common multiples* appropriate for the values of x? What name is given to the first multiple, 12?

3. In *Example 3b*, the inequalities could be written as $8 < x$ and $x < 3$. Could we combine these into one inequality by writing $8 < x < 3$? Explain.

4. In *Example 6*, how could you use the diagram to determine the percent of the students who have:

   a) not seen any of the movies?

   b) seen exactly two of the movies?

   c) seen exactly one movie?

   d) seen only one or two movies?

   e) seen two or three movies?

A **1.** There are two treasure chests. What can you conclude from the information on the sign?

There is a treasure in chest #1 or in chest #2.

The treasure is not in chest #1.

2. Refer to the illustration above. What can you conclude if the second statement on the sign is changed as shown?

a)
There is a treasure in chest #1 or in chest #2.

The treasure is not in chest #2.

b)
There is a treasure in chest #1 or in chest #2.

The first statement is false.

3. What is the negation of each statement?

a) $x = 7$

b) $x < 10$

c) x is a multiple of 2.

d) x is not a factor of 24.

e) $\angle B$ is a right angle.

f) $\angle B$ is an acute angle.

4. There are 25 students in a class. All the students take theatre arts or industrial arts. Fifteen students take theatre arts and 21 take industrial arts.

a) How many students take both subjects?

b) How many students take theatre arts only?

c) How many students take industrial arts only?

5. The list of ingredients on a box of cereal may contain this phrase:
"… sugar and/or glucose …"
Why are the words "and" and "or" combined as "and/or"?

B **6.** On a number line, indicate the numbers corresponding to each statement.

a) $x \neq 7$ **b)** $x \leq 6$ **c)** $x \not> 4$ **d)** $x \not\leq -3$

7. Which numbers x satisfy each statement?

a) x is a multiple of 5 and x is a multiple of 4.

b) x is divisible by 2 and x is divisible by 5.

c) x is a factor of 32 and x is not a perfect square.

8. Choose one part of exercise 7. Illustrate the solution on a Venn diagram. Write to explain how you completed the diagram.

9. Which numbers x satisfy each statement?

   a) x is a factor of 12 or x is a factor of 18.

   b) x is a factor of 25 or x is a factor of 21.

   c) x is a multiple of 4 or x is a multiple of 10.

10. Choose one part of exercise 9. Illustrate the solution on a Venn diagram. Write to explain how you completed the diagram.

11. On a number line, indicate the location of the numbers described by each statement.

    a) $x > 6$ and $x < 12$ b) $x \geq -2$ and $x \leq 7$

    c) $x \geq -5$ and $x \geq 2$ d) $x < 4$ and $x > 9$

12. On a number line, indicate the location of the numbers described by each statement.

    a) $x < 5$ or $x > 8$ b) $x \leq 7$ or $x \geq -2$

    c) $x < -4$ or $x > 4$ d) $x < 8$ or $x > 8$

13. On a number line, indicate the location of the numbers described by each statement.

    a) $x < 3$ or $x > 6$ b) $x < 3$ and $x > 6$

    c) $x < 6$ or $x > 3$ d) $x < 6$ and not $x > 2$

14. On a number line, indicate the location of the numbers described by each statement.

    a) $x \geq -2$ and $x \leq 4$ b) $x \geq -2$ or $x \leq 4$

    c) $x \geq 4$ and $x \leq -2$ d) $x \geq 4$ or not $x \leq -2$

15. On a number line, indicate the location of the numbers described by each statement.

    a) $3 < x < 9$ and $6 < x < 12$ b) $3 < x < 9$ or $6 < x < 12$

    c) $3 < x < 6$ and $9 < x < 12$ d) $3 < x < 6$ or $9 < x < 12$

16. For this diagram, which statement is true?

    a) ABCD is a square and ABCD is a rhombus.

    b) ABCD is a square or ABCD is a rhombus.

    c) ABCD is a rectangle and ABCD is a rhombus.

    d) ABCD is a rectangle or ABCD is a rhombus.

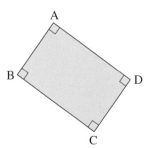

17. For this diagram, which statements are true?

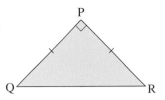

a) $\triangle PQR$ is isosceles and $\triangle PQR$ is a right triangle.

b) $\triangle PQR$ is isosceles or $\triangle PQR$ is a right triangle.

c) $\triangle PQR$ is isosceles and $\triangle PQR$ is equilateral.

d) $\triangle PQR$ is isosceles or $\triangle PQR$ is equilateral.

18. Andrea, Betty, and Carol have red, white, and blue caps, but not necessarily in that order. Only one of these statements is true.

• Andrea's cap is blue.

• Betty's cap is not blue.

• Carol's cap is not red.

What colour is each girl's cap?

19. The students in a class must take at least one of these subjects: French, music, or geography. Here is a list of how many students take these subjects.

14 students take French.   7 students take French and music.

18 students take music.   3 students take French and geography.

10 students take geography.   5 students take music and geography.

2 students take all three subjects.

How many students are in the class?

20. A school band has 40 members. Sixteen play the clarinet, 17 play the trumpet, and 23 play the flute. Five students play both the clarinet and the trumpet, 11 play both the trumpet and the flute, and 8 play both the clarinet and the flute. Three students play all three instruments. How many students in the band do not play any of these three instruments?

21. A sports club found that 95 of its members played at least one of these sports: football, tennis, or badminton. Thirty-six played football, 28 played tennis, and 51 played badminton. Of these, 5 played both football and tennis, 8 played both tennis and badminton, and 10 played both football and badminton. How many played:

a) all three sports?   b) exactly two sports?

c) only football?   d) football but not tennis?

22. Each member of a sports club plays at least one of these sports: soccer, rugby, or tennis. The following information is given:

163 play tennis; 36 play tennis and rugby; 13 play tennis and soccer; 6 play all three sports; 11 play soccer and rugby; 208 play rugby or tennis; 98 play soccer or rugby.

Use this information. Determine the number of members in the club.

23. a) John LeClair, Scott Mellanby, and German Titov played in the NHL in 1995-96. One played centre, one left wing, and the other right wing, but not necessarily in that order. Only one of these statements is true.

- Titov did not play centre.
- Mellanby played centre.
- LeClair did not play left wing.

What position did each athlete play? Check your answer using the *Hockey* database.

b) Use the database to make up a similar exercise for a classmate to complete.

24. Only one statement below is the negation of the statement "All teenagers like mathematics." Which statement is it?

a) All teenagers do not like mathematics.

b) Some teenagers like mathematics.

c) Some teenagers do not like mathematics.

25. In this section, the word "or" has been used in the *inclusive* sense, which allows for the possibility that both parts of a statement involving "or" are true. Unless stated otherwise, we always use it this way in mathematics. Another way to use "or" is in the *exclusive* sense, in which only one part of a statement involving "or" is true. Suppose we had used "or" in the exclusive sense in *Example 4*. How would the answer change? How would the Venn diagram change?

26. You are marooned on an island, where there are only liars and truth-tellers. You meet a couple and ask the wife, "Are you liars or truth-tellers?" She replies, "Either my husband is a truth-teller or I am a liar." What can you conclude?

C **27.** A condemned prisoner is given a choice. The judge says, "You must make a statement. If it is true, you will be shot. If it is false, you will be hanged." The prisoner made one of the following statements, and the judge had no choice but to set him free. Which statement did the prisoner make? Why did the judge set him free?

a) "I will be shot."   **b)** "I will be hanged."

COMMUNICATING THE IDEAS

Look up the word "or" in the dictionary. Do the meanings correspond to both the inclusive and the exclusive sense of this word? Write some examples from everyday speech that illustrate both the inclusive and the exclusive meaning of "or." Include examples that show how ambiguity in meaning can be avoided.

6.5 Statements Involving "If ... then"

After a practice, the coach announced:

"If it is raining, then the game is cancelled." ...①

Statement ① is formed by combining the statements "it is raining" and "the game is cancelled" with the words "If ... then." This is an example of an "If ... then" statement.

Example 1

Write each statement in an "If ... then" form. Is the statement true or false? Explain.

a) An isosceles triangle is equilateral.

b) All factors of odd numbers are odd.

Solution

a) If a triangle is isosceles, then it is equilateral. This is false. The triangle shown is a counterexample.

b) If a number is odd, then all its factors are odd. This is true because an odd number cannot have an even factor.

Starting with any "If ... then" statement, we can write other statements related to it. In the following examples, we start with statement ①.

Converse

Here is a statement related to statement ①.

> If the game is cancelled, then it is raining.   …②

Statement ② is the *converse* of statement ①. It does not have the same meaning as statement ① because the game could be cancelled for a different reason.

To form the converse of an "If … then" statement, interchange the two parts of the statement:

Statement: If … it is raining   then … the game is cancelled.

Converse: If … the game is cancelled   then … it is raining.

Contrapositive

Here is another statement that is related to statement ①.

> If the game is not cancelled, then it is not raining.   …③

Statement ③ is the *contrapositive* of statement ①. It *does* have the same meaning as statement ①. To see why, observe that it could not be raining because, if it were, then the game would be cancelled.

To form the contrapositive of an "If … then" statement, interchange the two parts of the statement, then form their negations.

Statement: If … it is raining   then … the game is cancelled.

Contrapositive: If … the game is *not* cancelled   then … it is *not* raining.

Example 2

Consider the statement: " Multiples of 6 are always even numbers."

a) Write the statement in an "If … then" form. Is it true or false? Explain.

b) Write the converse and the contrapositive of the statement. Is each new statement true or false? Explain.

Solution

a) If a number is a multiple of 6, then it is an even number.
This is true, because any number that is a multiple of 6 can be divided by 2.

b) Converse — If a number is even, then it is a multiple of 6.
This is false. The number 10 is a counterexample. It is an even number that is not a multiple of 6.

Contrapositive — If a number is not even, then it is not a multiple of 6. This is true, because only even numbers can be multiples of 6.

Example 3

Consider the statement: "All quadrilaterals with 4 equal sides are squares."

a) Write the statement in an "If … then" form. Is it true or false? Explain.

b) Write the converse and the contrapositive of the statement. Is each new statement true or false? Explain.

Solution

a) If a quadrilateral has 4 equal sides, then it is a square.

This is false. The quadrilateral at the right is a counterexample. It has 4 equal sides, but it is not a square.

b) Converse — If a quadrilateral is a square, then it has 4 equal sides.

This is true, from the definition of a square.

Contrapositive — If a quadrilateral is not a square, then it does not have 4 equal sides.

This is false. The quadrilateral above is a counterexample. It is not a square, but it has 4 equal sides.

DISCUSSING THE IDEAS

1. What is the converse of the converse of a statement? Explain.

2. What is the contrapositive of the contrapositive of a statement? Explain.

3. Is it possible to form the converse of the contrapositive of a statement? Explain.

6.5 EXERCISES

A 1. There are two treasure chests. What can you conclude from the information on the sign?

> If the treasure is in chest #1, it is worth at least $1 000 000.
>
> The treasure is in chest #1.

2. Refer to the illustration above. What can you conclude if the second statement on the sign is changed as shown?

a)
> If the treasure is in chest #1, it is worth at least $1 000 000.
>
> The treasure is in chest #2.

b)
> If the treasure is in chest #1, it is worth at least $1 000 000.
>
> The treasure is worth only $100.

3. Decide whether each statement is true or false. Write the converse of each statement, then decide if it is true or false.

   a) If you live in Red Deer, then you live in Alberta.

   b) If the ground is wet, then it is raining.

   c) If you study, you will pass the test.

   d) If you don't buy a ticket, you won't win the lottery.

   e) If you can't swim, you can't go out in a canoe.

4. Write the contrapositive of each statement in exercise 3, then decide if it is true or false.

B 5. Decide whether each statement is true or false. Write the converse of each statement, then decide if it is true or false.

   a) If two rectangles have the same area, then they have the same length.

   b) If two positive numbers have a product greater than 10, then both of the numbers are greater than 10.

   c) If a right triangle contains a 30° angle, then it also contains a 60° angle.

   d) If a triangle does not have three equal sides, then it is not equilateral.

   e) All rectangles are parallelograms.

6. Write the contrapositive of each statement in exercise 5, then decide if it is true or false.

7. a) Write the statement in the cartoon in an "If … then" form.

   b) Write the converse of the statement in part a.

   c) Write the contrapositive of the statement in part a.

8. Write each sentence in an "If … then" form.

   a) Anything that can go wrong will go wrong.

   b) Where there's a will, there's a way.

   c) It isn't over until it's over.

   d) It is impossible to eat only one potato chip.

© Jim Unger/dist. by LaughingStock Licensing Inc.

"I can start work Monday if I don't win the lottery."

9. Which of these statements are true for all real numbers?

   a) If x is even, then $x + 1$ is odd. **b)** If $x > 1$, then $x^2 > 1$

   c) If $x^2 > 0$, then $x > 0$ **d)** If $x^2 \leq 1$, then $0 \leq x \leq 1$

   e) If $xy = 0$, then $x = 0$ or $y = 0$ **f)** If $xy = 12$, then $x = 4$ and $y = 3$

   g) If $xy = 2$, then $x = 2$ or $y = 2$ **h)** If $xy < 0$, then $x < 0$ or $y < 0$

10. Decide whether each statement is true or false. Write the converse of each statement, then decide if it is true or false.

   a) Multiples of 6 are always multiples of 3.

   b) If x is a prime number, then $2x$ is not a prime number.

   c) A multiple of 4 is a multiple of 2.

   d) The square of a positive number is greater than 1.

   e) The square root of any number is positive.

11. Write the contrapositive of each statement in exercise 10, then decide if it is true or false.

12. Only one statement below has the same meaning as the statement "If a quadrilateral is a square, then it is a rectangle." Which statement is it? Write to explain your answer.

   a) If a quadrilateral is not a square, then it is not a rectangle.

   b) If a quadrilateral is a rectangle, then it is a square.

   c) If a quadrilateral is not a rectangle, then it is not a square.

13. A detective was investigating a crime. She interviewed the musician, the manager, and the server. Here are their statements.

Manager's statement: The musician is guilty.

Musician's statement: The manager or the server is guilty.

Server's statement: If the musician is innocent, the manager is guilty.

Two people were not telling the truth. Who committed the crime?

14. This sign is in an airport terminal.

    a) Write the statement on the sign in an "If … then" form.

    b) Why do you think an "If … then" form was not used on the sign?

> **NO SMOKING
> OTHER THAN IN
> DESIGNATED AREAS**

15. Match each statement in Column 2 with one in Column 1 that means the same.

    Column 1

    a) If you are early, you are lucky.

    b) If you are early, you are unlucky.

    c) If you are late, you are unlucky.

    Column 2

    1) If you are lucky, you are late.

    2) If you are lucky, you are early.

    3) If you are unlucky, you are late.

C 16. You are marooned on an island, where there are only liars and truth-tellers. You meet two people and ask one of them, "Are you liars or truth-tellers?" She replies, "At least one of us is a liar." What can you conclude?

17. Suppose you come to a fork in the road that leads either to danger or to safety. Two people are standing there. One of them always lies and the other always tells the truth, but you don't know which is which. You can ask one of them only one question. Which of these questions should you ask to be sure you know which road to take? Which road is it?

    a) "Which road leads to safety?"

    b) "Which road would your friend tell me leads to safety?"

    c) "If I asked you if the road on the right leads to safety, would you say yes?"

COMMUNICATING THE IDEAS

Explain what is meant by the converse of an "If … then" statement. Illustrate your explanation with an example of a true statement that has a true converse, and a true statement that has a false converse.

Explain what is meant by the contrapositive of an "If … then" statement. Illustrate your explanation with an example. Use your example to explain why the contrapositive has the same meaning as the original statement.

Logical Connectives in Sports Rules

The words "not," "and," "or," and "If … then" are called *logical connectives*. Many sports rules involve logical connectives. Here are some examples.

Tennis

- If the first serve misses the correct service court, then the server gets a second serve.
- A "let" is called if the ball hits the net on a serve and falls into the correct service court.
- After the serve, players must hit the ball either before it bounces or after it has bounced only once.

Baseball

- If the game is tied after 9 innings, then extra innings are played.
- The game is over if the home team is ahead after $8\frac{1}{2}$ innings.
- A strike is called if the batter hits a foul ball and there are fewer than 2 strikes.

Basketball

- If there is a technical foul on a coach, then two free throws are awarded.
- A team must try for a basket within 30 s of gaining control of the ball.
- When a violation is called before a shot is attempted the ball becomes dead, and if a basket is made it is not counted.

1. Choose one sport above. The first rule is written in an "If … then" form. Write the second rule in an "If … then" form.

2. Choose one sport above, or another sport. Use your knowledge of that sport to write some additional statements containing the logical connectives "not," "and," "or," and "If … then" involving the sport.

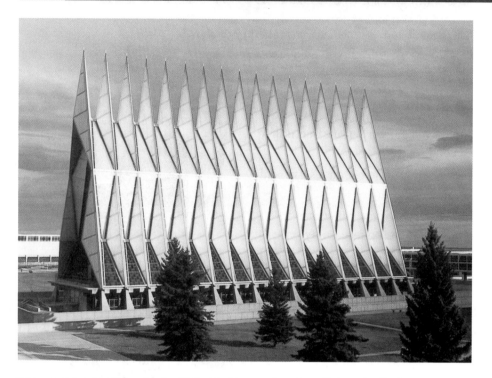

This striking building is the U.S. Air Force chapel at Colorado Springs, Colorado. It appears to consist of many triangular sections that all have the same size and shape. For this reason, we say that the triangular sections are *congruent*.

If $\triangle ABC$ and $\triangle PQR$ are congruent, we write: $\triangle ABC \cong \triangle PQR$. This means that:

Corresponding angles are equal.	Corresponding sides are equal.
$\angle A = \angle P$	$AB = PQ$
$\angle B = \angle Q$	$BC = QR$
$\angle C = \angle R$	$AC = PR$

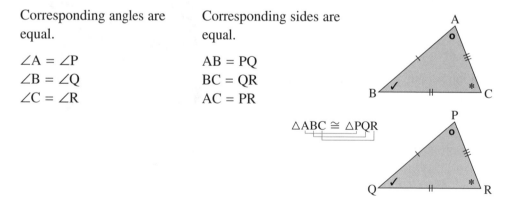

$\triangle ABC \cong \triangle PQR$

However, it is not necessary to know this much information to be certain that two triangles are congruent. The minimum conditions under which two triangles are congruent are called *congruence theorems*.

SSS Congruence Theorem

If three sides of one triangle are equal to three sides of another triangle, then the triangles are congruent.

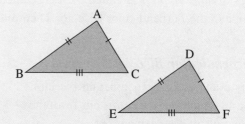

SAS Congruence Theorem

If two sides and the contained angle of one triangle are equal to two sides and the contained angle of another triangle, then the triangles are congruent.

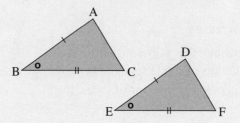

ASA Congruence Theorem

If two angles and the contained side of one triangle are equal to two angles and the contained side of another triangle, then the triangles are congruent.

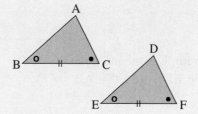

Example 1

In the diagram, P is any point on the perpendicular bisector of line segment AB.

a) Prove that △PAM ≅ △PBM.

b) Prove that PA = PB .

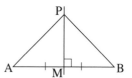

Solution

Think...

Both triangles are right triangles. Two corresponding legs in each triangle are given equal. The other legs are common.

a) In △PAM and △PBM,

$$AM = BM$$
$$\angle PMA = \angle PMB$$
$$PM = PM$$

Therefore, △PAM ≅ △PBM (SAS)

b) Since the triangles are congruent, PA = PB

The result of *Example 1* is called the Perpendicular Bisector Theorem. The converse of the Perpendicular Bisector Theorem is also true (see exercise 23).

Perpendicular Bisector Theorem

Any point on the perpendicular bisector of a line segment is equidistant from the endpoints of the segment.

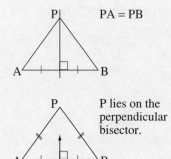

PA = PB

Converse

Any point that is equidistant from the endpoints of a line segment is on the perpendicular bisector of the segment.

P lies on the perpendicular bisector.

Example 2

In the diagram, E is the midpoint of both AC and BD. Prove that AB = CD.

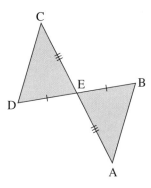

Solution

Think...

Two corresponding sides of each triangle are given equal.
The angles between these sides are equal by the Opposite Angles Theorem.

By the Opposite Angles Theorem, $\angle AEB = \angle CED$

In $\triangle ABE$ and $\triangle CDE$,
$$AE = CE$$
$$\angle AEB = \angle CED$$
$$BE = DE$$

Therefore, $\triangle ABE \cong \triangle CDE$ (SAS)

Since the triangles are congruent, AB = CD

Example 3

In the diagram, BC = ED, ∠OBA = ∠OEF, and ∠OCB = ∠ODE

Prove that ∠BOC = ∠EOD.

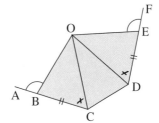

Solution

> *Think...*
>
> If we can show that ∠OBC = ∠OED, we can use ASA to prove that △OBC ≅ △OED.

Since ABC and DEF are straight lines and ∠OBA = ∠OEF, we know that ∠OBC = ∠OED.

In △OBC and △OED,   ∠OBC = ∠OED

BC = ED

∠OCB = ∠ODE

Therefore, △OBC ≅ △OED (ASA)

Since the triangles are congruent, ∠BOC = ∠EOD

DISCUSSING   THE IDEAS

1. a) In *Example* 1, what kind of triangle is △PAB ?

   b) Explain how this example can be used to prove the following theorem.

   ### Isosceles Triangle Theorem

   In any isosceles triangle, the angles opposite the equal sides are equal.

   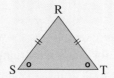

2. Some people might say that the ASA Congruence Theorem should be called the "ASA or AAS Congruence Theorem."

   a) Explain why two triangles are congruent if two angles and a non-contained side of one triangle are equal to two angles and a non-contained side of another triangle.

   b) Do you agree that the theorem should be renamed? Explain.

A **1.** For each diagram, state why the triangles are congruent; then list pairs of equal angles and equal sides.

a)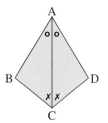

b)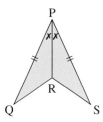

c)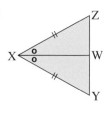

2. In each diagram, find two congruent triangles; then state their congruence theorem.

a)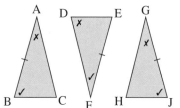

b)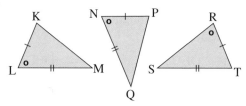

3. For each diagram, state a third condition necessary for the triangles to be congruent.

a)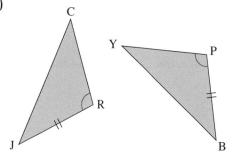

b)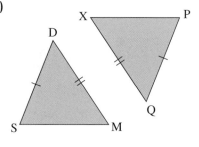

4. Give counterexamples to show that the following conditions are not conditions for congruent triangles.

   a) the measures of three angles, AAA

   b) the lengths of two sides and the measure of an angle not between them, SSA

5. a) In the diagram (below left), explain why $\triangle PQS \cong \triangle RQS$.

  b) Explain why $\angle P = \angle R$.

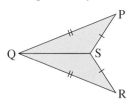 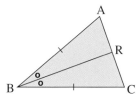

6. a) In the diagram (above right), explain why $\triangle ABR \cong \triangle CBR$.

  b) Explain why $\angle A = \angle C$.

7. a) In the diagram (below left), explain why $\triangle ABD \cong \triangle CBD$.

  b) Explain why $AB = CB$.

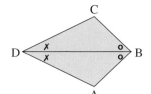 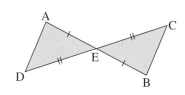

8. In the diagram (above right), explain why $AD = BC$ and $\angle D = \angle C$.

9. In the diagram (below left), explain why $CA = CD$ and $\angle A = \angle D$.

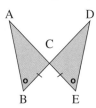 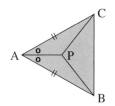

10. In the diagram (above right), explain why $\triangle PBC$ is isosceles.

11. In the diagram (below left), AB and CD are chords of equal length in a circle with centre O. Prove that $\angle AOB = \angle COD$.

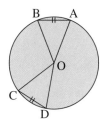 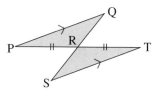

12. In the diagram (above right), $PQ \parallel ST$ and R is the midpoint of PT. Prove that R is also the midpoint of QS.

13. In the diagram (below left), the angles and segments marked are equal. Prove that AB = AD.

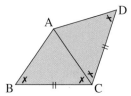
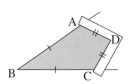

14. To bisect an angle (above right), a carpenter places a try-square as shown. Prove that BD bisects ∠ABC.

15. In quadrilateral PQRS (below left), PQ = RQ, PS = RS, and T is any point on the diagonal QS. Prove that PT = RT.

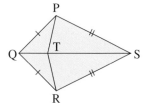
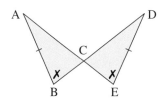

16. In the diagram (above right), use the information given to prove that AE = DB.

17. In the diagram (below left), use the information given to prove that KL = NM.

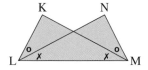
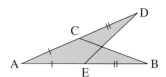

18. In the diagram (above right), use the information given to prove that BC = DE.

19. In the diagram (below left), AB = AC and AB ∥ DE. Prove that BC bisects ∠ACE.

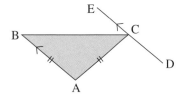
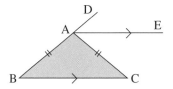

20. In the diagram (above right), AB = AC and AE ∥ BC. Prove that AE bisects ∠CAD.

21. In the diagram (below left), use the information given to prove that AD = AE.

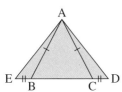 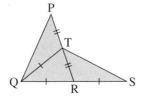

22. In the diagram (above right), use the information given to prove that ∠PTQ = ∠TRS and PQ = TS.

23. a) In the diagram (below left), C is the midpoint of AB and PA = PB. Prove that PC is perpendicular to AB.

b) Explain why the result of part a proves the converse of the Perpendicular Bisector Theorem.

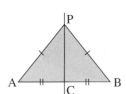 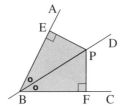

24. Use the diagram (above right). Prove the ***Angle Bisector Theorem***: Any point on the bisector of an angle is equidistant from the arms of the angle.

25. State and prove the converse of the Angle Bisector Theorem.

C **26.** In △PQR, S is the midpoint of QR and PS ⊥ QR. Prove that △PQR is isosceles.

27. In quadrilateral PQRS, PQ = QR and the diagonal QS bisects ∠Q. Prove that PS = RS.

28. Segments AB and CD bisect each other at M. Prove that AC = BD.

29. Given XY and WZ are diameters of a circle, prove that XW = YZ.

30. Circles with centres A and B intersect at C and D. Prove that ∠ACB = ∠ADB.

COMMUNICATING   THE IDEAS

Explain the difference between similar triangles and congruent triangles. Use diagrams to illustrate your explanation.

Logic Puzzles

From time to time, certain kinds of logic puzzles appear in books and magazines. Here is a sample of these puzzles. Try to solve as many of them as you can.

1. A farmer offered to cut a log into 3 pieces for $5.00. How much should she charge to cut the log into 6 pieces?

2. A drawer contains 10 brown socks and 10 blue socks. Suppose you want to be sure you have at least two socks of the same colour. What is the least number of socks you need to remove from the drawer?

3. There are 3 treasure chests. Each chest has a statement above it. Only 1 statement is correct. Which chest contains the treasure?

The treasure is in this chest.   The treasure is not in this chest.   The treasure is not in the red chest.

4. All 3 statements above these treasure chests are incorrect. Which chest contains the treasure?

The treasure is not in the red chest.   The treasure is in this chest.   The treasure is in the yellow chest.

5. All the labels on these boxes are incorrect. You may select only 1 fruit from 1 box. How can you relabel the boxes correctly?

6. The first names of King, Laird, and Port are Ken, Louise, and Max, but not necessarily in that order. Port's first name is not Max. Laird is Max's uncle. What are the first and last names of each person?

7. Ms. Vrentzos points to a family photograph in her home. She says, "I have no brothers and sisters, but that woman's mother is my mother's daughter." How is Ms. Vrentzos related to the person in the photograph?

8. Each card below has a colour on one side and a geometric figure on the other. Which cards should you pick up and turn over to find out if every green card has a circle on the other side?

9. You are marooned on an island, where there are only liars and truth-tellers. You meet a couple and the husband says, "My wife told me that she is a liar." Is he a liar or a truth-teller? How do you know?

10. Suppose that, in Puzzle 9, you meet another couple and ask the wife, "Are you liars or truth-tellers?" She replies, "My husband and I are both liars." What can you conclude?

11. Although the book *Anno's Hat Tricks* is written for young children, it contains several thought-provoking logic puzzles involving coloured hats. This is one of those puzzles.

Two children close their eyes, while their teacher puts coloured hats on their heads. The teacher has two red hats and one white hat from which to choose. This is what happens when the children open their eyes:

Assume you are the other child.

a) Before you heard Andy's response, could you tell the colour of your hat? Explain.

b) After you heard Andy's response, could you tell the colour of your hat? Explain.

c) How would this situation change if the teacher asked you the question first?

d) Could Andy have said his hat was red? Explain.

e) Could Andy have said his hat was white? Explain.

6.7 | Indirect Proof

In a previous grade, you probably drew two parallel lines and a transversal, measured the angles formed, and concluded that the alternate angles are equal. But you cannot do this for all parallel lines and all transversals. To explain why the alternate angles are equal, we can reason as follows.

l_1 and l_2 are two parallel lines.
A transversal meets these lines at A and B, forming alternate angles with measures x and y.

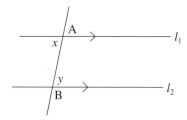

Think...

Either the alternate angles are equal, or they are not equal.
What would happen if these angles are *not* equal?

Assume that the alternate angles are not equal: $x \neq y$

Then, there must be another line l_3 through A such that the alternate angles are equal:

$$w = y \qquad \dots \text{①}$$

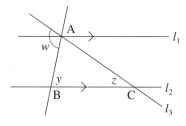

l_3 cannot be parallel to l_2 because l_1 is parallel to l_2, and there can only be one line through A parallel to l_2. Suppose l_3 meets l_2 at C.

Then, the angle marked w is an exterior angle in $\triangle ABC$.
Hence, according to the Exterior Angle Theorem, $w = y + z$.

This is impossible, because we already know from ① above that $w = y$.

Hence, the assumption that the alternate angles are not equal must be incorrect.

Therefore, the alternate angles must be equal.

Parallel Lines Theorem

If a transversal intersects two parallel lines, the alternate angles are equal.

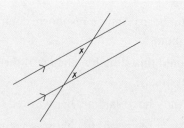

The Parallel Lines Theorem was proved above using a method called *indirect proof*. To use indirect proof, follow these steps.

Step 1. State that the result to be proved is either true or false.

Step 2. Assume that the result to be proved is false.

Step 3. Show that a conclusion can be reached that contradicts the known facts.

Step 4. Since there is a contradiction, the assumption in *Step 2* is incorrect. Using *Step 1*, we conclude that the result to be proved is true.

Example

In $\triangle ABC$, M is a point on BC such that BM $\neq$ CM, and AM bisects $\angle A$. Prove that AB $\neq$ AC.

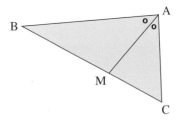

Solution

Use indirect proof.

Step 1. Either AB $\neq$ AC or AB = AC.

Step 2. Assume that AB = AC.

Step 3. **Think ...**

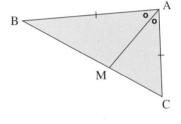

We could show that $\triangle ABM$ and $\triangle ACM$ are congruent.

In $\triangle ABM$ and $\triangle ACM$,   AB = AC   (assumption)

$\angle BAM = \angle CAM$ (given)

AM = AM

Therefore, $\triangle ABM \cong \triangle ACM$ (SAS)

Since the triangles are congruent, BM = CM

This contradicts the given information that BM $\neq$ CM.

Step 4. The assumption that AB = AC is incorrect. Therefore, AB $\neq$ AC

1. What would be the easiest way to prove that when a transversal intersects two parallel lines, the corresponding angles are equal?

2. When you use the method of indirect proof, are you using inductive reasoning, deductive reasoning, or another kind of reasoning? Explain.

3. The method of indirect proof is sometimes called *reductio ad absurdum*, which is a Latin phrase.

   a) What do you think this means?

   b) Why do you think this name is appropriate?

6.7 EXERCISES

A **1.** In $\triangle PQR$, $\angle Q = 50°$ and $\angle R = 60°$ Use the method of indirect proof to explain why $PQ \neq PR$.

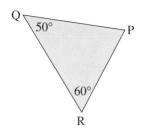

2. Use indirect proof.

   a) Prove that a triangle cannot have two right angles.

   b) Prove that a triangle cannot have two obtuse angles.

B **3.** Use the diagram at the right. Prove that only one perpendicular can be drawn from the point P to the line l.

4. Use the diagram at the right. Prove the converse of the Parallel Lines Theorem:

   If a transversal intersects two lines so that the alternate angles are equal, then the lines are parallel.

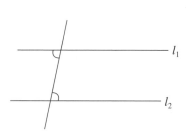

5. If n is an integer and n^2 is odd, prove that n is odd.

6. If n is an integer and n^2 is even, prove that n is even.

7. If m and n are positive integers and their product mn is odd, prove that both m and n are odd.

8. In $\triangle ABC$ (below left), AM is the median from A to BC, and $\angle AMC = 60°$. Prove that $AB \neq AC$.

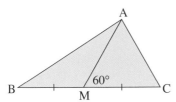 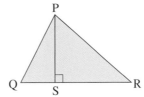

9. In $\triangle PQR$ (above right), PS is the altitude from P to QR, and $QS \neq RS$. Prove that $PQ \neq PR$.

10. Prove the converse of the Isosceles Triangle Theorem:

If two angles in a triangle are equal, then the sides opposite those angles are equal.

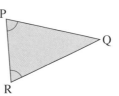

11. In $\triangle DEF$, G and H are the midpoints of DE and DF, respectively. The line segments EH and FG are medians. Prove that these medians cannot bisect each other.

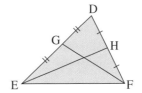

C **12.** The ages of Anjanee, Blair, and Concetta are three consecutive numbers. Only *one* of the following statements is true. Prove that Anjanee is the oldest of the three.

- Blair is 2 years older than Anjanee.
- Blair is 1 year older than Concetta.
- Anjanee is 1 year older than Concetta.
- Concetta is 1 year younger than Anjanee.

COMMUNICATING THE IDEAS

Suppose your friend was absent from today's mathematics class. How would you explain how to use the method of indirect proof? Choose one of the exercises above that is simple enough that you could discuss it on the telephone to illustrate your explanation.

Measuring Earth's Circumference

How do you think you could measure the distance around Earth? This problem was first tackled by the ancient Greeks more than 2000 years ago.

Eratosthenes (c. 276-192 B.C.) was an expert mathematician, astronomer, and geographer. He had a clever idea for estimating the circumference of Earth. He knew that at a certain time in Syene, a town south of Alexandria, vertical objects had no shadows. This meant that the sun was directly overhead. At the same time in Alexandria, vertical objects *did* have shadows. By measuring, Eratosthenes found that the inclination of the sun's rays was approximately 7.5°.

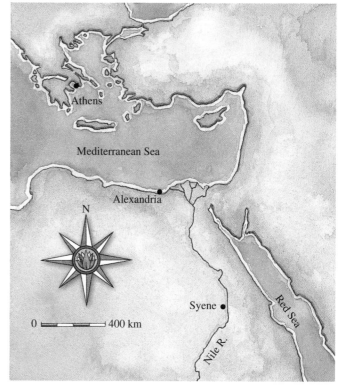

How is it possible that, at the same time, there are shadows at Alexandria but not at Syene? Eratosthenes realized that the only possible answer is that Earth is spherical. Further, the shadows at Alexandria were the key to calculating Earth's circumference.

Eratosthenes assumed that the sun was so far away that its rays at Syene and Alexandria are parallel. He drew a diagram like the one on page 417.

From his diagram, Eratosthenes concluded that ∠O at the centre of Earth is 7.5°.
Since a complete rotation is 360°, he calculated that ∠O represented about $\frac{7.5}{360}$, or $\frac{1}{48}$ of a complete rotation.
He then reasoned that the distance from Alexandria to Syene was about $\frac{1}{48}$ of the circumference of Earth.

Eratosthenes knew the distance from Alexandria to Syene, and therefore was able to calculate the circumference of Earth. Although he did not know it, his estimate was surprisingly close to the correct value. This was a remarkable achievement, accomplished over 2000 years ago, using only simple instruments and a clever idea.

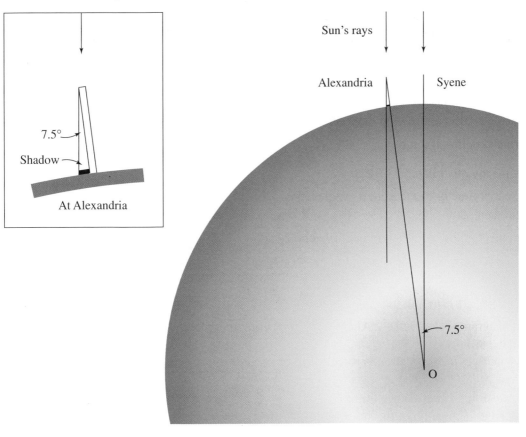

At Alexandria

1. Look at the diagram above. What property of parallel lines did Eratosthenes use to conclude that $\angle O = 7.5°$?

2. How do we know that $\frac{7.5}{360} = \frac{1}{48}$?

3. In Eratosthenes' time, distances were measured in a unit called "stades". Eratosthenes knew that the distance from Alexandria to Syene was approximately 5000 stades.

   a) Calculate the circumference of Earth in stades.

   b) One stade is about 157 m. Calculate Earth's circumference in metres and in kilometres.

4. The kilometre was originally defined so that the distance from the equator to the North Pole is 10 000 km.

   a) According to this definition, what is Earth's circumference in kilometres?

   b) How close was Eratosthenes' estimate of the circumference of Earth to the correct value?

5. How do you think Eratosthenes was able to determine information about shadows in two widely separated locations at the same time?

1. Suppose you multiply two odd numbers.

   25×41 5×7 101×203

   All the products are odd numbers.

   a) Check this by multiplying pairs of odd numbers.

   b) What conclusion might you make?

2. **a)** Draw any quadrilateral ABCD.

   b) Use a protractor to measure its angles, then find their sum.

   c) Do this for several different quadrilaterals.

   d) What conclusion might you make?

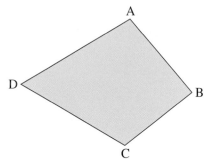

3. In the diagram (below left), $\angle ABD = \angle CBD$; write to explain why $\angle ABD = 90°$.

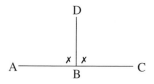

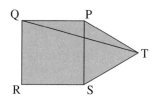

4. In the diagram (above right), PQRS is a square and $\triangle PST$ is an equilateral triangle. Explain why $\triangle PQT$ is isosceles.

5. Each example below suggests a conjecture. Find a counterexample to show each conjecture is false.

   a) $2^2 - 1 = 3$ $2^5 - 1 = 31$

   $2^3 - 1 = 7$ $2^{19} - 1 = 524\ 287$

   Conjecture: If n is a natural number, then $2^n - 1$ is a prime number.

   b)

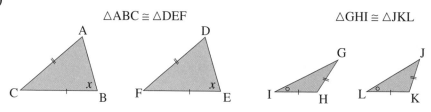

$\triangle ABC \cong \triangle DEF$ $\triangle GHI \cong \triangle JKL$

   Conjecture: If two pairs of corresponding sides and a pair of corresponding non-contained angles in two triangles are equal, then the triangles are congruent.

6. Thirty people are surveyed at a mall. All the people shopped at the grocery store or the department store. Eighteen people shopped at the grocery store and 17 shopped at the department store.

   a) How many people shopped at both stores?

   b) How many people shopped only at the grocery store?

   c) How many people shopped only at the department store?

7. Which numbers x satisfy each statement?

   a) x is a multiple of 2 and x is a factor of 54.

   b) x is a factor of 24 and x is a factor of 25.

   c) x is a factor of 35 or x is a factor of 56.

   d) x is a multiple of 7 or x is a multiple of 8.

8. Choose one part of exercise 7. Illustrate the solution on a Venn diagram. Write to explain how you completed the diagram.

9. Decide whether each statement is true or false.

   a) If a quadrilateral is a square, then it is a rhombus.

   b) If a quadrilateral has two pairs of parallel sides, then it is a parallelogram.

   c) If you water a plant, then it will grow.

   d) If it is noon, then it is light outside.

10. Write the converse of each statement in exercise 9, then decide if it is true or false.

11. Write the contrapositive of each statement in exercise 9, then decide if it is true or false.

12. In the diagram (below left), use the information given to prove that $\angle M = \angle K$.

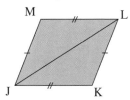 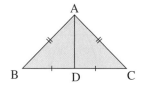

13. In the diagram (above right), use the information given to prove that AD bisects $\angle A$.

14. In the proof on page 412, C is on the right side of B. Modify the proof that the alternate angles are equal, assuming that C is on the left side of B.

15. Use the method of indirect proof to prove that when a transversal intersects two parallel lines, the corresponding angles are equal.

1. What is the present value of each investment?

   a) It yields $10 000 after 4 years at 3.75% interest compounded annually.

   b) It yields $5000 after 6 years at 6.5% interest compounded annually.

2. Determine each accumulated amount.

   a) $500 invested for 5 years at 4% compounded semi-annually

   b) $1200 invested for 2 years at $5\frac{1}{2}$% compounded semi-annually

3. Select one part of exercise 2. Write to explain how you determined the accumulated amount.

4. Antonia is planning to make regular contributions to her son's education fund, starting on January 1. Which of these plans would you recommend and why?

   a) Invest $300 on January 1, April 1, July 1, and October 1 at 7.5% compounded quarterly.

   b) Invest $100 at the beginning of each month at 7.25% compounded monthly.

5. Use the payroll deduction tables on pages 78 to 81.

   a) Alberto has gross biweekly earnings of $1127. He pays $200 per month into an RRSP. His claim code is 4. Determine his net earnings.

   b) Alberto gets a raise of $50 biweekly. He decides to increase his RRSP deduction to $225 per month. Recalculate his net earnings.

6. Identify each component of each quadratic function.

   i) the equation of the axis of symmetry　　　ii) the coordinates of the vertex

   iii) the x- and y-intercepts　　　iv) the domain and range

   a)
   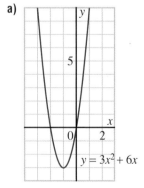
   $y = 3x^2 + 6x$

   b)
   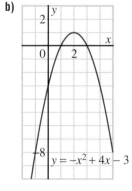
   $y = -x^2 + 4x - 3$

7. A producer of synfuel from coal estimates that the cost, C dollars, per barrel for a production run of x thousand barrels is given by $C = 9x^2 - 180x - 940$.

   a) How many thousand barrels should be produced each run to keep the cost per barrel at a minimum?

b) What is the minimum cost per barrel of synfuel?

8. Sketch a graph of each function. State its domain and range.

 a) $y = (x - 1)^2(x + 2)(x - 3)$ **b)** $y = (x - 3)^2(x + 4)^2$ **c)** $y = x^3(x - 1)$

9. A packaging company makes a box from cardboard 30.0 cm long and 25.0 cm wide. Determine the size of squares to be cut from the corners for each of these boxes. Determine the dimensions of each box.

 a) a box with volume 750 cm³   **b)** a box with the maximum possible volume

10. Determine $f(g(x))$ and $g(f(x))$ for each pair of functions.

 a) $f(x) = x^2$; $g(x) = 2x + 3$ **b)** $f(x) = \frac{1}{x}$; $g(x) = \frac{1}{2x}$

11. Choose one part of exercise 10. Write to explain how you determined $f(g(x))$ and $g(f(x))$.

12. Determine the remainder when $x^3 + 4x^2 - 7x + 10$ is divided by each binomial.

 a) $x - 1$ **b)** $x - 2$ **c)** $x - 3$ **d)** $x + 1$ **e)** $x + 2$ **f)** $x - 4$

13. Which polynomials have $x + 2$ as a factor?

 a) $2x^3 - 5x^2 - 2x + 7$ **b)** $x^3 - x^2 - 10x - 8$
 c) $3x^3 - 2x^2 - 4x - 3$ **d)** $2x^3 - x^2 - 7x + 6$

14. Solve each equation. Give the answers to 2 decimal places.

 a) $1.5x^2 + 2.0x - 3.2 = 0$ **b)** $\frac{1}{2}x^2 - \frac{3}{4}x - \frac{1}{8} = 0$
 c) $3x^2 + \sqrt{5}x - \sqrt{7} = 0$ **d)** $3.7x^2 - 5.7x + 1.1 = 0$

15. Choose one equation from exercise 14. Write to explain how you solved it.

16. a) Algeria, Benin, Cameroon, Djibouti, and Ethiopia are five countries in Africa. All the following statements about their 1990 populations are true. List the counties in order of decreasing population. Use the *Air Pollution* database to check your answer.

 • At least three of the countries had greater populations than Benin.
 • Algeria did not have the greatest population.
 • Cameroon had a smaller population than Algeria.
 • Djibouti had a smaller population than Benin.

 b) Use the database to make up a similar exercise for a classmate to complete.

17. Use the diagram at the right. Prove that if $x + y = 180°$, then lines l_1 and l_2 are parallel.

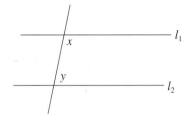

How Can We Map Earth's Surface?

 CONSIDER THIS SITUATION

Since Earth is spherical, it is impossible to represent its surface on a flat satellite image without distortions. These images were obtained from a site on the Internet that displays a variety of images generated from satellite data. The larger image represents Earth as it would appear from the sun at the moment the site was visited. It shows only half Earth's surface. The smaller image is a composite of many images arranged in a rectangle to represent Earth's entire surface. Both images contain distortions.

• Identify parts of Earth's surface that are distorted on each image.

• How was the rectangular image obtained from Earth's round surface?

On pages 446-447, you will develop the mathematical model that was used to generate the rectangular image from the satellite data. You will also use the model to visualize Earth's day and night regions.

 FYI Visit www.awl.com/canada/school/connections

For information related to the above problem, click on MATHLINKS followed by *Mathematics 11*. Then select a topic under How Can We Map Earth's Surface?

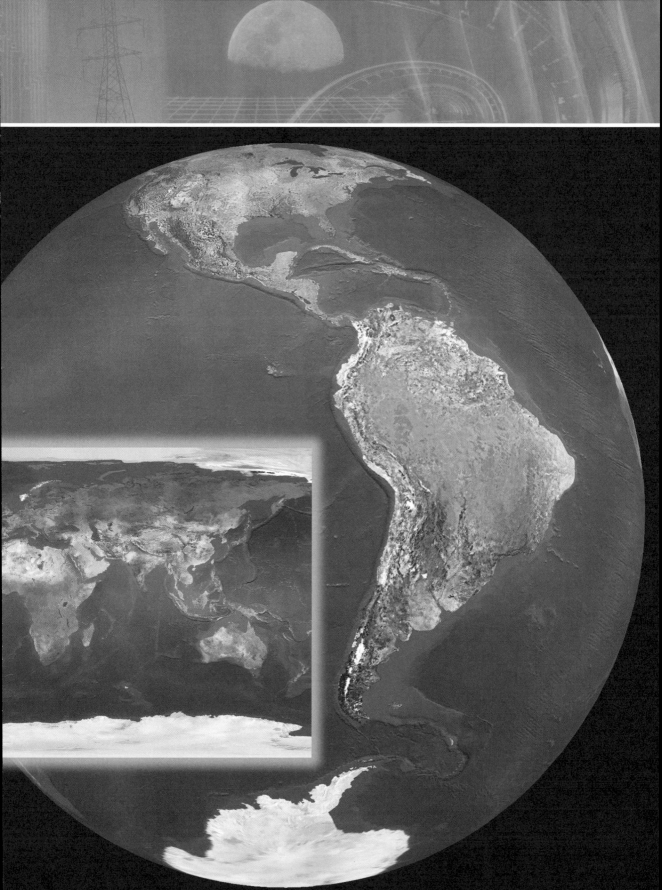

Conjectures and Proof in Geometry

In geometry, we construct figures and draw conclusions using inductive reasoning. For example, suppose we investigate the distances from the midpoint of the hypotenuse of a right triangle to the three vertices.

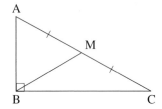

Making Conjectures

You can use dynamic geometry software such as *The Geometer's Sketchpad, The Geometric SuperSupposer*, and *Cabri,* or the *TI-92* calculator, to construct a right ΔABC and the midpoint M of its hypotenuse. You can measure the distances MA, MB, and MC, and display them on the screen. There are several ways you can change the diagram and create many others like it.

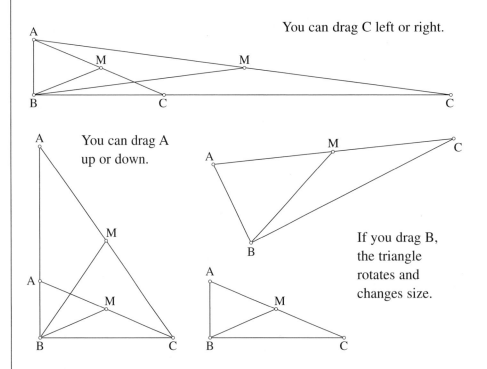

You can drag C left or right.

You can drag A up or down.

If you drag B, the triangle rotates and changes size.

At all times, B is a right angle and M remains the midpoint of AC. The distances MA, MB, and MC are always equal. Based on these examples, you would conjecture that the midpoint of the hypotenuse of a right triangle is equidistant from all three vertices.

You can use ruler and compasses, protractor, plastic triangles, or other instruments to construct a right triangle and the midpoint of its hypotenuse. By measuring, you would find that the distances MA, MB, and MC are equal. Before conjecturing that they are equal, you should make additional drawings, and compare the results with those of others.

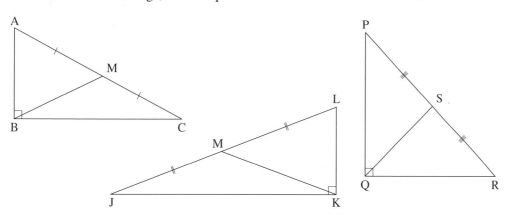

This symbol will accompany some of the activities and exercises in this chapter and the next. When you see it, you should use either a computer with dynamic geometry software, or geometric instruments.

Proving Conjectures

When you use geometric instruments, it is not feasible to consider more than a few examples. Although you can consider many more examples with a computer, the only way to be certain that the result is true in all cases is to use deductive proof.

You can prove the midpoint of the hypotenuse of a right triangle is equidistant from the three vertices as follows.

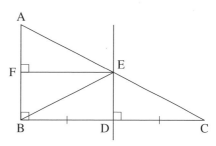

- Draw a right ΔABC in which ∠B = 90°. Draw the perpendicular bisector of BC, through D and meeting AC at E.

- Explain why ΔEBD ≅ ΔECD, and why EB = EC.

- Draw EF ⊥ AB.

- Explain why ΔEAF ≅ ΔEBF, and why EB = EA.

- Explain why the midpoint of AC is equidistant from A, B, and C.

If you were to open the plastic case of a floppy disk, you would find a thin, flexible disk. Information is stored on the disk as a vast number of magnetized spots, organized in tracks. The magnetized regions on any given track are all the same distance from the centre of the disk. A *circle* is defined as the set of all points on a plane that are equidistant from another point, the centre. Hence, the tracks on a floppy disk are examples of circles. Since all the tracks on the disk have the same centre, they are *concentric* circles.

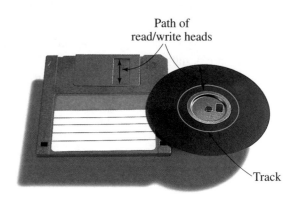

Path of read/write heads

Track

To read information from a floppy disk or write information on it, a magnetic read/write head moves across the disk surface along a radius. A *radius* is any line segment with one endpoint at the centre and the other on the circle. As the read/write head moves, the disk rotates around its centre. At each track, the read/write head pauses to decode or record information. When the disk has completed one rotation, the read/write head moves to the next track, and so on.

The disk completes one rotation in about 0.2 s. For time intervals less than 0.2 s, the read mechanism scans a fraction of a track. In 0.05 s, for example, only $\frac{1}{4}$ of the track will be read, as shown. The fractional part of the circle is called a *minor arc* if it is less than half the circle and a *major arc* if it is more than half the circle. The figure comprising the arc, the radii from the endpoints, and all enclosed points is a *sector* of the circle. The angle formed by the radii at the centre is the *sector angle*.

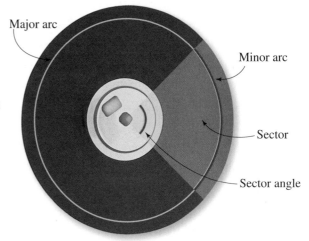

Major arc

Minor arc

Sector

Sector angle

A line segment that has both endpoints on the circle is a *chord*. You could model a chord on the disk by looping a rubber band around it. The part of the band that crosses one side of the disk represents a chord. It separates the disk into two *segments*.

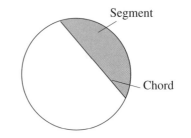

Segment

Chord

Example

Identify each item in the circle, centre O: radius, diameter, chord, minor arc, major arc, sector, segment

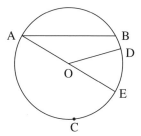

Solution

An example of each item is labelled in the diagram below. Other examples are possible for some items.

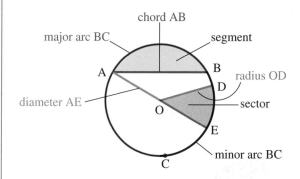

In the circle at the right, O is the centre and line segment PQ is a chord of the circle.
The line PQ is a *secant*. A secant is a line that intersects a circle at two different points.

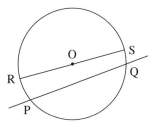

Some terms can have more then one meaning. In the diagram on the right,

- *radius* can refer either to the line segment OS or to its length.
- *diameter* can refer either to the line segment RS or to its length.
- *circumference* can refer either to the circle or to the distance around it.

1. Does a circle have a line of symmetry? If so, how many does it have and how would you construct them? If not, explain why not.

2. Why do you think the circle is the shape of choice for computer storage media?

3. Suggest meanings for the terms *minor segment, major segment, minor sector,* and *major sector.*

4. Classify each statement as true or false. Give reasons.

   a) All radii in a circle are equal.

   b) Every radius is a chord.

   c) Every chord is a diameter.

   d) A secant intersects a circle at just one point.

   e) A chord contains exactly two points on the circumference.

7.1 EXERCISES

A 1. Identify each item in the diagram (below left). Point O is the centre of the circle.

   a) 3 radii   b) a diameter   c) 3 chords   d) 2 secants   e) 5 minor arcs
   f) 4 major arcs   g) 2 segments   h) 2 sectors   i) 2 sector angles

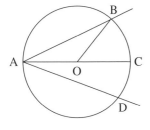

2. Identify each item in the photo (above right): radius, diameter, concentric circles, arc, sector.

3. The information storage pattern on a floppy disk is organized by sector and track.

   a) One common format uses 80 tracks and 18 sectors. Explain what this means.

   b) Draw a diagram to show what the storage pattern would look like on a floppy disk containing 10 tracks and 18 sectors.

4. Get an ordinary floppy disk. Slide the shutter aside so that you can see the curved edge of the disk inside the plastic case.

   a) Visualize the inner and outer tracks. As accurately as you can, measure the radius of each track.

   b) Assume there are 80 tracks on the disk. How far would the read/write head have to move to go from one track to an adjacent track?

   c) Calculate the area of the magnetic surface between the inner and outer tracks.

   d) Assume there are 18 sectors on the disk. Calculate the area of each sector.

5. Decide which items could exist. If an item could exist, sketch an example. If it cannot exist, write to explain why not.

   a) a chord that is also a diameter   b) a sector that is also a segment

6. Visualize two intersecting circles. At how many different points can they intersect? Draw diagrams to illustrate the different possibilities.

B 7. Visualize three intersecting circles, each with a different radius.

   a) What is the greatest number of points at which three circles can intersect?

   b) What is the least number of points at which three circles can intersect?

   c) Sketch diagrams to illustrate three circles intersecting at all possible numbers of points from the least to the greatest.

8. What is the point of intersection of any two diameters of a circle? Explain.

9. Two intersecting circles of equal radii are drawn. How many points on each circle are equidistant from both centres?

10. a) Calculate the perimeter of each inscribed regular polygon. Point O is the centre of each circle.

   i) diameter 17 cm   ii) radius 12 cm   iii) radius 6 cm

   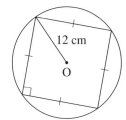
   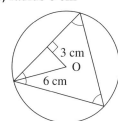

   b) Choose one circle from part a. Write to explain how you determined the perimeter of the polygon.

11. A plate with a diameter of 24 cm is placed on a square mat, with no overhang. Calculate the length of the diagonal of the mat.

12. a) Construct a circle. Mark two points on the circumference. Use the points to construct a chord. Measure the length of the chord.

b) Visualize keeping one endpoint of the chord fixed and moving the other endpoint around the circumference. What would you expect to happen to the chord length? Check your prediction.

c) Check your prediction with other circles.

d) Make a conjecture about the longest chord you can construct in any circle.

13. a) Construct a circle and two diameters BD and CA. Join the endpoints of the diameters to form quadrilateral ABCD.

b) Measure to determine what type of quadrilateral ABCD is. Make a conjecture about the type of quadrilateral that is formed by joining the endpoints of two diameters of a circle.

c) Visualize moving the endpoints of the diameters or changing the radius of your circle. Would your conjecture still hold?

d) What special case occurs when the two diameters are perpendicular?

C **14.** Repeat exercise 7, assuming that all three circles have the same radius.

15. All the circles have radius *r*. The dots indicate the centres of the circles. Determine the perimeter of each figure as a function of *r*.

a)

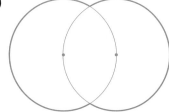

b)

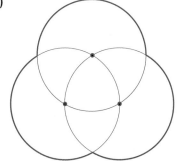

COMMUNICATING   THE IDEAS

In a dictionary, look up the words *radius, circumference, segment,* and *sector.* Examine the meanings carefully. For those words that have two mathematical meanings, are both included? Are other meanings included? Are any meanings given that would help you remember the difference between a segment and a sector?

Circles in Space

1. List as many examples of circles as you can find in the photographs. For each example, explain why it occurs.

2. Why do so many examples of circles occur in space?

3. Suggest an example of a circle that occurs in space that would be difficult or impossible to photograph.

Sometimes the setting sun looks like a large orange ball dropping below the horizon. Visualize the sun doing this as you look at the pictures. The chord formed by the horizon and the sun is horizontal. The centre of the sun moves straight down. What properties of chords of a circle do you think this illustrates? In this section, you will learn about the properties of chords.

INVESTIGATE

You will use a computer or geometric instruments to develop conjectures about chords. You will prove your conjectures in Section 7.3.

1. Construct a circle with centre O and a chord AB (not a diameter). Construct the midpoint, C, of AB. Construct the line OC.

   a) Measure ∠OCA.

   b) Visualize repeating this construction for other chords, and for chords in other circles. What would you expect to observe about ∠OCA? Check your prediction with other examples.

   c) Make a conjecture based on your findings.

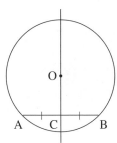

2. Construct a circle with centre O and a chord AB. Construct a line through O perpendicular to AB. Mark point C where AB and the perpendicular intersect.

   a) Measure AC and BC.

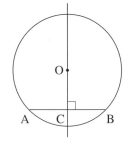

b) Visualize repeating this construction for other chords, and for chords in other circles. What would you expect to observe about AC and BC? Check your prediction with other examples.

c) Make a conjecture based on your findings.

3. Construct a circle with centre O and chord AB. Construct the midpoint, C, of AB and the line through C that is perpendicular to AB.

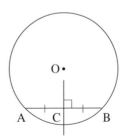

a) The line you constructed is the perpendicular bisector of chord AB. Through what other point does it pass?

b) Visualize repeating this construction for other chords, and for chords in other circles. What would you expect to observe about the perpendicular bisector of the chord? Check your prediction with other examples.

c) Make a conjecture based on your findings.

In *Investigate,* exercises 1 to 3, you should have conjectured the following properties of chords.

Chord Properties

1. A line through the centre of a circle that bisects a chord (not a diameter) is perpendicular to the chord.

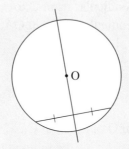

The line is perpendicular to the chord.

2. The perpendicular from the centre of a circle to a chord bisects the chord.

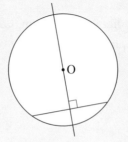

The line bisects the chord.

3. The perpendicular bisector of any chord contains the centre of the circle.

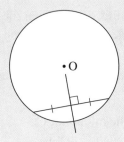

The line passes through the centre.

You can use these properties to solve problems involving chords.

Example

A circle has diameter 12 cm. A chord is 8 cm long. How far is the chord from the centre of the circle?

Solution

Draw a diagram.

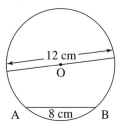

Think...

The distance from the centre to the chord is measured along the perpendicular from O to AB. If we construct the perpendicular and join OA, we will have a right triangle. If we can determine the lengths of two sides, we could use the Pythagorean Theorem.

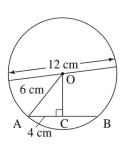

Drop the perpendicular from O onto AB at C.
According to Chord Property 2, C is the midpoint of AB.
Hence, AC = 4 cm

Draw radius OA. Since the diameter is 12 cm, OA = 6 cm

Apply the Pythagorean Theorem to $\triangle OCA$.

$OC^2 = OA^2 - AC^2$
$\qquad = 6^2 - 4^2$
$\ OC = \sqrt{6^2 - 4^2}$
$\qquad = \sqrt{20}$
$\qquad \doteq 4.47$

The chord is about 4.5 cm from the centre of the circle.

1. In *Investigate,* exercise 1, why can't the chord be a diameter? Can the chord be a diameter in the other exercises? Explain.

2. In *Investigate,* exercise 1, would it be possible for C to be the midpoint of AB, with OC not perpendicular to AB? Explain.

3. In *Investigate,* exercise 2, would it be possible for OC to be perpendicular to AB, with C not the midpoint of AB? Explain.

4. In *Investigate,* exercise 3, would it be possible for the perpendicular bisector of AB not to pass through the centre? Explain.

5. In the *Example,* the length was expressed in decimal form to 1 decimal place. What other ways are there to express this length?

7.2 EXERCISES

A 1. Determine each value of *x*. Point O is the centre of each circle.

a)

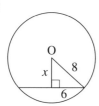

b)

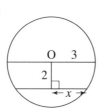

c)

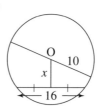

d)

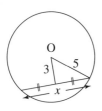

e)

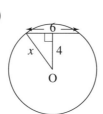

f)

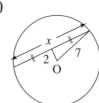

2. Determine each value of *x*. Point O is the centre of each circle.

a)

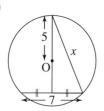

b)

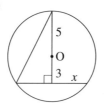

c)

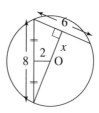

3. Choose one circle from exercise 1 or 2. Write to explain how you determined the value of *x*.

4. A circle (below left) has diameter 14 cm. A chord is 7 cm long. How far from the centre of the circle is the chord?

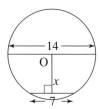 

5. Triangle PQR is inscribed in a circle with centre O (above right). Chord QR = 8.0 cm and PO extended is perpendicular to QR at M. The radius of the circle is 5.0 cm. Calculate each item.

a) the distance from O to QR   **b)** the length of PM

c) the length of PQ   **d)** the distance from O to PQ

6. a) Mark two points P and Q (below left).

b) Use a plastic or cardboard right triangle. Position the triangle so the points P and Q lie on the two shorter sides of the triangle. Mark the position of the right angle.

c) Repeat part b several times. Mark the different positions of the right angle.

d) What do you notice about the dots marking the positions of the right angle?

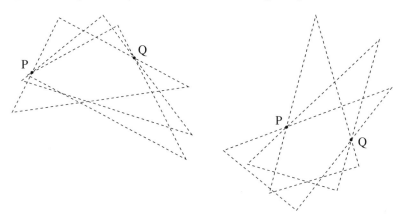

7. a) Repeat exercise 6. Position the triangle so points P and Q lie on the arms of one acute angle. Mark the different positions of the acute angle (above right).

b) How would the result change if the angle were smaller?

c) How would the result change if the angle were larger?

B 8. Determine each value of *z*. Point O is the centre of each circle.

a)

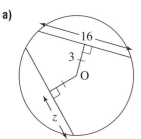

b)

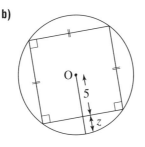

c)

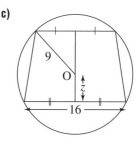

9. **a)** Determine each distance from the centre O of the circle to the chord.

i)

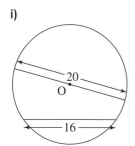

ii)

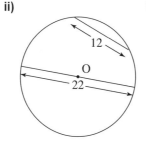

iii)

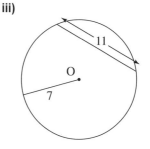

b) Choose one circle from part a. Write to explain how you determined the distance of the chord from the centre of the circle.

10. You are given a diagram of a circle with two chords. The centre of the circle is not marked. How could you locate the centre? Explain.

11. Visualize drawing a circle with a pencil, using a pen to draw two chords, then erasing the circle.

   a) Describe a method someone could use to re-create the circle using just the chords.

   b) Would it be possible to draw two chords like this in such a way that the circle could not be re-created? Explain.

12. This sketch shows a fragment of a plate that was unearthed by an archaeologist.

   a) Copy the sketch.

   b) Devise a method to sketch the remainder of the plate.

   c) Use your completed sketch to determine the plate's diameter.

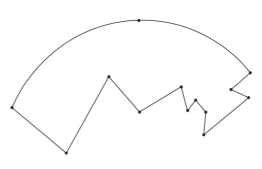

13. A circle has diameter 16 cm. A chord is 10 cm long. How far from the centre of the circle is the chord?

14. A circle has diameter 20 cm. Which of the following measures could be lengths of chords in this circle? For each possible length, determine the distance from the chord to the centre.

    a) 2 cm   **b)** 10 cm   **c)** 18 cm   **d)** 24 cm

15. A circle has diameter 12 cm. A chord is 4 cm from the centre. How long is the chord?

16. A circle has diameter 20 cm. Which of the following measures could be distances of chords from the centre? For each possible distance, determine the length of the chord.

    a) 2 cm   **b)** 10 cm   **c)** 18 cm   **d)** 24 cm

17. What is the diameter of a circle in which a chord 8.0 cm long is 5.3 cm from the centre?

18. What is the diameter of a circle in which a chord 16 cm long is 15 cm from the centre?

19. Choose exercise 17 or 18. Write to explain how you calculated the diameter.

20. The base of a large hemispherical dome (below left) is a circle with diameter 80 m. How far apart are two 20-m parallel support beams that form "chords" of the circular base?

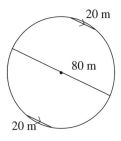
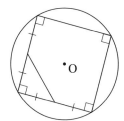

21. A square (above right) is inscribed in a circle, centre O and diameter 20 cm. What is the distance between the midpoints of adjacent sides of the square?

22. In the diagram, O is the centre of the circle. Determine the length of the line segment in each case. If not enough information is given, note this as your answer.

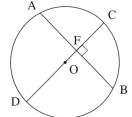

a) AF = 5 cm, AB = ▨

b) OB = 7 cm, CD = ▨

c) AC = 9 cm, OB = ▨

d) OB = 13 cm, OF = 5 cm, AB = ▨

e) CD = 30 cm, AB = 24 cm, AC = ▨

23. This diagram represents a fuel line on an F-18 aircraft. Some measurements are taken with a Vernier caliper: diameter EF = 0.357 cm and chord CD = 0.140 cm. A is the centre of the circle and H is the midpoint of segment CD. If the segment GH is longer than 0.010 cm, the fuel line is not within safety specifications and it must be replaced. Calculate to determine whether the fuel line must be replaced.

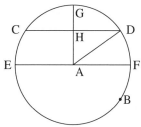

24. Construct a circle with centre O. Construct any chord AB. Construct another chord CD that has the same length as AB.

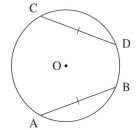

a) Measure the distance from each chord to the centre. What do you notice?

b) Visualize repeating this construction for other equal chords on the circle, and for other circles. What would you expect to observe about the distances from the chords to the centre? Check your prediction with other examples.

c) Make a conjecture based on your findings.

25. In exercise 24, find out if any special cases occur if:

a) the chords intersect

b) the chords are parallel

c) the chords are perpendicular

26. Repeat exercise 24. Use chords AB and CD with different lengths. In part b, visualize repeating the construction for other chords with different lengths.

27. Assume your conjecture from exercise 24 is correct. Determine each value of z. Point O is the centre of each circle.

a)

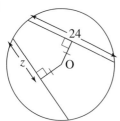

b)

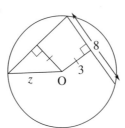

c)

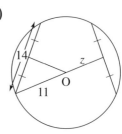

C **28.** Chord Property 1 states that a line through the centre of a circle that bisects a chord (not a diameter) is perpendicular to the chord. You can prove this property as follows.

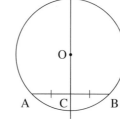

a) Draw a circle with centre O and a chord AB with midpoint C as shown. You want to prove that OC is perpendicular to AB.

b) Draw two triangles. Explain why they are congruent.

c) Explain how you can use the result of part b to conclude that OC is perpendicular to AB.

29. Chord Property 2 states that the perpendicular from the centre of a circle to a chord bisects the chord. You can prove this property as follows.

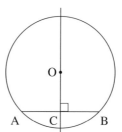

a) Draw a circle with centre O and a chord AB. Draw OC ⊥ AB. You want to prove that C is the midpoint of AB.

b) Draw two triangles, and explain why the congruence axioms cannot be used to conclude that they are congruent.

c) How else can you conclude that C is the midpoint of AB?

COMMUNICATING THE IDEAS

Write a brief summary of the chord properties you discovered in exercises 1 to 3 of *Investigate*. Include diagrams in your summary.

Chord Property 3 on page 434 is known as the Chord Perpendicular Bisector Theorem.

Chord Perpendicular Bisector Theorem

The perpendicular bisector of any chord contains the centre of the circle.

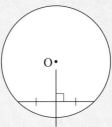

The line passes through the centre.

This theorem follows immediately from the converse of the Perpendicular Bisector Theorem. Since the centre is equidistant from the endpoints of the chord, the centre must be on the perpendicular bisector of the chord.

A theorem that is closely related to another theorem and is a self-evident consequence of that theorem is a *corollary* of that theorem. For example, here is a corollary of the Chord Perpendicular Bisector Theorem.

Corollary 1

The centre of a circle is the point of intersection of the perpendicular bisectors of any two non-parallel chords.

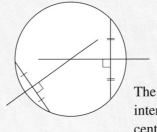

The point of intersection is the centre.

Chord Properties 1 and 2 on page 433 are also corollaries of the Chord Perpendicular Bisector Theorem.

Corollary 2

A line through the centre of a circle that bisects a chord (not a diameter) is perpendicular to the chord.

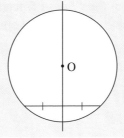

The line is perpendicular to the chord.

To see why this is true, consider a line through the centre and the midpoint of the chord. The perpendicular bisector passes through the same two points. Therefore, the line is the perpendicular bisector of the chord.

Corollary 3

The perpendicular from the centre of a circle to a chord bisects the chord.

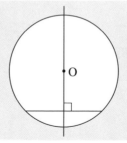

The line bisects the chord.

To see why this is true, consider a line through the centre that is perpendicular to a chord. It must be parallel to the perpendicular bisector. Both this line and the perpendicular bisector pass through the centre. Therefore, this line is the perpendicular bisector of the chord.

We can use Corollary 1 to construct a circle passing through the three vertices of any triangle. Such a circle is called the *circumcircle* of the triangle.

Example 1

A triangle with vertices A, B, and C is given.

a) Construct the circumcircle of △ABC.

b) Prove that your construction is correct.

Solution

a) Construct any △ABC.
Construct the perpendicular bisector of side AB.
Construct the perpendicular bisector of side BC.
Mark point O where the perpendicular bisectors intersect.
With centre O and radius OA, construct a circle.
The circle will pass through all three vertices of △ABC.

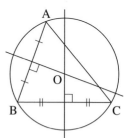

b) According to Corollary 1 of the Chord Perpendicular Bisector Theorem, O is the centre of a circle that passes through A, B, and C.

We can use the Chord Perpendicular Bisector Theorem to prove an important property of two equal chords in a circle.

Example 2

Two chords of a circle have the same length. Prove that they are the same distance from the centre.

Solution

Draw a circle with centre O and two chords AB and CD that have the same length. Drop perpendiculars OM and ON as shown. We must show that OM = ON.

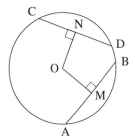

> **Think...**
>
> Suppose we join OA and OC. We know that OA = OC because O is the centre.
>
> Since OM ⊥ AB and ON ⊥ CD, we know that M and N are midpoints of the chords.
>
> Since ΔOAM and ΔOCN are right triangles, we can use the Pythagorean Theorem.

Join OA and OC. Since these are radii,

$$OA = OC \quad ①$$

Since OM ⊥ AB, according to the Chord Perpendicular Bisector Theorem, M is the midpoint of AB.
Similarly, N is the midpoint of CD.

Since it is given that AB = CD,

$$AM = CN \quad ②$$

Apply the Pythagorean Theorem in ΔOAM and ΔOCN to determine OM and ON.

In ΔOAM: $OM = \sqrt{OA^2 - AM^2}$

In ΔOCN: $ON = \sqrt{OC^2 - CN^2}$

$= \sqrt{OA^2 - AM^2}$ **Using ① and ②**

Since the two expressions for OM and ON are equal, OM = ON

We shall call the result of *Example 2* the Two Chords Theorem.

Two Chords Theorem

If two chords in a circle have the same length, they are equidistant from the centre.

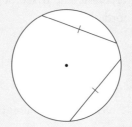

The chords are the same distance from the centre.

DISCUSSING **THE IDEAS**

1. a) What does it mean to say that a corollary is a self-evident consequence of a theorem?

   b) Explain why Corollary 1 is self-evident.

   c) Do you think Corollary 2 and Corollary 3 are self-evident? Explain.

2. The centre of the circumcircle of a triangle is called the *circumcentre*. In the solution of *Example 1,* the circumcentre was inside △ABC.

   a) Could the circumcentre of a triangle ever be outside the triangle? Explain.

   b) Could the circumcentre of a triangle ever be on one of the sides of the triangle? Explain.

3. An important step in the solution of *Example 2* was to draw the radii OA and OC.

   a) Why did drawing the radii help us solve the problem?

   b) What other radii could we have drawn instead?

7.3 EXERCISES

B **1.** Here is a way to prove Corollary 2 of the Chord Perpendicular Bisector Theorem that is different from the explanation on page 442.

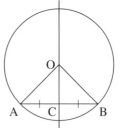

   a) Draw a circle with centre O and a chord AB. Join OA and OB. Mark the midpoint, C, of AB. Draw the line through O and C.

   b) Explain why ΔOAC ≅ ΔOBC.

   c) Explain how you can conclude that ∠OCA = ∠OCB = 90°

2. Find a way to prove Corollary 3 of the Chord Perpendicular Bisector Theorem that is different from the explanation on page 442. Write to explain your proof.

3. State and prove the converse of the Two Chords Theorem.

4. Two chords PQ and RS in a circle with centre O have the same length (below left). The chords intersect at T. Prove that PT = ST and QT = RT.

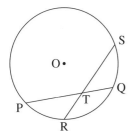
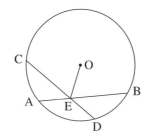

5. Two chords AB and CD in a circle with centre O have the same length (above right). The chords intersect at E. Prove that OE bisects ∠CEB.

C **6.** In the diagram (below left), diameter CD is perpendicular to chord AB. The diameter intersects the chord at E. Prove that ΔABC is isosceles.

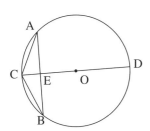
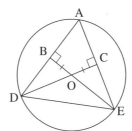

7. In the diagram (above right), EB is perpendicular to AD, DC is perpendicular to AE, and OB = OC. Prove that ΔADE is isosceles.

8. Use the method of indirect proof.

a) Prove the Chord Perpendicular Bisector Theorem.

b) Prove Corollary 1.

c) Prove Corollary 2.

COMMUNICATING THE IDEAS

Your friend is designing a logo. She has drawn a triangle and wants to draw a circle that passes through its three vertices. Write a series of steps she can follow. Include diagrams in your description, and explain why your method is correct.

How Can We Map Earth's Surface?

On page 422, you considered the problem of representing Earth's round surface in a rectangular image. This can be done in many ways.

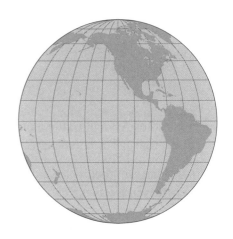

DEVELOP A MODEL

Visualize the lines of longitude on Earth's surface, perpendicular to the Equator. In the diagram, they are 15° apart. The lines of latitude are parallel to the Equator, also 15° apart.

We draw a rectangle, and use vertical and horizontal lines for the lines of longitude and latitude, respectively. Each small rectangle represents a region on Earth's surface determined by the corresponding lines of longitude and latitude.

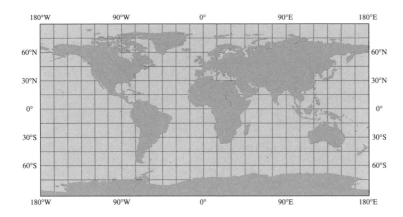

1. Does it matter what dimensions are used for the length and width of the rectangle? Explain.

2. To represent lines of longitude and latitude 15° apart, how many vertical and horizontal lines must be drawn in the rectangle? Explain how these numbers are determined.

3. Which two points on Earth's surface are not represented as single points on the rectangle? Explain.

4. Which parts of the surface are more distorted than others? Explain.

 **REVISIT THE SITUATION**

The site on the Internet that generated the images on page 423 also generated this image. This image shows the night and day regions of Earth at the moment the site was visited.

These exercises refer to the image above and the large image on page 423. These two images were made at the same time.

5. **a)** Check that both images show the same day region.

   b) What season was it in Canada when these images were made? Explain how you can tell from the images.

   c) Suppose the images had been made at the same time of day six months later. How would they differ from the ones in this book? Explain.

6. Find your location on each image. At your location, approximately what time of day was it when the images were obtained? Explain.

7. Suppose the images had been obtained later in the day. How would they differ from the images in this book?

All of these five photographs show the same building taken from different directions. The building takes up the entire width of each photograph. Where would the photographer have to stand to take photographs like these? You will discover the answer to this question in this section.

INVESTIGATE

You will use a computer or geometric instruments to develop conjectures about angles in circles. You will prove your conjectures in Section 7.5.

1. Construct a circle with centre O. Draw a diameter AB. Mark point C on the circle. We say that arc ACB is a *semicircle,* and that ∠C is inscribed in the semicircle.

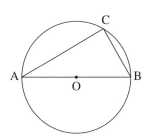

 a) Measure ∠C. What do you notice?

 b) Visualize repeating this construction for other positions of C on the semicircle, and for other semicircles. What would you expect to observe about the measure of ∠C? Check your prediction with other examples.

 c) Make a conjecture based on your findings.

2. Construct a circle with centre O. Mark points A and B on the circle. Mark points C and D on major arc AB. Join AC, BC, AD, and BD. We say that ∠C and ∠D are *inscribed angles subtended by* arc AB.

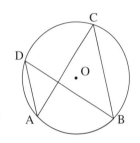

a) Measure ∠C and ∠D. What do you notice?

b) Visualize repeating this construction for other positions of C and D on major arc AB, for other positions of A and B, and for other circles. What would you expect to observe about the measures of ∠C and ∠D? Check your prediction with other examples.

c) In parts a and b, points C and D were on major arc AB. What would you expect to observe about the measures of ∠C and ∠D if points C and D were on minor arc AB? Check your prediction with examples.

d) Make a conjecture based on your findings.

3. Construct a circle with centre O. Mark points A and B on the circle. Mark point C on major arc AB. Join AC, BC, AO, and BO. We say that ∠AOB is the *central angle subtended by* arc AB.

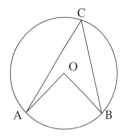

a) Measure ∠AOB and ∠ACB, and compare the results. What do you notice?

b) Visualize repeating this construction for other positions of C on major arc AB, for other positions of A and B, and for other circles. What would you expect to observe about the measures of ∠AOB and ∠ACB? Check your prediction with other examples.

c) In part b, point C was on major arc AB. In the diagram at the right, C is on minor arc AB. In this case, the central angle is greater than 180°. It is called a *reflex angle*. What would you expect to observe about the measures of ∠ACB and reflex ∠AOB in this case? Check your prediction with some examples.

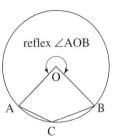

d) Make a conjecture based on your findings.

In exercises 1 to 3, you should have conjectured the following properties of angles in a circle.

Angle Properties

1. The angle inscribed in a semicircle is a right angle.

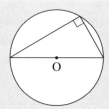

2. Inscribed angles subtended by the same arc of a circle are equal.

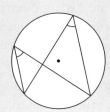

3. The measure of the central angle is twice the measure of the inscribed angle subtended by the same arc.

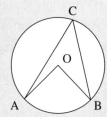

$$\angle AOB = 2\angle ACB$$

We can use these properties to solve problems involving angles in a circle.

Example

Use the information in the diagram to calculate the measures of P and Q.

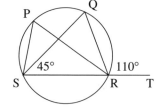

Solution

Think...

Angle P and ∠Q are equal, so we need only determine the measure of ∠P or ∠Q.

Triangle QSR contains a 45° angle and has an exterior angle of 110°.

We can determine ∠QRS, then we will know two angles in ΔQSR.

Since SRT is a straight line,

$$\angle QRS = 180° - 110°$$
$$= 70°$$

Since the sum of the angles in $\triangle QSR$ is 180°,

$$\angle Q = 180° - 70° - 45°$$
$$= 65°$$

According to Angle Property 2,

$$\angle P = \angle Q$$
$$= 65°$$

DISCUSSING THE IDEAS

1. In exercise 1 of *Investigate,* visualize joining C to the centre and extending the line to point D on the circle, to form another diameter. Join AD and BD. What figure is formed? Explain.

2. In exercise 2 of *Investigate,* visualize points C and D moving around the circle.

   a) What special case occurs when both AC and BD pass through O? Explain.

   b) What happens if C is on major arc AB and D is on minor arc AB? Explain.

3. In exercise 3 of *Investigate,* visualize C moving around the circle.

   a) What special case occurs when AC passes through O? Explain.

   b) Could C move past A or B so that it is on the minor arc AB? Would your conjecture still apply in this case? Explain.

4. Where would the photographer have to stand to take the photographs on page 448? Explain.

A **1.** Determine each value of *x*. Point O is the centre of each circle.

a)

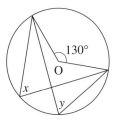

b)

c)

d)

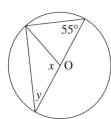

e)

f)

2. Determine each value of *x* and *y*. Point O is the centre of the circle.

a)

b)

c)

d)

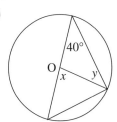

e)

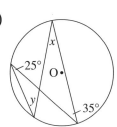

f)

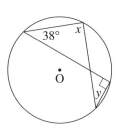

3. Choose one circle from exercise 1 or 2. Write to explain how you determined the value of *x* or *y*.

4. During hockey practice, the goalies go to their nets. Half the team lines up at one blue line and the other half lines up at the other blue line. As a warm up for the goalies, the players take turns shooting at the goal from the blue line.

  a) Explain why it is easier for some players to score than for others.

  b) How could the players be arranged so the shooting angle is the same for each player?

5. Determine the value of x (below left).

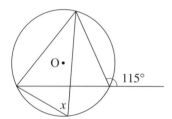

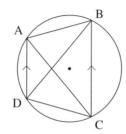

6. Trapezoid ABCD (above right) is inscribed in a circle and AD is parallel to BC. Angle ABD = 35° and ∠DAC = 40°, determine the measures of ∠DAB and ∠ADC.

 7. Chord AB has a length equal to the radius of the circle.

a) Determine the measure of ∠P.

b) Suppose P were on the minor arc AB. Would the answer to part a be the same? Explain.

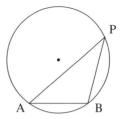

8. a) Determine the values of x, y, and z. Point O is the centre of each circle.

i)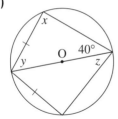

ii)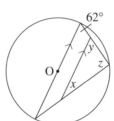

iii)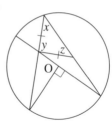

iv)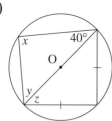

v)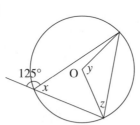

vi)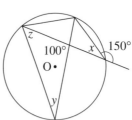

b) Choose one circle from part a. Write to explain how you determined the values of x, y, and z.

9. Isosceles △ABC is inscribed in a circle with diameter 12 cm. Side AB is a diameter. Determine the length of BC.

10. Quadrilateral ABCD is inscribed in a circle with diameter 20 cm. Diagonals AC and BD are both diameters.

   a) What type of quadrilateral is ABCD?

   b) The length of AB is 16 cm. Determine the length of BC.

   c) Calculate the area of ABCD.

11. Carpenters use an instrument called a square (sometimes called a carpenter's square) for laying out and testing right angles. Write to describe how a carpenter can find the centre of a circle using this instrument (below left).

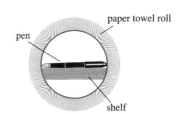

12. Get the tube from a roll of paper towels. Place a pen on a shelf at eye level and view it through the tube. Adjust your distance from the pen until it fits exactly across the opening (above right). Move to another position near the pen and repeat several times. Visualize the points on the floor where you were standing when you saw the pen framed by the tube. What geometric figure do they form? Write to explain why the points form this figure.

13. Lori's camera has a 50° field of view, as shown in this diagram. She is taking a picture of a house for a real estate advertisement. Since the neighbouring houses are in need of repair, Lori wants the house to take up the entire field of view.

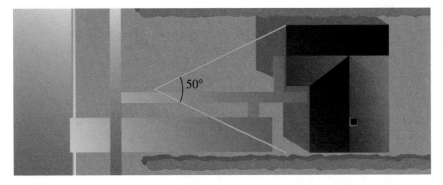

   a) Draw a diagram to show the possible places from which Lori can take the picture.

   b) Gita's camera has a wide-angle lens with a 70° field of view. Draw another diagram to show the possible places from which she could take a picture of the house.

   c) Compare and contrast your two diagrams.

14. a) For the circle with centre O, explain why RO is perpendicular to SO.

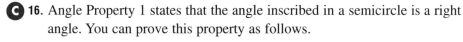

b) Visualize point Q moving around the circle on major arc RS. Describe the diagram for each special case.

    i) when S, O, and Q are collinear

    ii) when Q, O, and R are collinear

    iii) when Q coincides with R or with S   **iv)** when QS = QR

c) What happens to the angles and line segments on the diagram if Q moves along minor arc RS? Explain.

15. Construct a circle and two chords AB and CD that intersect inside the circle at P. Join AC and BD.

a) Measure ∠C and ∠D. Measure ∠A and ∠B. How are △PAC and △PBD related?

b) Visualize repeating this construction for other positions of the chords AB and CD. What would you expect to observe about △PAC and △PBD? Check your prediction with other examples.

c) Make a conjecture based on your findings.

d) Describe the diagram for each special case.

    i) when AB = CD   **ii)** when AB ⊥ CD

    iii) when AB = CD and AB ⊥ CD

C 16. Angle Property 1 states that the angle inscribed in a semicircle is a right angle. You can prove this property as follows.

a) Draw a circle with centre O, then draw a diameter AB. Mark point C on the circumference. You want to prove that ∠C = 90°.

b) Join CO, CB, and CA. What kind of triangle is △OAC? What does this tell you about ∠OAC and ∠OCA?

c) What kind of triangle is △OBC? What do you know about ∠OBC and ∠OCB?

d) What is the sum of all four angles in parts b and c? Explain.

e) Explain how you can conclude that ∠C = 90°.

COMMUNICATING THE IDEAS

Write a brief summary of the angle properties you discovered in exercises 1 to 3 of *Investigate.* Include diagrams in your summary.

Paradoxes and Proof

About 2500 years ago, a new civilization flourished in the eastern Mediterranean. Known as the ancient Greeks, these people recognized that humans had the ability to reason, or think logically. They knew that successful reasoning would produce new knowledge. And, since they were interested in discovering new ideas, they tried to develop laws of reasoning.

In their work with logic, the ancient Greeks encountered some strange paradoxes such as this:

> This sentence is false.

Is this sentence true or is it false? Suppose it is true. Then, by what it says, it must be false. Next, suppose it is false. Then what it says is false, so the sentence must be true!

The sentence above involves the idea of *self-reference*. This means that the sentence refers to itself. Many other paradoxes also involve self-reference. Some of them are presented on the next page.

Until the 20th century, mathematicians considered paradoxes like these to be merely riddles. But, in 1931, the Austrian mathematician Kurt Gödel used a similar paradox to prove a surprising result. He showed that mathematics contains "undecidable" statements that can never be proved true or false. His proof uses self-reference and is very clever. In effect, Gödel proved a theorem something like this:

> This theorem can never be proved.

Mathematicians were astonished to learn that there are some true statements that can never be proved. Perhaps Goldbach's Conjecture on page 383 is one of them.

1. In a booklet of test questions, the page (below left) was found. Explain how self-reference is involved in this example.

> This page is intentionally left blank.

2. You have probably seen statements like the one (above right). Explain how self-reference is involved in this example.

3. The first sentence below applies to this book. The second sentence applies to this page in this book. Is the third sentence true or is it false? Explain.

> 1. This book has 1000 pages.
> 2. This page is in Chapter 4.
> 3. Sentences 1, 2, and 3 are all false.

4. Are these sentences true or are they false? Explain.

> The sentence in the box on the right is true.

> The sentence in the box on the left is false.

5. Does the rule in the box have exceptions? Explain.

> All rules have exceptions.

6. A certain village has only one barber, who is clean-shaven. The barber shaves all men and only those men who do not shave themselves. Who shaves the barber?

7. Gödel proved that mathematics contains "undecidable" statements that can never be proved to be true or false. Explain why this means that there are true statements that can never be proved.

In Section 7.4, you developed some properties of angles in a circle. Each property was based on a conjecture made by examining diagrams of angles in circles.

Angle Property 3 on page 450 describes how the central angle and an inscribed angle subtended by the same arc are related. This relationship is called the Angles in a Circle Theorem.

Angles in a Circle Theorem

The measure of the central angle is twice the measure of the inscribed angle subtended by the same arc.

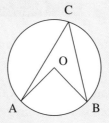

$\angle AOB = 2\angle ACB$

We can prove this theorem as follows.

Join CO and extend to point D.

We must prove that $\angle AOB = 2\angle ACB$.

Think...

Triangle OAC and ΔOBC are isosceles.

We can use the Isosceles Triangle Theorem.

The sum of the angles in ΔOBC is 180°.

Angle DOC is a straight angle.

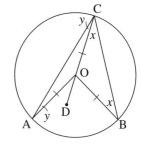

Since OB and OC are radii, OB = OC

Hence, ΔOBC is isosceles.

According to the Isosceles Triangle Theorem, $\angle OBC = \angle OCB$

Let x represent the measure of each of these angles, in degrees.

Similarly, let y represent the measure, in degrees, of each equal angle in ΔOAC.

According to the Exterior Angle Theorem,

$$\angle DOB = 2x \quad ①$$
$$\angle DOA = 2y \quad ②$$

Hence, $\angle DOB + \angle DOA = 2x + 2y$
$$= 2(x + y)$$
$$= 2\angle ACB$$

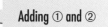

Adding ① and ②

Angle Properties 1 and 2 on page 450 are corollaries of the Angles in a Circle Theorem.

Corollary 1

Inscribed angles subtended by the same arc of a circle are equal.

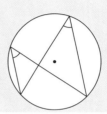

Corollary 2 Semicircle Theorem

The angle inscribed in a semicircle is a right angle.

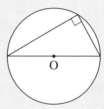

DISCUSSING THE IDEAS

1. Look at the proof of the Angles in a Circle Theorem on page 458.

   a) What was the reason for joining CO?

   b) What was the reason for extending CO to D?

2. In the diagrams on page 458, observe that point C is on major arc AB. Does the proof of the Angles in a Circle Theorem apply when C is on minor arc AB, as shown in the diagram? Explain.

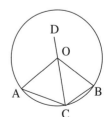

3. Does the proof of the Angles in a Circle Theorem on page 458 apply to all positions of C on major arc AB? Explain.

4. Recall that a corollary is a self-evident consequence of a theorem.

   a) Explain why Corollary 1 is self-evident.

   b) Explain why Corollary 2 is self-evident.

B **1.** Two chords AB and CD of a circle intersect at P inside the circle (below left). Prove that ΔPAD ~ ΔPCB.

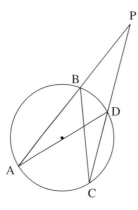

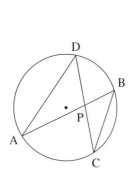

2. The lines containing two non-parallel chords AB and CD of a circle meet outside the circle at P (above right). Prove that ΔPAD ~ ΔPCB.

3. Here is one way to construct the perpendicular at a point P on a line. Write to explain why this method works.

Step 1 Mark a point O and draw a circle with radius OP that intersects the line again at A.

Step 2 Join AO and extend it to meet the circle again at B. Join BP. Then BP is perpendicular to the line.

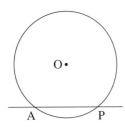

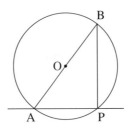

4. Segment BC is a diameter of a circle with centre O. Point A is on the circle. Segments OD and OE bisect chords AB and AC. Prove that OD is perpendicular to OE.

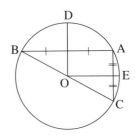

5. Quadrilateral PQRS is inscribed in a circle. Side PQ is parallel to side SR. The diagonals intersect at T. Prove that ΔTSR and ΔTPQ are isosceles.

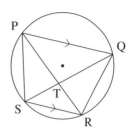

6. Two circles intersect at A and B. A line is drawn through A, to intersect the circles at P and Q.

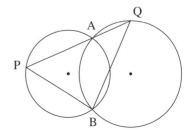

a) Prove that for all possible positions of line PAQ, the measure of ∠PBQ is constant.

b) What special case occurs when the circles have equal radii? Explain why it occurs.

c) What special case occurs when each circle passes through the centre of the other circle? Explain why it occurs.

7. In exercise 6, what are the possible positions of the line PAQ?

8. In exercise 3 of *Discussing the Ideas* on page 459, you should have noted that the proof of the Angles in a Circle Theorem on page 458 does not apply for all positions of C on major arc AB. It applies only when O is in the interior of inscribed ∠ACB.

a) Draw a diagram similar to the one on page 458 with C located so that O is not in the interior of ∠ACB.

b) Prove the Angles in a Circle Theorem in this case.

COMMUNICATING THE IDEAS

In this chapter, you made some conjectures about chords in Section 7.2 and you proved them in Section 7.3. You made some conjectures about angles in a circle in Section 7.4 and you proved them in this section. Explain why it is important to prove the conjectures. Use an example to illustrate your explanation.

Dynamic Circle Designs

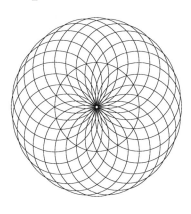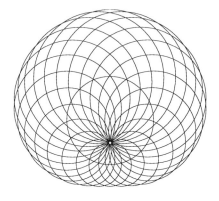

 These designs were made using a computer with dynamic geometry software. Follow these steps to make the designs.

Step 1

Construct a circle.
Construct a vertical line through the centre.
Hide the centre.
Mark a point on the vertical line.
Move the point to the centre.
Mark one point where the line intersects the circle.

Step 2

Hide the vertical line.
Construct equally spaced points every 15° around the circumference.
Construct circles with centres at the points on the circumference, all of which pass through the point at the centre of the circle.
Hide the points.

Step 3

The design will look like the first one above.
To make the other designs, move the point at the centre up or down.

Although the point at the centre of the first design is at the centre of the circle, it is not the centre. It can be moved anywhere along an imaginary vertical line without changing the original circle. As it moves, all the

LINKING IDEAS

Mathematics & Technology

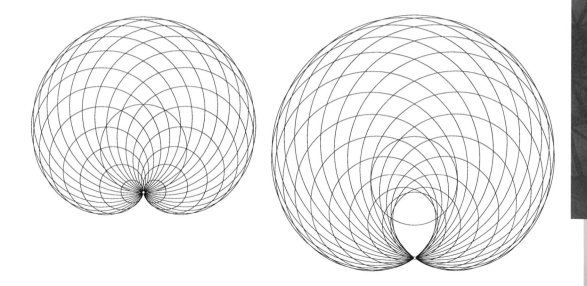

circles that were constructed in Step 2 will adjust automatically to pass through it, while their centres remain fixed on the original circle.

1. Look at the four designs. Visualize how the circles change as the point at the centre moves each way.

   a) down to meet the original circle, then continues far beyond it

   b) back up to pass through the original circle, then continues far beyond it

2. The diameter of the original circle is 2.3 cm.

   a) What is the diameter of all the other circles in the first design? Explain.

   b) Estimate the diameter of the smallest and the largest circles in each of the other three designs.

   c) Let x centimetres represent the distance that the point at the centre has moved from its original position. Write the diameters of the smallest and largest circles as functions of x.

3. In exercise 3 on page 460, a method to construct a perpendicular at a point P on a line is described. The method begins by marking a point O. Visualize repeating the construction many times using different points O that all lie on a circle. Describe what the result would look like in each case.

   a) The circle is so large that P is inside it.

   b) The circle is so small that P is outside it.

   c) The circle passes through P.

7 Review

1. Determine each value of *x*. Point O is the centre of each circle.

a)

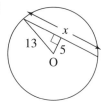

b)

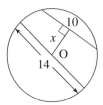

c)

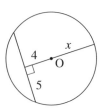

2. Determine each value of *x*. Point O is the centre of each circle.

a)

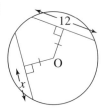

b)

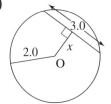

c)

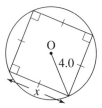

3. Determine each value of *x*. Point O is the centre of each circle.

a)

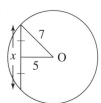

b)

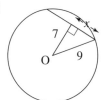

c)

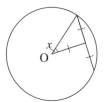

4. Choose one circle from exercise 1, 2, or 3. Write to explain how you determined the value of *x*.

5. Two chords JK and LM of a circle intersect at Q (below left). Point O is the centre of the circle and OQ bisects ∠JQM. Prove that JK = LM.

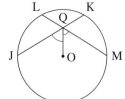

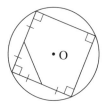

6. A square is inscribed in a circle with diameter 15 cm (above right). What is the distance between the midpoints of adjacent sides of the square?

7. Calculate the radius of the semicircle (below left).

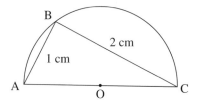

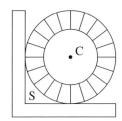

8. A plate is supported so it touches two sides of a shelf (above right). The plate has centre C and diameter 18.0 cm. What is the distance between the centre of the plate and the inside corner of the shelf, S?

9. Chords AB and AD are equal chords in a circle, centre O (below left). Prove that ∠OAD = ∠OBA.

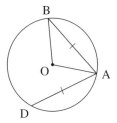

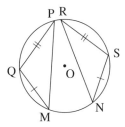

10. Equal chords PQ and RS are in a circle, centre O (above right). Points M and N lie on the circumference so that QM = SN. Prove that ∠QMP = ∠RNS.

11. Chord AB is in a circle, centre O and radius 12.0 cm. Points C and D lie on the circumference so that ∠ACB and ∠ADB are on opposite sides of chord AB. Angle ACB and ∠ADB are equal. Determine the length of chord AB.

12. Prove that any trapezoid inscribed in a circle is an isosceles trapezoid; that is, at least one pair of opposite sides are equal.

13. Quadrilateral WXYZ is inscribed in a circle, with WX = YZ. Prove that the diagonals of WXYZ are equal.

14. Construct three overlapping circles. Construct chords where the circles overlap, as shown. Repeat with other circles having different radii or positions.

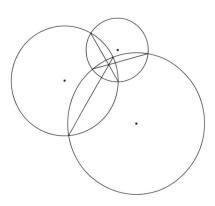

a) Assume the three circles intersect. What is always true about the chords?

b) What special case occurs when each circle passes through the centres of the other two?

1. What is the present value of an investment that yields $6400 after 8 years at 7% interest compounded annually?

2. Determine each accumulated amount.

   a) $2100 for 20 months at $6\frac{1}{4}\%$ compounded monthly

   b) $855 for 3 years at 5% compounded monthly

3. Select one part of exercise 2. Write to explain how you determined the accumulated amount.

4. Which is the greater rate: $6\frac{3}{4}\%$ compounded annually or 6% compounded monthly?

5. Patricia is offered two different sales jobs. One pays a salary of $1500 per week, with no commission. The other pays $800 per week, plus a 5% commission on sales. How much would Patricia have to sell at the second job to make as much money as at the first job?

6. Determine the monthly payment and the total cost of a $110 000 mortgage at 5.5% interest amortized over 20 years.

7. Write a quadratic function that has each pair of zeros.

   a) $-\frac{1}{7}, 3$ **b)** $-\frac{1}{4}, -\frac{1}{2}$ **c)** $2, -\frac{2}{3}$

8. For the parabola defined by $y = 5 - 3x^2$, state:

   a) the maximum or minimum value of y

   b) whether it is a maximum or minimum

   c) the value of x when it occurs

   d) the domain and range of the function

9. Solve for x.

   a) $x^3 - 13x + 12 = 0$ **b)** $x^3 - 3x^2 - 4x + 12 = 0$ **c)** $x^3 + 27 = 0$

10. Choose one part of exercise 9. Write to explain how you solved for x.

11. For each function, write the equation of the corresponding reciprocal function.

    a) $y = \sqrt{x^2 - 5}$ **b)** $y = -4x^3 + 2x - 5$ **c)** $y = x^4 + 1$

12. Choose one part of exercise 11. Write to explain how you wrote the reciprocal function.

13. Solve each equation. Give the answers to 2 decimal places.

    a) $9x^2 - 6x - 143 = 0$ **b)** $12x^2 - 29x + 14 = 0$ **c)** $20x^2 + x - 12 = 0$

14. Determine the nature of the roots in each equation.

a) $2x^2 + 36x - 25 = 0$ **b)** $4x^2 - 24x + 36 = 0$

c) $x^2 - 6x + 10 = 0$ **d)** $3x^2 + 8x - 20 = 0$

15. Select one equation from exercise 14. Write to explain how you determined the nature of the roots.

16. Solve each equation.

a) $\dfrac{3x - 11}{x - 5} = x - 2$ **b)** $\dfrac{6}{5 - x} = x + 2$ **c)** $\dfrac{4x + 2}{x - 1} = x + 4$

17. Solve each inequality.

a) $\dfrac{8}{x} > -2$ **b)** $-\dfrac{15}{x + 1} \le 3$ **c)** $\dfrac{6}{2x + 1} \ge 1$

18. a) Sketch the graph of $f(x) = \dfrac{7}{x + 1}$.

b) Determine the values of x for which $f(x) \ge 0$.

19. Solve by graphing.

a) $x - 3y = 6$ **b)** $3x - 2y = -10$ **c)** $y = 7 - 2x$
   $4x + 3y = -21$ $6x + 8y = -2$ $3x - 2y = 7$

20. It costs $26.00 for movie tickets for one adult and 8 children. On another occasion, it cost $34.00 for 4 adults and 4 children. What are the prices of the movie tickets for adults and children?

21. Graph the region defined by each system of inequalities. Describe the shape of the region, then determine its area.

a) $x \ge 0$ **b)** $x \le 0$
   $y \le 0$ $y \ge -5$
   $x \le 3$ $y \ge -3x$
   $2x - y \le 5$ $y \le 3x + 5$

22. Use deductive reasoning. Copy and complete each conclusion.

a) All cats have four paws. Fluffy is a cat. Therefore, …

b) All months have more than 27 days. February is a month. Therefore, …

23. Which numbers x satisfy each statement?

a) x is a multiple of 2 and x is a factor of 54.

b) x is a factor of 35 or x is a factor of 56.

24. Choose one part of exercise 23. Illustrate the solution on a Venn diagram. Write to explain how you completed the diagram.

How Long Is the Chain?

 CONSIDER THIS SITUATION

A gear chain on a bicycle transfers the motion of the pedals to the rear wheel. A belt on a car engine transfers the motion of the engine to another component, such as the alternator. The gear chain and alternator belt each encircle two wheels. The lengths of the chain and belt depend on the sizes of these wheels and the distance between their centres.

- How could you determine the length of the bicycle chain?
- How could you determine the length of the alternator belt on the car engine?
- Would your method give more accurate results for the chain or for the belt? Explain.

On pages 494-496, you will develop mathematical models to determine the length of a chain or a belt encircling two wheels. These models will involve estimation and scale diagrams.

 FYI Visit www.awl.com/canada/school/connections

For information relating to the above problem, click on
<u>MATHLINKS</u>, followed by *Mathematics 11*. Then select a topic under How Long Is the Chain?

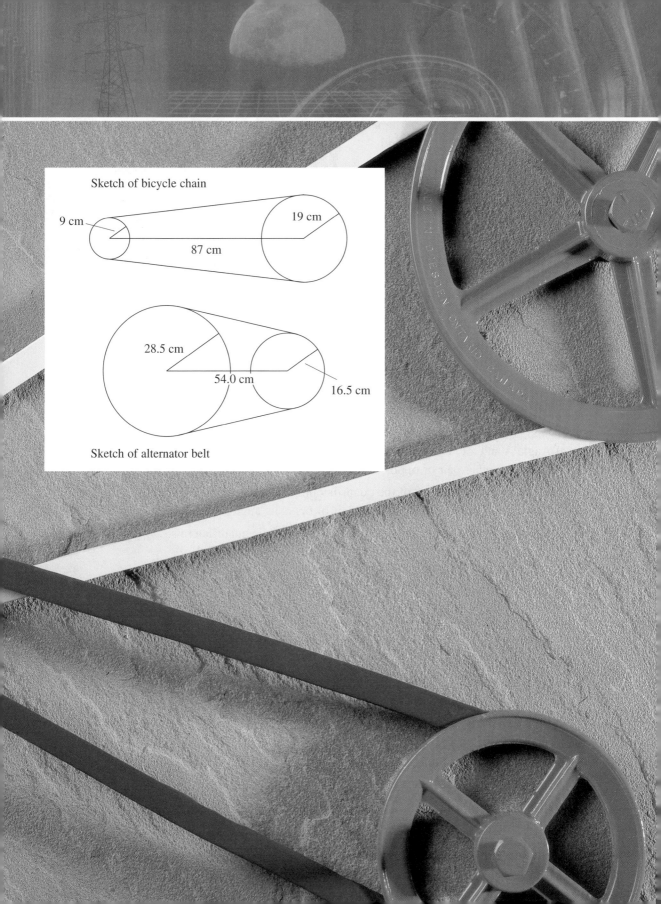

Sketch of bicycle chain

9 cm

19 cm

87 cm

28.5 cm

54.0 cm

16.5 cm

Sketch of alternator belt

St. Paul's Cathedral in London, England, is noted for its large hemispherical dome. Inside the cathedral, a walkway around the circumference of the dome is known as the Whispering Gallery. If you stand on the walkway, you can hear other people on the walkway.

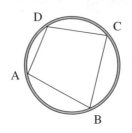

The diagram shows the positions of four people on the walkway. We say that quadrilateral ABCD is a *cyclic quadrilateral* because all its vertices lie on a circle.

INVESTIGATE

You will use a computer or geometric instruments to develop conjectures about cyclic quadrilaterals. You will prove your conjectures in Section 8.2.

1. Construct a circle. Construct cyclic quadrilateral ABCD.

   a) Angle A and ∠C are *opposite angles* in the quadrilateral. Name the other pair of opposite angles.

   b) Measure the four angles of the quadrilateral. What do you notice about the measures of each pair of opposite angles?

   c) Visualize repeating this construction for other cyclic quadrilaterals. What would you expect to observe about the opposite angles? Check your predictions with other examples.

   d) Make a conjecture based on your findings.

2. Construct cyclic quadrilateral ABCD. Angle CBE formed by extending one side, as shown, is an *exterior angle* of the quadrilateral.

   a) Construct exterior ∠CBE and measure it. Measure the angles of the quadrilateral.

   b) Compare the measure of ∠CBE with the measures of the other angles. What do you notice?

   c) Visualize repeating this construction for other exterior angles and for other cyclic quadrilaterals. What would you expect to observe about the exterior angles? Check your predictions.

   d) Make a conjecture based on your findings.

In exercises 1 and 2, you should have conjectured these properties of cyclic quadrilaterals.

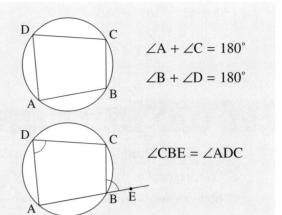

Cyclic Quadrilateral Properties

1. The opposite angles of a cyclic quadrilateral are supplementary.

$\angle A + \angle C = 180°$

$\angle B + \angle D = 180°$

2. Each exterior angle of a cyclic quadrilateral is equal to the opposite interior angle.

$\angle CBE = \angle ADC$

You can use these properties to solve problems involving cyclic quadrilaterals.

Example

Determine the values of x, y, and z. Point O is the centre of each circle.

a)

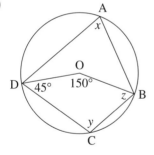

b)

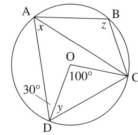

Solution

a) From the Angles in a Circle Theorem,

$x = \frac{1}{2}(150°)$

$= 75°$

According to Cyclic Quadrilateral Property 1,

$x + y = 180°$

$75° + y = 180°$

$y = 105°$

b) From the Angles in a Circle Theorem,

$x = \frac{1}{2}(100°)$

$= 50°$

Since OC and OD are radii,

$\angle OCD = \angle ODC = y$

By the Angles in a Triangle Theorem in $\triangle ODC$,

$100° + 2y = 180°$

$y = 40°$

Since the angle sum of quadrilateral OBCD is 360°,

$$z + 150° + 45° + 105° = 360°$$
$$z + 300° = 360°$$
$$z = 60°$$

According to Cyclic Quadrilateral Property 1,

$$\angle ABC + \angle ADC = 180°$$
$$z + 30° + 40° = 180°$$
$$z = 110°$$

DISCUSSING THE IDEAS

1. For the Cyclic Quadrilateral Properties, describe each special case.

   a) One diagonal of the quadrilateral is a diameter of the circle.

   b) Both diagonals are diameters of the circle.

2. In the solution of the *Example* part a, we stated that the angle sum of quadrilateral OBCD is 360°. Explain why this is true.

3. Suppose you construct an exterior angle at each vertex of a cyclic quadrilateral. Determine the sum of the four exterior angles. Explain.

8.1 EXERCISES

A 1. Determine each value of x and y. Point O is the centre of each circle.

a)

b)

c)

2. Choose one part of exercise 1. Write to explain how you determined the values of x and y.

B 3. Determine each value of x and y. Point O is the centre of each circle.

a)

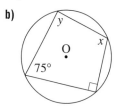

b)

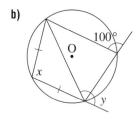

c)

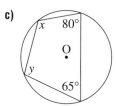

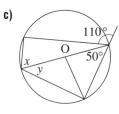

4. Choose one part of exercise 3. Write to explain how you determined the values of x and y.

5. For each circle, describe how the value of y can be found if the value of x is known. Write an equation relating x and y.

a)

b)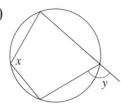

6. Construct any quadrilateral ABCD that is obviously not cyclic. Construct the perpendicular bisectors of sides CD and AD. Mark point O where the lines intersect. Construct a circle with centre O that passes through A.

a) Why does the circle pass through C and D?

b) Measure the four angles of the quadrilateral. Is each pair of opposite angles supplementary?

c) Visualize moving B toward the circle.

   i) Predict what will happen to the measures of ∠B and ∠D.

   ii) Predict what will happen to the measures of ∠A and ∠C.

d) Check your predictions in part c.

e) Do your observations support Cyclic Quadrilateral Property 1? Explain.

 7. Cyclic Quadrilateral Property 1 states that the opposite angles of a cyclic quadrilateral are supplementary. You can prove this property as follows.

a) In the diagram, how does ∠DAB compare with ∠DOB?

b) How does ∠DCB compare with reflex ∠DOB?

c) Use the results of parts a and b to prove that ∠DAB + ∠DCB = 180°.

COMMUNICATING THE IDEAS

Write a brief summary of the properties of cyclic quadrilaterals you discovered in *Investigate*. Include diagrams in your summary.

8.2 Proving Properties of Cyclic Quadrilaterals

Cyclic Quadrilateral Property 1, on page 471, is known as the Cyclic Quadrilateral Theorem.

Cyclic Quadrilateral Theorem

The opposite angles of a cyclic quadrilateral are supplementary.

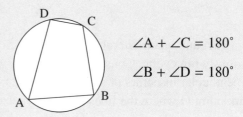

$\angle A + \angle C = 180°$

$\angle B + \angle D = 180°$

To prove this theorem, consider quadrilateral ABCD inscribed in a circle with centre O. We must show that $\angle A + \angle C = 180°$ and $\angle B + \angle D = 180°$.

Think...

There are two central angles, one for $\angle DAB$ and one for $\angle DCB$. Apply the Angles in a Circle Theorem to each central angle.

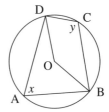

Let x and y represent the measures of $\angle DAB$ and $\angle DCB$, respectively.

According to the Angles in a Circle Theorem,

$$\angle DOB = 2x \qquad ①$$

and reflex $\angle DOB = 2y \qquad ②$

Hence, $\angle DOB + \text{reflex } \angle DOB = 2x + 2y$ ——— Adding ① and ②

$$360° = 2(x + y)$$
$$180° = x + y$$

Therefore, $\angle A + \angle C = 180°$

In a similar way, by considering central angles for $\angle ADC$ and $\angle ABC$, we can prove that $\angle B + \angle D = 180°$

Hence, the opposite angles of cyclic quadrilateral ABCD are supplementary.

Cyclic Quadrilateral Property 2, on page 471, is a corollary of the Cyclic Quadrilateral Theorem.

Corollary

Each exterior angle of a cyclic quadrilateral is equal to the opposite interior angle.

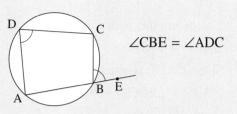

∠CBE = ∠ADC

The converse of the Cyclic Quadrilateral Theorem is also true.

Converse of Cyclic Quadrilateral Theorem

If the opposite angles of a quadrilateral are supplementary, then the quadrilateral is cyclic.

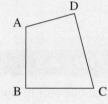

If ... ∠A + ∠C = 180°
∠B + ∠D = 180°

then ... ABCD is cyclic.

To prove the converse, draw quadrilateral ABCD in which:

∠BAD + ∠BCD = 180°   ①

∠ABC + ∠ADC = 180°   ②

Construct a large arc of a circle passing through A, B, and C. Do not draw the part of the circle near D. We must prove D lies on the circle.

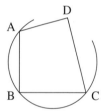

Think...

We could use indirect proof. We could assume that D is not on the circle and try to arrive at a contradiction.

Use indirect proof.

Step 1. Either D is on the circle or D is not on the circle.

Step 2. Assume D is not on the circle.

Step 3. There must be some other point on the line through A and D that is on the circle. Let this point be E.

Then, ABCE is a cyclic quadrilateral.

According to the Cyclic Quadrilateral Theorem, ∠E and ∠B are supplementary.

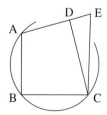

$\angle ABC + \angle AEC = 180°$

Comparing this with equation ② above,

$\angle ADC = \angle AEC$ ③

Since $\angle ADE$ is a straight angle,

$\angle ADC + \angle EDC = 180°$ ④

Comparing equations ③ and ④,

$\angle AEC + \angle EDC = 180°$

But this is impossible in $\triangle EDC$.

Step 4. The assumption that D is not on the circle is incorrect. Therefore, D lies on the circle.

Hence, quadrilateral ABCD is cyclic.

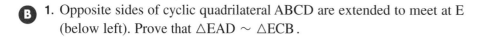

DISCUSSING THE IDEAS

1. Why is it enough to show that one pair of opposite angles in a quadrilateral is supplementary to establish that both pairs are supplementary?

2. We proved the Cyclic Quadrilateral Theorem using the Angles in a Circle Theorem. You can also prove it using Corollary 1 of that theorem. How could you do this?

3. In the proof of the converse theorem, how did we know we could draw an arc through A, B, and C? Explain.

4. In the proof of the converse theorem, point E was located on the opposite side of D from A. What changes, if any, occur in the proof if E is between A and D?

8.2 EXERCISES

B 1. Opposite sides of cyclic quadrilateral ABCD are extended to meet at E (below left). Prove that $\triangle EAD \sim \triangle ECB$.

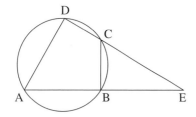
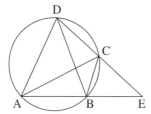

2. Opposite sides of cyclic quadrilateral ABCD are extended to meet at E (above right). Prove that $\triangle EAC \sim \triangle EDB$.

3. Adjacent sides of cyclic quadrilateral PQRS are equal (below left).

   a) Prove that △PQR and △PSR are right triangles.

   b) What special case occurs when the centre of the circle lies on SQ?

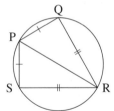
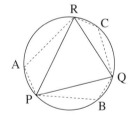

4. Triangle PQR is inscribed in a circle (above right). Points A, B, and C are any three points on the three arcs defined by the sides of the triangle.

   a) Prove that ∠A + ∠B + ∠C = 360°.

   b) What special case occurs when both △PQR and △ABC are equilateral triangles?

5. The diagram in exercise 4 suggests that a similar result could be obtained for a cyclic quadrilateral. Create a problem similar to the one in exercise 4, starting with a quadrilateral PQRS inscribed in a circle. Solve your problem.

6. For exercise 1, describe each special case (if any).

   a) One side of quadrilateral ABCD is a diameter of the circle.

   b) One diagonal is a diameter.

   c) Both diagonals are diameters.

C 7. Quadrilateral ABCD is cyclic, with diagonals AC and BD perpendicular, and intersecting at E. Point M is the midpoint of CD. Prove that the line through M and E is perpendicular to AB at N.

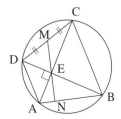

COMMUNICATING THE IDEAS

A student used the following argument to prove that any quadrilateral with vertices A, B, C, and D is cyclic: "A, B, and C lie on a circle. Likewise, B, C, and D lie on a circle. Therefore, A, B, C, and D lie on a circle." How would you explain to the student that the argument is wrong? Write out your explanation for someone to follow your thinking.

Eclipses

An eclipse of the sun occurs when the moon passes between Earth and the sun. During this time, the moon's shadow falls on Earth. People living in the shadow see the sun partially or completely hidden by the moon.

The moon's shadow has the shape of a cone. During an eclipse of the sun, the tip of this cone touches Earth in a small region called the *umbra*. People in the umbra will observe a total eclipse of the sun, like the one in the photograph.

As you look at the diagram, visualize the moon moving in its orbit around Earth. Visualize Earth moving in its orbit around the sun, and also rotating on its axis. Since the sun is very far away, the umbra sweeps rapidly across Earth's surface. A total eclipse lasts just a few minutes.

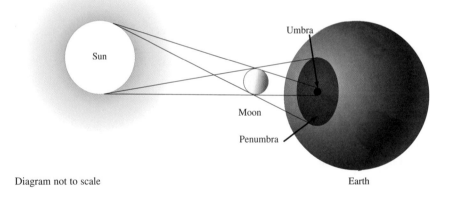

Diagram not to scale

Earth

Facts about total solar eclipses
- The diameter of the umbra is never more than 268 km.
- A total solar eclipse lasts no longer than 7.5 min.

1. Use the information above. Imagine you are about to experience a total eclipse of the sun. Visualize the moon's circular shadow approaching your location. Visualize how long it takes to pass over your location. Calculate its approximate speed in kilometres per hour.

Mathematics & Science

2. People outside the umbra but inside the larger shaded region, called the *penumbra*, observe a partial solar eclipse that can last up to 4 h. To them, only part of the sun is obscured by the moon. Estimate the diameter of the penumbra.

3. Explain the significance of the lines on the diagram on page 478.

Look at the full moon in the night sky. Visualize a gigantic screen placed in the sky, slightly farther away than the moon. On this screen, the umbra of Earth's shadow would be a large dark circle. A total eclipse of the moon occurs when the moon passes through the umbra. The diagram below shows the moon moving from right to left in front of the screen and across the umbra.

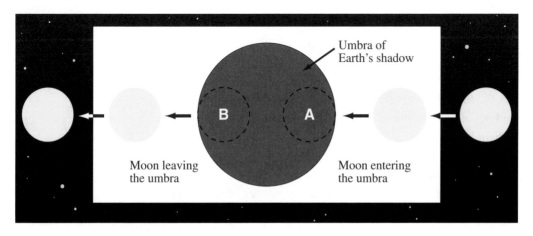

Umbra of Earth's shadow

B   A

Moon leaving the umbra

Moon entering the umbra

4. Draw a diagram like that on the facing page to show the relative positions of the moon, Earth, and sun during a lunar eclipse.

5. Use the information on the right. Imagine the moon has just passed completely into the umbra at A.

   a) The diameter of the moon is 3480 km. How far does it travel until it begins to leave the umbra at B?

   b) Calculate the moon's approximate speed in kilometres per hour.

   Facts about total lunar eclipses

   - The diameter of the umbra is 9200 km.

   - A total lunar eclipse can last up to 1 h 40 min.

6. Which type of eclipse do you think most people are more likely to see, a solar eclipse or a lunar eclipse? Explain your reasoning.

What Is a Tangent?

The word *tangent* comes from the Latin verb "tangere," which means to touch. We think of a tangent as a line that touches a curve, as in these examples.

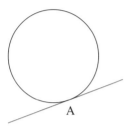

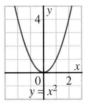

This line is a tangent to the circle at A.

The *x*-axis is a tangent to the parabola $y = x^2$ at the origin.

However, not all lines that touch a curve are tangents.

All three lines touch this curve at A, but none of them is a tangent.

We define a tangent to a circle as a line that intersects the circle at exactly one point. This definition is fine for circles, but not for other curves.

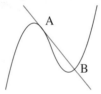

The axis of symmetry of a parabola intersects it at exactly one point, but it is not a tangent.

This line is a tangent to the curve at A, although it intersects the curve again at B.

A tangent to any curve is defined as follows. Let P be a point on the curve. Let Q be another point on the curve, on either side of P. Construct the line PQ. This line is called a *secant*. Visualize what happens to this line when Q approaches P from either side. If the secant PQ approaches the same line as Q approaches P from either side, this line is called the tangent at P.

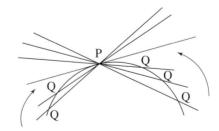

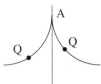

Visualize Q approaching A from either side to define the tangent shown.

This curve has no tangent at A because the secants from either side do not approach the same line.

1. Use an example from page 480 to illustrate each answer.

   a) In the definition of a tangent, does it matter how far Q is from P at the beginning?

   b) Does it matter if Q is on the other side of P at the beginning?

2. a) Could a line be a tangent to a curve at two different points on the curve? Explain.

   b) Could a line be a tangent to a curve and intersect the curve in another point where the line is not a tangent to the curve? Explain.

3. In the third diagram on page 480, three lines that touch a curve but are not tangents are shown. Sketch a curve similar to this one. Draw the tangent.

4. To demonstrate the definition of a tangent, construct a circle. Mark points P and Q on the circle. Construct the line PQ.

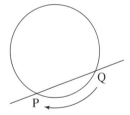

   a) If you are using a computer, move Q closer to P. Otherwise, draw more lines PQ where Q is closer to P. What happens to the line PQ as Q approaches P? What happens when Q coincides with P?

   b) Repeat part a with Q on the other side of P.

5. Construct a circle with centre O. Mark points P and Q on the circle. Construct the line PQ. Join OP.

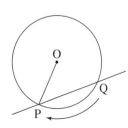

   a) Measure ∠OPQ. Visualize how this angle changes as Q moves around the circle.

   b) What happens to ∠OPQ as Q moves away from P?

   c) What happens to ∠OPQ as Q approaches P?

The diagram shows a satellite in orbit. The lines represent the limit of the paths of radio signals transmitted between Earth and the satellite. Points A and B represent the places on Earth that are on the edge of the satellite's area of coverage.

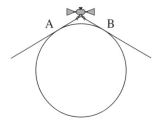

Each line is a tangent to the circle because it touches the circle in exactly one point.

A *tangent to a circle* is a line that intersects the circle in exactly one point. The point where the tangent intersects the circle is called the *point of tangency*.

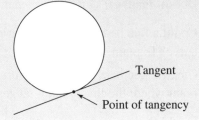

Tangent

Point of tangency

I N V E S T I G A T E

You will use a computer or geometric instruments to develop conjectures about tangents to circles. You will prove your conjectures in Section 8.4.

1. Construct a circle with centre O. Mark a point A on the circle. Join OA.

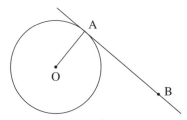

a) Use visual estimation to construct a line through A that is a tangent to the circle. Mark any other point B on this line. Measure $\angle$OAB.

b) Visualize repeating this construction for other positions of A on the circle, and for other circles. What would you expect to observe about the angle between the tangent and the radius? Check your predictions with other examples.

c) Make a conjecture based on your findings.

2. Construct a circle. Mark a point P outside the circle.

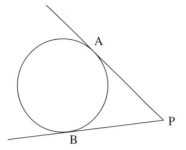

a) Use visual estimation to construct two tangents from P to the circle. Mark the points of tangency A and B. The line segments PA and PB are called *tangent segments*. Measure PA and PB. What do you notice?

b) Visualize repeating this construction for other positions of point P outside the circle, and for other circles. What would you expect to observe about the lengths of the tangent segments PA and PB? Check your predictions with other examples.

c) Make a conjecture based on your findings.

In exercises 1 and 2, you should have conjectured these properties of tangents to a circle.

Tangent Properties

1. A tangent to a circle is perpendicular to the radius at the point of tangency.

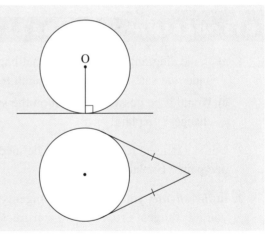

2. The tangent segments to a circle from an external point are equal.

You can use these properties to solve problems involving tangents.

Example

In the diagram, O is the centre of the circle, and AB and AC are tangent segments. Determine the lengths of AC and AO.

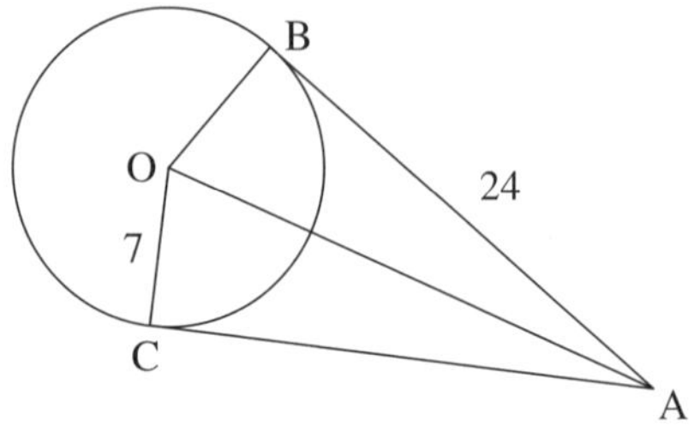

Solution

According to Tangent Property 2, AC = AB

Hence, AC = 24

According to Tangent Property 1, $\angle ACO = 90°$

Use the Pythagorean Theorem in $\triangle ACO$.

$$
\begin{aligned}
AO &= \sqrt{AC^2 + OC^2} \\
&= \sqrt{24^2 + 7^2} \\
&= \sqrt{625} \\
&= 25
\end{aligned}
$$

$AO = 25$

DISCUSSING THE IDEAS

1. a) In the diagram on page 482, what happens to the area of coverage as Earth rotates under the satellite, and as the satellite revolves around Earth?

   b) Would it be possible to position the satellite so that the area of coverage never changes? Explain.

2. Could a tangent to a parabola be defined as a line that intersects the parabola in exactly one point? Explain.

3. In *Investigate*, you constructed tangents to a circle using visual estimation. How could you use Tangent Property 1 to construct a tangent to a circle at a point on the circle?

4. In the *Example:*

   a) Explain why $\triangle ABO \cong \triangle ACO$.

   b) What kind of quadrilateral is ABOC? Explain.

In each diagram below, assume that lines that look like tangents or tangent segments are tangents or tangent segments unless stated otherwise.

A 1. Determine each value of *x*. Point O is the centre of each circle.

a)

b)

c)

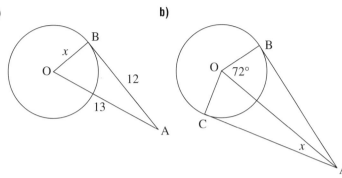

2. Determine each value of *x* and *y*. Point O is the centre of each circle.

a)

b)

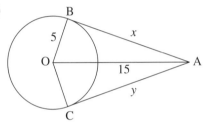

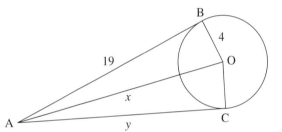

c)

d)

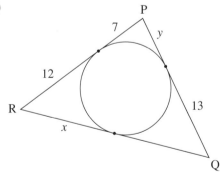

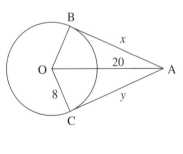

3. Choose one diagram from exercise 1 or 2. Write to explain how you determined the value of *x* or *y*.

4. Determine each value of *x* and *y*. Points O and M are the centres of the circles.

a)

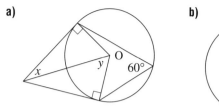

b)

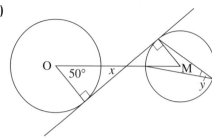

B **5.** Determine the perimeter of △LMN (below left).

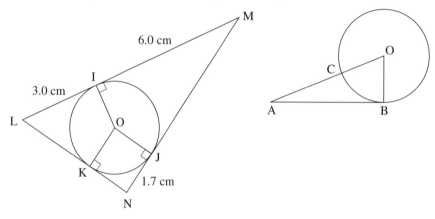

6. The circle (above right) has radius 2.0 cm. Tangent segment AB is 5.0 cm long. Determine the length of AC.

7. A circle is inscribed in an equilateral triangle. The length of XY is 3.2 cm. What is the perimeter of the triangle?

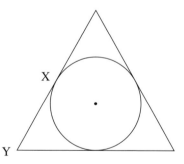

8. A circle is inscribed in a square with side length 15 cm. How far is the centre of the circle from a vertex of the square?

9. Determine each value of x and y. Point O is the centre of each circle.

a)

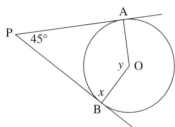

b)

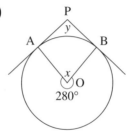

c)

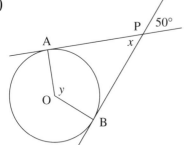

d)

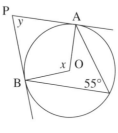

10. In each diagram, PR and PQ are tangents to the circle. Point O is the centre of each circle. Express y as a function of x. What is the domain of each function?

a)

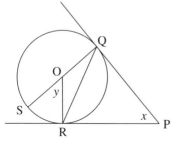

b)

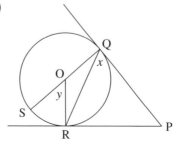

11. The distances between the centre O of the inscribed circle and the vertices of △ ABC are 14 cm, 33 cm, and 16 cm, respectively. The diameter of the circle is 18 cm. What is the perimeter of △ ABC?

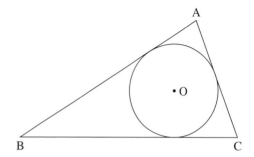

12. The distance from point A to the centre of a circle is 20 cm. The radius of the circle is 5 cm. Calculate the lengths of the two tangent segments from A to the circle.

13. Construct a circle with centre O. Mark points A and B on the circle. Construct tangents at points A and B. Label point P where the two tangents intersect.

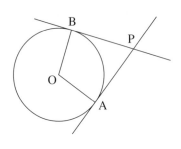

a) Measure ∠AOB and ∠APB. What do you notice about the measures of the angles?

b) Suppose you were to move points A and B or vary the radius of the circle. Would your observation in part a still hold? Check.

c) Make a conjecture based on your findings.

14. Construct a circle with centre O. Mark four points A, B, C, and D on the circle. Construct a tangent at each point. Construct points E, F, G and H where the tangents intersect. Construct segments FE, GF, HG and EH to form a quadrilateral.

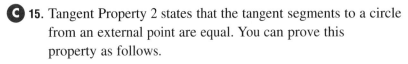

a) Measure the lengths of the sides of the quadrilateral. The lengths of the sides have a certain property. What is it?

b) Suppose you were to move the points A, B, C, or D, or vary the radius of the circle. Would your observation from part a still hold? Check.

c) Make a conjecture based on your findings.

C 15. Tangent Property 2 states that the tangent segments to a circle from an external point are equal. You can prove this property as follows.

a) Draw a circle with centre O. Draw tangent segments from an external point P to the circle at A and B. You want to prove that PA = PB.

b) Join OA and OB. What kind of triangles are △PAO and △PBO? Explain.

c) What do you know about the sides of △PAO and △PBO? How can you use this to prove that PA = PB?

16. A spherical glass globe just fits inside a cubical box with edge length 42 cm. Determine the distance from the centre of the globe to a vertex of the box.

COMMUNICATING THE IDEAS

Write a brief summary of the properties of tangents you discovered in *Investigate*. Include diagrams in your summary.

8.4 Proving the Tangent Properties

Tangent Property 1, on page 483, is known as the Tangent-Radius Theorem.
The converse theorem is also true.

Tangent-Radius Theorem

A tangent to a circle is perpendicular to
the radius at the point of tangency.

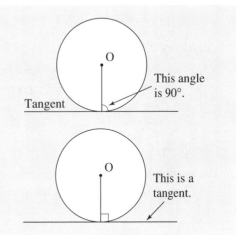

Converse

A straight line drawn at right angles to a radius
of a circle at a point on the circle is a tangent to
the circle.

We can prove the Tangent-Radius Theorem using indirect proof.

Suppose that line *l* is a tangent to a circle at A.

Either *l* is perpendicular to OA or *l* is not perpendicular to OA.

Assume that *l* is not perpendicular to OA.

Then there must be some other point, B, on *l* so that *l*
is perpendicular to OB.

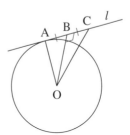

Let C be a point on *l* so that CB = BA, where C is
on the opposite side of B from A.

In △OBC and △OBA,

$$OB = OB$$
$$\angle OBC = \angle OBA = 90°$$
$$CB = AB$$

Therefore, △OBC ≅ △OBA   (SAS)

Since the triangles are congruent, OC = OA

Since OA is a radius, OC must also be a radius. Hence, C lies on the circle.
But then *l* intersects the circle at two points. This is impossible because *l* is a
tangent to the circle.

Our assumption that *l* is not perpendicular to OA must be false.
Therefore, *l* is perpendicular to OA.

Corollary 1 Tangent-Diameter Theorem

A straight line drawn at right angles to
a diameter of a circle at a point on the circle
is a tangent to the circle.

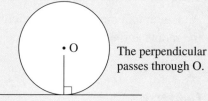

This is a
tangent.

Corollary 2

A line perpendicular to a tangent of a circle
at the point of tangency passes through the centre.

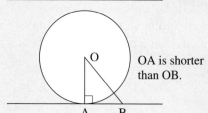

The perpendicular
passes through O.

Corollary 3

The minimum distance from the centre of
a circle to a tangent occurs at the point of
tangency.

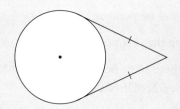

OA is shorter
than OB.

Tangent Property 2, on page 483, is known as the Equal Tangents Theorem.

Equal Tangents Theorem

The tangent segments to a circle from
an external point are equal.

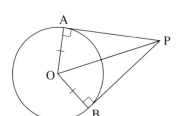

We can prove this theorem as follows.

Let P be any point outside a circle with centre O.
Let PA and PB be tangent segments to the circle.
We must prove that PA = PB.
Join OP.
According to the Tangent-Radius Theorem,
∠OAP = 90° and ∠OBP = 90°
Since OA and OB are radii of the circle, OA = OB
According to the Pythagorean Theorem,

$$PA^2 = OP^2 - OA^2$$
and $$PB^2 = OP^2 - OB^2$$

Since OA = OB, the right sides of the above equations are equal.
Therefore, $PA^2 = PB^2$ and PA = PB

Example

Point P is outside a circle with centre O.

a) Construct the two tangents from P to the circle.

b) Prove that the construction is correct.

Solution

a) Construct a circle with centre O (below left). Mark any point P outside the circle.

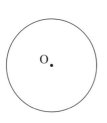 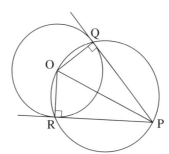

> ***Think...***
>
> Visualize a circle with diameter OP (above right). This circle intersects the given circle at Q and R, forming right angles at Q and R. PQ and PR are tangents to the circle.

Join OP.

Construct the midpoint, M, of OP.

With centre M and radius MO, construct a circle.

The circle intersects the given circle at Q and R.

Draw a line through PQ and a line through PR.
These are the required tangents.

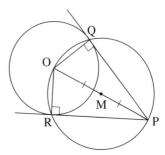

b) According to the Semicircle Theorem,
   ∠OQP = 90° and ∠ORP = 90°

According to the converse of the Tangent-Radius Theorem,
PQ and PR are tangents to the circle.

1. Why was an indirect proof used to prove the Tangent-Radius Theorem?

2. Explain why each of Corollaries 1, 2, and 3 is self-evident.

3. Could we have proved the Equal Tangents Theorem using congruent triangles? Explain.

4. In the last step of the proof of the Equal Tangents Theorem, why did we write, after $PA^2 = PB^2$, $PA = PB$ instead of $PA = \pm PB$?

8.4 EXERCISES

1. In the diagram (below left), PQ and PR are tangents to the circle. Prove that $\angle PQR = \angle PRQ$.

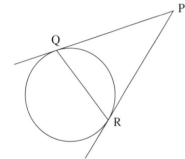
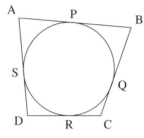

2. In the diagram (above right), the sides of quadrilateral ABCD are all tangents to the circle.

   a) Prove that AB = AS + BQ   **b)** Prove that CD = CQ + DS

3. Two tangents are drawn from an external point P to points A and B on a circle with centre O (below left). Prove that PAOB is a cyclic quadrilateral.

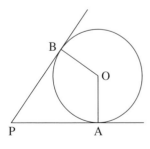
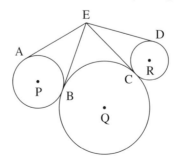

4. In the diagram (above right), EA, EB, EC, and ED are tangent segments. The centre of each circle is labelled. Prove that ABCD is a cyclic quadrilateral.

5. Parallel tangents AB and CD intersect a circle at points M and N (below left). Prove that MN is a diameter of the circle.

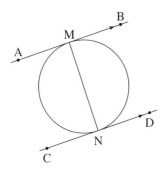

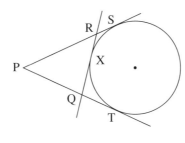

6. Tangents PS, PT, and QR intersect a circle at points S, T, and X respectively (above right). Prove that the perimeter of △PQR is equal to 2PS.

7. In the diagram (below left), PQ is a tangent segment to a circle with centre O. Given that QO = QR, prove that OR = RP.

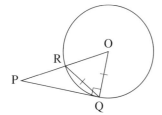

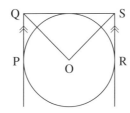

8. Tangents PQ and RS intersect a circle at points P and R (above right). The circle has centre O. QS is a tangent segment to the circle, and PQ ∥ RS. Prove that △OQS is a right triangle.

 9. AB is a diameter of a circle with centre O, and C is any other point on the circle. MN is the tangent at C.

a) Given AD ⊥ MN, prove that AC bisects ∠BAD.

b) E is a point on MN so that BC bisects ∠ABE. Prove that BE ⊥ MN.

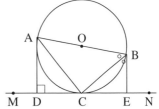

10. AB is a diameter of a circle and PQ is a tangent to that circle. Prove that the sum of the distances of A and B from PQ is a constant.

COMMUNICATING THE IDEAS

Suppose a circle and a point outside the circle are given. Write to describe how you can construct the two tangents from the point to the circle. Include diagrams in your description, and explain why your method is correct.

How Long Is the Chain?

On page 468, you considered the problem of determining the length of a chain or a belt that encircles two wheels. This problem involves common tangent segments to two circles.

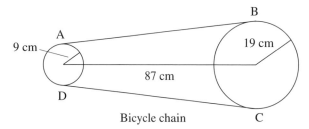

Bicycle chain

DEVELOP A MODEL

An *estimation model* for determining the length of the bicycle chain is based on two assumptions.

- The length of the chain along the common tangent segments AB and CD is approximately the same as the distance between the centres of the circles.

- The length of the chain around the circles from B to C and from D to A is approximately half the circumference of each circle.

1. **a)** Use the estimation model to determine the length of the bicycle chain. Express your answer to the nearest hundredth of a centimetre.

   b) The length of the chain can be accurately calculated using trigonometry. The result is 263.12 cm. Calculate the difference between your estimated length in part a and the actual length.

   c) Express your answer in part b as a percent of the actual length.

2. **a)** Suppose you use the estimation model to determine the length of the alternator belt in the diagram on page 495. Would you expect the result to be as accurate as it was for the bicycle chain? Explain.

   b) Verify your prediction by repeating exercise 1 for the alternator belt. The actual length of the belt is 252.05 cm.

3. What are some sources of error in the estimation model?

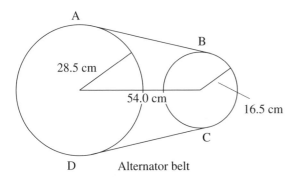

28.5 cm

B

54.0 cm

16.5 cm

Alternator belt

The accuracy of the estimation model decreases when the common tangent segments between the two circles become more inclined with respect to the line joining their centres. To improve the accuracy in these situations, another model is needed. The *measurement model* is based on making a scale diagram and taking measurements from it.

4. **a)** Choose a convenient scale. Make a large scale drawing to show the two circles in the diagram above.

   b) Use visual estimation and a ruler to draw the two common tangent segments. Measure the lengths of these segments. Use your scale to determine the lengths of these parts of the alternator belt.

   c) Join the points of tangency to the centres of the circles. Measure the angles formed at the centres. The length of each arc is proportional to the angle it subtends at the centre. Use each ratio to write a proportion involving the circumference of the circle and the 360° angle it subtends at the centre. Solve each proportion to determine the length of the part of the belt that encircles each wheel.

   d) Calculate the length of the alternator belt.

   e) Use the length in exercise 2b. Calculate the difference between your measured length and the actual length. Express this difference as a percent.

5. What are some sources of error in the measurement model?

REVISIT THE SITUATION

6. Suppose a bicycle chain or an alternator belt is manufactured to fit the wheels in the diagrams on these pages. What are some additional factors to take into account when the length of the chain or belt is determined?

When a bicycle chain or an alternator belt is manufactured, its exact length must be known. For this purpose, a different model is needed. The *calculation model* follows.

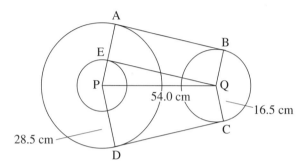

Visualize this construction. Construct circles with centres P and Q, as shown. With centre P, construct a third circle whose radius is the *difference* of the radii of the given circles. From Q, construct a tangent segment to this circle, intersecting it at E. Join PE, and extend it to meet the first circle at A. Construct AB perpendicular to PA, to meet the second circle at B.

7. a) Explain how you would construct the tangent segment from Q to the third circle.

   b) Explain why the perpendicular drawn at A is a tangent segment to both circles.

8. a) Look at △PEQ. What kind of triangle is it? How long is PE? Calculate the length of EQ.

   b) Explain why your answer in part a is also the length of AB.

9. a) In right △PEQ, you know the length of the hypotenuse PQ and side PE, which is adjacent to ∠P. Recall from trigonometry that $\cos P = \frac{PE}{PQ}$. Use this ratio to determine ∠P to the nearest hundredth of a degree.

   b) Determine ∠APD and ∠BQC. Explain why these two angles are equal.

   c) Calculate the fraction of the second circle represented by the minor arc from B to C. Use this fraction to calculate the length of the belt from B to C.

   d) Calculate the fraction of the first circle represented by the major arc from D to A. Use this fraction to calculate the length of the belt from D to A.

10. Use your results from exercises 8 and 9. Calculate the length of the alternator belt to the nearest hundredth of a centimetre.

11. Consider the construction you used to solve this problem. Explain why the radius of the third circle had to be the difference of the radii of the given circles.

According to the Tangent-Diameter Theorem, the angle between a tangent to a circle and the diameter at the point of tangency is 90°. Since each diameter of a circle is a chord, this suggests that there might be a relation between the tangent to a circle and other chords.

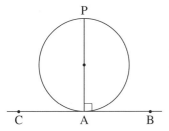

$\angle PAB = 90°$
$\angle PAC = 90°$

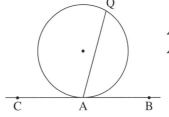

$\angle QAB = ?$
$\angle QAC = ?$

INVESTIGATE

You will use a computer or geometric instruments to make a conjecture about the angles between a tangent to a circle and chords in the circle. You will prove your conjecture later in this section.

1. Construct a circle with centre O. Mark a point A on the circle. Construct a line passing through A perpendicular to OA. This line is a tangent to the circle. Mark another point B on the tangent.

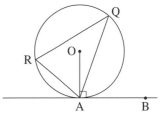

   a) Mark a point Q on the circle. Join A to Q to form chord AQ. Measure $\angle QAB$.

   b) Mark a point R on the circle, on the side of the chord opposite B. Measure $\angle QRA$. What do you notice?

   c) Visualize repeating this construction for other positions of Q on the circle, and for other circles. What would you expect to observe about the angle between the tangent and the chord, and the inscribed angle on the opposite side of the chord? Check your prediction with other examples.

   d) Make a conjecture based on your findings.

The conjecture you made in *Investigate* is known as the Tangent-Chord Theorem.

Tangent-Chord Theorem

The angle between a tangent to a circle and a chord of the circle is equal to the inscribed angle on the opposite side of the chord.

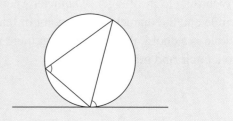

We can prove this theorem as follows.

In the diagram, QA is a chord in a circle with centre O, and AB is the tangent at A. R is a point on the circle, on the side of chord QA opposite B. We must prove that $\angle QAB = \angle QRA$.

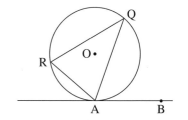

Think...

In some way, we must use the fact that AB is a tangent. We know AB is perpendicular to a diameter.

Join AO and extend it to meet the circle at P.

Then, $\angle PAB = 90°$

Join RP.

According to the Semicircle Theorem, $\angle PRA = 90°$

From Corollary 1 of the Angles in a Circle Theorem, $\angle PAQ = \angle PRQ$

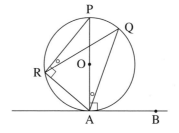

Then, $\angle QAB = \angle PAB - \angle PAQ$
$= 90° - \angle PAQ$ ①

and, $\angle QRA = \angle PRA - \angle PRQ$
$= 90° - \angle PRQ$ ②

Since $\angle PAQ = \angle PRQ$, expressions ① and ② are equal.
Therefore, $\angle QAB = \angle QRA$

Corollary

The Tangent-Chord Theorem applies to both angles formed by a tangent and a chord.

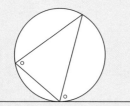 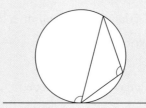

VISUALIZING

The Tangent-Diameter Theorem and the Tangent-Chord Theorem

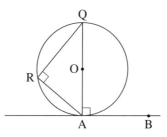 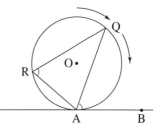

R is a fixed point on the circle. As Q moves around the circle, ∠QAB is always equal to ∠QRA.

Example

PA and PB are tangent segments from a point P to a circle. C is a point on the circle on the side of AB that is opposite P. Given ∠P = 30°, determine the measure of ∠C.

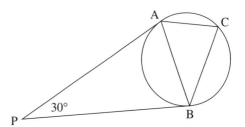

Solution

> **Think...**
>
> We can determine ∠C if we can determine ∠ABP. We can determine ∠ABP from △PAB since we know the two tangent segments PA and PB have the same length.

According to the Equal Tangents Theorem, PA = PB

Hence, △PAB is isosceles, and ∠PAB = ∠PBA

By the Angles in a Triangle Theorem in △PAB,

30° + 2∠PBA = 180°

∠PBA = 75°

According to the Tangent-Chord Theorem, ∠C = 75°

DISCUSSING THE IDEAS

1. Explain why the Tangent-Diameter Theorem is a special case of the Tangent-Chord Theorem.

2. Explain why the corollary of the Tangent-Chord Theorem is self-evident.

3. In the second diagram in *Visualizing* on page 499, suppose Q is a fixed point on the circle. Which two angles are always equal if R moves around the circle?

4. In the *Example*, how could you prove that ∠PAB = ∠PBA without using the Equal Tangents Theorem or the Isosceles Triangle Theorem?

8.5 EXERCISES

A 1. Determine each value of *x*. AB and AC are tangent segments. Point O is the centre of each circle.

a)

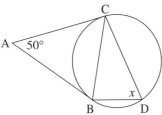

b)

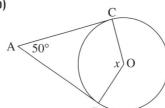

c)

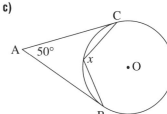

2. Determine each value of *x*. PQ and PR are tangents.

a)

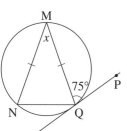

b)

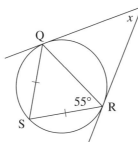

c)

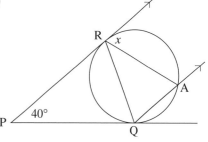

3. Choose one part of exercise 2. Write to explain how you determined the value of *x*.

4. Determine each value of *x* and *y*. AB and CD are tangents. Points O and M are the centres of the circles.

a)

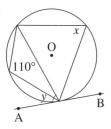

b)

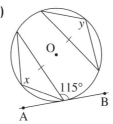

c)

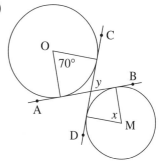

5. Choose one part of exercise 4. Write to explain how you determined the values of *x* and *y*.

6. In the diagram (below left), DCE is a tangent to the circle.

a) Explain why ∠ABC = ∠ACE.

b) Find two different ways to explain why ∠BAC = ∠BCD.

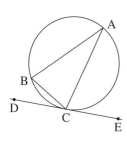

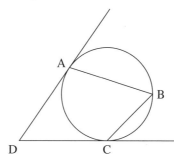

7. In the diagram (above right), DA and DC are tangents to the circle. Is it possible for ABCD to be a cyclic quadrilateral? Explain.

B **8.** Tangent segments are drawn from an external point P to points A and B on a circle (below left). AD is a chord parallel to tangent segment PB. Given ∠PAB = 60° and AP = 7 cm, determine each value.

a) the measure of ∠ADB

b) the measure of ∠BAD

c) the length of AD

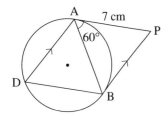

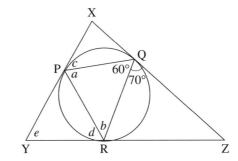

9. XZ, YX, and ZY are tangent segments to a circle at points Q, P, and R, respectively (above right). Determine the values of *a, b, c, d,* and *e.*

10. Triangle ABC is inscribed in a circle. Given ∠A = 50° and ∠B = 70°, determine the measures of the three angles of the triangle formed by the tangents to the circle at A, B, and C.

11. A circle is inscribed in △ABC. Given ∠A = 40° and ∠B = 80°, determine the measures of the three angles of the triangle formed by the points of tangency of △ABC and the circle.

12. In each diagram, AB and AC are tangents to the circle. Express *y* as a function of *x*. What is the domain of each function?

a)

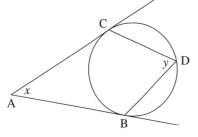

b)

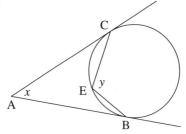

13. In the diagram (below left), MN is parallel to tangent ABC. Prove that △ MBN is isosceles.

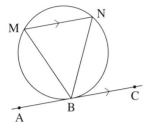

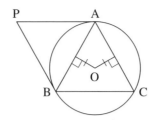

14. AB and AC are two chords that are equidistant from the centre of a circle (above right). P is any point external to the circle, so that PA and PB are tangent segments to the circle. Prove that ∠APB = ∠BAC.

15. PA and PB are two tangents from a point P to a circle with centre O (below left). Prove that ∠APB = $\frac{1}{2}$(reflex ∠AOB − ∠AOB).

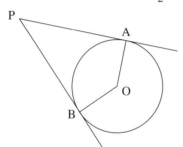

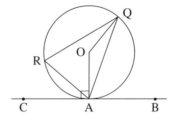

16. One way to prove the Tangent-Chord Theorem was shown on page 498.

  a) Use the diagram (above right) to prove the Tangent-Chord Theorem a different way.

  b) Compare your proof with the one on page 498. Which proof do you prefer? Why?

C **17.** One leg of a right triangle is a diameter of a circle. Prove that the tangent drawn at the point of intersection of the circle with the hypotenuse bisects the other leg of the triangle.

18. State and prove the converse of the Tangent-Chord Theorem.

COMMUNICATING THE IDEAS

Your friend has missed today's mathematics lesson. How would you explain the Tangent-Chord Theorem and its relationship to the Tangent-Diameter Theorem? Write your ideas in your notebook.

Sweeping a Circle with Lines

P is a point outside a circle. Visualize
a line through P that rotates and sweeps
across the circle. The turning line touches
the circle at Q and leaves it at R. As it
moves from Q to R, it intersects the
circle in two points A and B that move
around the two arcs of the circle,
approaching R.

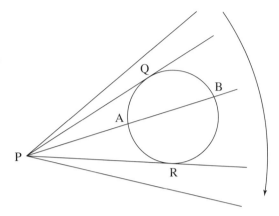

In the following exercises, you will investigate how the lengths of the
segments PQ, PA, and PB are related.

1. In the diagram (below left), PQ is a tangent to the circle at Q. PA is a
line that intersects the circle at A and B.

    a) Prove that $\triangle PAQ \sim \triangle PQB$.

    b) Use the result of part a to establish a relation among the lengths PQ,
PA, and PB.

    c) Visualize what happens for other positions of the line PB. In the
diagram (below right), how are the lengths of PA_1, PB_1, PA_2, and
PB_2 related? Explain.

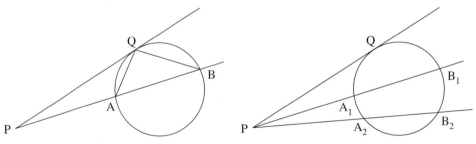

2. Use the diagram at the top of the page. Describe how the lengths of
segments PQ, PA, PB, and PR are related as the line sweeps from Q to R.

3. Construct a large circle. Mark a point P outside the circle. Construct a tangent to the circle at Q from P. Construct a line PA that also intersects the circle at B.

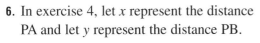

a) Measure PQ, PA, and PB.

b) Move A to other positions on the circle, or construct other lines through P to intersect the circle. Measure PA and PB for each line. Record the measurements in a table with headings for the lengths of PA and PB.

c) Use your measurements to verify the relation you discovered in exercise 1.

4. The data you recorded in exercise 3b consist of ordered pairs.

a) Visualize graphing these ordered pairs on a grid, where the first coordinate is the distance PA and the second coordinate is the distance PB. What do you think the graph would look like?

b) Construct the graph to verify your prediction.

5. For exercise 4, explain how the graph would change in each situation.

a) The circle is larger or smaller.

b) Point P is farther from the circle or closer to the circle.

c) Point P is on the circle.

6. In exercise 4, let x represent the distance PA and let y represent the distance PB.

a) Observe that y is a function of x. What kind of function do you think it might be?

b) Use your graphing calculator to determine an approximate equation for the function.

c) What are the domain and range of the function?

7. For the diagram at the top of page 504, visualize a similar diagram with P inside the circle. Do you think you might get similar results in exercises 1 to 6 if you had started with P inside the circle? Investigate to find out.

8.6 Angles and Polygons

This is a dissection puzzle called Supernova. There are 24 pieces forming a star. To solve the puzzle, you must use all the pieces to form three smaller stars that have the same shape as the star formed by all the pieces.

Each piece of the puzzle has the shape of a polygon. A *polygon* is a closed figure that consists of line segments.

In this section, you will develop a formula for the sum of the angles in any polygon.

A *regular polygon* has all sides equal and all angles equal. Consider the regular polygon formed by joining the outer vertices of the Supernova puzzle. We can determine the measures of its angles in different ways.

Method 1

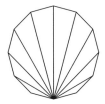

Visualize joining one vertex to all the other vertices, creating 10 triangles. The sum of the angles in the polygon is the same as the sum of the angles in these triangles.

The sum of the angles in the polygon is $10 \times 180° = 1800°$.

Hence, each angle is $\frac{1800°}{12} = 150°$.

Method 2

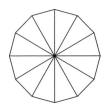

Visualize joining the centre to each vertex, creating 12 congruent triangles. The sum of the angles in the polygon is the sum of the angles in the triangles minus the angle at the central vertex of each triangle. These central angles add to 360°.

The sum of the angles in the polygon is

$12 \times 180° - 360° = 1800°$.
Hence, each angle is $\frac{1800°}{12} = 150°$.

Method 3

Visualize extending each side of the polygon, forming 12 exterior angles. Visualize $\angle OAX$ rotating about O. As it rotates to $\angle OBY$, segment AX rotates through the same number of degrees as segment OA. Hence, $\angle YBX = \angle BOA$. Angle YBX is one of the exterior angles of the polygon. The rotation can be repeated all around the polygon.

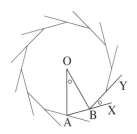

Hence, the sum of all the exterior angles is the same as the sum of all the angles formed at the centre, or 360°. Since there are 12 equal exterior angles, each exterior angle measures $\frac{360°}{12} = 30°$.

Therefore, each interior angle of the polygon measures $180° - 30° = 150°$.

I N V E S T I G A T E

1. Choose one of the methods above. Determine the measure of each angle in each regular polygon.

   a)
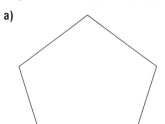
   b)
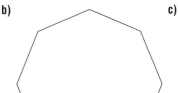
   c)
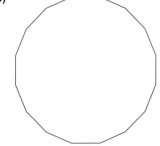

2. Suppose a regular polygon has n sides. Use your method in exercise 1. Determine a formula for the measure of one angle of a regular polygon with n sides.

Many polygons are not regular polygons. Although you cannot determine the angles in these polygons without further information, you can determine the sum of the angles if you know the number of sides. For example, you know the sum of the angles in any triangle is 180°.

3. Determine the sum of the angles in each polygon. Use one of the three methods. You may need to modify the method slightly for some of these polygons.

a)

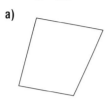

b)

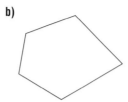

c)

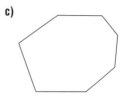

d)

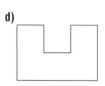

e)

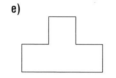

f)

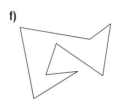

4. Suppose a polygon has n sides. Use your method in exercise 3. Determine a formula for the sum of the angles of a polygon with n sides.

In exercise 4 of *Investigate*, you developed a formula for the sum of the angles of a polygon with n sides.

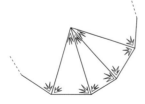

A polygon with n sides can be divided into n triangles. The sum of the angles in n triangles, in degrees, is $180n$.

There is an extra $360°$ at the central point. The sum of the angles in the polygon, in degrees, is $180n - 360$, or $180(n - 2)$.

For a polygon with many sides, the sum of the angles can be quite large. Hence, as well as using degrees as a unit of angle measure, we will also use the equivalent number of right angles. For example, since $\frac{180}{90} = 2$, we say that the sum of the angles in a triangle is 2 right angles. To change from degree measure to right-angle measure, divide by 90. Hence, we can express the above formulas in right-angle measure.

In right angles, the sum of the angles in a polygon with n sides is $\frac{180n - 360}{90} = 2n - 4$, or $2(n - 2)$.

This establishes the following theorem.

Angle Sum Theorem for Polygons

In any polygon with n sides, the sum of the angles is:

in degrees:	$180n - 360$
in right angles:	$2n - 4$

In exercise 2 of *Investigate*, you developed a formula for the measure of one angle of a regular polygon with n sides.

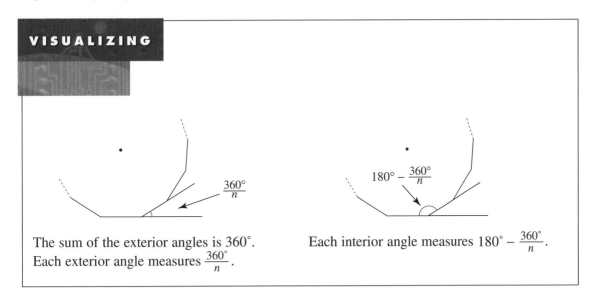

VISUALIZING

The sum of the exterior angles is 360°. Each exterior angle measures $\frac{360°}{n}$.

Each interior angle measures $180° - \frac{360°}{n}$.

This establishes the following corollary of the Angle Sum Theorem for Polygons.

Corollary   Angle Theorem for Regular Polygons

In a regular polygon with n sides, each angle measures $180° - \frac{360°}{n}$.

Example

Construct a regular 7-sided polygon, with each side 2.5 cm long.
Then measure the lengths of its diagonals.

Solution

Draw a rough sketch of the polygon.

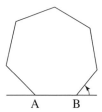

> **Think...**
>
> We can determine the measure of one exterior
> angle. We can draw one side AB that is
> 2.5 cm long. Then we can rotate the arm of
> the exterior angle at B to get the direction of
> the adjacent side of the polygon.

The measure of each exterior angle of the polygon is $\frac{360°}{7} \doteq 51.4°$.

Draw a base line. Mark two points A and B that are 2.5 cm apart
(below left).

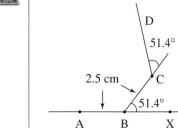

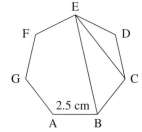

With a protractor, construct a 51.4° angle at B.

With a computer, rotate segment BX through 51.4° about centre B.

On the new arm, mark point C that is 2.5 cm from B.

Repeat until the polygon has been formed (above right).

A diagonal of a polygon is a line segment joining any two non-adjacent
vertices. The regular 7-sided polygon has diagonals of two different lengths.
By measuring, EB $\doteq$ 5.6 cm and EC $\doteq$ 4.5 cm

1. Why do we say that the sum of the angles in a triangle is 2 right angles, when we know that a triangle cannot contain 2 right angles?

2. a) Explain how you can convert an angle measure from degrees to a number of right angles.

   b) Explain how you can convert an angle measure from a number of right angles to degrees.

3. a) Does the explanation in *Visualizing* on page 508 apply to all polygons? Explain.

   b) In *Visualizing* on page 508, how else could we have explained the sum of the angles in an *n*-sided polygon?

   c) Does the explanation in part b apply to all polygons? Explain.

4. a) Explain why the corollary on page 509 is self-evident.

   b) Why was right-angle measure not included in the statement of the corollary?

8.6 EXERCISES

A 1. What is the sum of the angles in a quadrilateral, in degrees and in number of right angles?

2. What is the sum of the angles in each polygon?

   a)

   b)

   c)

B 3. What is the sum of the angles in a polygon with each number of sides?

   a) 5   b) 8   c) 10   d) 100

4. What is the measure of each angle in a regular polygon with each number of sides?

   a) 5   b) 8   c) 10   d) 100

5. Choose one polygon from exercises 3 and 4. Write to explain how you determined the sum of the angles and the measure of each angle.

6. The formulas in the statement of the Angle Sum Theorem for Polygons are functions of *n*.

a) What kind of functions are they?

b) Choose one function.

  i) What is the domain of the function?   **ii)** Graph the function.

7. The formula in the statement of the Angle Theorem for Regular Polygons is a function of *n*.

a) What kind of function is it?

b) What is the domain of the function?

c) Graph the function.

Exercises 8 to 12 involve the Supernova puzzle on page 506. All polygons in the puzzle are formed using angles of 90°, 60°, 45°, and 30°. For this reason, the angle measures are not shown on these diagrams. You can determine the angle measures by estimation.

8. a) These polygons are pieces from the Supernova puzzle. Determine the angle sum for each polygon. Do this in two different ways.

i)

ii)

b) Choose one polygon from part a. Write to explain the two ways you determined the angle sum.

9. Two adjacent pieces from the Supernova puzzle form these polygons. Identify the pieces, then determine the angle sum for each polygon.

a)

b)

10. In the solution of the Supernova puzzle, eight pieces are arranged to form a star that has the same shape as the star in the puzzle. Two adjacent pieces from the solution form the polygon (below left). Identify the pieces, then determine the angle sum for this polygon.

          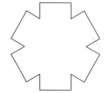

11. Six congruent pieces in the middle of the puzzle form the polygon (bottom right on page 512). Determine the angle sum for the polygon in two different ways.

12. All the pieces of the puzzle form a star-shaped polygon. Determine the angle sum of this polygon.

C 13. Each exterior angle of a regular octagon measures 45°. Each interior angle measures 135°. Observe that $135° = 3 \times 45°$. That is, the interior angle is a multiple of the exterior angle. Find other regular polygons whose exterior angles are multiples of their interior angles. Can you find any patterns in the results? Explain.

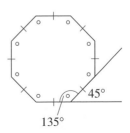

14. About 100 years ago, mathematicians devised some strange figures that have been revived by modern computer graphics. The *snowflake curve* is one example. The first four steps in the construction of the snowflake curve are shown below. Visualize these steps continuing forever. After each step, the figure comes closer and closer to the snowflake curve.

a) How was each polygon formed from the previous polygon?

b) How many sides does each polygon have?

c) What are the measures of the angles in each polygon?

d) What is the sum of the angles in each polygon?

e) Is the snowflake curve a polygon? Explain.

   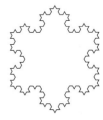

COMMUNICATING THE IDEAS

Your friend has trouble memorizing the formulas for the sum of the angles in a polygon and for the angle measure in a regular polygon. What would you say to help your friend remember these formulas? Write your ideas in your notebook.

A Bestseller from Way Back

In 300 B.C., the city of Alexandria was the intellectual and cultural centre of ancient Greek civilization. Euclid was the head of the mathematics department at the university.

Euclid's *Elements* is one of the world's most famous books. More than 1000 editions have appeared in the last 500 years. In the *Elements,* Euclid organized the results in geometry, number theory, and algebra of mathematicians from 600 B.C. to 300 B.C.

Because it was the first organized source of mathematical knowledge, Euclid's *Elements* has probably had more influence on mathematical thinking than any other book. It has dominated the study of geometry for more than 2000 years.

1. Euclid used the word "elements" to refer to the most important theorems. From Chapters 6, 7, 8, and your previous work in geometry, list the theorems and other results that you think might be considered the "elements" of geometry.

1. Determine each value of *x* and *y*. Point O is the centre of each circle.

a)

b)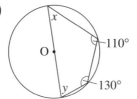

c)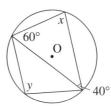

d)

e)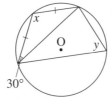

f)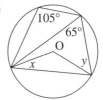

2. Construct a circle with centre O and with points A, B, and C on the circle. Construct line segment AB. Construct the line through C that is parallel to AB. Construct line segment BC. Construct the line through A that is parallel to BC. Label point D where the two lines intersect.

   a) What kind of quadrilateral is formed by the points A, B, C, and D? How can you tell?

   b) Move the points A, B, and C to make the quadrilateral cyclic. Other than being cyclic, what word best describes the quadrilateral? How do you know? Make appropriate measurements to check your description.

   c) Make a conjecture based on your findings.

3. Segments AB and AC are two equal chords of a circle with centre O (below left). The midpoints of AB and AC are D and E, respectively. Prove that quadrilateral ADOE is cyclic.

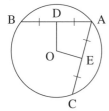

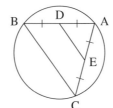

4. Segments AB and AC are two equal chords of a circle (above right). The midpoints of AB and AC are D and E, respectively. Prove that quadrilateral BCED is cyclic.

5. Quadrilateral ABCD is cyclic, with AB = AD and CB = CD. Prove that the quadrilateral formed by joining the midpoints of adjacent sides is a rectangle.

6. Two circles have the same centre O. A tangent to the smaller circle intersects the larger circle at P and Q. Another tangent to the smaller circle intersects the larger circle at R and S. Prove that PQ = RS .

7. See page 489. Prove the converse of the Tangent-Radius Theorem.

8. Tangent segments PA and PB are drawn from an external point P to a circle (below left). The bisector of ∠P intersects the circle at Q and C. Prove that ∠CAQ = 90° .

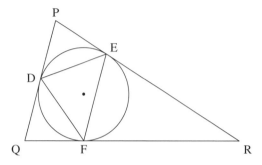

9. In the diagram (above right), D, E, and F are the points where the sides of △PQR are tangent to its inscribed circle. Use the Tangent-Chord Theorem. Prove that the two tangent segments from each vertex of △PQR are equal.

10. Carpenters sometimes use set squares together, as shown (below left).

a) Why do you think carpenters might place set squares this way?

b) Suppose AD is extended to meet BC extended. At what angle do these lines intersect?

c) What other angles can be formed by placing two set squares together?

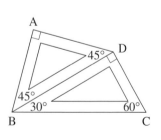

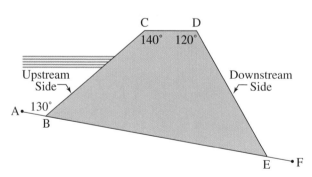

11. The diagram (above right) shows a cross section of a dam. Use the information in the diagram to calculate ∠DEF.

1. At 6% annual simple interest, how long would it take each investment to grow to the given amount?

   a) $400 to $448   **b)** $1000 to $1300   **c)** $750 to $885

2. Determine each accumulated amount.

   a) $1200 earns 3.25% annual interest for 6 months.

   b) $875 earns 5% annual interest for 45 days.

3. At an interest rate of 7% compounded annually, how much would you have to invest today to earn each indicated amount?

   a) $6470 in 7 years   **b)** $14 000 in 10 years

4. Determine the interest rate for an investment of $7500 that grows to $8203.55 in one year, with interest compounded monthly.

5. Show that the rates in each pair are equivalent.

   a) 8% compounded quarterly; 8.243% compounded annually

   b) 12% compounded monthly; 12.6825% compounded annually

6. The table shows some expenses for the Mak family.

Date	Expense	Amount ($)
January 1	Property taxes	2950.00
January 15	Car insurance	525.00
January 31	Alberta Health	404.00
March 31	Alberta Health	404.00
June 30	Alberta Health	404.00
July 15	Car insurance	525.00
September 30	Alberta Health	404.00
October 30	Car license and registration	125.00

   How much money should the Maks put aside each month to pay these expenses?

7. Determine the monthly payment and the total cost of an $80 000 mortgage at 6.75% interest amortized over 15 years.

8. Determine the equation of each parabola.

   a) with vertex $(-1, 4)$ and y-intercept 16

   b) with y-intercept 10, x-intercept 2, and axis of symmetry $x - 3 = 0$

9. Choose one part of exercise 8. Write to explain how you found the equation of the parabola.

10. Sketch the parabola defined by $v = \frac{1}{2}t^2 + 10t + 21$. Label the vertex with its coordinates. Label the axis of symmetry with its equation. Label two points on the graph with their coordinates.

11. Use a graph to approximate the zeros of each function.

a) $f(x) = x^3 - 10x + 12$ b) $q(x) = -x^4 + 2x^2 + 12$

12. a) Use a graphing tool or grid paper. Graph each rational function.

i) $y = \frac{x}{x^2 + 9}$ ii) $y = \frac{x^2 + 9}{x}$

b) Choose one function in part a. Determine the following information about this function, where possible.

i) its domain and range   ii) its zeros

iii) the equations of any asymptotes   iv) a description of any symmetry

v) the coordinates of any maximum or minimum points

13. Given $f(x) = x + 2$ and $g(x) = 3x + 2$, determine each value.

a) $f(2)$ b) $g(f(2))$ c) $g(2)$ d) $f(g(2))$

14. Determine which of these trinomials can be factored.

a) $x^2 + x + 1 = 0$ b) $3x^2 - 4x + 7 = 0$ c) $2x^2 + 7x - 1 = 0$

15. Solve each inequality. Show each solution on a number line.

a) $x^2 - 2x - 15 > 0$ b) $3x - 18 + x^2 \le 0$ c) $7x^2 - 7 < 0$

16. Solve.

a) $\frac{2 - 4x}{x - 3} \ge x + 4$ b) $\frac{7}{x + 5} < x - 1$ c) $\frac{15 - x}{x - 6} > x + 1$

17. Choose one inequality from exercise 16. Write to explain how you solved it.

18. Determine each value of k.

a) When $x^3 - kx^2 + 3x - 4$ is divided by $x - 3$, the remainder is 5.

b) When $2x^3 - 3x^2 + kx + 7$ is divided by $x + 2$, the remainder is 7.

19. Solve each system by graphing.

a) $y = x^2$ b) $y = x^2 - 4$ c) $y = |x| - 3$ d) $y = 2\sqrt{x}$
   $y = 2x$ $y = 4 - x^2$ $y = 2x$ $y = 3x - 1$

20. Choose one non-linear system from exercise 19. Write to explain how you solved it.

21. Solve by addition or subtraction.

a) $\frac{1}{3}x + \frac{1}{4}y = 0$ b) $\frac{2}{3}x + \frac{1}{5}y = -2$ c) $\frac{1}{3}x + \frac{1}{2}y = -\frac{1}{2}$
   $x + y = -1$ $\frac{1}{3}x - \frac{1}{2}y = -7$ $\frac{1}{5}x - \frac{1}{3}y = \frac{27}{5}$

22. Solve by substitution.

a) $x + 2y = 4$
$3x + 2y = 0$

b) $2x + y = 9$
$x - y = 3$

c) $2y + x + 10 = 0$
$y - 4x = 13$

23. For the school play, adult tickets cost $5.00 and student tickets cost $3.00. Twice as many student tickets as adult tickets were sold. The total receipts were $1650. How many of each kind of ticket were sold?

24. Graph the region defined by each system of inequalities. Describe the shape of the region, then determine its area.

a) $x \leq 0$
$x \geq -4$
$y \leq 0$
$y \geq -4$

b) $x \geq 0$
$y \geq 0$
$x + y \leq 5$

25. What is the negation of each statement?

a) $x \neq 4$

b) $x > 9$

c) x is a factor of 25.

26. Twenty-six people are surveyed at a mall. All the people shopped at the hardware store or the dollar store. Sixteen people shopped at the hardware store and 20 shopped at the dollar store.

a) How many people shopped at both stores?

b) How many people shopped only at the hardware store?

c) How many people shopped only at the dollar store?

27. Determine each value of x. Point O is the centre of each circle.

a)

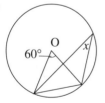

b)

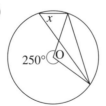

c)

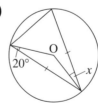

28. Equal chords AB and AD are in a circle, centre O, diameter AE. Prove that $\angle BAE = \angle DAE$.

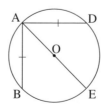

29. Construct each regular polygon, with each side 2.5 cm long. Then measure the lengths of its diagonals.

a) a regular pentagon

b) a regular 9-sided polygon

What Comes after the Cube?

 CONSIDER THIS SITUATION

We live in a three-dimensional world. We can hold a three-dimensional object such as a cube and count its faces, edges, and vertices. The faces are two-dimensional squares, which we can draw on paper. The edges are one-dimensional segments, and the vertices are points, which have zero dimension.

Visualize starting with a point and increasing the dimension by one:

Figure	• point	— segment	square	cube	?
Dimensions	0	1	2	3	4

• How are the figures below related to those above?

• What comes next, after the cube?
• How many faces, edges, and vertices does the next object have?

On pages 542–544, you will develop mathematical models for visualizing the next object.

 FYI Visit www.awl.com/canada/school/connections

For information related to the above problem, click on MATHLINKS followed by AWMath. Then select a topic under What Comes after the Cube?

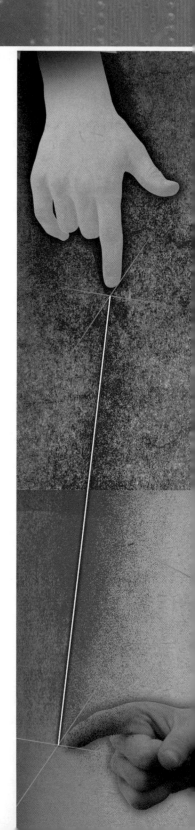

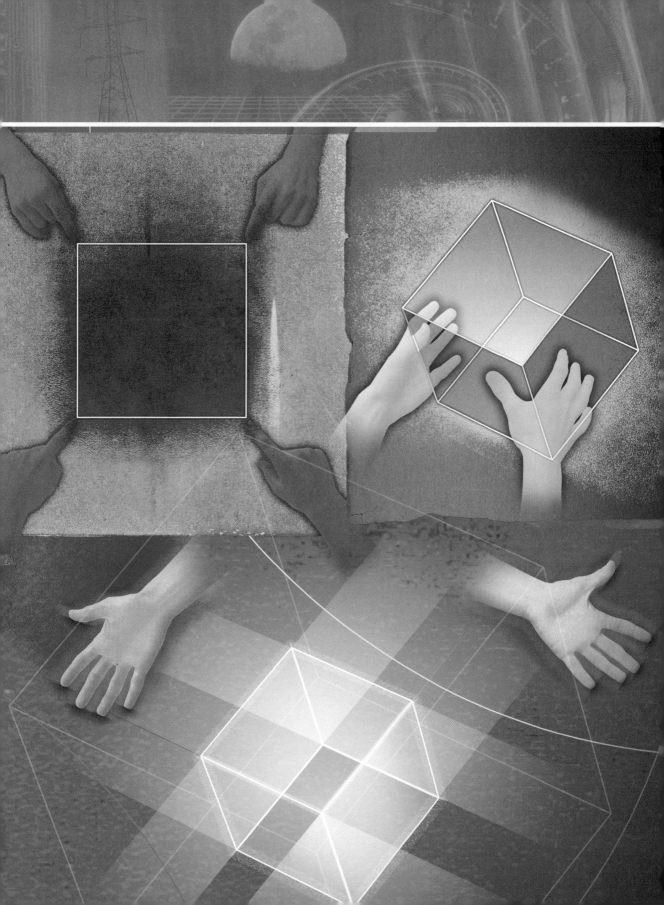

9.1 The Equation of a Circle

Many farms in western North America use an
automated centre-pivot irrigation system.
A long pipe sprays water as it rotates about
the centre. Distinctive circular traces are left
by the wheels and, since the end of the pipe
is always the same distance from the centre,
the area watered forms a circle.

On a coordinate grid, we can determine the
equation of a circle, just as we can determine
equations of lines and parabolas.

Example 1

Consider the circle with centre C(2, −1) and radius 5 units.

a) Graph the circle.

b) Determine the equation of the circle.

Solution

a) Draw a circle with centre C(2, −1) and radius 5.

b) Let P(x, y) be any point on the circle.

Since CP is a radius, CP = 5

Use the distance formula.

$\sqrt{(x-2)^2 + (y+1)^2} = 5$

Square each side to eliminate the radical.

$(x-2)^2 + (y+1)^2 = 25$

This is the equation of the circle.

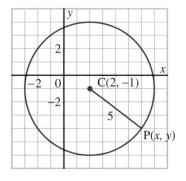

We can use the method of *Example 1* to
determine the equation of any circle with
centre C(h, k) and radius r.
Let P(x, y) be any point on the circle.
Then,

$$CP = r$$
$$\sqrt{(x-h)^2 + (y-k)^2} = r$$
$$(x-h)^2 + (y-k)^2 = r^2$$

Square each side.

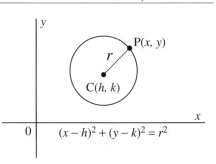

$(x-h)^2 + (y-k)^2 = r^2$

Equation of a Circle

The equation of a circle with centre (h, k) and radius r is:

$$(x - h)^2 + (y - k)^2 = r^2$$

When the centre is $O(0, 0)$, the equation is:

$$x^2 + y^2 = r^2$$

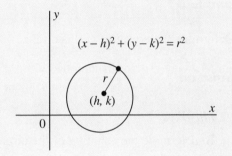

Example 2

Graph the circle defined by $(x - 5)^2 + (y + 1)^2 = 20$.

Solution

The equation $(x - h)^2 + (y - k)^2 = r^2$ represents a circle with centre (h, k) and radius r.

Hence, $(x - 5)^2 + (y + 1)^2 = 20$ represents a circle with centre $(5, -1)$ and radius $\sqrt{20}$.

Plot the point $C(5, -1)$ on a grid.

With centre C, draw a circle with radius $\sqrt{20} \doteq 4.47$.

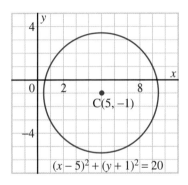

VISUALIZING

Consider the graph in *Example 2*. Observe that $r^2 = 20 = 2^2 + 4^2$.

Start at the centre. Points on the circle can be located by moving:

2 left, 4 up	2 right, 4 up
4 left, 2 up	4 right, 2 up
2 left, 4 down	2 right, 4 down
4 left, 2 down	4 right, 2 down

This method can only be used when the square of the radius is the sum of two squares.

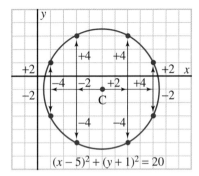

Example 3

Determine the equation of this circle.

Solution

> **Think...**
>
> To determine the equation of the circle, we need
> to know the radius and the coordinates of the
> centre. The coordinates of the centre are given,
> and we can calculate the radius using the distance
> formula.

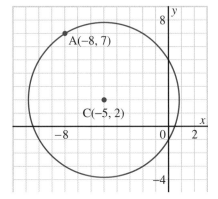

The radius is the length of CA.

$$CA = \sqrt{(-8 + 5)^2 + (7 - 2)^2}$$
$$= \sqrt{9 + 25}$$
$$= \sqrt{34}$$

The radius of the circle is $\sqrt{34}$. The centre is $(-5, 2)$.

Use the equation $(x - h)^2 + (y - k)^2 = r^2$.

The equation of the circle is $(x + 5)^2 + (y - 2)^2 = 34$.

DISCUSSING THE IDEAS

1. In *Example 1*, the equation of the circle was written as $(x - 2)^2 + (y + 1)^2 = 25$. Why did we not expand the left side of the equation?

2. For the equation of the circle in *Example 3*:
   a) Why did we write +5 in $(x + 5)^2$, but −2 in $(y - 2)^2$?
   b) What happened to the radical sign in $\sqrt{34}$?

3. a) Does the graph of a circle represent a function? Explain.
   b) Does the equation of a circle represent a function? Explain.

A 1. Determine if each point is on the circle defined by $x^2 + y^2 = 85$.

    a) A(9, −2)   **b)** B(−5, 8)   **c)** C(−7, −6)   **d)** D(4, 8)

2. State the radius and the coordinates of the centre of the circle defined by each equation.

    a) $(x - 3)^2 + (y + 4)^2 = 81$ **b)** $(x + 2)^2 + (y - 1)^2 = 5$

    c) $(x + 4)^2 + y^2 = 15$ **d)** $x^2 + (y - 6)^2 = 48$

    e) $x^2 + y^2 = 64$ **f)** $x^2 + y^2 = 12$

3. **a)** Write the equation of the circle with each given centre and radius.

    i) O(0, 0), 3   **ii)** O(0, 0), 7   **iii)** A(5, 3), 4   **iv)** B(−2, 6), 5

    v) C(4, 0), 6   **vi)** D(0, −3), 9   **vii)** E(2, 0), $\sqrt{5}$ **viii)** F(3, −5), $\sqrt{10}$

    b) Choose one circle from part a. Write to explain how you determined its equation.

4. Write the equation of each circle.

a) 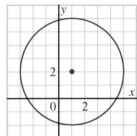   **b)** 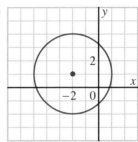   **c)**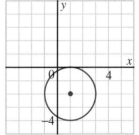

5. What does each equation represent? Explain.

    a) $x^2 + y^2 = 1$ **b)** $(x - 3)^2 + (y - 2)^2 = 2$

B 6. State which of these circles have:

    a) radius less than 5

    b) centre on the x-axis

    c) centre on the line $y = x$

        i) $(x - 3)^2 + (y - 3)^2 = 1$ **ii)** $(x + 2)^2 + (y - 4)^2 = 9$

        iii) $x^2 + (y + 7)^2 = 13$ **iv)** $(x - 5)^2 + y^2 = 20$

        v) $(x + 1)^2 + (y + 1)^2 = 25$ **vi)** $x^2 + y^2 = 32$

7. On the same grid, graph the equations in each list. Write to explain how the patterns in the equations are related to the patterns in the graphs.

a) $(x - 2)^2 + (y - 5)^2 = 10$
   $(x + 2)^2 + (y - 5)^2 = 10$
   $(x - 2)^2 + (y + 5)^2 = 10$
   $(x + 2)^2 + (y + 5)^2 = 10$

b) $(x - 2)^2 + (y + 5)^2 = 5$
   $(x - 2)^2 + (y + 5)^2 = 10$
   $(x - 2)^2 + (y + 5)^2 = 20$
   $(x - 2)^2 + (y + 5)^2 = 40$

8. a) Graph the circle $(x - 3)^2 + (y + 5)^2 = 100$ and the points R(10, 2), S(9, −13), and T(−6, 0).

   b) Which, if any, of the points R, S, and T are on the circle?

9. Determine the equation of each circle.

   a) The line segment with endpoints A(−2, 0) and B(6, −6) is a diameter of the circle.

   b) The circle passes through C(1, −6) and has centre D(4, 2).

   c) The circle just touches the x-axis and has centre E(5, 4).

10. Determine the equation of each circle.

   a) Its centre is C(3, −2), and R(−1, 1) is a point on the circle.

   b) The endpoints of a diameter are M(5, 1) and N(−3, 3).

   c) The circle passes through A(2, 2) and B(5, 3), and its centre is on the line defined by $y = x + 1$.

11. The centre of a circle lies on the x-axis, 3 units from the origin. The circle passes through A(6, −4). Determine the possible equations of the circle.

12. A circle has centre O(0, 0) and radius 6 units.

   a) Write the equation of the circle.

   b) The point B(4, k) is on the circle. Determine the value of k.

13. To graph any equation on a graphing calculator, the equation must be expressed in this form: "y ="

   a) Solve the equation $x^2 + y^2 = 16$ for y. Write two equations.

   b) Graph the equations from part a on the same grid. Describe the graph.

   c) By tracing and zooming, determine to two decimal places:
      i) the value(s) of m when (2, m) is on the circle
      ii) the value(s) of k when (k, k) is on the circle.

14. A treasure map has this information.
   The treasure is buried 3 m from point A and 8 m from point B.
   Point A is located at O(0, 0). Point B has coordinates (5, 0).
   What are the possible locations of the treasure?

15. a) A circle has centre A(3, 2) and radius 5 units.

    i) The point C(m, 3) is on the circle. Determine the value of m.

    ii) The point D(2, n) is on the circle. Determine the value of n.

  b) Choose one point from part a. Write to explain how you determined the value of the variable.

16. A circle has x-intercepts 0 and 4, and y-intercepts 0 and 6. Determine the equation of the circle.

17. Describe the graph of each equation.

  a) $x^2 + y^2 = 0$ **b)** $x^2 + y^2 = -9$

  c) $y = \sqrt{25 - x^2}$ **d)** $x = \sqrt{25 - y^2}$

C **18.** Determine if the circles defined by each pair of equations intersect.

  a) $(x - 3)^2 + (y + 1)^2 = 5$; $(x - 1)^2 + (y + 2)^2 = 32$

  b) $(x - 5)^2 + (y + 1)^2 = 8$; $(x + 3)^2 + (y + 7)^2 = 49$

19. Determine the equation of the circle that passes through the points J(−3, 2), K(4, 1), and L(6, 5).

20. Four circles, each with radius 5, touch the x- and y-axes and a smaller circle with centre O(0, 0). Determine the equation of the smaller circle.

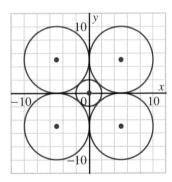

21. The equation $x^2 + y^2 = 25$ represents a circle with centre O(0, 0) and radius 5. Use a calculator to determine what each equation represents. Sketch the graph of each equation on plain paper.

  a) $x^2 - y^2 = 9$ **b)** $x^2 + 2y^2 = 36$

  c) $x^4 + y^4 = 625$ **d)** $\sqrt{x} + \sqrt{y} = 4$

COMMUNICATING THE IDEAS

Write to explain how the equation of a circle and the Pythagorean Theorem are related. Assume the centre of the circle is any point, not necessarily the origin.

Circles and Regular Polygons

A regular pentagon ABCDE is inscribed in a circle with radius 3 cm. The centre of the circle is the origin. The coordinates of A are (3, 0). We can use trigonometry to determine the coordinates of the other vertices. To determine the coordinates of B, construct BM perpendicular to the *x*-axis.

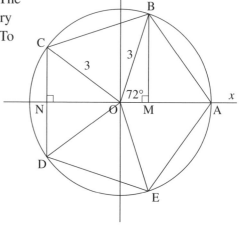

Then, $\angle BOM = \frac{360°}{5}$, or 72°

In $\triangle BOM$, $\sin 72° = \frac{BM}{OB}$

$$\sin 72° = \frac{BM}{3}$$
$$BM = 3 \sin 72°$$
$$\doteq 2.853\ 17$$

Similarly, $\cos 72° = \frac{OM}{OB}$
$$OM = 3 \cos 72°$$
$$\doteq 0.927\ 05$$

Hence, the coordinates of B are approximately (0.93, 2.85).

1. **a)** What are the coordinates of E?

   b) In $\triangle CON$, what is the measure of $\angle CON$? Explain.

   c) Determine the coordinates of C and D.

2. Determine the coordinates of the vertices of each regular polygon.

   a)
   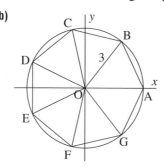

   b)
   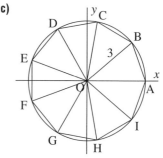

   c)

3. A regular polygon has its centre at the origin. One vertex has coordinates (3, 0). Another vertex has coordinates approximately (2.427, 1.763). How many sides could the polygon have?

9.2 Problems Involving Circles and Lines

Problems about circles and lines usually involve their points of intersection. The coordinates of these points are found by solving the non-linear system that consists of the equation of the line and the equation of the circle.

Example 1

a) Graph the circle $(x - 2)^2 + (y + 1)^2 = 25$ and the line $2x + y = 5$.

b) Determine the exact coordinates of the points of intersection.

Solution

a) The circle has centre C(2, −1) and radius 5. The line passes through A(2.5, 0) and B(0, 5).

b) To determine the exact coordinates of the points of intersection, we solve this system.

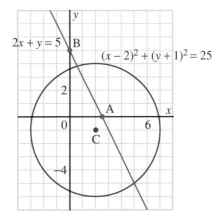

$$2x + y = 5 \qquad ①$$
$$(x - 2)^2 + (y + 1)^2 = 25 \qquad ②$$

Use the substitution method.
Solve equation ① for y.

$$y = 5 - 2x \qquad ③$$

Substitute this expression for y in equation ②, then solve for x.

$$(x - 2)^2 + (6 - 2x)^2 = 25$$
$$x^2 - 4x + 4 + 36 - 24x + 4x^2 = 25$$
$$5x^2 - 28x + 15 = 0$$
$$(5x - 3)(x - 5) = 0$$

Either $5x - 3 = 0$ or $x - 5 = 0$
$$x = \frac{3}{5} \qquad x = 5$$

Substitute each value of x in equation ③ to determine y.

When $x = 5$, $y = 5 - 2(5)$ When $x = \frac{3}{5}$, $y = 5 - 2\left(\frac{3}{5}\right)$
$$= -5 \qquad\qquad\qquad\qquad = \frac{19}{5}$$

The coordinates of the points of intersection are (5, −5) and $\left(\frac{3}{5}, \frac{19}{5}\right)$, or (0.6, 3.8).

When two lines intersect, the angle between them is relevant. To work with angles, we use trigonometry.

Example 2

A line is drawn through the origin at an angle of 32° to the positive x-axis. R is a point on this line and R is 8 cm from the origin. With centre R, a circle is drawn with radius 5 cm. This circle intersects the x-axis at S and T. Calculate the length of chord ST, to two decimal places.

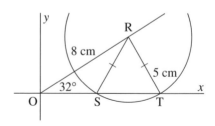

Solution

> ### Think...
>
> Visualize the perpendicular RN from R to the x-axis.
> This creates a right triangle whose height we can determine.
> The perpendicular also bisects chord ST.
> We can apply the Pythagorean Theorem to $\triangle$RSN to determine the length of SN.

Construct the perpendicular RN from R to the x-axis.

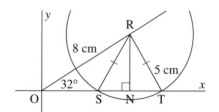

In $\triangle$ORN, $\sin 32° = \dfrac{RN}{OR}$

$$RN = 8 \sin 32°$$
$$\doteq 4.239\,35$$

RN is approximately 4.24 cm long.

In $\triangle$RSN, $SN = \sqrt{RS^2 - RN^2}$

$$\doteq \sqrt{5^2 - 4.239\,35^2}$$
$$\doteq 2.651\,02$$

According to the Chord Perpendicular Bisector Theorem, N is the midpoint of ST.

Hence, ST $\doteq 2 \times 2.651\,02$, or $5.302\,04$

Chord ST is approximately 5.30 cm long.

Recall that the SAS Congruence Theorem states that if two sides and the contained angle of one triangle are equal to two sides and the contained angle of another triangle, then the triangles are congruent. In this theorem, the angles must be contained by the corresponding equal sides. The triangles in *Example 2* illustrate the reason why. Two sides and a non-contained angle of $\triangle$ORS are equal to two sides and a non-contained angle of $\triangle$ORT, and the triangles are clearly not congruent.

DISCUSSING THE IDEAS

1. In the graph in *Example 1*, one point of intersection appears to be $(5, -5)$. If we use only the graph, can we be certain this is correct? Explain.

2. When we found the values of x in the solution of *Example 1b*, why did we substitute in equation ③? Could we have substituted in equation ①? Could we have substituted in equation ②? Explain.

3. For *Example 2*, give some reasons why we should think of constructing the perpendicular from R to ST at the beginning of the solution.

4. For *Example 2*, how could you determine each item?
   a) the coordinates of R   b) the angles in $\triangle$RSN
   c) the angles in $\triangle$ORS   d) the length of OS

9.2 EXERCISES

For each exercise for which there is no diagram, include a diagram as part of your solution.

A 1. Visualize the circle $x^2 + y^2 = 16$.
   a) What are the coordinates of its centre?
   b) What is the radius of the circle?
   c) How far apart are its two horizontal tangents? Explain.

2. For what values of k is the line $y = k$ a tangent to the circle $x^2 + y^2 = 100$?

3. The endpoints of a diameter of a circle are A(1, 2) and B(5, 4).
   a) Graph the circle.   b) Determine the equation of the circle.

4. a) Determine the coordinates of the points of intersection of each line and circle.

    i) $y = 2x$ **ii)** $y = 2x - 5$ **iii)** $y = 2x - 5$ **iv)** $x - y = 8$

        $x^2 + y^2 = 5$ $x^2 + y^2 = 5$ $x^2 + y^2 = 10$ $x^2 + y^2 = 25$

b) Choose one system from part a. Write to explain how you determined the points of intersection.

B **5. a)** Graph the circle $(x - 3)^2 + (y - 1)^2 = 5$.

b) Determine the coordinates of the points of intersection of the circle in part a and each line.

    i) $y = 2x - 2$ **ii)** $y = 2x$ **iii)** $y = 2x + 2$

c) Graph the lines in part b on the graph you drew in part a. Use your graph to explain the results you obtained in part b.

6. a) Graph the circle $x^2 + (y - 2)^2 = 20$.

b) Show that the point M(2, 6) lies on the circle.

c) Graph the tangent to the circle at the point M(2, 6).

d) Determine the equation of the tangent.

7. a) Graph the circle $(x - 3)^2 + (y + 5)^2 = 25$.

b) Use your graph to determine the coordinates of the endpoints of a diameter that has slope $-\frac{4}{3}$.

c) Determine the equations of the tangents to the circle with slope $\frac{3}{4}$.

8. a) Graph the circle $(x + 1)^2 + (y - 3)^2 = 25$.

b) Determine the equations of the tangents to the circle with slope $-\frac{3}{4}$.

9. Point P is on the line $3x + y = 26$ and is 10 units from the origin. Determine the coordinates of P. Illustrate your solution with a graph.

10. From a lighthouse, the range of visibility on a clear day is 40 km. On a coordinate system, where O(0, 0) represents the lighthouse, a ship is travelling on a course represented by $y = 2x + 80$. Between which two points on the course can the ship be seen from the lighthouse?

11. A line is drawn through the origin at an angle of $40°$ to the positive x-axis. R is a point on this line, and is 6 cm from the origin. With centre R, a circle is drawn. This circle intersects the x-axis at S and T.

a) Calculate the length of chord ST for each radius of the circle.

    i) 4 cm   **ii)** 8 cm

b) What is the minimum possible radius for the circle?

12. Line segment AB is 11 cm long. It is drawn at an angle of 44° to a horizontal line AE. A circle centre B and radius 9 cm is drawn, intersecting the horizontal line at C and D. Calculate the length of chord CD.

13. The endpoints of a line segment are O(0, 0) and C(8, 6).

a) Calculate the length of OC.

b) Determine the angle formed by OC and the positive x-axis.

c) Determine the coordinates of the points on the x-axis that are 9 units from C.

d) Check your answer by determining the intersection points of the circle $(x - 8)^2 + (y - 6)^2 = 81$ and the line $y = 0$.

14. Streetlights A, B, C, D, and E are placed 50 m apart on the main road, as shown below. The light from a streetlight is effective up to a distance of 60 m.

a) Determine the distance from A to the farthest point on the side road that is effectively illuminated.

b) Determine the length of the side road that is effectively illuminated by both streetlights C and D.

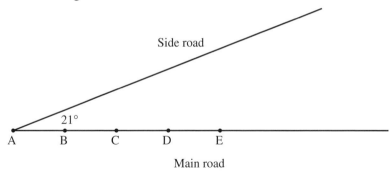

15. A dog is on a 20-m leash in a yard. The leash is staked to the ground. The yard is not fenced and is bounded by a sidewalk. Suppose you are standing on the sidewalk. The angle between the sidewalk and the line joining you and the stake is 67°. You are 21 m from the stake. What length of sidewalk can the dog reach? Give the answer to the nearest tenth of a metre.

Ⓒ 16. A square is inscribed in a circle. Two vertices of the square are A(3, 2) and B(6, 5). Determine the possible equations of the circle.

17. Points Q(−3, 4) and A(3, 4) lie on the circle $x^2 + y^2 = 25$ (below left). Point P(5, 10) is outside the circle.

a) Show that line PQ intersects the circle at only one point.

b) Line PA intersects the circle again at B. Determine the coordinates of B.

c) Show that $PQ^2 = PA \times PB$.

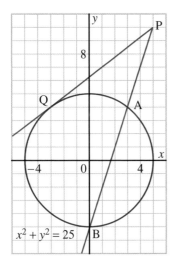
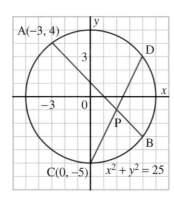

18. Chords AB and CD of the circle $x^2 + y^2 = 25$ (above right) intersect at P(2, −1).

a) Points A and C have the coordinates shown. Determine the coordinates of B and D.

b) Show that $PA \times PB = PC \times PD$.

19. Pyramid ABCDE has a rectangular base ABCD. Calculate the possible lengths of edge EC.

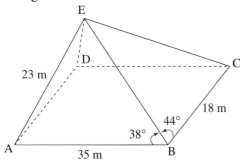

Write to explain the importance of using a diagram to solve a problem involving circles and lines. Use an example to illustrate your explanation.

Crop Circles

Designs like the one in the photograph below have been appearing in farmers' fields in great numbers worldwide since the 1970s. Some people believe that the earliest formation, in England, dates back to 1678. There are many theories as to how the circles are formed. The designs are not limited to circles and often include spirals and lines.

Probably the easiest way to learn about the mathematics in this crop circle is to try to replicate it.

1. Draw an equilateral triangle.

2. Construct the centre of the inscribed circle. Recall that the incentre is the point of intersection of the bisectors of the angles. Draw the inscribed circle.

3. Construct the centre of the circumscribed circle. Recall that the circumcentre is the point of intersection of the perpendicular bisectors of the sides. Draw the circumscribed circle.

4. One theory is that the designs are made by people at night. Write a set of instructions for someone in a field to replicate the design. The diameter of the circumscribed circle is 100 m.

If you are interested in viewing more crop circles and reading about them, search the Internet for crop circles.

1. This diagram shows the line defined by $2x + 3y = 18$. Line segment OM is perpendicular to this line.

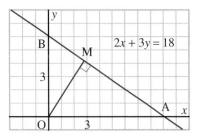

   a) Draw this diagram on grid paper. There are some right triangles in the diagram. Use your knowledge of: the equation of a line, geometry, and trigonometry. Determine the lengths of all the sides and the measures of all the angles in the triangles.

   b) Explain why the distance OM is the shortest distance from the origin to the line.

2. Mark the point P(7, 5) on your graph. Calculate to determine the shortest distance from P to the line.

3. Write to describe how you calculated the shortest distances in exercises 1 and 2.

The shortest distance from a point to a line is measured along the perpendicular from the point to the line.

PM is shorter than PC.
The length, d, is the shortest distance from P to AB.

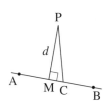

We can use a graph to determine the shortest distance from a point to a line.

Example 1

Determine the shortest distance from the origin to the line defined by
$x - 2y = -8$. Give the answer to two decimal places.

Solution

Graph the line on grid paper.

The line defined by $x - 2y = -8$ intersects the axes
at A(−8, 0) and B(0, 4). This line has slope $\frac{1}{2}$.
Hence, any line perpendicular to it has slope −2.
Draw a line through O with slope −2 that intersects
this line at M. Then OM is the shortest distance
from O to the line.

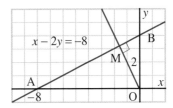

Think ...

We can determine the lengths of OA and OB.

We can use $\triangle AOB$ to determine $\angle BAO$.

We can use $\triangle AOM$ to determine the length of OM.

Since the coordinates of A are (−8, 0), the length of AO is 8 units.

Since the coordinates of B are (0, 4), the length of OB is 4 units.

In $\triangle AOB$, $\tan A = \dfrac{OB}{AO}$

$\qquad\qquad = \dfrac{4}{8}$

$\qquad\qquad = 0.5$

$\qquad\angle A \doteq 26.565\ 05°$

$\angle A$ is approximately 26.6°.

In $\triangle AOM$, $\sin A = \dfrac{OM}{OA}$

$\qquad\quad OM \doteq 8 \sin 26.565\ 05°$

$\qquad\qquad\quad \doteq 3.577\ 71$

The shortest distance from the origin to the line is approximately 3.58 units.

We can use the same method to determine the distance from a point, other than
the origin, to a line.

Example 2

Determine the shortest distance from R(7, 6) to the line defined by $3x + 4y = 28$. Give the answer to one decimal place.

Solution

Graph the point R(7, 6) and the line $3x + 4y = 28$.
This line has slope $-\frac{3}{4}$. Draw a line through R
with slope $\frac{4}{3}$ that intersects the line at S.
Then RS is the shortest distance.

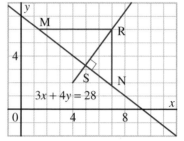

> ### Think ...
>
> Draw vertical and horizontal lines through R to form a triangle like $\triangle$ABO in *Example 1*. Then $\triangle$MRN is a right triangle.
>
> To use the method of *Example 1*, we need to know the lengths of MR and NR.

Draw a vertical line through R to intersect the given line at N.
Draw a horizontal line through R to intersect the given line at M.

To determine the coordinates of M, substitute 6 for y in the equation.

$$3x + 4(6) = 28$$
$$3x = 4$$
$$x = \frac{4}{3}$$

The coordinates of M are $\left(\frac{4}{3}, 6\right)$. The length of MR is $7 - \frac{4}{3} = \frac{17}{3}$

To determine the coordinates of N, substitute 7 for x in the equation.

$$3(7) + 4y = 28$$
$$4y = 7$$
$$y = \frac{7}{4}$$

The coordinates of N are $\left(7, \frac{7}{4}\right)$. The length of NR is $6 - \frac{7}{4} = \frac{17}{4}$

In $\triangle$MRN, $\tan M = \dfrac{NR}{MR}$

$$= \dfrac{\frac{17}{4}}{\frac{17}{3}}$$

$$= \frac{3}{4}, \text{ or } 0.75$$

$$\angle M \doteq 36.869\ 89°$$

$\angle M$ is approximately $36.9°$.

In $\triangle$MSR, $\sin M = \dfrac{SR}{MR}$

$$SR \doteq \frac{17}{3} \times \sin 36.869\ 89°$$

$$\doteq 3.399\ 999$$

Point R is 3.4 units from the line.

1. Explain why the shortest distance from a point to a line occurs along the perpendicular from the point to the line.

2. In *Example 1:*

   a) How did we know that AO = 8 and OB = 4?

   b) Why is AO positive, when the *x*-intercept is negative?

   c) Why did we use 26.565 05° in the second calculation, and not 26.6°?

3. In *Example 2:*

   a) How did we know that the length of MR is $7 - \frac{4}{3}$?

   b) How did we know that the length of RN is $6 - \frac{7}{4}$?

   c) What is another way to determine $\tan M$?

The shortest distance from a point to a line can be calculated in other ways. Two methods are described in exercises 4 and 5.

4. In *Example 1,* you could calculate the area of $\triangle AOB$ using AO as the base and OB as the height. You can express the same area using AB as the base and OM as the height. Since you can determine the length of AB and you know the area, you can find the length of OM.

   a) Use this method to solve *Example 1.*

   b) Use this method to solve *Example 2.*

5. In *Example 1,* you could determine the equation of OM, then solve it with the given equation to determine the coordinates of M. Then you could use the distance formula to determine the length of OM.

   a) Use this method to solve *Example 1.*

   b) Use this method to solve *Example 2.*

6. Which of the three methods do you think is the simplest for calculating the shortest distance from a point to a line?

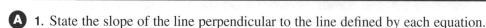

A 1. State the slope of the line perpendicular to the line defined by each equation.

a) $y = \frac{2}{3}x + 6$ b) $5x + 7y = 12$ c) $3x - 2y = 6$ d) $4x - 2y + 8 = 0$

2. Determine the equation of the line through the given point, with the given slope.

a) $(1, -3)$, slope 2

b) $(1, 7)$, slope $\frac{4}{5}$

c) $(4, -5)$, slope $-\frac{2}{3}$

d) $(-2, 9)$, slope $-\frac{10}{7}$

3. For each pair of points A and B, graph the line segment AB. Use your graph to determine the coordinates of the midpoint of AB.

a) A(0, 3), B(8, 1)

b) A(−5, 0), B(3, 7)

c) A(−2, −5), B(5, 5)

d) A(6, −3), B(−2, 4)

4. For each pair of points C and D, graph the line segment CD. Use your graph to determine the coordinates of the three points that divide line segment CD into four equal parts.

a) C(0, 2), D(8, 6)

b) C(−3, 0), D(5, −6)

c) C(0, 0), D(−4, 4)

d) C(6, −2), D(−2, 6)

B 5. a) Determine the shortest distance from each point to the line.

i) A(2, 5) to the line $3x - 4y = 16$

ii) B(−1, 4) to the line $5x + 12y = 4$

iii) the origin to the line $6x + 8y - 30 = 0$

b) Choose one point and corresponding line from part a. Write to explain how you determined the shortest distance.

6. Graph $y = \frac{4}{3}x + 4$ and $y = \frac{4}{3}x - 4$.

a) Draw a horizontal line segment connecting the two lines. What is the horizontal distance between the two lines?

b) From one endpoint of the horizontal segment in part a, draw a vertical line segment connecting the two lines. What is the vertical distance between the two lines?

c) What is the area of the triangle formed by the horizontal and vertical segments?

d) What is the length of the hypotenuse of this triangle?

e) What is the shortest distance between the lines defined by $y = \frac{4}{3}x + 4$ and $y = \frac{4}{3}x - 4$?

7. Graph $y = -3x + 4$, then plot the point C(2, −11).

    a) Draw a line through C(2, −11) parallel to $y = -3x + 4$.

    b) Determine the vertical distance between the two lines.

    c) Determine the horizontal distance between the two lines.

    d) Determine the shortest distance between the two lines.

    e) What is the shortest distance between $y = -3x + 4$ and C(2, −11)?

8. What is the distance between each pair of parallel lines?

    a) $y = 2$ and $y - 5 = 0$

    b) $y = 4x - 9$ and $y = 4x + 6$

    c) $3x + 4y - 8 = 0$ and $3x + 4y + 2 = 0$

9. Use each diagram. Visualize two points dividing line segment AB into three equal parts. Determine the coordinates of these two points.

    a)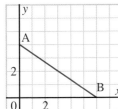
    b)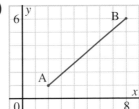

10. Choose one part of exercise 9. Write to explain how you determined the coordinates of the points.

11. Graph the points A(2, 1) and B(2, 7). Join AB to form one side of △ABC. Visualize point C moving so that △ABC is always a right triangle.

    a) Where are all the possible positions of C? Indicate these on your graph.

    b) Determine the equation(s) of any lines or circles you drew in part a.

12. Repeat exercise 11 for isosceles △ABC.

C 13. A line through B(0, 10) is 6 units from the origin. Determine its equation.

14. Refer to exercise 11. Write one equation that represents all the possible positions of C.

COMMUNICATING THE IDEAS

Write to explain how the shortest distance from a point to a line is defined. Explain how you can calculate this distance if you know the coordinates of the point and the equation of the line.

What Comes after the Cube?

On page 520, you considered what came next in the following sequence, where the dimension increased by one from each object to the next.

point, segment, square, cube, ...

DEVELOP A MODEL

A number line represents a 1-dimensional coordinate system. We call the segment from 0 to 1 the unit segment.

0   1

Visualize moving the unit segment in a direction perpendicular to itself to form a unit square. In a 2-dimensional coordinate system, each vertex of the unit square has 2 coordinates, listed below.

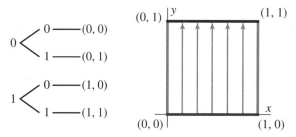

Visualize moving the unit square in a direction perpendicular to itself to form a unit cube. In a 3-dimensional coordinate system, each vertex of the unit cube has 3 coordinates, listed below.

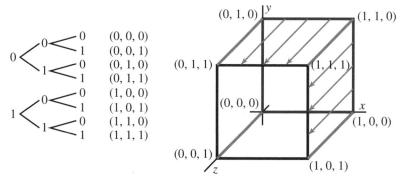

If we could move the unit cube in a direction perpendicular to itself, the next object would be a 4-dimensional object called a *hypercube*.

Since we live in a 3-dimensional world, we cannot move the unit cube in a direction perpendicular to itself. However, we can visualize doing this to form a unit hypercube. In a 4-dimensional coordinate system, each vertex of the unit hypercube has 4 coordinates.

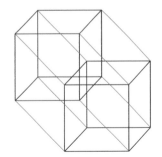

1. Use a tree diagram. List the coordinates of the vertices of the unit hypercube.

2. How many vertices does the unit hypercube have? Explain.

 LOOK AT THE IMPLICATIONS

3. The main diagonal of the unit square (below left) is the line segment joining (0, 0) and (1, 1). According to the Pythagorean Theorem, its length is $\sqrt{1^2 + 1^2}$, or $\sqrt{2}$.

  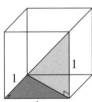

   a) The main diagonal of the unit cube (above right) is the line segment joining (0, 0, 0) and (1, 1, 1). Use the Pythagorean Theorem. Determine the length of this diagonal.

   b) The main diagonal of the unit hypercube is the line segment joining (0, 0, 0, 0) and (1, 1, 1, 1). Use the results for the unit square and the unit cube. Determine the length of the main diagonal of the unit hypercube.

4. a) How many edges does the unit hypercube have? Explain.

   b) Copy and complete this table. Describe the patterns in the last two rows.

Figure	point	segment	square	cube	hypercube
Dimensions	0	1	2	3	4
Number of vertices	1	2			
Number of edges	0	1			

   c) Extend the table a few more columns to the right. Use the names "5-cube," "6-cube," … for the objects.

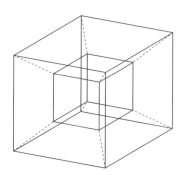

REVISIT THE SITUATION

The diagram at the top of page 543 represents a 3-dimensional model of a hypercube. Just as we can draw different 2-dimensional views of a cube on paper, we can also construct different 3-dimensional models of a hypercube.

5. This diagram shows another model of a hypercube. Compare this model with the one on page 543. How are they similar? How are they different?

We can construct another model of a hypercube using the idea of a fold-out pattern. A fold-out pattern of a 2-dimensional square is a 1-dimensional pattern showing 4 segments linked together (below left). A fold-out pattern of a 3-dimensional cube is a 2-dimensional pattern showing 6 squares joined together (below middle). Similarly, a fold-out pattern of a 4-dimensional hypercube is a 3-dimensional pattern showing 8 cubes joined together (below right).

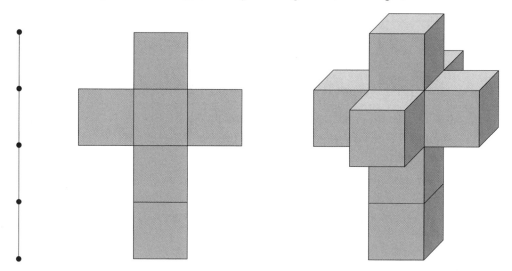

6. a) Visualize how the second diagram above forms a cube by joining some edges of the squares. How many pairs of edges must be joined?

   b) To form a hypercube, some faces of the cubes in the third diagram above must be joined. How many pairs of faces must be joined?

7. a) Identify some parts of the hypercube model at the top of this page that correspond to the cubes in the third fold-out diagram above.

   b) Repeat part a using the hypercube model at the top of page 543.

INVESTIGATE

Use this diagram in exercises 1 to 3.

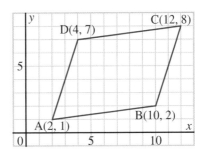

1. Recall that the slope of the line segment joining $P_1(x_1, y_1)$ and $P_2(x_2, y_2)$ is $\frac{y_2 - y_1}{x_2 - x_1}$.

   a) Calculate the slope of each side of quadrilateral ABCD.

   b) Explain why quadrilateral ABCD is a parallelogram.

2. a) Calculate the length of each side of parallelogram ABCD. What do you notice?

   b) Visualize repeating part a for other parallelograms. What would you expect to observe about the lengths of the sides? Check your prediction with another example.

   c) Make a conjecture based on your findings.

3. a) Determine the coordinates of the midpoints of the diagonals AC and BD. What do you notice?

   b) Visualize repeating part a for other parallelograms. What would you expect to observe about the diagonals? Check your prediction with another example.

   c) Make a conjecture based on your findings.

4. A rectangle is a parallelogram in which all the angles are right angles. Visualize the diagonals of a rectangle. Make a conjecture about the diagonals of a rectangle that is different from the conjecture you made in exercise 3. Verify your conjecture with some examples.

5. A rhombus is a parallelogram with all sides the same length. Visualize the diagonals of a rhombus. Make a conjecture about the diagonals of a rhombus that is different from the conjecture you made in exercise 3. Verify your conjecture with some examples.

In *Investigate* exercises 2 and 3, you should have conjectured these properties of a parallelogram.

Properties of a Parallelogram

1. The opposite sides of a parallelogram are equal.

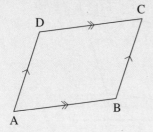

AB = DC
AD = BC

2. The diagonals of a parallelogram bisect each other.

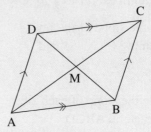

M is the midpoint of AC and BD.

You used coordinates to conjecture these properties. However, we cannot be certain they are true for all parallelograms. In Section 9.5, we will use coordinates to prove conjectures.

Example

a) Show that P(7, 7) is on the perpendicular bisector of the line segment joining A(6, −1) and B(0, 3).

b) Verify that P is equidistant from A and B.

Solution

a) Draw a diagram.

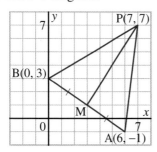

Let M be the midpoint of AB.

The coordinates of M are $\left(\frac{0 + 6}{2}, \frac{3 + (-1)}{2} \right)$, or (3, 1).

To show that P is on the perpendicular bisector of AB, show that PM is perpendicular to AB.

Slope of PM $= \dfrac{7-1}{7-3}$ Slope of AB $= \dfrac{3-(-1)}{0-6}$

$\qquad\qquad\quad = \dfrac{6}{4}$ $= \dfrac{4}{-6}$

$\qquad\qquad\quad = \dfrac{3}{2}$ $= -\dfrac{2}{3}$

Since the slopes are negative reciprocals, PM is perpendicular to AB.

b) Use the distance formula to calculate the lengths of PA and PB.

PA $= \sqrt{(7-6)^2 + (7+1)^2}$ PB $= \sqrt{(7-0)^2 + (7-3)^2}$

$\quad\; = \sqrt{1 + 64}$ $= \sqrt{49 + 16}$

$\quad\; = \sqrt{65}$ $= \sqrt{65}$

Since PA = PB, P is equidistant from A and B.

DISCUSSING THE IDEAS

1. In *Investigate,* you verified some properties of parallelograms. Explain why we cannot be certain that these properties are true for all parallelograms.

2. Does the *Example* prove the Perpendicular Bisector Theorem? Explain.

9.4 EXERCISES

A **1.** Consider A(−1, 3), B(1, 7), and C(5, 5).

    a) Verify that △ABC is a right triangle.

    b) Is △ABC an isosceles triangle? Justify your answer.

    c) M is the midpoint of AB and N is the midpoint of AC. Verify that MN is parallel to BC.

    d) Find a point D so that the four points A, B, C, and D are vertices of:
    i) a square   **ii)** a parallelogram that is not a square

2. Consider A(−5, −1), B(−2, 1), C(5, −6), and D(2, −8).

    a) Verify that AB is parallel to CD.

    b) Verify that BC = DA.

3. a) Verify that the triangle with vertices P(−3, 4), Q(−1, −2), and R(3, 2) is isosceles.

    b) Let M be the midpoint of PQ and let N be the midpoint of PR. Verify that RM = QN.

B 4. Consider $\triangle ABC$ with A(−1, 3), B(5, 7), and C(4, −2). D is the midpoint of AB and E is the midpoint of BC. Verify that the length of DE is one-half the length of AC.

5. Consider the circle defined by $(x − 3)^2 + (y + 7)^2 = 169$.

   a) Determine the coordinates of two points on this circle. Join these points.

   b) Determine the slope and the coordinates of the midpoint of the chord in part a.

   c) Determine the equation of the perpendicular bisector of the chord in part a.

   d) Does the perpendicular bisector pass through the centre of the circle? Explain.

6. A line segment has endpoints A(6, 2) and B(2, −6).

   a) Show that AB is a chord of the circle $x^2 + y^2 = 40$.

   b) Determine the equation of the perpendicular bisector of chord AB.

   c) Write to explain how you can tell from your answer in part b that the perpendicular bisector passes through the centre of the circle.

7. A line segment has endpoints M(8, 6) and N(−6, 8).

   a) Show that MN is a chord of the circle $x^2 + y^2 = 100$.

   b) Determine the equation of the line that passes through the centre of the circle and the midpoint of chord MN.

   c) Write to explain how you can tell from your answer in part b that the line is perpendicular to the chord MN.

8. A line segment has endpoints P(2, 5) and Q(5, −2).

   a) Show that PQ is a chord of the circle $x^2 + y^2 = 29$.

   b) Determine the equation of the line, perpendicular to the chord PQ, that passes through the centre of the circle.

   c) Show that the line in part b passes through the midpoint of chord PQ.

9. Verify that the line passing through the centre of the circle $(x − 3)^2 + (y + 4)^2 = 25$ and the midpoint of a chord in this circle is perpendicular to the chord.

10. Verify that the perpendicular, from the centre of the circle $(x + 8)^2 + (y − 7)^2 = 25$ to a chord in this circle, bisects the chord.

11. Consider the circle $x^2 + y^2 = 25$. Verify that the angle inscribed in a semicircle is a right angle. Begin by finding the coordinates of the endpoints of one diameter as well as one other point.

12. Heron of Alexandria was a Greek mathematician and physicist. He developed a formula to determine the area of any $\triangle ABC$ when the lengths of its three sides a, b, and c are known. The formula is $A = \sqrt{s(s-a)(s-b)(s-c)}$, where s is one-half the perimeter of the triangle (called the *semiperimeter*).

a) Consider the triangle with vertices A(2, 0), B(10, 0), and C(5, 4). Use the formula $A = \frac{1}{2}bh$ to determine the area of this triangle.

b) Calculate the lengths of the three sides of $\triangle ABC$.

c) Calculate the semiperimeter, s.

d) Verify the formula $A = \sqrt{s(s-a)(s-b)(s-c)}$ for the area of $\triangle ABC$.

In coordinate geometry, we need a tool to work with angles that are not right angles. This tool is the Cosine Law. Recall that in any $\triangle ABC$, $c^2 = a^2 + b^2 - 2ab\cos C$. You can use the Cosine Law to calculate the angles in a triangle if you know the lengths of the three sides. You will need to do this in exercises 13 to 16.

13. a) Plot the points A(1, 3), B(5, 5), C(4, 1), and D(10, 4). Verify that AB is parallel to CD.

b) Calculate the lengths of AB, BC, and AC.

c) Use the Cosine Law to calculate the measure of $\angle ABC$.

d) Calculate the lengths of CD and BD.

e) Verify that $\angle BCD = \angle ABC$.

14. a) Plot the points A(−8, −1), B(4, −7), and C(7, 4). Verify that these points lie on the circle $x^2 + y^2 = 65$.

b) Calculate the lengths of the three sides of $\triangle OAB$.

c) Use the Cosine Law to calculate the measure of $\angle AOB$.

d) Calculate the measure of $\angle ACB$.

e) Verify that $\angle AOB = 2\angle ACB$.

C 15. Verify that the opposite angles of a cyclic quadrilateral are supplementary. Begin by drawing the circle $x^2 + y^2 = 65$ and choosing four points on this circle.

16. Verify that the angle between a tangent and a chord of a circle is equal to the inscribed angle on the opposite side of the chord. Use the circle $x^2 + y^2 = 25$ and the line $y = -5$.

COMMUNICATING THE IDEAS

Write to explain how you can use coordinates to verify conjectures in geometry. Use some examples to illustrate your explanations. Include an explanation of whether this method proves the conjectures.

9.5 Using Coordinates to Prove Conjectures

One of the most useful problem-solving strategies in mathematics is credited to the great French mathematicians of the seventeenth century, Rene Descartes and Pierre de Fermat. Their idea was to use arithmetic and algebra to prove theorems in geometry. In this section, we will use their method to prove some new theorems, some conjectures you made in Section 9.4, and some theorems you encountered in Chapters 7 and 8.

In exercises 1 and 4 on pages 547 and 548, you made conjectures about how the line segment joining two sides of a triangle is related to the third side. This relationship is called the Side-Splitting Theorem.

Side-Splitting Theorem

The line segment joining the midpoints of two sides of a triangle is parallel to the third side and one-half as long as the third side.

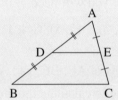

DE ∥ BC

DE = $\frac{1}{2}$BC

We can use coordinates to prove this theorem.

Draw any △ABC. Points D and E are midpoints of AB and AC, respectively.

We must prove that DE ∥ BC and DE = $\frac{1}{2}$BC.

Think...

Draw coordinate axes on the figure.

We can use any point as the origin, and we can use any line as one axis.

Assign general coordinates to the vertices. Solve the problem using the coordinates.

Draw coordinate axes on the figure.

Let B be the origin.

Draw the x-axis along side BC.

Draw the y-axis through B perpendicular to BC.

Let the coordinates of the vertices of △ABC be:

A(a, b), B(0, 0), C(c, 0).

Since D is the midpoint of AB, the coordinates of D are $\left(\frac{a}{2}, \frac{b}{2}\right)$.

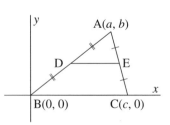

Since E is the midpoint of AC, the coordinates of E are $\left(\frac{a+c}{2}, \frac{b}{2}\right)$.

Since the y-coordinates of D and E are equal, DE is parallel to the x-axis. Hence, DE is parallel to BC.

Since DE is parallel to the x-axis, its length is the difference in the x-coordinates of D and E. Therefore,

$$\begin{aligned} DE &= \frac{a+c}{2} - \frac{a}{2} \\ &= \frac{a}{2} + \frac{c}{2} - \frac{a}{2} \\ &= \frac{c}{2} \end{aligned}$$

Since the length of BC is c, then DE is one-half as long as BC. That is, $DE = \frac{1}{2}BC$

In Chapter 7, we proved the Semicircle Theorem. The example below shows how we can prove this theorem using coordinates.

Example

Use coordinates to prove that the angle in a semicircle is a right angle.

Solution

Draw a circle with diameter AB. Let C be any point on the circle. We must prove that $\angle ACB = 90°$.

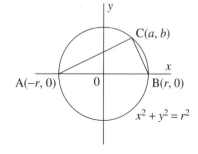

Think...

Use coordinates where the centre of the circle is the origin. Show that the slopes of AC and BC are negative reciprocals.

Draw coordinate axes on the figure.

Let the centre of the circle be the origin.

Let the x-axis coincide with the diameter AB.

Draw the y-axis through the centre perpendicular to AB.

Let the equation of the circle be $x^2 + y^2 = r^2$.

Then the coordinates of the endpoints of the diameter AB are A$(-r, 0)$ and B$(r, 0)$.

Let the coordinates of C be (a, b).

Since C is on the semicircle, its coordinates satisfy the equation.

Hence, $a^2 + b^2 = r^2$ ①

The slope of AC is $\dfrac{b-0}{a+r} = \dfrac{b}{a+r}$

The slope of BC is $\dfrac{b-0}{a-r} = \dfrac{b}{a-r}$

The product of the slopes of AC and BC is

$$\dfrac{b}{a+r} \times \dfrac{b}{a-r} = \dfrac{b^2}{a^2-r^2}$$

Substitute $b^2 = r^2 - a^2$ from ①.

$$= \dfrac{r^2 - a^2}{a^2 - r^2}$$

$$= -1$$

The slopes of AC and BC are negative reciprocals.

Therefore, $\angle ACB = 90°$

DISCUSSING THE IDEAS

1. Why does the proof of the Side-Splitting Theorem apply to all triangles, and not just to those with one vertex at the origin and one side along the x-axis?

2. In the proof of the Side-Splitting Theorem, the axes were placed to coincide as much as possible with parts of the given figure.

   a) Why was this done?

   b) What other positions might be good choices for the origin and the axes? Explain.

3. In the proof of the Side-Splitting Theorem:

   a) Why did we let the coordinates of C be $(c, 0)$, where c is a variable, and not something such as $(5, 0)$, where the first coordinate is a constant?

   b) Why did we let the coordinates of A be (a, b), where a and b are variables?

4. In the proof of the Side-Splitting Theorem, suppose we let the vertices of $\triangle ABC$ be A($2a$, $2b$), B($0, 0$), and C($2c, 0$).

   a) Follow the same steps as those used previously to prove the theorem.

   b) What is the advantage of using the 2 in the coordinates of A and C?

   c) How would you know in advance that using a 2 in the coordinates of A and C would simplify the proof?

5. In the solution of the *Example,* one reason was given for writing $a^2 + b^2 = r^2$. What is another reason why $a^2 + b^2 = r^2$?

B **1.** Use the diagram (below left). Prove that the diagonals of a square are perpendicular.

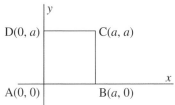

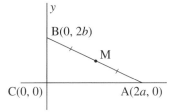

2. Use the diagram (above right). Prove that the midpoint of the hypotenuse of a right triangle is equidistant from the three vertices.

Use coordinate proofs in exercises 3 to 6, with coordinate axes drawn in convenient positions.

3. Point P is any point on the perpendicular bisector of a line segment AB (below left). Prove that PA = PB .

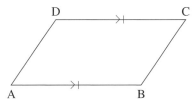

4. In quadrilateral ABCD, AB = DC and AB ∥ DC (above right).

a) Prove that AD = BC and AD ∥ BC.

b) What theorem does this prove about a quadrilateral in which two sides are parallel and equal?

5. Prove that the line segments joining the midpoints of adjacent sides of a rectangle form a rhombus (below left).

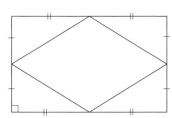

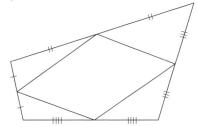

6. Prove that the line segments joining the midpoints of adjacent sides of a quadrilateral form a parallelogram (above right).

7. Use coordinates to prove that the diagonals of a parallelogram bisect each other.

8. Use coordinates to prove that the diagonals of a rectangle are equal.

9. M and N are the midpoints of the equal sides of an isosceles triangle. Prove that the medians to M and N are equal in length.

10. Three or more lines that intersect at a common point are called *concurrent*. Use the diagram of △ABC.

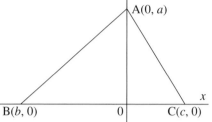

  a) Let BN be the altitude from B to AC. Determine the equation of the line containing BN.

  b) Determine the *y*-intercept of the line in part a.

  c) Let CM be the altitude from C to AB. Determine the *y*-intercept of the line containing CM.

  d) Compare the results in parts b and c. Explain why this proves that the altitudes of a triangle are concurrent.

11. Prove that the perpendicular bisectors of the sides of a triangle are concurrent.

12. Prove that the medians of a triangle are concurrent.

C 13. In △ABC, ∠B = 90° and P is the centre of the square on AC (below left). Prove that ∠PBC = 45°.

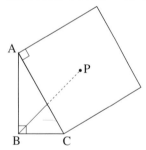
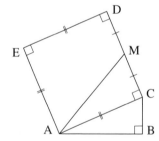

14. In △ABC, ∠B = 90° and side AB is three times as long as side BC. Square ACDE is drawn on the hypotenuse AC as shown (above right). Point M is the midpoint of CD. Prove that ∠MAB = 45°.

COMMUNICATING THE IDEAS

Your friend missed class today and asks you what she missed. How would you explain the method of proving conjectures using coordinate geometry? Write what you would say to her.

Creative Problem Posing

What happens if two angles of a triangle are the coordinates of a point?

If you know two angles of a triangle, you can find the third angle, since the sum of the three angles is 180°. Therefore, if the angles of a triangle are listed in order from smallest to largest, the first two angles form an ordered pair (x, y) that can be plotted as a point on a coordinate grid.

For example, in △ABC, the smallest angle is 40°, and the next larger angle is 55°. Hence, the point (40, 55) on the grid below represents △ABC. In a similar manner, you can plot a point to represent any given triangle.

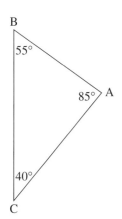

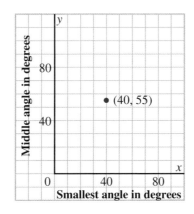

1. **a)** Draw a grid similar to the one above. Mark the point (40, 55) to represent the triangle above.

   b) Suppose you know the angles of a triangle. How do you determine the ordered pair (x, y)? Can you do this for any triangle? What happens if two or three of the angles are equal?

   c) Draw examples of other triangles, marking the measures of their angles. For each triangle, plot a point on the grid to represent it.

2. On your graph, plot several points that represent only right triangles. Find out as much as you can about the points on the grid that correspond to right triangles.

3. Repeat exercise 2 for each type of triangle.

   a) equilateral triangles   **b)** isosceles triangles

   c) acute triangles   **d)** obtuse triangles

Up to now in your study of mathematics, most or all of the problems you solved have probably been in your textbook or provided by your teacher. In other words, somebody created the problems for you. But mathematicians usually do not solve problems that have been created by others. They try to solve problems they have created for themselves.

How do mathematicians create problems? They may look for a pattern and try to extend patterns they find. They may try to think of new problems that might be related to a problem they have already solved. They may try to think about certain topics in new and different ways.

One way to do this is suggested by the last sentence of the quotation on page 557 and the problem on page 555. How did the person who created this particular problem happen to think of it? This problem involves two topics in mathematics — angles of a triangle and the coordinates of a point. Suppose we write these down, like this:

Angles of a triangle: A   C   B

                                     ↓   ↓

Coordinates of a point: x y

The arrows suggest that the person who created the problem thought about linking the two topics and asking questions:

- What happens if two angles of a triangle are the coordinates of a point?
- Which points correspond to the different kinds of triangles?

You can create your own problems if you think about other topics in this way.

4. Think of these topics.

    Quadratic equation: $x^2 + px + q = 0$

                                  ↓   ↓

    Coordinates of a point: x y

    Create a problem based on these two topics, then solve the problem. To create your problem it may help to think, "What happens if …?"

5. Create a problem based on these topics, then solve the problem.

    a) consecutive numbers and coordinates of a point

    b) the parabola $y = ax^2 + bx + c$ and the line $y = mx + b$

Some of the topics you studied in mathematics are listed in the table on page 557. Each cell in the table links two topics — one by reading down to the right, and the other by reading up to the right. For example, cell A links the topics *angles of a triangle* and *coordinates of a point*. Hence, this cell suggests the problem on page 555. Similarly, cells B, C, and D suggest the problems in exercises 4 and 5 above.

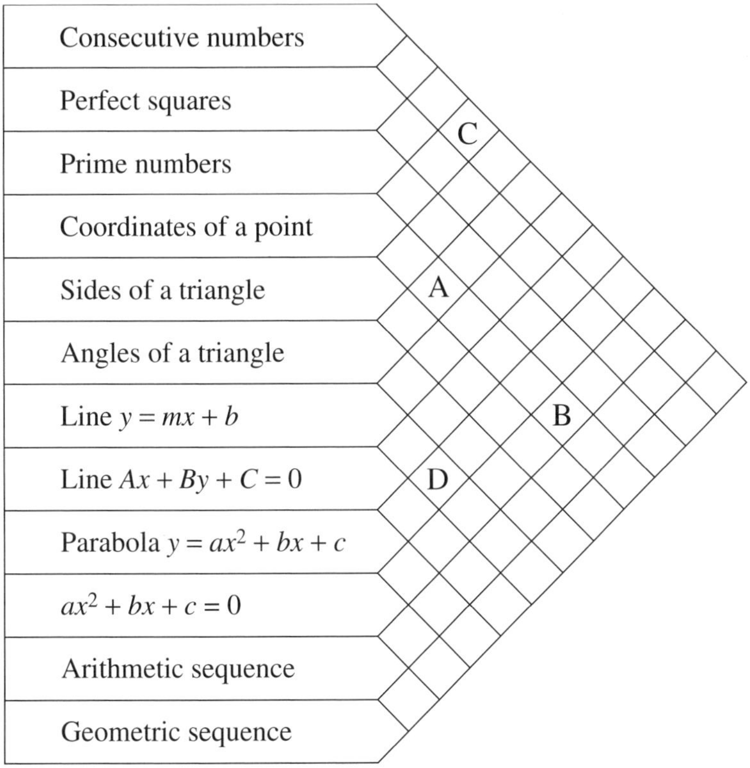

Consecutive numbers

Perfect squares

Prime numbers

C

Coordinates of a point

Sides of a triangle

A

Angles of a triangle

Line $y = mx + b$

B

Line $Ax + By + C = 0$

D

Parabola $y = ax^2 + bx + c$

$ax^2 + bx + c = 0$

Arithmetic sequence

Geometric sequence

6. Choose one of the empty cells in the table. Create a problem based on the two topics linked by the cell, then solve the problem.

7. The quotation below is from an essay entitled *How Much Mathematics Can There Be?*, published in 1981.

> All experience so far seems to show that there are two inexhaustible sources of new mathematical questions. One source is the development of science and technology, which makes ever new demands on mathematics for assistance. The other source is mathematics itself. As it becomes more elaborate and complex, each new, completed result becomes the potential starting point for several new investigations. Each pair of seemingly unrelated mathematical specialties pose an implicit challenge: to find a fruitful connection between them.

Explain how the quotation suggests the method of creating problems illustrated on pages 555, 556.

1. State the radius and the coordinates of the centre of the circle defined by each equation.

    a) $(x - 1)^2 + y^2 = 25$

    b) $(x + 2)^2 + (y - 3)^2 = 40$

    c) $(x + 3)^2 + (y - 4)^2 = 36$

    d) $x^2 + (y + 5)^2 = 55$

    e) $x^2 + y^2 = 42$

    f) $x^2 + (y + 1)^2 = 64$

2. Write the equation of the circle with each given centre and radius.

    a) O(0, 0), 5

    b) A(1, 3), 3

    c) B(2, −2), 1

    d) C(−4, 2), $\sqrt{3}$

    e) D(−1, −3), $2\sqrt{2}$

    f) E(2, 0), 6

3. Choose one circle from exercise 2. Write to explain how you determined its equation.

4. In this diagram, the circle has centre C(5, 0) and radius 3. Calculate the length of chord AB.

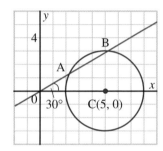

5. For what values of k is the line $y = k$ a tangent to the circle $x^2 + y^2 = 36$?

6. a) Graph the circle $(x - 1)^2 + (y + 3)^2 = 100$.

    b) Use your graph. Determine the coordinates of the endpoints of a diameter that has slope $\frac{3}{4}$.

    c) Determine the equations of the tangents to the circle with slope $-\frac{4}{3}$.

7. Determine the shortest distance from each point to the line.

    a) A(3, −4) to the line $x + y = 5$

    b) B(−4, 5) to the line $y = 3x - 2$

    c) the origin to the line $2x - 4y - 8 = 0$

8. Choose one point and corresponding line from exercise 7. Write to explain how you determined the shortest distance.

9. Determine the distance between the parallel lines in each pair.

   a) $y = 2x + 1$ and $y = 2x - 5$

   b) $y = 3$ and $y + 5 = 0$

   c) $2x - 3y - 5 = 0$ and $2x - 3y - 3 = 0$

10. Choose one pair of parallel lines from exercise 9. Write to explain how you determined the distance between the lines.

11. For each pair of points M and N, graph the line segment MN. Determine the coordinates of the points that divide line segment MN into the indicated number of equal parts.

   a) M(0, 3), N(9, 0)   three equal parts

   b) M(–4, 6), N(3, 7)   three equal parts

   c) M(–2, 0), N(6, 2)   five equal parts

   d) M(4, 7), N(–3, 8)   five equal parts

12. Visualize a circle and a line that does not intersect the circle.

   a) Explain how you would define the shortest distance from the circle to the line.

   b) Determine the shortest distance from the circle $(x - 5)^2 + (y + 1)^2 = 5$ to the line $x + 3y = 18$.

13. Consider points A(–3, 1), B(1, 5), and C(–2, 8).

   a) Verify that △ABC is a right triangle.

   b) Is △ABC an isosceles triangle? Justify your answer.

   c) Verify the Pythagorean Theorem for △ABC.

14. Consider the circle $x^2 + y^2 = 4$. Verify that a tangent is perpendicular to the radius of the circle at the point of tangency.

15. Draw the circle $x^2 + y^2 = 9$. Verify that the angle subtended by a diameter at the circumference is a right angle.

16. Use the diagram. Prove that the diagonals of a rectangle bisect each other.

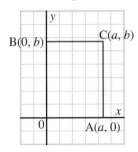

1. When Susana turned 18, her parents gave her $8000 toward her university education. This is the accumulated amount of an investment her parents made when she was born. Calculate the principal for each interest rate.

    a) 6% compounded semi-annually   b) 6% compounded quarterly

2. Stephen's monthly net income is $2500. Use the spending guidelines on page 63. Determine the range of money he should budget monthly in each category.

    a) housing and utilities   b) savings

    c) transportation   d) recreation and education

3. Determine if one function is the inverse of the other function.

    a) $y = 3x - 4$, $y = 4x - 3$ b) $y = \frac{1}{3}x + 5$, $y = 3x - 15$

4. Determine the inverse of each function.

    a) $y = 4x^2 - 9$ b) $y = (x - 5)^2 + 9$ c) $f(x) = \frac{1}{4}(x - 7)^2 - 2$

5. Choose one function from exercise 4. Write to explain how you found its inverse.

6. Sketch an example of a function with the given zeros.

    a) $-1, -1, 3, 4$ b) $1, 2, 2, 3$ c) $2, 2, 2, -5$

7. Graph each pair of functions on the same grid.

    a) $y = x^2 - 1$ and $y = \dfrac{1}{x^2 - 1}$ b) $y = (x - 5)^2$ and $y = \dfrac{1}{(x - 5)^2}$

8. Determine two functions whose composite function is $k(x)$.

    a) $k(x) = \dfrac{1}{\sqrt{3x - 4}}$ b) $k(x) = 2(x - 1)^2 + 4(x - 1) + 3$ c) $k(x) = 4^{x+2}$

9. Solve each equation. Give the answers to 2 decimal places, where necessary.

    a) $5x^2 + 11x - 12 = 0$ b) $3x^2 + 10x - 32 = 0$ c) $5x^2 - 15x + 11 = 0$

10. Select one equation from exercise 9. Write to explain how you solved it.

11. Solve for x.

    a) $-5\sqrt{x} = -10$ b) $\sqrt{2x} + 3 = 8$ c) $\sqrt{7x} - 2 = 3$

    d) $\sqrt{x} < 5$ e) $2\sqrt{x} \geq 4$ f) $3\sqrt{x} \leq 9\sqrt{2}$

12. Point P lies on the line $y = x$. The coordinates of M and N are $(-5, 5)$ and $(5, -5)$ respectively. The sum of the distances MP and NP is 20 units. Determine the coordinates of P.

13. Solve for x.

    a) $|x| = 3$ b) $|x| = 0$ c) $|x - 4| = 5$

14. Factor each polynomial completely.

a) $x^3 - 4x^2 + x + 6$

b) $2x^3 + 11x^2 + 17x + 6$

c) $x^3 - 3x^2 - 6x + 8$

d) $2x^3 - 3x^2 - 8x + 12$

15. Solve by graphing.

a) $x - 4y = 12$
$3x + 2y = 8$

b) $x - 4y = 3$
$2x + 3y = 6$

c) $2x + y = 4$
$x - 2y = -3$

16. Solve by substitution.

a) $2x + 3y = 9$
$x - y = 3$

b) $2x + 5y = -5$
$x + y = 2$

c) $4x + y = -6$
$-2x + 3y = 24$

17. Use deductive reasoning. Copy and complete each conclusion.

a) All the students in Raphael's class have at least one pet at home. Julie is in Raphael's class.
Conclusion:

b) All quadrilaterals have four vertices. A parallelogram is a quadrilateral.
Conclusion:

18. Determine each value of x and y.

a)
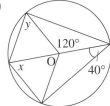

b)
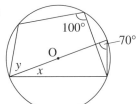

c)
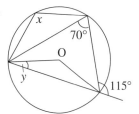

19. A circle, centre O, intersects parallelogram ABOD at points D, E, and F (below left). Prove that quadrilateral BOEA is cyclic.

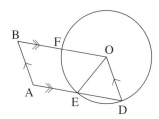

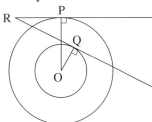

20. A tangent is drawn to each of two concentric circles with centre O (above right). Points P and Q are points of tangency, and R is the point of intersection of the tangents. Prove that $\angle PRQ = \angle POQ$.

21. Determine the coordinates of the points of intersection of each line and circle.

a) $y = 3x$
$x^2 + y^2 = 9$

b) $y = x - 2$
$x^2 + y^2 = 36$

c) $y = 2x + 1$
$x^2 + y^2 = 10$

Answers

Chapter 1

Investigate, page 4

1. a) $8 b) $16 c) $24 d) $40 e) $80
   f) $12 g) $4 h) $0.67 i) $0.66

2. Multiply the investment by the interest rate expressed as a decimal and by the time in years.

1.1 Exercises, page 9

1. Answers may vary.

2. a) $2.09 b) $10.51 c) $1.68 d) $72.91

3. a) $2.09 b) $1107.59 c) 18.94%
   d) 31 e) $10 086.18 f) $0.29

4. Explanations may vary.

5. $437.67

6. a) $2181.16 b) $2526.47

7. 146 days

8. 8.75%

9. 1.39%/month

10. $682.67

11. Explanations may vary.

12. a) $525.00 b) $1657.50 c) $2670.50 d) $3776.06 e) $2593.78

13. a)

Month	Previous balance ($)	Payment ($)	Unpaid balance ($)	Credit charge ($)	New purchases ($)	New balance ($)
January	782.50	250.00	532.50	11.98	531.87	1076.35
February	1076.35	300.00	776.35	17.47	425.76	1219.58
March	1219.58	300.00	919.58	20.69	723.05	1663.32
April	1663.32	500.00	1163.32	26.17	673.12	1862.61
May	1862.61	400.00	1462.61	32.91	358.29	1853.81
June	1853.81	375.00	1478.81	33.27	206.38	1718.47
				Total credit charge:	142.49	

b) No; explanations may vary.
c) No; explanations may vary.
e) $240.75 f) $302.58

14. a) $37.92 b) $246.78 c) $250.67
   d) Answers may vary.

15. a) $250.67 b) Explanations may vary.

16. $621.51

Investigate, page 13

1.

Year	0	1	2	3	4	5	6	7	8	9
Value($)	500.00	531.25	562.50	593.75	625.00	656.25	687.50	718.75	750.00	781.25

3. Descriptions may vary. The value increases by the same amount every year. It is an arithmetic sequence.

4. Descriptions may vary. The graph would be a straight line with a greater slope.

5.

Year	0	1	2	3	4	5	6	7	8	9
Value($)	500.00	531.25	564.45	599.73	637.21	677.04	719.36	764.32	812.09	862.85

7. Descriptions may vary. The graph would curve up faster for a higher interest rate, and slower for a lower interest rate.

8. Descriptions may vary. Value of bond = Face value of bond
   $(1 + \text{annual interest rate expressed as a decimal})^{\text{number of years}}$

1.2 Exercises, page 17

1. a)

Year	0	1	2	3	4	5
Value ($)	500.00	526.25	553.88	582.96	613.56	645.77

b)

Year	0	1	2	3	4	5	6
Value ($)	300.00	314.25	329.18	344.81	361.19	378.35	396.32

c)

Year	0	1	2	3	4
Value ($)	1000.00	1055.00	1113.03	1174.24	1238.82

2. a) $1500.73 b) $1458.88 c) $261.54 d) $1284.10

3. a) $1268.48 b) $1486.41 c) $1885.49

4. a) $174.18 b) $870.89 c) $4354.47

5. a) $5200.00 b) $2282.33 c) $13 225.19 d) $8943.89

6. Explanations may vary.

7. a) $3009.41 b) $2497.45 c) $10 000 d) $4999.89

8. Explanations may vary.

9. $194 530.68

10. a)

Years	Accumulated amount for 5% interest ($)	Accumulated amount for 10% interest ($)
0	100.00	100.00
5	127.63	161.05
10	162.89	259.37
15	207.89	417.72
20	265.33	672.75
25	338.64	1083.47

c) i) 23 years ii) 12 years
d) No; explanations may vary.

11. Between 10 and 11 years

12. a) 10.5% b) 8.5%

13. a) $543.09 b) $2092.67 c) $721.26 d) $2981.87

14. $3736.29

15. b) 1974 c) Answers may vary.
   d) Descriptions may vary.

16. 12.9%

17. a) $7107.90 b) $6838.43 c) $6581.62
   d) $6336.75 e) $6103.16

18. 10.4%

19. 6 years

20. a) i) 7 years ii) 5 years
   b) i) 12% ii) 7%

22. $30 million

Investigate, page 21

1. a) Explanations may vary. 1%
   b) $10; $1010 c) $10.10; $1020.10

2. Explanations may vary.

3. Explanations may vary.

4. $1020.18; explanations may vary

1.3 Exercises, page 24

1. a) $716.44 b) $167.90 c) $298.71 d) $2610.82
2. a) $172.92 b) $1079.08 c) $348.53 d) $4023.20
3. a) $45.51 b) $45.94
4. a) $290.37 b) $289.92
5. Explanations may vary.
6. a) $875.02 b) $873.94
7. 7.7%
8. $1073.25
9. b) Descriptions may vary.
   c) 15 years d) $9171.78
10. $1703.18
11. a) $1520.39 b) $1499.04
12. b) Explanations may vary.
    c) Between annually and semi-annually
    d) Answers may vary.
13. $33.20
14. Answers may vary.
15. a) $695.63 b) $557.67 c) $841.03
16. Explanations may vary.
17. 25 years
18. 24 years 2 months
19. Investment part b; explanations may vary
20. a) About 10 years b) About 13 years c) About 8 years
21. Explanations may vary.
22. a) $816.48; $816.61; $816.63
    b) $812.36 c) Suggestions may vary.
23. $473.06
24. $1133.98

Problem Solving: Investigating Financial Calculations on the TI-83, page 28

1. $6190.43
2. a) 6.705% b) Explanations may vary.
3. About 14 years
4. a)

Investment period (years)	Accumulated amount ($)
1	1 239.72
5	7 011.89
10	16 469.87
15	29 227.28
20	46 435.11
25	69 645.89
30	100 953.76
35	143 183.39
40	200 144.82

c)

Investment period (years)	Amount invested ($)
1	1 200
5	6 000
10	12 000
15	18 000
20	24 000
25	30 000
30	36 000
35	42 000
40	48 000

5. $1295

Investigate, page 30

1. a) $107.12 b) $107.25
   c) $7\frac{1}{4}$% compounded annually
2. a) $1 071 225 b) $1 072 500
   c) Part b earns $1275 more d) Part b earns $209.92 more
3. Explanations may vary.

Investigate, page 32

1. a) $2852 b) $345
2. b) Answers may vary.
   c) $93, with a final payment of $82.76
   d) $2221.76 e) $214.76
3. 15.7%
4. Answers may vary.

1.4 Exercises, page 34

1. Answers may vary.
2. a) i) $109.20 ii) $109.25; part ii is the greater rate
   b) i) $102.26 ii) $102.50; part ii is the greater rate
   c) i) $111.02 ii) $110.25; part i is the greater rate
3. a) $11\frac{3}{4}$% b) $6\frac{1}{2}$% c) $9\frac{3}{4}$% d) $11\frac{1}{4}$%
4. a) $12\frac{1}{2}$% by $97.97 b) $10\frac{1}{4}$% by $9.65 c) 17% by $117.94
5. Explanations may vary.
6. a) 8.16% b) 12.625% c) 16.075% d) 9.925%
7. Explanations may vary.
8. Part b; explanations may vary
9. a) $752.00 b) 17.1%
10. a) $188.80 b) Descriptions may vary.
    c) The finance charge is $752.00 in exercise 9, and $530.15 in exercise 10. The interest rate is 17.1% in exercise 9, and $12\frac{1}{4}$% in exercise 10. Explanations may vary.
11. a) $353.54; $785 b) $191.61; $1497.56
    c) Explanations may vary.
12. a) $520.86; $406.70; $338.52
    b) $18 750.96; $19 521.60; $20 311.20
    c) Explanations may vary.
13. a) 6.152% b) 5.83%
14. No; 20.262%
15. a) $r = \dfrac{i(4 + i)}{4}$
    c) i) 10.25% ii) 17.3%

Investigate, page 37

1. After 40 h; answers may vary
2. 6; 9; the wage per hour in dollars
3. $510; explanations may vary.

1.5 Exercises, page 40

1. Answers may vary. The employee will work harder to earn a greater salary. Good salespeople will earn more money.

2. Biweekly is being paid 26 times a year, semi-monthly is being paid 24 times a year.

3. Explanations may vary. If the first pay of a year is on January 1, there will be 53 paydays. It is not possible for there to be fewer than 52 paydays.

4. Descriptions may vary. $2216.20

5. $12/h; answers may vary, four arrangements per hour is only $10/h.

6. a) $336.50; wage b) $598.00; wage
   c) $681.00; commission d) $1087.50; wage
   e) $1307.50; graduated commission
   f) $391.20; commission g) $2739.20; commission

7. Explanations may vary.

8. $390.30

9. For an 8-h day, the gross pay is the same. For less than 8 h, the Fish House pays more. For more than 8 h, the Tea House pays more.

10. Explanations may vary.

11. b) Answers may vary. c) Answers may vary.

12. $23.75/h

13. Base salary: $750; commission rate: 6%

Mathematical Modelling: Should You Buy or Lease a Car?, page 42

For Alberta:

1. a) $19 530.71 b) $21 160.60 c) $1629.89 d) $0.35

2. a) $100.10; $12 403.60 b) $9128.17
   c) $21 531.77 d) $0.36

3. a) Buying b) Answers may vary.

4. a) $1370.70 b) $22 902.47 c) Answers may vary.

5. Answers may vary.

6. Answers may vary.

1.6 Exercises, page 48

1. $29.25; $46.07

2. Answers may vary.

3. Different provinces have different income tax rates.

4. a) Increases b) Increases c) Increases d) Decreases

5. a) $25.40 b) $32.78 c) $24.15
   d) $251.90 e) $295.70 f) $33.79

6. Explanations may vary.

7. a) Jackie b) Jackie c) Jackie d) Answers may vary.

8. a) $803.13 b) $884.28

9. a) $839.59 b) $842.36 c) Explanations may vary.

10. a) Estimates may vary. $734.20 b) Higher
    c) Yes; answers may vary.

11. a) Estimates may vary. $1369.87
    b) Answers may vary. Lower

12. Explanations may vary.

13. a) $412.50 b) Estimates may vary; $313.50
    c) 76%

14. a)

Claim code	Income tax deducted ($)	Percent of taxable income deducted (%)
1	501.10	28.8
2	492.55	28.3
3	475.40	27.3
4	458.30	26.3
5	441.15	25.3
6	424.05	24.4
7	407.05	23.4
8	390.70	22.5
9	374.35	21.5
10	358.00	20.6

c) Decreases d) Functions may vary; $y = -17x + 500$

Linking Ideas: Mathematics and Technology
The Future Is Now!, page 51

1. About 2.8 times as great

2. Conclusions may vary.

3. $5569

4. Answers may vary.

1.7 Exercises, page 54

1. a) $1.39 for 907 mL b) $2.98 for 400 g
   c) $2.19 for 450 g
   d) First 6 for $0.69 each and next 6 for $0.96 each

2. a) 12 for $49.99 b) Answers may vary. No

3. a) Suggestions may vary. b) $97.31
   c) Explanations may vary.

4. a) $366.25 b) Answers may vary. No

5. a) $2350.02 b) $221.70 c) $214.31
   d) Suggestions may vary.

6. $187.69

7. a) No; explanations may vary. b) $386.11 Can.

8. $312.00 Can.

9. Explanations may vary. Yes

10. Explanations may vary.

11. 63.39 pounds

12. Explanations may vary. No

13. a) $6.31 b) $84.16 c) $231 431.80
    d) $1 051 962.74

14. $2518.44

15. b) $259.83 c) $1039.32 at 6%

Mathematics File: Keeping Track of Your Money, page 58

1. His balance after the deposit should be $752.56. The cheque to Clothing Express should be $137.64. He forgot to record the $60.00 withdrawal.

2. Chris subtracted the $58.72 cheque incorrectly. The Record Bin cheque should be $33.15.

3. Answers may vary.

4. Explanations may vary. Chris saves enough to support his spending habits.

5. Explanations may vary.

6. Answers may vary.

1.8 Exercises, page 66

1. a) Housing and utilities: 32.5%; food and clothing: 17.5%; health and personal care: 2.5%; transportation: 30%; recreation and education: 7.5%; savings: 2.5%; miscellaneous: 7.5%
 b) Overspending: transportation; underspending: food and clothing, health and personal care, savings, miscellaneous; within guidelines: housing and utilities, recreation and education

2. a) $720 to $880　　　　**b)** $533.33 to $693.33
 c) $320 to $373.33　　　**d)** $160 to $266.67

3. Budgets may vary.

4. Explanations may vary.

5. Budgets may vary.

6. Budgets may vary.

7. Answers may vary.

8. Answers may vary.

9. Budgets may vary.

1.9 Exercises, page 71

1. a) $895.20; $214 848.00　　**b)** $658.80; $79 056
 c) $694.80; $208 440.00

2. Explanations may vary.

3. a) $636.30　　　　　　　**b)** $755.10

4. Lists may vary.

5. a) $655.27
 b) ii) $562.28　　　**iii)** Explanations may vary. Yes

6. a) i) $552.68　　　**ii)** $611.72　　　**iii)** $635.50
 b) Reasons may vary.

7. a) $519.45　　　**b)** $737.73　　　**c)** $1244.23

8. a) $64 978.40　　　　　**b)** $75 621.67
 c) $83 223.39　　　　　**d)** Factors may vary.

9. a)

Amortization period (years)	Monthly payment ($)	Total cost ($)	Percent of original repaid (%)
5	1584.00	95 040.00	118.80
10	928.80	111 456.00	139.32
15	719.20	129 456.00	161.82
20	620.00	148 800.00	186.00
25	565.60	169 680.00	212.10

b) As the amortization period increases, the monthly payment decreases.
 c) As the amortization period increases, the total cost increases.

10. a)

Interest rate (%)	Monthly payment ($)	Total cost ($)	Percent of original repaid (%)
4	666.00	119 880.00	133.20
6	759.60	136 728.00	151.92
8	860.40	154 872.00	172.08
10	967.50	174 150.00	193.50
12	1080.00	194 400.00	216.00

b) Increases　　　　　　**c)** Increases
 d) Explanations may vary. No

11. a)

Payment period	Payment ($)	Total cost ($)	Percent of original repaid (%)
Monthly	745.57	178 937.55	178.94
Semi-monthly	372.28	178 695.89	178.70
Biweekly	343.61	178 677.32	178.68
Weekly	171.70	178 565.95	178.57

b) As the payment period decreases, the payment decreases.
 c) As the payment period decreases, the total cost of the mortgage decreases.

12. a) $771.82　　　**b)** $98 690.90　　　**c)** 14%

d)

Number of payments	12	36	60	120	180	240	300
Unpaid balance ($)	98 690.90	95 737.70	92 273.94	80 763.31	63 614.24	38 064.77	0

f) Descriptions may vary.

13. a) $449.09　　　**b)** $63 966.81　　　**c)** $60 966.81
 d) 257 payments; about 21 years
 e) 163 payments; about 14 years
 f) Earlier; reasons may vary.

1 Review, page 77

1. a) $17.53　　**b)** $8.86　　**c)** $168.84　　**d)** $116.83

2. a) $584.93　　　**b)** $1431.13　　　**c)** $1068.60

3. Explanations may vary.

4. a) 7.186%　　**b)** 8.681%　　**c)** 9.381%　　**d)** 12.747%

5. The second job

6. Estimates may vary. $466.85

7. a) $4803.20　　　　　**b)** $400.27

8. Answers may vary.

Chapter 2

Linking Ideas: Mathematics and Technology
Graphing a Function, page 84

1. a) 122.1 m　　**b)** 122.6 m　　**c)** 87.3 m　　**d)** 31.4 m

2. a) 0 s, 3.5 s　　**b)** 4.4 s　　**c)** 5.4 s　　**d)** 6.1 s

3. 124.8 m, 1.8 s

4. 6.8 s

5. $0 \le x \le 6.8$

6. a) 17

  b) i) The graph would have a greater maximum and a greater x-intercept.
    ii) The graph would have a greater maximum and a greater x-intercept.
    iii) The graph would have a lesser maximum and a lesser x-intercept.
    iv) The graph would have a lesser maximum and a lesser x-intercept.

7. a) 141.9 m, 2.6 s   **b)** 7.9 s

8. a) 28 m/s   **b)** 38 m/s

9. a) 110

  b) i) The graph would have a greater vertical intercept and a greater x-intercept.
    ii) The graph would have a greater vertical intercept and a greater x-intercept.
    iii) The graph would have a lesser vertical intercept and a lesser x-intercept.
    iv) The graph would have a lesser vertical intercept and a lesser x-intercept.

2.1 Exercises, page 94

1. a) i) Yes   **ii)** No   **iii)** Yes   **iv)** Yes   **v)** Yes
  vi) No   **vii)** No   **vii)** Yes   **ix)** No
  b) Explanations may vary.

2. a) i) $x = -1$ **ii)** $(-1, -2)$ **iii)** $-2, 0; 0$
  iv) D: all real numbers; R: $y \geq -2$
  b) i) $x = 3$ **ii)** $(3, 4)$ **iii)** $1, 5; -5$
  iv) D: all real numbers; R: $y \leq 4$
  c) i) $x = 1$ **ii)** $(1, 3)$ **iii)** $0, 2; 0$
  iv) D: all real numbers; R: $y \leq 3$
  d) i) $x = -4$ **ii)** $(-4, -9)$ **iii)** $-7, -1; 7$
  iv) D: all real numbers; R: $y \geq -9$
  e) i) $x = 6$ **ii)** $(6, -4)$ **iii)** $4, 8; 32$
  iv) D: all real numbers; R: $y \geq -4$
  f) i) $x = -5$ **ii)** $(-5, 4)$ **iii)** $-7, -3; -21$
  iv) D: all real numbers; R: $y \leq 4$

3. The x-coordinate of the vertex of a parabola defines the axis of symmetry.

4. Explanations may vary.

  a) 0, 1, or 2   **b)** 1   **c)** Yes

6. a) i) $x = 0$ **ii)** $(0, -8)$ **iii)** $-2, 2; -8$
  iv) D: all real numbers; R: $y \geq -8$
  b) i) $x = 3$ **ii)** $(3, -1)$ **iii)** $2, 4; 8$
  iv) D: all real numbers; R: $y > -1$
  c) i) $x = 0.5$ **ii)** $(0.5, 6.25)$ **iii)** $-2, 3; 6$
  iv) D: all real numbers; R: $y \leq 6.25$
  d) i) $x = 1.5$ **ii)** $(1.5, 0.5)$ **iii)** 5
  iv) D: all real numbers; R: $y \geq 0.5$
  e) i) $x = 2$ **ii)** $(2, 5)$ **iii)** $5.2, -1.2; 5$
  iv) D: all real numbers; R: $y \leq 5$
  f) i) $x = 0.5$ **ii)** $(0.5, 6.75)$ **iii)** $-1, 2; 6$
  iv) D: all real numbers; R: $y \leq 6.75$

7. b) D: $0 \leq t \leq 4.1$; R: $0 \leq h \leq 82$ **c)** 51.4 m

8. a) Xmin $= -10$, Xmax $= 40$, Xscl $= 10$, Ymin $= -1200$,
  Ymax $= 1200$, Yscl $= 200$, Xres $= 1$; explanations may vary.
  b) 643.5 m   **c)** $9.3 \text{ s} \leq t \leq 22.0$ s
  d) $(15.6, 1197.5)$
  e) D: $0 \leq t \leq 31.3$ s; R: $0 \leq h \leq 1197.5$ m; explanations may vary.

9. a) $(5x + 4)(x + 1)$, $(5x + 2)(x + 2)$, $(5x + 1)(x + 4)$, $(5x - 4)(x - 1)$, $(5x - 2)(x - 2)$, $(5x - 1)(x - 4)$
  b) Explanations may vary; k can only be $-9, 9, -12, 12, -21$ or 21
  c) Answers may vary. $3x^2 + 9x + 6 = (3x + 6)(x + 1)$,
  $3x^2 + 11x + 6 = (3x + 2)(x + 3)$,
  $3x^2 + 19x + 6 = (3x + 1)(x + 6)$

10. Explanations may vary.
  a) $-6, 6, -9$ or 9   **b)** $-7, 7, -11$ or 11
  c) $-7, 7, -8, 8, -13$ or 13   **d)** $-7, 7, -8, 8, -13$ or 13
  e) $-8, 8, -16$ or 16   **f)** $-9, 9, -11, 11, -19$ or 19

11. a) -2 or 2   **b)** $-4, 4, -5$ or 5
  c) $-6, 6, -8, 8, -9, 9, -10, 10, -17$ or 17
  d) $-5, 5, -7$ or 7

12. a) $-7, 3$ **b)** $-4, \frac{3}{2}$ **c)** $-3, \frac{3}{4}$ **d)** $-\frac{4}{5}, 3$
  e) $-\frac{1}{2}, 6$ **f)** $\frac{1}{3}, 4$

13. Answers may vary.
  a) $x^2 + 2x - 15 = 0$ **b)** $3x^2 - 20x + 12 = 0$
  c) $5x^2 + 4x = 0$ **d)** $4x^2 + 4x + 1 = 0$
  e) $4x^2 + 19x + 12 = 0$ **f)** $16x^2 + 62x + 21 = 0$

14. a) $-6, 2$ **b)** $-1, \frac{5}{2}$ **c)** $-8, \frac{1}{4}$
  d) $\frac{1}{2}, 5$ **e)** $\frac{2}{3}, 3$ **f)** $-1, 2$

15. Answers may vary.
  a) $f(x) = x^2 - 2x$ **b)** $f(x) = 2x^2 - 3x - 9$
  c) $f(x) = x^2 + 12x + 32$ **d)** $f(x) = 8x^2 + 26x + 21$
  e) $f(x) = x^2 + 8x - 180$ **f)** $f(x) = 36x^2 - 60x + 25$

16. a) Predictions may vary.
  b) Explanations may vary; $x = 0.628, 6.372$

17. a) -3.31 **b)** Answers may vary.

18. Each part has an infinite number of answers. For example,
  a) $-2, -6, -12, \dots$ **b)** $1, -3, -8, \dots$
  c) $2, -4, -10, -18, \dots$

Exploring with a Graphing Calculator: The Parabola of Best Fit, page 98

1. a) A quadratic function would best describe the data.
  b) Answers may vary; $P(t) = 0.0036t^2 - 14.0t + 13701.11$
  c) 17.9% in 2011, 29.8% in 2036

2. a) A quadratic function would best describe the data.
  Answers may vary; $y = -4.9t^2 + 3.00$
  b) 0.78 s   **c)** 0.55 s; time to hit $= 1.4 \times$ time halfway

2.2 Exercises, page 105

1. a) Both dimensions are between 5 m and 15 m.
  b) Both dimensions are between 2.9 m and 17.1 m.
  c) Both dimensions are between 1.3 m and 18.7 m.

2. a) $n = 110 - 0.1s$; $R = -0.1s^2 + 110s$; $s = \$550.00$;
$R = \$30\ 250.00$; $500 < s < 600$
b) $n = 120 - 0.12s$; $R = -0.12s^2 + 120s$; $s = \$500.00$;
$R = \$30\ 000.00$; no new values, $s = 500$

3. a) $A = -x^2 + 8x$
c) Both dimensions are between 2 m and 6 m.

4. a) $n = 225 - 0.5s$ **b)** $R = -0.5s^2 + 225s$
c) $s = \$225.00$, $R = \$25\ 312.50$
d) $\$120.00 < s < \$320.00$

5. a) $R = -0.4s^2 + 190s$ **b)** 50 at \$350 each
c) $130.70 < s < 344.30$

6. a) $R = -4s^2 + 300s$ **b)** \$5625
c) $\$30 < s < \$45$

7. a) 171   **b)** $p = 351 - 0.24t$
c) $R = -0.24t^2 + 351t$
d) A ticket price of \$731.25 will produce the maximum revenue of \$128 334.38
e) There will be about 176 passengers. This is within the seating capacity of 256.

8. a) \$25; \$5780   **b)** No, maximum profit is \$5780.

9. b) 4.84 m   **c)** 0.61 s   **d)** About 1.2 s

10. The graph would be a straight line, $l = 20 - w$ for $0 < w < 10$.

11. $b^2 = 4ac$ and $a < 0$, $c < 0$

Linking Ideas: Mathematics and Sports

The Rising Fastball, page 108

1. 0.44 s

2. 95 cm

3. 76 cm

Investigate, page 109

2.

Function	Value of p	Direction of opening	Vertex	Axis of symmetry	Congruent to y = x²?
$y = x^2$	0	up	(0, 0)	$x = 0$	yes
$y = (x - 2)^2$	2	up	(2, 0)	$x = 2$	yes
$y = (x - 4)^2$	4	up	(4, 0)	$x = 4$	yes
$y = (x + 2)^2$	−2	up	(−2, 0)	$x = -2$	yes
$y = (x + 4)^2$	−4	up	(−4, 0)	$x = -4$	yes

3. a) x-coordinate of vertex, when equated to x, the equation of the axis of symmetry
b) If p is positive, the graph has been translated to the right. If p is negative, the graph has been translated to the left.
c) Explanations may vary.

4. b)

Function	Value of q	Direction of opening	Vertex	Axis of symmetry	Congruent to y = x²?
$y = x^2$	0	up	(0, 0)	$x = 0$	yes
$y = x^2 - 1$	−1	up	(0, −1)	$x = 0$	yes
$y = x^2 - 3$	−3	up	(0, −3)	$x = 0$	yes
$y = x^2 + 2$	2	up	(0, 2)	$x = 0$	yes
$y = x^2 + 4$	4	up	(0, 4)	$x = 0$	yes

q provides information about the vertex. If q is positive, the graph has been translated up. If q is negative, the graph has been translated down. Explanations may vary.

5. b)

Function	Value of a	Direction of opening	Vertex	Axis of symmetry	Congruent to y = x²?
$y = x^2$	1	up	(0, 0)	$x = 0$	yes
$y = 2x^2$	2	up	(0, 0)	$x = 0$	no
$y = 0.5x^2$	0.5	up	(0, 0)	$x = 0$	no
$y = -x^2$	−1	down	(0, 0)	$x = 0$	yes
$y = -2x^2$	−2	down	(0, 0)	$x = 0$	no
$y = -0.5x^2$	−0.5	down	(0, 0)	$x = 0$	no

a provides information about the direction of opening and the expansion or compression of the parabola. If a is positive, the parabola opens up. If a is negative, the parabola opens down. Explanations may vary.

6. a) i) $a = -1$, $p = 2$, $q = 3$
ii) Vertex has coordinates $(2, 3)$; axis of symmetry, $x = 2$; opens down

2.3 Exercises, page 115

1. a) $a = 1$, $p = -3$, $q = 0$ **b)** $a = 5$, $p = 0$, $q = 0$
c) $a = 1$, $p = 0$, $q = 4$ **d)** $a = -1$, $p = 2$, $q = 0$
e) $a = 1$, $p = 3$, $q = -5$ **f)** $a = 0.5$, $p = 0$, $q = 2$

2. Explanations may vary.

3. a) Reflected in the x-axis   **b)** Translated down 3 units
c) Compressed vertically   **d)** Translated right 2 units
e) Translated left 4 units, up 3 units
f) Translated right 5 units, down 3 units, reflected in the x-axis

4. a) i   **b)** iii   **c)** iv   **d)** ii

5. Explanations may vary.

6. Descriptions may vary.

7. a) i) $(5, 2)$ ii) $x = 5$ iii) 27
b) i) $(-3, -8)$ ii) $x = -3$ iii) 10
c) i) $(-1, 4)$ ii) $x = -1$ iii) 0
d) i) $(2, -8)$ ii) $x = 2$ iii) −6

11. a) As a increases, the parabola becomes thinner. Eventually it approaches the y-axis.
b) Predictions may vary. The graphs approach the y-axis in the first case, and the x-axis in the second case.

12. a) $y = 2(x - 4)^2 - 1$ **b)** $y = -\frac{1}{3}(x + 2)^2 + 3$
c) $y = -\frac{1}{2}(x + 3)^2 + 2$ **d)** $y = (x - 3)^2 - 4$

13. Statements 2 and 3 are true.

14. a) $y = x^2 + 2$ **b)** $y = x^2 + 5$
c) $y = (x - 4)^2$ **d)** $y = (x + 3)^2$

15. a) $y = -3(x - 5)^2 + 2$ **b)** $y = 2(x - 4)^2 - 3$
c) $y = \frac{1}{3}(x + 2)^2 + 6$ **d)** $y = -\frac{1}{2}(x + 3)^2 - 1$

16. a) i) $y = (x - 3)^2 - 1$ ii) $y = -2(x + 1)^2 + 4$
iii) $y = 3(x - 2)^2 - 27$
b) Explanations may vary.

17. a) $y = x^2 - 1$, $y = (x - 1)^2 - 1$, $y = (x - 2)^2 - 1$,
$y = (x - 3)^2 - 1$, ...; $y = (x + 1)^2 - 1$, $y = (x + 2)^2 - 1$,
$y = (x + 3)^2 - 1$, ...
b) $y = \frac{4}{9}x^2 - 2$, $y = -\frac{4}{9}x^2 + 2$, $y = \frac{2}{9}x^2$, $y = -\frac{2}{9}x^2$

18. Patterns may vary.

19. a) $y = x^2 + 5$ **b)** $y = -\frac{1}{2}x^2 + 3$

20. $h = -4.8(x - 5)^2 + 120$

21. Yes; explanations may vary.

Linking Ideas: Mathematics and Technology
Dynamic Graphs Part I, page 120

1. b) From right to left, $a \doteq -2, -\frac{1}{2}, -\frac{1}{4}, -\frac{1}{9}, \frac{1}{9}, \frac{1}{4}, \frac{1}{2}, 2$

2. a) Sixth graph from left, $y = (x - 1)^2 - 1$ **b)** b

c) From right to left: $y = (x - 2)^2 - 1$, $y = \left(x - \frac{3}{2}\right)^2 - 1$,

$y = (x - 1)^2 - 1$, $y = \left(x - \frac{1}{2}\right)^2 - 1$, $y = \left(x + \frac{1}{2}\right)^2 - 1$,

$y = (x + 1)^2 - 1$, $y = \left(x + \frac{3}{2}\right)^2 - 1$, $y = (x + 2)^2 - 1$

3. Each graph along the bottom of both pages would be reflected in the y-axis.

2.4 Exercises, page 124

1. a) $f(x) = (x - 3)^2 - 1$ **b)** $f(x) = (x + 5)^2 - 11$
c) $f(x) = 2(x + 1)^2 + 5$ **d)** $f(x) = -2(x - 1)^2 + 7$
e) $f(x) = 3(x - 4)^2 - 8$ **f)** $f(x) = -5(x + 2)^2 - 10$

3. a) The maximum value of y is 4 when x is 1.
b) The minimum value of y is -4 when x is 3.
c) The minimum value of y is -1 when x is -2.
d) The maximum value of y is 2 when x is -3.
e) The minimum value of y is -8 when x is 6.
f) The maximum value of y is 3 when x is 2.

4. a) Yes; $y = -8$ when $x = -5$ **b)** No
c) Yes; $y = 7.5$ when $x = 3$
d) Yes; $y = 5$ when $x = 0$ **e)** No
f) Yes; $y = -19$ when $x = -4$

5. a) i) $y = 2(x - 2)^2 + 7$ **ii)** $y = 3(x + 2)^2 - 19$
iii) $y = (x - 3)^2 - 2$ **iv)** $y = -2(x - 1.5)^2 + 15.5$
v) $y = -1(x + 1.5)^2 - 0.75$ **vi)** $y = 1.5(x - 3)^2 - 3.5$
b) i) i) 7, minimum   **ii)** 2
ii) i) -19, minimum   **ii)** -2
iii) i) -2, minimum   **ii)** 3
iv) i) 15.5, maximum   **ii)** 1.5
v) i) -0.75, maximum   **ii)** -1.5
vi) i) -3.5, minimum   **ii)** 3

6. The values of a are equal. Explanations may vary.

9. Explanations may vary.

10. a) i) 5   **ii)** Minimum   **iii)** 3
iv) D: all real numbers; R: $y \geq 5$
b) i) -3 **ii)** Minimum   **iii)** -1
iv) D: all real numbers; R: $y \geq -3$
c) i) 4   **ii)** Maximum   **iii)** 1
iv) D: all real numbers; R: $y \leq 4$
d) i) -6 **ii)** Maximum   **iii)** -2
iv) D: all real numbers; R: $y \leq -6$
e) i) -9 **ii)** Minimum   **iii)** 0
iv) D: all real numbers; R: $y \geq -9$
f) i) 7   **ii)** Maximum   **iii)** 0
iv) D: all real numbers; R: $y \leq 7$

11. b) Descriptions may vary.

c) The parabola has its vertex on the y-axis.

12. b) Descriptions may vary.
c) The graph is a straight line.

13. Answers may vary.
a) $y = (x + 2)^2 + 4$ **b)** $y = (x - 3)^2 - 7$ **c)** $y = -(x - 5)^2 - 1$
d) $y = -2x^2 + 14$ **e)** $y = 3x^2 - 6$

14. $a > 0$ and $c \leq 0$, or $a < 0$ and $c \geq 0$

15. a) $f(x) = a\left(x + \frac{b}{2a}\right)^2 - \frac{b^2}{4a} + c$

b) $x + \frac{b}{2a} = 0$, $\left(\frac{-b}{2a}, -\frac{b^2}{4a} + c\right)$, c

c) $x = \frac{-b + \sqrt{b^2 - 4ac}}{2a}$, $x = \frac{-b - \sqrt{b^2 - 4ac}}{2a}$

2.5 Exercises, page 130

1. 4, -4

2. 6, 6

3. 30, 30

5. 10, -10

6. 8, 8

7. 7, -9

8. 45 000 m^2

9. a) 400 m^2 **b)** 20 m by 20 m

10. 150 m by 100 m

11. \$30.00

12. 70¢

13. 12.5 cm

14. 28.125 cm^2

15. 200 m by 300 m

16. 14.1 cm, 15.9 cm

17. $\frac{1}{2}$

18. $\frac{p^2}{16}$ units^2, where p is the perimeter

20. 3.33 units

Problem Solving: Elegance in Mathematics, page 132

1. Answers may vary.

2. Suggestions may vary.

2.6 Exercises, page 136

1. a) Yes   **b)** No   **c)** Yes

2. a) $y = x - 3$ **b)** $y = \frac{x + 1}{4}$ **c)** $y = \frac{1}{2}x$
d) $y = \frac{x + 4}{3}$ **e)** $y = 2x - 12$ **f)** $y = \frac{3}{2}x + \frac{3}{2}$

4. a) Yes   **b)** No   **c)** No   **d)** Yes

5. a) $f^{-1}(x) = x - 6$ **b)** $f^{-1}(x) = \frac{1}{2}x$
c) $f^{-1}(x) = 3 - x$ **d)** $f^{-1}(x) = 2(x + 3)$
e) $f^{-1}(x) = \frac{x - 1}{5}$ **f)** $f^{-1}(x) = \frac{x - 2}{2}$

6. a) No   **b)** No   **c)** Yes   **d)** Yes

7. a) iii   **b)** v   **c)** iv

9. $f(x) = 2x + 5$; explanations may vary.

10. Any function whose graph is perpendicular to the line $y = x$ is its own inverse.

2.7 Exercises, page 140

1. a) $y = \pm\sqrt{x}$ **b)** $y = \pm\sqrt{x + 1}$

c) $y = \pm\sqrt{x - 3}$ **d)** $y = \pm\sqrt{\dfrac{x - 5}{2}}$

e) $y = \pm\sqrt{4x + 8}$ **f)** $y = \pm\sqrt{4x + 2}$

3. a) $y = \pm\dfrac{1}{2}\sqrt{x}$ **b)** $y = \pm\sqrt{1 - x}$

c) $y = \pm\sqrt{\dfrac{2 - x}{3}}$ **d)** $y = \pm\sqrt{x} - 3$

e) $y = \pm\sqrt{\dfrac{x}{5} + 2}$ **f)** $y = \pm\sqrt{3(x + 3)} - 1$

4. Restrictions may vary.

a) $x \geq 0$, $y = \sqrt{x + 1}$ **b)** $x \geq 0$, $y = \sqrt{x - 2}$

c) $x \geq -1$, $y = \sqrt{x} - 1$ **d)** $x \geq 2$, $y = \sqrt{x - 1} + 2$

e) $x > 0$, $y = \sqrt{\dfrac{3 - x}{2}}$ **f)** $x \geq 1$, $y = \sqrt{3x + 6} + 1$

5. a) No   **b)** Yes   **c)** No   **d)** No

6. a) ii   **b)** v   **c)** iv

7. a) $y = (x - 1)^2$, $x \geq 1$ **b)** $y = \left(\dfrac{3 - x}{2}\right)^2$, $x \leq 3$

Mathematical Modelling: Modelling Golf Ball Trajectories, page 142

1. a) 26.5 m   **b)** 4.65 s

2. 135.87 m

3., 4. Time is rounded to the nearest hundredth of a second, and horizontal distance is rounded to the nearest metre.

Club	Maximum height (m)	Time to hit the ground (s)	Horizontal distance (m)
Driver	10.29	2.90	246
3 wood	11.18	3.02	210
3 iron	15.45	3.55	170
5 iron	21.23	4.16	153
7 iron	26.52	4.65	136
9 iron	31.63	5.08	118
Pitching wedge	35.83	5.41	97
Sand wedge	40.29	5.73	75

5. a) All answers are given in metres.

Time, t seconds	x (from ④)	y (from ⑤)
0.0	0	0
0.5	14.6	10.2
1.0	29.2	17.9
1.5	43.8	23.2
2.0	58.4	26.0
2.5	73.0	26.4
3.0	87.6	24.3
3.5	102.2	19.8
4.0	116.8	12.8
4.5	131.4	3.4

c) $y = -4.9\left(\dfrac{x}{29.2}\right)^2 + \dfrac{22.8}{29.2}x$

6. Answers may vary.

7. a) Explanations may vary.
    b) Explanations may vary.

8. Answers may vary.

9. Answers may vary.

10. Answers may vary.

Linking Ideas: Mathematics and Technology
Dynamic Graphs Part II, page 146

1. Explanations may vary.
    a) a **b)** b

2. a) Second graph from left   **b)** a

c) From right to left: $y = 3x^2 + 2x + 1$, $y = x^2 + 2x + 1$,
$y = \dfrac{1}{2}x^2 + 2x + 1$, $y = \dfrac{1}{4}x^2 + 2x + 1$,
$y = -\dfrac{1}{4}x^2 + 2x + 1$, $y = -\dfrac{1}{2}x^2 + 2x + 1$,
$y = -x^2 + 2x + 1$, $y = -3x^2 + 2x + 1$

3. a) Sixth graph from left   **b)** b

c) $y = x^2 + 3x - 1$, $y = x^2 + 2x - 1$, $y = x^2 + \dfrac{3}{2}x - 1$,
$y = x^2 + \dfrac{3}{4}x - 1$, $y = x^2 - \dfrac{3}{4}x - 1$, $y = x^2 - \dfrac{3}{2}x - 1$,
$y = x^2 - \dfrac{9}{4}x - 1$, $y = x^2 - 3x - 1$

4. a) c **b)** Descriptions may vary.

5. Answers may vary.

2 Review, page 148

1. a) iv   **b)** ii   **c)** i   **d)** iii

2. a) $2.00   **b)** $160.00

3. b) 40.1 m   **c)** 1.02 s
    d) $0 \leq t \leq 3.88$ s; $0 \leq h \leq 40.1$ m

4. 14, 14

5. a) Predictions may vary.   **b)** 149.05 m, 4.6 s, 13.6 s

6. a) i) $(-7, 0)$ **ii)** $x = -7$ **iii)** 147
    b) i) $(3, 4)$ **ii)** $x = 3$ **iii)** -14
    c) i) $\left(\dfrac{1}{2}, -\dfrac{3}{4}\right)$ **ii)** $x = \dfrac{1}{2}$ **iii)** -1

7. a) $y = \dfrac{1}{9}x^2 + 3$ **b)** $y = \dfrac{3}{16}(x - 4)^2$

9. Explanations may vary.

10. 9, 3

11. a) $f^{-1}(x) = x - 3$ **b)** $f^{-1}(x) = \dfrac{x}{3}$
    c) $f^{-1}(x) = -\dfrac{1}{3}x + 2$

12. Explanations may vary.

14. a) Yes, for $y = 4x^2$ restricted to $x \geq 0$
    b) No   **c)** Yes, for $y = 2x^2 + 10$ restricted to $x \geq 0$
    d) Yes, for $y = 4x^2 + 9$ restricted to $x \leq 0$

2 Cumulative Review, page 150

1. a) 14.38   **b)** 525.11   **c)** 7.5   **d)** 730

2. $102.67

3. a) $3242.30   **b)** $1113.63

4. Explanations may vary.

5. a) 6%   **b)** 8%

6. Answers may vary.

7. a) $592.00 **b)** 16.9%

8. a) $1922.75 **b)** $405.00

9. a) $4803.20 **b)** $400.27

10. $900.00 U.S.

11. Explanations may vary.
   a) Yes **b)** No **c)** No
   d) No **e)** Yes **f)** No

12. b) 9.0 A **c)** 40.5 W

13. $1.45 < s < $3.80, where s is the fare.

14. a) $-3, -4$ **b)** $1, \frac{3}{2}$ **c)** $-\frac{2}{3}, 3$ **d)** $-\frac{4}{3}, \frac{1}{2}$
   e) $\frac{2}{3}, -3$ **f)** $4, -2$

15. Explanations may vary.

16. a) i) -8 **ii)** Maximum **iii)** -5
   iv) D: all real numbers; R: $y \leq -8$
   b) i) -9 **ii)** Minimum **iii)** 2
   iv) D: all real numbers; R: $y \geq -9$
   c) i) 7.5 **ii)** Maximum **iii)** 3
   iv) D: all real numbers; R: $y \leq 7.5$

17. Explanations may vary.

18. 5000 m^2

19. a) No **b)** Yes

Chapter 3

Exploring with a Graphing Calculator: Graphing Polynomial Functions, page 154

1. Conclusions may vary.

2. Conclusions may vary.

3. Conclusions may vary.

4. b),c)

Degree	Sign of leading coefficient	Extends from quadrant...	...to quadrant	Greatest number of hills and valleys
3	+	3	1	2
3	–	2	4	2
4	+	2	1	3
4	–	3	4	3
5	+	3	1	4
5	–	2	4	4

5. a) Odd degree polynomials
   b) Even degree polynomials

6. The greatest number of hills and valleys is one less than the degree of the function.

3.1 Exercises, page 160

1. Conclusions may vary.

   a) Polynomial **b)** Polynomial **c)** Polynomial
   d) Other **e)** Other **f)** Polynomial
   g) Polynomial **h)** Other

2. Explanations may vary. Parts a, c, d, and f

3. a) Graphs d, e, and f; descriptions may vary.
   b) Graphs c and d; descriptions may vary.

4. a) Polynomials of even degree
   b) No; explanations may vary.

5. a) $g(x) = x^4 - 10x^2 - 5x + 5$ **b)** $k(x) = 5x^4 - 14x^3$
   c) $f(x) = x^3 - 3x^2 - 5x + 16$

6. b) The inverse of $f(x) = x^3 - 2x^2 + 2x + 3$ is a function, since there is exactly one y-value for each x-value.

7. a) Functions with no maximums or minimums have inverses that are functions. Otherwise, the inverse will have more than one y-value for some x-values.
   b) No; explanations may vary.

8. a) Answers may vary.
   b) Predictions may vary.
   c) Descriptions may vary.

9. b) Predictions may vary.
   c) Descriptions may vary.

3.2 Exercises, page 166

1. a) False, $y = x^3$, no relative maximum or minimum
   b) True. Explanations may vary.
   c) True. Explanations may vary.
   d) False, $y = x^4$, D: all real numbers, R: $y \geq 0$

2. a) Point symmetry; descriptions may vary.
   b) Maximum at $(1.37, 8.71)$, minimum at $(4.63, -8.71)$
   c) D: all real numbers, R: all real numbers

3. a) Point symmetry; descriptions may vary; maximum at $(-3, 26)$, minimum at $(1, -6)$;
   D: all real numbers, R: all real numbers
   b) Point symmetry; descriptions may vary; no maximum or minimum; D: all real numbers, R: all real numbers
   c) No symmetry; descriptions may vary; maximum at $(0, 2)$, minimum at $(1.09, -0.05)$, minimum at $(-1.84, -6.31)$;
   D: all real numbers, R: $y \geq -6.31$
   d) No symmetry; descriptions may vary; minimum at $(2.30, -7.82)$; D: all real numbers, R: $y \geq -7.82$

4. a) D: all real numbers, R: all real numbers
   b) D: all real numbers, R: all real numbers
   c) D: all real numbers, R: $y \geq -9.86$
   d) D: all real numbers, R: $y \geq 1.79$

5. b) $x = y^3 - 6y^2 + 6y + 3$; D: all real numbers, R: all real numbers
   c) No, fails vertical line test

6. a) i) Point symmetry; descriptions may vary.
   ii) Line symmetry; descriptions may vary.
   b) i) D: all real number, R: all real numbers
   ii) D: $x > -11$, R: all real numbers
   c) Neither function has an inverse; both graphs fail vertical line test.

7. No; explanations may vary.

3.3 Exercises, page 173

1. a) 3 **b)** 2 **c)** 3
   d) 4 **e)** 3 **f)** 5

2. Answers may vary.
  a) $f(x) = (x - 1)(x - 2)(x - 3)$
  b) $f(x) = (x - 2)^2(x - 5)$
  c) $f(x) = x(x + 4)(x - 1)$
  d) $f(x) = (x - 2)^3$

3. Answers may vary.
  a) $y = (x - 1)(x - 2)(x - 3)(x - 4)$
  b) $y = (x + 2)(x - 1)(x - 2)^2$
  c) $y = (x + 3)(x - 1)^3$
  d) $y = x^2(x + 1)^2$

4. Explanations may vary.

5. a) True   **b)** False   **c)** True   **d)** False

6. No equation is unique. Explanations may vary.
  a) $y = (x + 1)(x - 1)(x - 3)$
  b) $y = x(x - 3)(x - 5)$
  c) $y = (x + 2)^2(x - 2)$
  d) $y = (x - 1)^2(x - 6)$

7. Explanations may vary.
  a) $y = (x + 2)(x - 1)(x - 3)(x - 6)$
  b) $y = \frac{1}{4}x(x + 3)^2(x - 5)$
  c) $y = (x + 1)^2(x - 3)^2$
  d) $y = (x + 1)(x - 3)^3$

9. Explanations may vary.

10. a) $y = \frac{3}{2}(x + 1)(x - 4)$

11. a) $y = -2(x + 3)(x + 1)(x - 2)$

12. a) $y = 3(x - 2)^2$
  b) $y = 3(x + 2)(x - 1)(x - 4)$
  c) $y = -2(x + 2)(x - 2)^2$
  d) $y = -3x(x - 2)(x - 4)$

13. a) $-7, -3, 0$ **b)** $0, 1, 6$ **c)** $-4, 0, 5$
  d) $-3, 0, 2$ **e)** $0, \frac{1}{2}, 3$ **f)** $-\frac{5}{3}, 0, 1$
  g) $0, 1$ **h)** $0, 1, -1$ **i)** 1 **j)** -1

14. a) $2, 8$ **b)** $0, 3, 4$ **c)** $0, 7, -2$
  d) $0, -5, 3$ **e)** $-1, 0, \frac{3}{2}$ **f)** $-7, 0, \frac{2}{3}$
  g) $-5, 0, 5$ **h)** $0, \frac{9}{4}$ **i)** $0, 8, -8$ **j)** 4

15. Explanations may vary.

16. a) ± 15 **b)** $\pm\frac{3\sqrt{14}}{2}$ **c)** ± 3 **d)** 1

17. a) 9 **b)** ± 20 **c)** ± 6 **d)** 49

18. a) 1 **b)** ± 24 **c)** 48 **d)** ± 48

19. a) Estimates may vary; $k = 0, 1.2$
  c) $k = \frac{32}{27}$

20. a) $y = x^3 - 9x^2 + 26x - 18$, $y = x^3 - 9x^2 + 27x - 19$,
  $y = x^3 - 9x^2 + 28x - 20$
  c) Descriptions may vary.
  d) $y = x^3 - 9x^2 + 22x - 14$, $y = x^3 - 9x^2 + 21x - 13$,
  $y = x^3 - 9x^2 + 20x - 12$
  e) Descriptions may vary.

21. a) i) $x^3 - 12x + 10 = 0$
  ii) $x^3 - 12x + 20 = 0$, $x^3 - 12x - 20 = 0$
  b) i) $-16 < k < 16$ **ii)** $k = \pm 16$
  iii) $k > 16, k < -16$

Problem Solving: Visual Proofs, page 176

1. Explanations may vary.

2. $a^2 - b^2 = (a + b)(a - b)$

3. a) 25 **b)** n^2

4. a) Explanations may vary.
  b) $\frac{n(n + 1)}{2}$
  c) $1, 3, 6, 10, 15, 21, \ldots, 5050$

5. a) Explanations may vary.
  b) 9 by 10; 90
  c) 30
  d) $2n + 1$ by $\frac{n(n + 1)}{2}$; $\frac{(2n + 1)n(n + 1)}{2}$
  e) $\frac{n(n + 1)(2n + 1)}{6}$
  f) $1, 5, 14, 30, 55, 91, \ldots, 338\ 350$

6. a) i) $A = (a + b)^2$ **ii)** $A = c^2 + 2ab$

3.4 Exercises, page 180

1. Estimates may vary.
  a) 1.7 **b)** $-2.3, 2.3$
  c) $-2.7, -1, 0.5, 3.3$ **d)** $-2.8, -1.4, 0, 1.5, 2.9$

2. a) 1.85 **b)** $0, 1.7, -1.7$
  c) -2.6 **d)** $2.4, 0.7, -3.1$

3. a) $0, \pm 3.2$ **b)** $-3.7, 1.7, 2$
  c) $-2, -1.7, 3.7$ **d)** 4

4. a) $-3.5, -0.7, 4.2$ **b)** 2.3
  c) $-3.7, -1.2, 1.2, 3.7$ **d)** $-2.3, 3.2$

5. a) 1.6 **b)** 3.5
  c) $-3.3, -0.6, 1.3, 2.7$ **d)** $-2.1, 3.2$

6. a) No; explanations may vary.
  b) Two unequal negative zeros, or no zeros in this region.
  c) Explanations may vary.

7. a) $-3.58, 5.58$ **b)** $-3.59, 5.58$
  c) i) No. The two give the same results.
  ii) The second method may be more reliable, since only one graph is drawn.

8. a) $0.73, 2.26$ **b)** $0.73, 2.26$
  c) i) No. The two give the same results.
  ii) The second method may be more reliable, since only one graph is drawn.

9. a) Explanations may vary.
  b) Positive. Explanations may vary.
  c) No; explanations may vary.

10. $-3.97, 4.03, 249.94$

3.5 Exercises, page 185

1. a) 1.28-cm squares, 22.44 cm by 17.44 cm by 1.28 cm;
  6.78-cm squares, 11.44 cm by 6.44 cm by 6.78 cm
  b) 3.68-cm squares, 17.64 cm by 12.64 cm by 3.68 cm

2. a) 0.93-cm squares, 23.13 cm by 23.13 cm by 0.93 cm;
  8.71-cm squares, 7.58 cm by 7.58 cm by 8.71 cm
  b) 4.17-cm squares, 16.67 cm by 16.67 cm by 4.17 cm

3. a) $V = x(21 - 2x)(14 - x)$
c) 1.75-cm squares, 17.50 cm by 12.25 cm by 1.75 cm; 6.66-cm squares, 7.68 cm by 7.34 cm by 6.66 cm
d) 3.96-cm squares, 13.08 cm by 10.04 cm by 3.96 cm

4. a) $V = x(28 - 2x)(10.5 - x)$
c) 1.75-cm squares, 24.50 cm by 8.75 cm by 1.75 cm; 6.66-cm squares; 14.68 cm by 3.84 cm by 6.66 cm
d) 3.96-cm squares, 20.08 cm by 6.54 cm by 3.96 cm

5. 25

6. 88.7 m

7. 7.8 cm by 5.4 cm by 11.9 cm

8. Descriptions may vary.

9. a) $V = \dfrac{250r - \pi r^3}{2}$
c) i) $r = 1.66$ cm, $h = 23.14$ cm; $r = 7.98$ cm, $h = 1.00$ cm
ii) $r = h = 5.15$ cm, $V = 429.2$ mL

10. Answers may vary; rectangle: 80, 22; triangle: 80, 80

Mathematics File: Function Operations, page 188

1. a) $18x - 21$ **b)** $3x^2 - 6x + 12$
c) $3(x - 5)^2$ **d)** 3

2. a) $3x - 2$ **b)** $x^2 + 2x - 1$
c) $2x^2 - 2x + 1$ **d)** $5x^2 - 8x + 5$

3. a) $4x - 1$ **b)** $6x + 2$
c) $4x$ **d)** $-5x^2 - 24x + 5$

4. a) $6x + 3$ **b)** $7x^2 - 26x - 8$
c) $x^4 - 1$ **d)** $2x^3 - 3x^2$

5. No; answers may vary.
a) $f(x) = (x + 1)^2$, $g(x) = x + 4$
b) $f(x) = x^3 + 4x^2 + 2$, $g(x) = -3x^2 - x - 2$
c) $f(x) = (2x + 2)^2$, $g(x) = 2x^2 + 7x + 5$
d) $f(x) = x^2$, $g(x) = 4x$
e) $f(x) = x^2 - 3x + 4$, $g(x) = 1$
f) $f(x) = x - 3$, $g(x) = x + 3$

6. Yes
a) $f(x) = x^2 + 10$, $g(x) = 7x$
b) $f(x) = x^2$, $g(x) = -1$
c) $f(x) = x^2 + x$, $g(x) = x + 1$

7. No; explanations may vary.

Exploring with a Graphing Calculator: Graphing Rational Functions, page 189

1. c) The y-coordinate disappears.
d) Predictions may vary.

2. b) The y-coordinate changes, but the x-coordinate stays the same.

3. a) y approaches 0
b) y increases to very large positive values, for example, 100 000
c) y approaches 0
d) y increases to very large negative values, for example, $-100\ 000$

4. Conclusions may vary.

5. a) $y = \dfrac{1}{x - 1}$ **b)** $y = \dfrac{1}{x^2 + 1}$
c) $y = \dfrac{x}{x^2 - 1}$ **d)** $y = \dfrac{x^2 - 1}{x - 1}$
e) $y = \dfrac{x^2 + 1}{x^2 - 1}$ **f)** $y = \dfrac{x^2 - 1}{x}$

Investigate, page 191

1. b)

x	-10	-5	-2	-1	-0.5	-0.2	-0.1	0	0.1	0.2	0.5	1	2	5	10
$\frac{1}{x}$	-0.1	-0.2	-0.5	-1	-2	-5	-10	$-$	10	5	2	1	0.5	0.2	0.1

2. a) $\dfrac{1}{x}$ is a negative number with a very small absolute value.
b) i) $\dfrac{1}{x}$ decreases from -1
ii) $\dfrac{1}{x}$ decreases from -1 to numbers with large absolute values.
iii) $\dfrac{1}{x}$ decreases from very large positive values to 1.
iv) $\dfrac{1}{x}$ decreases to a very small positive number.

3.6 Exercises, page 194

1. a) $y = \dfrac{1}{3x - 7}$ **b)** $y = \dfrac{1}{5x^2 - 2x}$
c) $y = \dfrac{1}{(x - 2)^2 - 1}$ **d)** $y = \dfrac{1}{\sqrt{x + 1}}$
e) $y = \dfrac{1}{-2x^3 + 3x^2 - 7x}$ **f)** $y = 2^{-x}$, or $y = \dfrac{1}{2^x}$

2. a) Asymptotes: $x = -1$, $y = 0$
b) Asymptotes: $x = 4$, $y = 0$
c) Asymptotes: $x = 4$, $y = 0$
d) Asymptotes: $x = -\dfrac{3}{2}$, $y = 0$
e) Asymptotes: $x = 2$, $y = 0$
f) Asymptotes: $x = -1$, $y = 0$

4. Explanations may vary.

5. a) Asymptotes: $x = 0$, $y = 0$
b) Asymptotes: $x = \pm 3$, $y = 0$
c) Asymptotes: $x = \pm 4$, $y = 0$
d) Asymptote: $y = 0$
e) Asymptotes: $x = 2$, $y = 0$
f) Asymptotes: $x = -3$, $y = 0$

8. Explanations may vary.

9. b) $y = \dfrac{1}{2x + 6}$
i) D: $x \neq -3$; R: $y \neq 0$
ii) No zeros
iii) $x = -3$; $y = 0$
iv) Point symmetry; line symmetry about $y = x + 3$ and $y = -x - 3$
v) No maximum or minimum points
$y = \dfrac{1}{x + 3}$
i) D: $x \neq -3$; R: $y \neq 0$
ii) No zeros
iii) $x = -3$; $y = 0$
iv) Point symmetry; line symmetry about $y = x + 3$ and $y = -x - 3$
v) No maximum or minimum points

$y = \frac{1}{3-x}$

   i) D: $x \neq 3$; R: all real numbers

   ii) No zeros

   iii) $x = 3$; $y = 0$

   iv) Point symmetry line symmetry about $y = 3 - x$ and
      $y = x - 3$

   v) No maximum or minimum points

$y = \frac{1}{x^2 - 25}$

   i) D: $x \neq \pm 5$; R: $y > 0$ or $y < -0.04$

   ii) No zeros

   iii) $x = 5$, $x = -5$

   iv) Line symmetry about $x = 0$

   v) Maximum at $(0, -0.04)$

$y = \frac{1}{x^2 - 10}$

   i) D: $x \neq \pm\sqrt{10}$; R: $y > 0$ or $y < -0.1$

   ii) No zeros

   iii) $x = \sqrt{10}$, $x = -\sqrt{10}$

   iv) Line symmetry about $x = 0$

   v) Maximum at $(0, -0.1)$

$y = \frac{1}{(x-5)^2}$

   i) D: $x \neq 5$; R: $y > 0$

   ii) No zeros

   iii) $x = 5$

   iv) Line symmetry about $x = 5$

   v) No maximum or minimum points

10. Explanations may vary.
   a) No   b) No   c) No

11. Descriptions may vary.

12. b) $y = \frac{1}{x} + 2$ c) Yes, passes vertical line test
   d) D: $x \neq 0$; R: $y \neq 2$

13. b) i) D: $0 < x \leq 1$, R: all real numbers
      ii) D: $x > 0$, R: $y \neq 0$
   c) Neither; the graphs of both inverses fail the vertical line
   test.

14. a) Predictions may vary.
   b) Descriptions may vary.

15. a) Predictions may vary.
   b) Descriptions may vary.

3.7 Exercises, page 200

1. Conclusions may vary.
   a) Polynomial   b) Rational   c) Polynomial
   d) Rational   e) Polynomial   f) Other
   g) Polynomial   h) Other

2. Explanations may vary.
   a) Rational   b) Polynomial   c) Rational
   d) Polynomial   e) Rational   f) Rational

3. Explanations may vary.
   a) Yes   b) No

4. $y = \frac{4x}{x^2 + 1}$
   a) D: all real numbers; R: $|y| \leq 2$
   b) 0   c) Point symmetry; descriptions may vary.
   d) None   e) $y = 0$

$y = \frac{x^2}{x^2 - 4}$

   a) D: $x \neq \pm 2$; R: $y \leq 0$, $y > 1$
   b) 0   c) Line symmetry; descriptions may vary.
   d) $x = \pm 2$ e) $y = 1$

$y = \frac{x^2}{x - 5}$

   a) D: $x \neq 5$; R: $y \leq 0$, $y \geq 20$ b) 0
   c) Point symmetry; descriptions may vary.
   d) $x = 5$ e) None

$y = \frac{x^2 - 9}{x - 3}$

   a) D: $x \neq 3$; R: $y \neq 6$ b) -3
   c) Point symmetry; line symmetry; descriptions may vary.
   d) None   e) None

9. Explanations may vary.

10. b) $y = \frac{x}{x^2 - 4}$
     i) D: $x \neq \pm 2$, R: all real numbers
     ii) $x = 0$
     iii) $x = \pm 2$, $y = 0$
     iv) Point symmetry; descriptions may vary.
     v) None

$y = \frac{x^2 - 4}{x}$

     i) D: $x \neq 0$, R: all real numbers
     ii) $x = \pm 2$
     iii) $x = 0$, $y = x$
     iv) Point symmetry; descriptions may vary.
     v) None

$y = \frac{x}{x^2 + 4}$

     i) D: all real numbers, R: $|y| \leq 0.25$
     ii) $x = 0$
     iii) $y = 0$
     iv) Point symmetry; descriptions may vary.
     v) Maximum at $(2.0, 0.25)$, minimum at $(-2.0, -0.25)$

$y = \frac{x^2 + 4}{x}$

     i) D: $x \neq 0$, R: $|y| \geq 4$
     ii) None
     iii) $x = 0$, $y = x$
     iv) Point symmetry; descriptions may vary.
     v) Maximum at $(-2, -4)$, minimum at $(2, 4)$

11. $y = \frac{x + 3}{x + 1}$
     i) D: $x \neq -1$, R: $y \neq 1$
     ii) $x = -3$
     iii) $x = -1$, $y = 1$
     iv) Point symmetry; descriptions may vary.
     v) None

$y = \frac{(x + 3)^2}{x + 1}$

     i) D: $x \neq -1$, R: $y \leq 0$, $y \geq 8$
     ii) $x = -3$
     iii) $x = -1$, $y = x + 5$
     iv) Point symmetry; descriptions may vary.
     v) Maximum at $(-3, 0)$, minimum at $(1, 8)$

$y = \frac{x + 3}{(x + 1)^2}$

     i) D: $x \neq -1$, R: $y \geq -0.125$
     ii) $x = -3$
     iii) $x = -1$, $y = 0$
     iv) No symmetry
     v) Minimum at $(-5, -0.125)$

$y = \dfrac{(x+3)^2}{(x+1)^2}$

    i) D: $x \neq -1$, R: $y \geq 0$
    ii) $x \neq -3$
    iii) $y = 1$, $x = -1$
    iv) No symmetry
    v) Minimum at $(-3, 0)$

12. b) Answers may vary. $2 + 2 = 2 \times 2$
  c) $y = \dfrac{x}{x-1}$
  d) Answers may vary; $(5, 1.25)$, $(-3, 0.75)$

14. b) i) D: all real numbers; R: all real numbers
    ii) D: $x \leq 0$, $x > 1$; R: $y \neq \pm 2$
  c) Both inverses fail the vertical line test.

15. b) i) $y = \dfrac{x}{x-1}$ ii) $y = \dfrac{x}{1-x}$
  c) i) D: $x \neq 1$; R: $y \neq 1$ ii) D: $x \neq 1$; R: $y \neq -1$
  d) Both inverses pass the vertical line test.

16. Conclusions may vary.

17. a) No; explanations may vary.
  b) Yes; explanations may vary.

3.8 Exercises, page 206

1. a) $v = \dfrac{900}{t}$ c) $t > 0$

2. a) Approximately 50 min   b) Approximately 28 min
  c) Approximately 6 min

3. b) i) $6.25   ii) 400

4. a) 3.86 cm   b) $h = \dfrac{1050}{(16.5 - x)^2} - 3.86$
  d) i) 0.8 cm   ii) 2.5 cm

5. a) $h = \dfrac{375}{(10.4 - x)^2} - 3.47$
  c) i) 0.4 cm   ii) 0.8 cm   iii) 1.3 cm
  d) i) 0.7 cm   ii) 1.2 cm   iii) 1.7 cm

6. a) Approximately 2.30 h, or 2 h 18 min
  b) $t = \dfrac{1025}{446 + s} - 2.30$
  c) ii) Approximately 0.12 h, or 7 min
    iii) Approximately 0.14 h, or 8 min
    iv) Approximately 500 km/h
  d) Answers may vary.

Mathematical Modelling: The Water and the Wine, page 208

1. a) Yes; answers may vary.   b) The piles are the same.

2. Yes; explanations may vary.

3. a) No; explanations may vary.
  b) The amount of wine in the water and the amount of water in the wine do not depend on the original amount of water and wine.

4. a) No   b) The contents do not need to be stirred.

5. a) Yes
  b) The volumes of liquid transferred must be equal.

6. a) i) $t - x$
    ii) Water: x; wine: n, total volume $= x + n$

    iii) Water: $\dfrac{x}{x+n}$; wine: $\dfrac{n}{x+n}$

b) i) $\dfrac{nx}{x+n}$ ii) $\dfrac{nx}{x+n}$

7. a) $y = \dfrac{x}{x+1}$, $y = \dfrac{2x}{x+2}$, $y = \dfrac{3x}{x+3}$, $y = \dfrac{4x}{x+4}$
  b) The volume of water in the jug does not affect the volume of wine.
  c) $x \leq 1$, $x \leq 2$, $x \leq 3$, $x \leq 4$

8. a) 0.33 L, 0.43 L   b) 0.55 L, 0.67 L
  c) 0.75 L, 1.00 L   d) 1.09 L, 1.33 L

Exploring with a Graphing Calculator: Graphing $f(x) + g(x)$, page 210

1. Predictions may vary.

2. c) The y-coordinate of any point on the line $y = 2x + 1$ is the sum of the y-coordinates for points with the same x-coordinate on the lines $y = x$ and $y = x + 1$.

3. b) Explanations may vary.

4. c) The line $y = 2x + 3$ touches the parabola defined by $y = x^2 + 2x + 3$ at the point where it intersects the y-axis.

5. b) They have the same y-intercept because when $x = 0$, $y = x^2$ has no effect on the sum.

6. c) The parabola $y = -4x^2 + 2x + 3$ and the graph of $y = x^3 - 4x^2 + 2x + 3$ intersect at the point where both curves intersect the y-axis.

7. b) They have the same y-intercept because when $x = 0$, $y = x^3$ has no effect on the sum.

8. c) The line $y = 2x + 3$ touches the graph of $y = x^3 - 4x^2 + 2x + 3$ at the point where they both intersect the y-axis.

9. b) They have the same y-intercept because when $x = 0$, $y = x^3 - 4x^2$ has no effect on the sum.

10. c) Explanations may vary.
  d) Its graph looks like the graph of a polynomial function.

3.9 Exercises, page 217

1. a) 7   b) 22   c) 10   d) 21

2. $6x + 4$; $6x + 3$

3. a) 5   b) 10   c) 4   d) 17

4. $4x^2 + 1$; $2x^2 + 2$

5. a) $-6x + 19$; $-6x - 3$
  b) $4x^2 + 14x + 6$; $2x^2 + 10x + 1$
  c) $32x^2 - 100x + 78$; $-8x^2 + 12x + 3$

6. a) $3x^2 + 12x + 11$; $3x^2 + 1$
  b) $12x^2 - 12x + 2$; $3 - 6x^2$
  c) $3x^4 - 1$; $9x^4 - 6x^2 + 1$
  d) $3x^4 + 6x^3 + 3x^2 - 1$; $9x^4 - 3x^2$
  e) $12x^4 - 36x^3 + 27x^2 - 1$; $18x^4 - 21x^2 + 5$

7. $A = \dfrac{\pi d^2}{4}$

8. $V = \dfrac{\pi d^3}{6}$

9. a) -11 b) -8 c) 5   d) 16

10. a) $1 - 6x$ b) $4 - 6x$ c) $4x - 3$ d) $9x - 2$

11. a) $\sqrt{4-2x}$; $4-2\sqrt{x}$; $\sqrt[4]{x}$; $4x-4$

b) $\sqrt{2+6x}$; $1+3\sqrt{2x}$; $\sqrt{2\sqrt{2x}}$; $4+9x$

c) $\dfrac{x^2-1}{x^2}$; $\dfrac{-2x-1}{(x+1)^2}$; $\dfrac{x}{2x+1}$; x^4-2x^2

d) $6x-8$; $6x-4$; $4x$; $9x-16$

12. $d = \dfrac{s^2}{51.84}+4$

13. $A = \dfrac{P^2}{16}$

14. $A = \dfrac{d^2}{2}$

16. b) No; explanations may vary.

17. $T(t) = 0.05t + 20$

18. $K(t) = 0.4(49-9.8t)^2$

19. a) $f(k(x))$ **b)** $g(e(x))$ **c)** $f(e(x))$ **d)** $k(e(x))$

20. Answers may vary.

a) $f(x) = x^3$, $g(x) = x^2+2x+1$, $k(x) = g(f(x))$

b) $f(x) = x^2+3x+4$, $g(x) = x-4$, $k(x) = f(g(x))$

c) $f(x) = \sqrt{x}$, $g(x) = 3x-2$, $k(x) = f(g(x))$

d) $f(x) = x+3$, $g(x) = \dfrac{1}{x}$, $k(x) = g(f(x))$

21. a) $\sqrt{x}-3$ **b)** $x \geq 0$ **c)** $y \geq -3$

d) $\sqrt{x-3}$ **e)** $x \geq 3$ **f)** $y \geq 0$

22. a) x **b)** All real numbers

c) All real numbers   **d)** x

e) All real numbers   **f)** All real numbers

23. $ad + b = cb + d$

Linking Ideas: Mathematics and Technology
Patterns in Graphs of Functions, page 220

1. b) i) $y = \dfrac{-6}{x^2-6}$ **ii)** $y = \dfrac{6}{x^2+6}$ **iii)** $y = \dfrac{-6}{x^2-6}$

2. $y = 0$

3. a) $(0, 1)$ **b)** $y = 0$; explanations may vary.

3 Review, page 221

1. a) $f(x)$ **b)** $g(x)$ **c)** $h(x)$ **d)** $k(x)$

2. Answers may vary.

a) $f(x) = (x-3)(x+2)(x-1)$

b) $f(x) = x^2(x-1)$

c) $f(x) = (x+4)(x-5)(x+1)$

d) $f(x) = (x+3)^3$

4. a) ± 4 **b)** 12   **c)** 4   **d)** ± 8

5. b) i) D: all real numbers, R: all real numbers; D: $x \neq 3$; R: $y \neq 0$

ii) D: all real numbers, R: all real numbers; D: $x \neq 0$; R: $y \neq 0$

c) Explanations may vary.

6. a) i) Polynomial   **ii)** Rational

iii) Polynomial   **iv)** Rational

v) Other   **vi)** Other

b) Explanations may vary.

7. b) No; y is undefined; 0.025

c) Explanations may vary.

8. a) $h = 13.1 - \dfrac{725}{\pi(r+4.2)^2}$

c) i) 2.7 cm   **ii)** 5.5 cm   **iii)** 6.0 cm

d) i) 0.08 cm   **ii)** 0.17 cm   **iii)** 0.26 cm

9. a) $l = \dfrac{7200}{w}$ **b)** $0 < w < 60\sqrt{2}$

c) $p = 2\left(\dfrac{7200}{w}+w\right)$ **d)** $0 < w < 60\sqrt{2}$

10. a) $\dfrac{x-1}{x}$ **b)** $-\dfrac{1}{x}$

3 Cumulative Review, page 223

1. a) 5%   **b)** 8%   **c)** 4.5%

2. a) i) $3242.30   **ii)** $1113.63

b) Explanations may vary.

3. $3247.98

4. a) $2   **b)** $745

5. b) Explanations may vary.

6. a) $y = \dfrac{7}{4}x^2$ **b)** $y = -3(x+3)^2 - 2$

7. 16.2 m, 1.8 s

8. 11, -13

Chapter 4
Investigate, page 226

1. a) $3 \pm \sqrt{3.5}$

2. $2, -\dfrac{1}{3}$ **b)** $-1 \pm \sqrt{10}$ **c)** $-2 \pm \sqrt{2.5}$ **d)** $-\dfrac{5}{2}, 1$

3. $x = \dfrac{-b \pm \sqrt{b^2-4ac}}{2a}$

4.1 Exercises, page 231

1. a) $3 \pm \sqrt{5}$ **b)** $\dfrac{-3 \pm \sqrt{13}}{2}$ **c)** $\dfrac{-7 \pm \sqrt{37}}{2}$

d) $\dfrac{5 \pm \sqrt{17}}{2}$ **e)** $-2 \pm \sqrt{5}$ **f)** $4 \pm \sqrt{22}$

2. a) i) $\dfrac{1}{2}, 2$ **ii)** $-3, -\dfrac{1}{2}$ **iii)** $\dfrac{14}{3}, -1$

iv) $1, \dfrac{5}{4}$ **v)** $-1, -\dfrac{2}{5}$ **vi)** $\dfrac{4}{3}, -\dfrac{5}{2}$

b) Explanations may vary.

3. a) $\dfrac{2}{3}, \dfrac{1}{2}$ **b)** $\dfrac{11}{2}, 7$ **c)** $-\dfrac{1}{3}, \dfrac{1}{2}$

d) $-\dfrac{9}{2}, 5$ **e)** $-\dfrac{3}{2}, 4$ **f)** $-\dfrac{1}{2}, \dfrac{2}{3}$

4. a) 0.89, 5.61   **b)** 0.72, 2.78   **c)** -1.14, 1.47

d) -0.11, 4.61   **e)** -0.24, 0.84   **f)** $-2.28, -0.22$

5. a) -1.35, 0.15   **b)** -0.16, 3.16   **c)** -1.90, 1.23

d) -1.15, 0.65   **e)** 0.18, 1.82   **f)** 0.18, 2.82

6. a) 2.8 s   **b)** 1.5 s

7. a) $0, \dfrac{4}{3}$ **b)** $4, -4$ **c)** $-\dfrac{7}{5}$

d) $\dfrac{3 \pm \sqrt{19}}{10}$ **e)** $\dfrac{5 \pm \sqrt{89}}{2}$ **f)** $\dfrac{5\sqrt{2} \pm \sqrt{82}}{4}$

8. a) $\dfrac{1}{4}, 4$ **b)** $\dfrac{1}{2}, \dfrac{4}{3}$ **c)** $-\dfrac{2}{3}, \dfrac{3}{4}$

d) $-9, \dfrac{11}{3}$ **e)** $-\dfrac{2}{5}, 3$ **f)** $-\dfrac{2}{3}, \dfrac{1}{5}$

9. a) $\dfrac{-4 \pm \sqrt{26}}{2}$ **b)** $-4, \dfrac{2}{3}$ **c)** $\dfrac{7 \pm \sqrt{33}}{2}$

d) $1, -\dfrac{4}{3}$ **e)** $-\dfrac{3}{2}, \dfrac{1}{2}$ **f)** $-\dfrac{2}{3}$

10. a) $(4x + 3)(2x + 1)$ **b)** $(3x - 4)(2x + 1)$
c) $(4x - 5)(3x - 4)$ **d)** $(9x + 4)(2x - 5)$
e) $(10x - 21)(2x + 3)$ **f)** $(15x - 8)(3x + 2)$

11. a) Descriptions may vary. A trinomial can be factored if $b^2 - 4ac$ is a perfect square.
b) i) Yes **ii)** No **iii)** Yes
iv) No **v)** Yes **vi)** Yes

12. a) i) 53 m **ii)** 80 m
b) i) 63 km/h **ii)** 110 km/h

13. a) i) 512.28 cm^2 **ii)** 5117.5 cm^2
b) i) 3.99 cm **ii)** 14.9 cm

14. BM = DN $\doteq$ 2.29 cm

15. a) i) $\dfrac{-1 \pm \sqrt{5}}{2}$ **ii)** $-2, 1$ **iii)** $\dfrac{-1 \pm \sqrt{13}}{2}$ **iv)** $\dfrac{-1 \pm \sqrt{17}}{2}$
b) 0, 2, 6, 12, 20; n is the product of two consecutive whole numbers

16. a) i) $-4, -45$ **ii)** $-5, \dfrac{21}{4}$ **iii)** $\dfrac{29}{6}, \dfrac{35}{6}$ **iv)** $\dfrac{6}{5}, -\dfrac{3}{5}$
b) $-\dfrac{b}{a}, \dfrac{c}{a}$

17. a) $b^2 = \dfrac{25}{6}ac$ **b)** $b^2 = \dfrac{(m + n)^2}{mn}ac$

18. Answers may vary. Two examples where the consecutive integers are in order are $-x^2 + 0x + 1$ and $-2x^2 - x + 0$.

Mathematics File: Deriving the Quadratic Formula, page 234

1. Answers may vary.

2. a) $\dfrac{-3 \pm \sqrt{45}}{2}$ **b)** $\dfrac{-3 \pm \sqrt{17}}{4}$ **c)** $\dfrac{-b \pm \sqrt{b^2 - 4ac}}{2a}$

3. a) $-3 \pm \sqrt{2}$ **b)** $\dfrac{-1 \pm \sqrt{17}}{4}$ **c)** $\dfrac{-b \pm \sqrt{b^2 - 4ac}}{2a}$

4. Answers may vary.

5. Either method will work. Explanations may vary.

Mathematical Modelling: How Can We Model a Spiral?, page 236

1. a) 1 unit **b)** $(x - 1)$ units

2. a) $\dfrac{x}{1} = \dfrac{1}{x - 1}$ **b)** $\dfrac{1 + \sqrt{5}}{2} \doteq 1.618\,033\,989$

3. a), 4., 5. d)

Square	Side length	Side length	Decimal form
1	$2\phi - 3$	ϕ^{-3}	0.236 067 977
2	$2 - \phi$	ϕ^{-2}	0.381 966 011
3	$\phi - 1$	ϕ^{-1}	0.618 033 989
4	1	1	1
5	ϕ	ϕ	1.618 033 989
6	$\phi + 1$	ϕ^2	2.618 033 989
7	$2\phi + 1$	ϕ^3	4.236 067 978

b) The length to width ratio for every triangle is the golden ratio.

5. a) Answers may vary.
b) Answers may vary.
c) Answers may vary.
e) There is a common ratio of ϕ.

6. a) The side lengths are in geometric sequence. The common ratio is the golden ratio.

Rectangle	Length, l	Width, w	$l : w$ ratio
1	$2\phi - 3$	$5 - 3\phi$	ϕ
2	$\phi - 1$	$2 - \phi$	ϕ
3	1	$\phi - 1$	ϕ
4	ϕ	1	ϕ
5	$\phi + 1$	ϕ	ϕ
6	$2\phi + 1$	$1 + \phi$	ϕ
7	$3\phi + 2$	$2\phi + 1$	ϕ

b) They are all golden rectangles.

7. a) $\dfrac{1 - \sqrt{5}}{2}$ **b)** Explanations may vary.

8. a) $\sqrt{5}$ units **b)** $(\sqrt{5} + 1)$ units
c) $\dfrac{\sqrt{5} + 1}{2}$, the golden ratio
d) Explanations may vary; the rectangles are similar.

9. a) $\dfrac{\sqrt{5} + 1}{2}$ **b)** 1.618 033 989
c) $\dfrac{1 - \sqrt{5}}{2}$, or −0.618 033 989; explanations may vary

10. a) $\dfrac{\sqrt{5} + 1}{2}$ **b)** 1.618 033 989
c) $\dfrac{1 - \sqrt{5}}{2}$, or −0.618 033 989; explanations may vary

11. a) Lists may vary. **b)** Lists may vary.

12. Explanations may vary.

13. Reasons may vary.

4.2 Exercises, page 244

1. a) 25 **b)** 8 **c)** 0
d) −39 **e)** 289 **f)** −24

2. a) $x^2 + 11x + 24 = 0$, $x^2 - 4x + 2 = 0$, $3x^2 + 13x - 10 = 0$
b) $4x^2 - 20x + 25 = 0$
c) $2x^2 - 5x + 8 = 0$, $7x^2 + 12x + 6 = 0$

3. a) i) Two different real roots
ii) Two equal real roots
iii) Two different real roots
iv) Two different real roots
v) No real roots
vi) Two equal real roots
b) Explanations may vary.

4. a) i) $-1, 3$ **ii)** 1 **iii)** No solution
b) i) $y = x^2 - 2x - 3$ intersects the x-axis twice.
ii) $y = x^2 - 2x + 1$ intersects the x-axis once.
iii) $y = x^2 - 2x + 5$ does not intersect the x-axis.

5. a) $q^2 - 4pr > 0$ **b)** $q^2 - 4pr = 0$ **c)** $q^2 - 4pr < 0$

6. a) $k < -2$, $k > 2$ **b)** $k > -\dfrac{4}{3}$
c) $k < -2\sqrt{6}$, $k > 2\sqrt{6}$

7. a) $\pm 2\sqrt{7}$ **b)** $\dfrac{5}{6}$

8. a) No values; that is, equation has real roots for all values of p.
b) $p > \dfrac{16}{9}$ **c)** $p < -\dfrac{5}{2}$, $p > \dfrac{5}{2}$

9. Explanations may vary.

10. a) i) Yes **ii)** No
b) i) For $t \geq 0$, the graph of $h = -2.75 + 250t - 4.9t^2$ intersects the t-axis once. The discriminant is greater than 0.

ii) The graph of $h = -4.0 + 250t - 4.9t^2$ does not intersect the t-axis. The discriminant is less than 0.

11. a) No, explanations may vary.   **b)** $0 \le x < 116$

12. a) $k = -8$ or $k = 4$ **b)** $k < -8$ or $k > 4$ **c)** $-8 < k < 4$

13. a) i) $-26, -24$ **ii)** -25 **iii)** No real roots

     b) The nature of the roots can change.

     c) $y = x^2 + 50x + 624$ intersects the x-axis twice.
       $y = x^2 + 50x + 625$ intersects the x-axis once.
       $y = x^2 + 50x + 626$ does not intersect the x-axis.

14. No solution, explanations may vary.

15. a) Sometimes true   **b)** Never true   **c)** Always true
     d) Sometimes true   **e)** Always true

16. a) Answers may vary.
     b) $b + c = -1$

Linking Ideas: Mathematics and History
A Short History of Number Systems, page 246

1. Explanations may vary.

2. a) $-2 \pm i$ **b)** $1 \pm i\sqrt{2}$

3. $1 + i$, $1 - i$

Exploring with a Graphing Calculator: Quadratic Equations and the Butterfly Effect, page 248

2. a) No. -2 is an initial value that leads to another root
     b) -2
     c) Initial values may vary. $-1.999\,999$, $-2.000\,001$; explanations may vary

3. b) There are some differences.
     d) Initial values may vary.
       Left-hand screen: 3.999 999 (Novel2)
       Right-hand screen: 3.999 999, 4.000 001 (Novel3)

4. a) Answers may vary.   **b)** Answers may vary.

Investigate, page 251

1. a) -18 **b)** -10 **c)** -30 **d)** -30

2. a) -18 **b)** -10 **c)** -30 **d)** -30

3. $f(k)$ is equal to the remainder when $f(x)$ is divided by $x - k$.

4. a) Polynomials may vary.
     b) The relation is still true.

4.3 Exercises, page 254

1. a) -5 **b)** 16 **c)** 27
     d) -9 **e)** 0 **f)** 7

2. a) 3 **b)** 14 **c)** 43
     d) 11 **e)** 18 **f)** 19

3. a) -4 **b)** 11 **c)** 0
     d) 16 **e)** 21 **f)** -11

4. Explanations may vary.

5. a) 4 **b)** 2 **c)** 3

6. a) -2 **b)** -1

7. a) 2 **b)** 6

8. a) -5 **b)** -23

9. a) $f(x) = (x - k)q(x) + r$ **b)** Proofs may vary.

Investigate, page 255

1. a) i) 0 **ii)** 30 **iii)** -12 **iv)** 0
     b) It has $x - 2$ and $x + 5$ as factors.

2. a) i) 4 **ii)** 0 **iii)** 0 **iv)** 10
     b) It has $x - 3$ and $x - 2$ as factors.

3. a) $x + 1$, $x - 2$, $x - 4$
     b) $x^3 - 5x^2 + 2x + 8 = (x + 1)(x - 2)(x - 4)$

4. a) $1, 3, -2$
     b) $x^3 - 2x^2 - 5x + 6 = (x - 1)(x - 3)(x + 2)$

5. Substitute $x = k$ in the polynomial. If $f(k) = 0$, then $x - k$ is a factor of $f(x)$.

4.4 Exercises, page 259

1. a) -11 **b)** 0 **c)** 165
     d) -3 **e)** 4 **f)** -35

2. $x - 2$

3. 0

4. a) $x + 2$ **b)** $x + 2$, $x - 1$
     c) $x + 1$, $x - 1$, $x - 2$

5. Parts a, c, and d

6. Part a

7. Explanations may vary.

8. According to the Factor Property, I choose factors of the constant, which in this case is 8. ± 1, ± 2, ± 4, ± 8

9. Factors may vary.
     a) $x - 1$ **b)** $x + 1$ **c)** $y + 1$
     d) $x - 3$ **e)** $y - 2$ **f)** $x + 1$

10. b) $2x - 1$

11. b) $x - 4$, $x - 1$ **c)** Explanations may vary.

12. a) $(x - 1)(x + 4)(x + 2)$ **b)** $(x + 1)(x + 3)(x + 5)$
     c) $(x + 1)(x + 5)(x - 4)$ **d)** $(x + 1)(x + 2)(x - 3)$
     e) $(x - 1)^2(5x + 3)$ **f)** $(x - 2)(x^2 - 7x + 3)$
     g) $(x + 1)(x + 2)(x + 5)$ **h)** $(x + 2)(x - 3)(2x + 1)$
     i) $(x - 3)(x^2 - 3x + 1)$

13. a) $1, -2, 4$ **b)** $-1, 2, -5$ **c)** $2, 2, 4$
     d) $-2, -2, 3$ **e)** $2, \dfrac{-1 \pm \sqrt{5}}{2}$ **f)** $2, 1 \pm i\sqrt{3}$

14. a) $1, 3, -\dfrac{1}{2}$ **b)** $-1, 2, \dfrac{1}{3}$ **c)** $2, -\dfrac{2}{3}, -\dfrac{1}{2}$
     d) $1, -2, -\dfrac{5}{2}$ **e)** $4, \dfrac{1 \pm \sqrt{5}}{2}$ **f)** $-2, \dfrac{-1 \pm \sqrt{17}}{4}$
     g) $3, \dfrac{3 \pm \sqrt{-23}}{4}$ **h)** $-1, 6, \dfrac{1}{3}$ **i)** $2, -1 \pm i\sqrt{2}$

15. a) 11 **b)** -8 **c)** -4

16. b) Explanations may vary; $\dfrac{1}{2}$

17. a) 11 **b)** -4

18. Yes, explanations may vary.

19. Proofs may vary.

20. Proofs may vary.

4.5 Exercises, page 264

1. a) i) $-2 < x < 3$ **ii)** $x \le -2$ or $x \ge 3$
  b) i) $x \le -1$ or $x \ge 3$ **ii)** $-1 < x < 3$
  c) i) $x \le -4$ or $-1 \le x \le 2$ **ii)** $-4 < x < -1$ or $x > 2$
  d) i) $x = -3$ or $x \ge 2$ **ii)** $-3 < x < 2$ or $x < -3$

2. a) $x < -2$ or $x > 2$ **b)** $-2 \le x \le -1$
  c) $x < 0$ or $x > 5$ **d)** $0 \le x \le 2$
  e) $2 < a < 3$ **f)** $n \le -5$ or $n \ge 3$

3. a) i) $x < -3$ or $x > -2$ **ii)** $m < -2$ or $m > 4$
  iii) $y < -6$ or $y > 3$ **iv)** $-1 < x < 3$
  v) $1 \le x \le 9$ **vi)** $-\frac{5}{2} < x < 3$
  b) Explanations may vary.

4. a) $x < 2$ or $x > 2$
  b) $x \le -\frac{1}{2}$ or $x \ge \frac{1}{2}$
  c) $x \le \frac{2 - \sqrt{19}}{3}$ or $x \ge \frac{2 + \sqrt{19}}{3}$
  d) $-\frac{5}{2} < x < \frac{4}{3}$
  e) All values of x **f)** No values of x
  g) $x = \frac{2}{3}$ **h)** $\frac{-5 - \sqrt{53}}{4} \le x \le \frac{-5 + \sqrt{53}}{4}$

5. b) $x < -1$ or $x > 3$

6. b) $-1 \le x \le 5$

7. a) Answers may vary.
  i) $2x^2 - 7x + 3 \ge 0$ **ii)** $x^2 - x - 12 < 0$
  iii) $x^2 + x - 2 > 0$ **iv)** $2x^2 + 9x + 7 \le 0$
  b) Explanations may vary.

8. a) $x < 1$ or $x > 5$ **b)** $-2 < x < 7$

9. a) $x > 3$ or $-3 < x < 1$ **b)** $x < -3$ or $1 < x < 3$
  c) $x \le -1$ or $2 \le x \le 5$ **d)** $-5 < x < 0$ or $x > 2$
  e) $-1 < x < 4$ or $x > 4$ **f)** $-3 \le x \le -1$ or $2 \le x \le 5$

10. a) $-1 < x < 1$ or $x > 4$ **b)** $x \le -1$ or $2 \le x \le 3$
  c) $x \le -3$ or $x \ge 2$ or $-2 \le x \le 1$
  d) $1 < x < 5$ or $-1 < x < 1$
  e) $x < \frac{-1 - \sqrt{37}}{2}$ or $-1 < x < \frac{-1 + \sqrt{37}}{2}$
  f) $x \ge -2$

11. Answers may vary.
  a) $x^3 + x^2 - 6x \ge 0$
  b) $4x^3 - 56x^2 + 225x - 225 < 0$
  c) $x^3 - 19x^2 + 111x - 189 > 0$
  d) $x^3 - 21x + 20 < 0$

12. $x > -1$

13. Explanations may vary.
  a) True   **b)** True   **c)** False
  d) False   **e)** False   **f)** True

14. $\frac{10x - x^2}{5} < y < 5$

4.6 Exercises, page 273

1. a) -2.5 **b)** -3 **c)** -4
  d) 13 **e)** -8 **f)** No solution

2. a) i) $-1, 5$ **ii)** $-2, 3$ **iii)** $1, 6$
  iv) $-3, -2$ **v)** $-9, -2$ **vi)** $-4, 2$
  b) Explanations may vary.

4. a) $-1, -\frac{8}{5}$ **b)** $-5, -\frac{3}{2}$ **c)** $-2, \frac{11}{7}$
  d) $-6, 4$ **e)** $-5, \frac{7}{3}$ **f)** $-\frac{1}{12}, \frac{2}{3}$

5. a) $x < -4$ or $x > 0$ **b)** $0 < x < 4$
  c) $x < 9$ or $x \ge 12$ **d)** $-3 < x < -2.5$
  e) $5 < x \le 7$ **f)** $x < -1.5$ or $x \ge -1$

6. a) i) $-4 < x < 0$, $x > 1$
  ii) $x \le -1$, $0 < x \le 2.5$ **iii)** $x \le -3$, $0 < x \le \frac{1}{3}$
  iv) $x < 0, \frac{1}{2} \le x \le \frac{5}{3}$ **v)** $-1 < x < 0, x > \frac{7}{5}$
  vi) $x < -\frac{9}{4}, 0 < x < 1$
  b) Explanations may vary.

7. a) $x < -9$ or $-3 < x < -2$ **b)** $x < \frac{4}{3}$ or $2 \le x \le \frac{10}{3}$
  c) $7 \le x < 9$ or $x \ge 11$ **d)** $x < -\frac{1}{2}$
  e) $x < -1$ or $0 < x < 4$ **f)** $x \le -\frac{7}{4}$ or $\frac{1}{4} < x \le \frac{1}{2}$

8. b) $x < -3$ or $x > 0$

9. b) $x < 0$ or $x \ge 1$

10. a) $-7, \frac{-11 \pm \sqrt{37}}{2}$ **b)** $5, 2 \pm \sqrt{3}$
  c) -3 **d)** $3, 7$

11. a) $\frac{3 \pm \sqrt{321}}{4}$ **b)** $2, 16$
  c) $2, 8$ **d)** $\pm 10\sqrt{5}$

12. a) $0, 8$ **b)** $\frac{3}{5}$ **c)** $\frac{-9 \pm \sqrt{97}}{4}$
  d) $\frac{1 \pm \sqrt{7}}{2}$ **e)** $1 \pm \sqrt{14}$ **f)** $\frac{-15 \pm \sqrt{217}}{4}$

13. a) $0, 4$ **b)** $0, -\frac{2}{3}$ **c)** $\frac{-5 \pm \sqrt{185}}{10}$
  d) $\frac{-1 \pm \sqrt{10}}{3}$ **e)** $\frac{-4 \pm \sqrt{61}}{3}$ **f)** No solution

14. a) All values of x **b)** $-2 < x < 2$
  c) $-1 < x < 1$ **d)** $-0.5 < x < 0.5$

15. a) $-4 < x < 0$ or $x > 2$ **b)** $x < -1$ or $0 < x < 3$
  c) $0 < x < 1$ or $x > 2$
  d) $x \le 2 - \sqrt{2}, 1 < x < 2$, or $x \ge 2 + \sqrt{2}$

16. a) $x < -5, -1 < x < \frac{5}{3}$, or $x > 3$
  b) $-\frac{21}{2} \le x < \frac{1}{2}$ or $2 \le x < 6$
  c) $-8 < x < -6, \frac{7 - \sqrt{137}}{2} < x < 8$, or $x > \frac{7 + \sqrt{137}}{2}$
  d) $x \le -\sqrt{14}, -3 < x \le 1$, or $2 < x \le \sqrt{14}$

Problem Solving: The Most Famous Problem in Mathematics, page 276

1. Answers may vary. 3, 4, 5; 5, 12, 13

2. a) Answers may vary. 19, 180, 181; 24, 32, 40; 8, 15, 17

3. Examples may vary.
  a) $1^2 + 4^2 + 8^2 = 9^2$ **b)** $3^3 + 4^3 + 5^3 = 6^3$
  c) $2^2 + 9^2 = 6^2 + 7^2$ **d)** $1^3 + 12^3 = 9^3 + 10^3$

4. Explanations may vary.

5. Examples may vary.

a) 6, 3, 2; 10, 15, 6

b) Examples may vary. $n = \frac{1}{2}$; $4\frac{1}{2} + 9\frac{1}{2} = 25\frac{1}{2}$

4.7 Exercises, page 284

1. a) 25   **b)** 36   **c)** 12

    d) $\frac{1}{16}$ **e)** 12   **f)** 6

2. a) $x > 16$ **b)** $0 \le x \le 25$ **c)** $0 \le x < \frac{27}{4}$

    d) $0 \le x \le \frac{9}{25}$ **e)** $x > 2$ **f)** $0 \le x \le \frac{81}{8}$

3. a) 16   **b)** 9   **c)** 16

    d) $\frac{25}{4}$ **e)** 1   **f)** 9

4. a) $-2 \le x < \frac{21}{2}$ **b)** $x > 34$ **c)** $x \ge 49$

    d) $0 \le x < 16$ **e)** $0 \le x \le 1$ **f)** $x > 16$

5. a) i) 27   **ii)** 2.8   **iii)** 4

    iv) $\frac{10}{3}$ **v)** No solution   **vi)** 4

    b) Explanations may vary.

6. a) i) $x \ge 53$ **ii)** $-\frac{7}{2} \le x < 37$ **iii)** $x > 4$

    iv) $x > \frac{49}{18}$ **v)** $\frac{3}{4} \le x \le \frac{19}{4}$ **vi)** $x \ge 3$

    b) Explanations may vary.

8. $\sqrt{x+3} + 5 = 0$, $4 + \sqrt{2x-7} = 0$, $7 + 5\sqrt{2x+3} = 4$

9. $a = \sqrt{d^2 - b^2}$

10. $P(-8, 0)$ or $P(8, 0)$

11. $P(0, -20)$ or $P(0, 20)$

12. $\left(\dfrac{648 \pm 36\sqrt{485}}{161}, \dfrac{648 \pm 36\sqrt{485}}{161} \right)$

13. a) 1440 min   **b)** 35 848.5 km

14. a) i) 3   **ii)** $\frac{1}{2}$ **iii)** 4

    iv) -11 **v)** No solution   **vi)** $-\frac{1}{2}$

    b) Explanations may vary.

16. a) $\frac{196}{9}$ **b)** $\frac{19 - \sqrt{189}}{2}$ **c)** 3.705   **d)** -2

17. BC = 20 cm, AB = 21 cm, AC = 29 cm

18. a) $x > 0$ **b)** $x > 1$ **c)** $x > 4$

    d) $x > 1$ **e)** $x > 1.64$ **f)** $x > -0.96$

19. a) $x > -2$ **b)** $x > -1$ **c)** $-2 \le x < 2$

    d) $x > 3$ **e)** $x > -2$ **f)** $x > -1$

21. a) i) $x > 12$ **ii)** $-\frac{5}{8} \le x < -\frac{1}{2}$ **iii)** $x \ge 3$

    iv) $-1 \le x \le 0$ **v)** $2 \le x \le 3$ **vi)** No solution

    b) Explanations may vary.

4.8 Exercises, page 294

1. a) ± 5 **b)** 0   **c)** $-5, 9$

    d) 2, 6   **e)** $-6, 4$ **f)** 5

2. a) $-3 < x < 3$ **b)** $x \le -4$ or $x \ge 4$

    c) $-3 < x < 7$ **d)** $-1 \le x \le 3$

    e) $x \le -8$ or $x \ge 6$ **f)** $-10 < x < 8$

3. a) 3, 7   **b)** $-3, 7$ **c)** $-3, 1$

    d) $-3, -1$ **e)** $-6, 12$ **f)** $-1, 7$

4. a) $2 \le x \le 4$ **b)** $x < 1, x > 5$

    c) $-8 < x < 4$ **d)** $x \le 1, x \ge 9$

    e) $x \le -11, x \ge 9$ **f)** $-14 \le x \le 4$

5. b) $x < -4, x > 4$

6. b) $-2 \le x \le 4$

7. a) No solution   **b)** 1   **c)** 1

    d) $0, \frac{1}{2}$ **e)** $\frac{1}{5}, \frac{1}{3}$ **f)** No solution

8. a) All values of x **b)** $x > \frac{1}{2}$ **c)** 2

    d) $\frac{1}{4} \le x \le \frac{3}{2}$ **e)** No solution   **f)** No solution

10. a) $\frac{5}{2}$ **b)** No solution   **c)** $x \ge 3$

11. a) $-\frac{1}{2}$ **b)** $-\frac{4}{3} < x < \frac{8}{3}$ **c)** $x < -\frac{7}{2}, x > -\frac{1}{2}$

    d) All values of x **e)** $x \ge -2$ **f)** $x > \frac{1}{3}$

12. Explanations may vary. $|2x+3| = -5$, $|x-2| + |2x+6| = 0$, $x = |2x+1|$

13. Answers may vary.

    a) $|x| \le 3$ **b)** $|x| \ge 3$ **c)** $|x-1| < 2$ **d)** $|x-1| > 2$

14. Answers may vary.

    a) $2x - 6 \ge |x - 3|$ **b)** $|x-2| \le 2$

    c) $|x - 2| > 2$ **d)** $|x - 2| > 0$

15. a) $-3, 14$ **b)** $-2, 3$

    c) No solution   **d)** $-1, 5$

16. a) $1 \le x \le 3$ **b)** $-4 < x < 4$

    c) $x < 1$ or $x > 3$ **d)** No solution

18. Answers may vary.

    a) $|x| < 0$ **b)** $|x| \ge 0$ **c)** $|x| \le 0$

19. a) $x < -2$ or $x \ge -1$ **b)** $x \le -4$ or $x > -2$

4 Review, page 296

1. a) $-10.13, 1.13$ **b)** 0.69, 12.31

    c) 0, 1.17   **d)** 0.05, 3.45

2. Explanations may vary.

3. a) 1   **b)** 148   **c)** 0   **d)** -152

4. a) $2x^2 - 3x + 1 = 0$, $3x^2 + 10x - 4 = 0$

    b) $x^2 - 2x + 1 = 0$ **c)** $7x^2 - 10x + 9 = 0$

5. a) 0   **b)** -1 **c)** 9

    d) 11   **e)** 2   **f)** -8

6. Explanations may vary.

7. a) $2, -\frac{1}{2}, -3$ **b)** 1

    c) $2, \dfrac{3 + \sqrt{29}}{2}, \dfrac{3 - \sqrt{29}}{2}$

    d) $1, 3, -\frac{2}{5}$

8. b) $0 < x < 6$

9. Answers may vary.

    a) $x^2 - 6x + 8 > 0$ **b)** $4x^2 - 4x - 3 \le 0$

    c) $x^2 + 3.9x + 3.74 \ge 0$ **d)** $x^2 - 10x + 21 < 0$

10. a) $-0.50, 0$ **b)** $-1.30, 2.30$ **c)** $-6.53, 1.53$ **d)** 0.5

11. a) 10   **b)** 34   **c)** 14   **d)** No solution

12. Explanations may vary.

13. $\sqrt{x+1} + 2 = 0$, $\sqrt{3x-1} - 7 = 0$

14. a) No solution   **b)** 5

    c) No solution   **d)** $\frac{625}{49}$

15. $h = \frac{2r \pm \sqrt{4r^2 - c^2}}{2}$

16. a) Substitute $c = r$ to get $h = r$.

    b) $c \le 2r$

17. a) $x < -1$ or $x > 1$ **b)** $-4 \le x \le 4$

    c) $-2 < x < 2$ **d)** $x < -3$ or $x > 1$

18. a) $x \le -\frac{1}{3}$ **b)** $0 < x < 1$

    c) $x \le 3$ **d)** $x > 1$

19. Explanations may vary.

20. $|3x + 1| = -2$, $2|x| = 4|x| + 3$, $|x - 3| + |3x - 2| = 0$;
explanations may vary.

21. Answers may vary.

    a) $4x - 2 \ge |3x + 2|$ **b)** $|x| \le 2$

    c) $|2x - 3| > 7$ **d)** $|x - 5| > 0$

22. a) i) 3, 4   **ii)** 2, 4   **iii)** 3, 4   **iv)** $x \ge 3$

    b) Answers may vary.

4 Cumulative Review, page 298

1. $725.62

2. $408

3. About 7 years

4. 3.5%

5. $13 327.31

6. a) $25.69   **b)** $31.79   **c)** $384.30

8. Explanations may vary.

9. Answers may vary.

    a) $f(x) = x^2 + 4x - 12$ **b)** $f(x) = x^2 - 5x + 4$

    c) $f(t) = 2t^2 - 9t + 9$

10. Explanations may vary.

11. a) $R = -5s^2 + 400s$ **b)** $8000

    c) $20 < s < $70

12. a) $y = \pm\frac{1}{2}\sqrt{x + 9}$ **b)** $y = 5 \pm \sqrt{x - 9}$

    c) $y = 7 \pm \sqrt{x + 2}$

13. Restrictions may vary.

    a) $y = x^2 + 3$, $x \ge 0$ **b)** $y = 2(x - 1)^2 + 4$, $x \ge 1$

    c) $y = 3x^2 + 7$, $x \ge 0$

14. a) Yes   **b)** No   **c)** Yes

    d) No   **e)** Yes   **f)** No

15. Explanations may vary.

17. a) $y = \frac{3}{2}(x + 3)(x - 4)^2$ **b)** $y = -\frac{1}{3}x^2(x - 1)(x + 2)$

18. a) 1, 1, −3   **b)** 0, 4   **c)** 3

19. a) 1.6   **b)** −1.1, 0

20. a) $y = \frac{1}{2x + 1}$ **b)** $y = \frac{1}{3x^2 - 5}$ **c)** $y = \frac{1}{3^x - 2}$

21. b) $y = \frac{2x}{x^2 - 9}$

    i) D: $x \ne \pm 3$, R: all real numbers

    ii) $x = 0$

    iii) $x = \pm 3$

iv) Point symmetry; descriptions may vary.

    v) No maximum, no minimum

$$y = \frac{x^2 - 9}{2x}$$

    i) D: $x \ne 0$, R: all real numbers

    ii) $x = 0, \pm 3$

    iii) $x = 0$

    iv) Line symmetry, point symmetry; descriptions may vary.

    v) No maximum, no minimum

22. Answers may vary.

    a) $f(x) = x^2$, $g(x) = 4x^2 + 4x + 1$, $k(x) = g(f(x))$

    b) $f(x) = x^2 + 2x + 1$, $g(x) = \sqrt{x}$, $k(x) = f(g(x))$

    c) $f(x) = \sqrt{x}$, $g(x) = 2x + 1$, $k(x) = f(g(x))$

23. Yes; explanations may vary.

24. a) i) $k < \frac{9}{28}$ **ii)** $k = \frac{9}{28}$ **iii)** $k > \frac{9}{28}$

    b) i) $k < -4\sqrt{21}$, $k > 4\sqrt{21}$ **ii)** $k = \pm 4\sqrt{21}$

    iii) $-4\sqrt{21} < k < 4\sqrt{21}$

    c) i) $k < 69$ **ii)** $k = 69$ **iii)** $k > 69$

25. a) $-3 \le x \le 1$ or $x \ge 2$ **b)** $x < -1$ or $2 < x < 4$

Chapter 5

5.1 Exercises, page 305

1. a) $(-1, 6)$ **b)** $(2, 4)$ **c)** $(4, 3)$ **d)** $(-3, -5)$

    e) $(0, -4)$ **f)** $(4, 5)$ **g)** $(-3, 1)$ **h)** $(-4, 2)$

2. a) $(-2, -6)$ **b)** $(-2, -3)$ **c)** $(-4, 3)$ **d)** $(-3, -5)$

3. Explanations may vary.

4. a) $(30, -20)$ **b)** $(0, 30)$ **c)** $(25, -30)$ **d)** $(-14, 4)$

5. a) $(2, 2)$ **b)** $(9, -2)$ **c)** $(-1, -10)$ **d)** $(3, 2)$

    e) $(6, -2)$ **f)** $(1, 0)$ **g)** $(-4, 7)$ **h)** $(6, -3)$

6. Answers may vary.

    a) $2x + y = 2$, $x - y = 10$ **b)** $x - y = -3$, $2x - y = 2$

    c) $x + y = 4$, $x - y = -10$ **d)** $x - y = 2$, $2x - y = 0$

7. i) $(-3, -4)$; part b   **ii)** $(-2, -3)$, $(6, 1)$; part d

    iii) $(-2, -2)$, $(4, 1)$; part c   **iv)** $(6, -2)$; part a

8. No, the lines could coincide or be parallel.

9. a) $(0, 0)$, $(1, 1)$ **b)** $(2.7, 3.3)$, $(-3.7, 9.7)$

    c) $(2, 2)$; $(-2, 2)$ **d)** $(-2, 2)$, $(6, 6)$

    e) $(1, 6)$, $(-3, -2)$ **f)** $(4, 2)$

10. Explanations may vary.

11. a) $(2, 3)$ **b)** $4x + y = 11$, line passes through $(2, 3)$

    c) $2x - 3y = -5$, line passes through $(2, 3)$

12. a) $(2, 4)$ **b)** $(1, 0)$ **c)** $(-3, -5)$ **d)** $\left(\frac{19}{3}, \frac{1}{3}\right)$

13. Explanations may vary.

    a) 0, 1, 2 or infinitely many points   **b)** 0, 1, or 2 points

    c) 0, 1, or 2 points

Exploring with a Graphing Calculator: Solving Systems of Equations, page 308

1. a) Plan A is better if your sales are less than $6607.14.

    Plan B is better if your sales are greater than $6607.14.

    b) Plan A is better if your sales are less than $5473.68.

    Plan B is better if your sales are greater than $5473.68.

2. a) $(1.71, 3.14)$ **b)** $(6.67, 5.33)$
c) $(4.83, 8.66)$, $(-0.83, -2.66)$ **d)** $(0.41, 1.41)$, $(-2.41, -1.41)$
e) $(-0.42, 0.75)$, $(3.72, 13.15)$ **f)** $(5.30, 2.30)$

3. Rewrite the equations to isolate y.
a) $(2.86, 1.29)$ **b)** $(2.37, 2.63)$, $(-3.37, 8.37)$
c) $(3, 4)$, $(-0.69, -3.38)$

4. Answers may vary.

Investigate, page 310

1. a) $(4, 1)$ **b)** $(4, 1)$
c) All the lines intersect at the same point.
d) Answers may vary; $9x - 3y = 33$, $3x + 6y = 18$

2. Multiplying both sides of either equation of a linear system by a constant does not change the solution.

3. b) It also passes through $(4, 1)$.
d) It also passes through $(4, 1)$.

4. Adding or subtracting the equations of a linear system does not change the solution.

5. Answers may vary.
a) $3x + 2y = 18$, $2x - 2y = 2$
b) $2x + y = 9$, $2x - 6y = 16$
c) $5x - y = 4$, $5x + 20y = 25$

6. Answers may vary.

5.2 Exercises, page 315

1. $3x + 4y = 1$, $5x - 3y = -8$; $2x + 3y = 1$, $4x + 6y = 2$

2. a) $(3, 4)$ **b)** $(-1, 3)$ **c)** $(2, -3)$ **d)** $(4, 3)$
e) $(-2, 3)$ **f)** $(2, -4)$ **g)** $(3, -1)$ **h)** $(3, 7)$

3. a) $(2, -3)$ **b)** $(-1, 5)$ **c)** $(3, -1)$ **d)** $(4, 3)$
e) $(-4, 2)$ **f)** $(-2, -3)$ **g)** $(3, 4)$ **h)** $(2, -1)$

4. Explanations may vary.

5. a) $(4, -2)$ **b)** $(3, 5)$ **c)** $\left(\frac{1}{2}, \frac{1}{3}\right)$ **d)** $(2, 3)$
e) $(-3, -7)$ **f)** $(-2, 1)$ **g)** $\left(\frac{5}{2}, -3\right)$ **h)** $(-6, 3)$

6. a) $\left(\frac{4}{5}, \frac{-3}{5}\right)$ **b)** $\left(\frac{23}{11}, \frac{8}{11}\right)$ **c)** $\left(\frac{3}{14}, \frac{15}{14}\right)$ **d)** $(2, 1)$
e) $\left(\frac{25}{11}, \frac{20}{11}\right)$ **f)** $(-3, 5)$ **g)** $(2, -1)$ **h)** $(4, -6)$

7. a) -0.0062, 27.5 **b)** $T = -0.0062h + 27.5$
c) m represents the temperature change in degrees per metre. b is the temperature at Earth's surface.
d) Domains may vary; $0 \le x \le 5000$. 0 m is sea level and 5000 m is a height high enough, to include most of Earth's surface.

8. a) 0.3, 1600 **b)** $C = 0.3d + 1600$
c) m represents the cost per kilometre to operate the car. b represents the fixed cost to operate the car.
d) Domains may vary. The domain could be 0 to 100 000 km.

9. a) $g + a = 161$, $g + 23 = a$; $g = 69$, $a = 92$ where g is number of goals and a is number of assists
b) Answers may vary.

10. a) The number of apples always depends on the number of trees, t.
b) $1140 = 3600m + 60b$, $1360 = 6400m + 80b$
c) -0.1, 25; $n = -0.1t^2 + 25t$

11. a) Wheat cannot have a value unless it is planted for more than one day.
b) -0.05, 7

12. a) $(2, 3)$ **b)** $(11, 3)$ **c)** $(3, 2)$ **d)** $(10, 4)$
e) $(3, 8)$ **f)** $\left(4, \frac{1}{2}\right)$ **g)** $(8, 2)$ **h)** $\left(\frac{116}{13}, \frac{106}{13}\right)$

13. a) $x + 4y = 7$, $3x + 6y = 9$ **b)** $(-1, 2)$
c) Answers may vary. $2x + 4y = 6$; $x + 3y = 5$, $3x + 5y = 7$ $(-1, 2)$
The solution is the same.

14. 2, 5

15. 5, -2

16. 2, -3

17. a) $(4, 1)$, $(6, 5)$, $(0, 4)$ **b)** $(4, 7)$, $(-2, 4)$, $(7, -2)$

18. a) i) 4, 3 ii) 3, 1 iii) -1, 2 iv) 2, 0
b) Explanations may vary.

19. a) $(3, 4)$ **b)** $(3.07, 3.95)$
c) In this case, an algebraic solution is more accurate than a graphical solution.

20. b) Answers may vary. $(11, 4)$, $(14, 6)$
c) The lines coincide.
d) Explanations may vary.

21. a) Explanations may vary. This is impossible. There are no ordered pairs that satisfy this system.
b) The lines are parallel. They do not intersect.
c) No, there is no solution because the lines do not intersect.

5.3 Exercises, page 321

1. a) No solutions **b)** Infinitely many solutions
c) No solutions **d)** Infinitely many solutions

2. Answers may vary.
a) $2x + y = 10$, $6x + 3y = 5$ **b)** $2x + y = 10$, $2x - 6y = 24$
c) $2x + y = 10$, $4x + 2y = 20$

3. Answers may vary.
a) $3x - 4y = 1$ **b)** $4x - 3y = 12$ **c)** $6x - 8y = 24$

4. a) i) Consistent ii) Consistent
iii) Inconsistent iv) Consistent
b) Explanations may vary.

5. Answers may vary.
a) $4x + 2y = 20$, $x - 3y = 12$; $4x + 2y = 20$, $5x - 15y = -60$
b) $x - 3y = 12$, $5x - 15y = -60$; $4x + 2y = 20$, $2x + y = 20$

6. $x - y = 5$, $3x + 4y = -6$; $4x - y = 11$, $-12x + 3y = -33$; $5x - 3y = 19$, $-2x + 4y = -16$

7. $x - y = 5$, $3x + 4y = -6$; $5x - 3y = 19$, $-2x + 4y = -16$; in parts a and d, the lines intersect at the point $(2, -3)$; in part c, the lines are coincident, so there are infinitely many solutions.

8. Explanations may vary.
  a) Yes   **b)** Yes

9. b) $(n, 9 - 3n)$
  c) Explanations may vary; n can be any real number.
  d) $(n, 4n - 11)$

10. a) All three lines can intersect at one point.
    A transversal can intersect two parallel lines.
    A line can intersect two coincident lines.
    A transversal can intersect two intersecting lines.
    All three lines can be coincident.
  b) i) Inconsistent   **ii)** Consistent   **iii)** Inconsistent

5.4 Exercises, page 325

1. $5x - 3y = -5$, $3x + 2y = 4$; $\frac{3}{2}x + \frac{2}{5}y = -1$,
  $\frac{5}{4}x - \frac{3}{10}y = -4$

2. a) $(2, 7)$ **b)** $(3, -2)$ **c)** $(-1, -8)$ **d)** $(1, 4)$
  e) $(-2, 5)$ **f)** $(-2, 3)$ **g)** $(3, 5)$ **h)** $(4, -2)$

3. Explanations may vary.

4. a) $(-1, 3)$ **b)** $(2, -2)$ **c)** $(2, 1)$ **d)** $(3, 1)$

5. a) $\left(\frac{7}{6}, \frac{1}{12}\right)$ **b)** $(2, -1)$ **c)** $\left(\frac{23}{2}, \frac{-15}{2}\right)$ **d)** $(19, -30)$

6. a) $(-2, 4), (3, 9)$ **b)** $\left(\frac{1}{2}, \frac{1}{4}\right)$ **c)** No solution

7. a) $(1, 1)$; $(1 + \sqrt{2}, 3 + 2\sqrt{2})$, $(1 - \sqrt{2}, 3 - 2\sqrt{2})$; $(3, 9)$,
    $(-1, 1)$; $(1 + \sqrt{6}, 7 + 2\sqrt{6})$, $(1 - \sqrt{6}, 7 - 2\sqrt{6})$
  b) Explanations of patterns may vary. The next two
    equations are $y = 2x + 7$ and $y = 2x + 9$.
    The solutions are $(1 + 2\sqrt{2}, 9 + 4\sqrt{2})$,
    $(1 - 2\sqrt{2}, 9 - 4\sqrt{2})$ and $(1 + \sqrt{10}, 11 + 2\sqrt{10})$,
    $(1 - \sqrt{10}, 11 - 2\sqrt{10})$.
  c) Explanations may vary.

8. a) $(2, 4), (-2, 4)$ **b)** $(1, 1)$ **c)** $(-2, 4), \left(\frac{5}{2}, \frac{25}{4}\right)$

9. a) $(-1, -4), \left(\frac{4}{3}, 3\right)$ **b)** $(-1, -4), (2, 2)$ **c)** $(-1, -4), (4, 1)$

10. a) $(9, 3)$ **b)** $\left(\frac{1}{4}, \frac{1}{2}\right)$ **c)** No solution

11. a) $(-1, 1)$ **b)** No solution   **c)** $\left(-\frac{8}{3}, \frac{8}{3}\right), (8, 8)$

12. Explanations may vary.

13. $(9.6, -2.8)$ or $(6, 8)$

14. 14 cm, 2.5 cm

15. $\sqrt{10}$ cm, $\sqrt{10}$ cm

16. a) About 15 s   **b)** About 3940 m

17. a) i) $(0, 2)$ **ii)** $(0, 3)$
  b) The property in each case is the y-coordinate of the point
    of intersection is c. It occurs because the x-coordinate of
    the point of intersection is 0. The solution is $(0, c)$.

18. -5

19. a) i) $(-2, -3)$ **ii)** $(0.62, 9.61), (2.80, 2.14), (-3.43, -1.75)$
  b) No, explanations may vary.
  c) Yes, explanations may vary.

Mathematical Modelling: Will There Be Enough Food? (Part I), page 328

1. Population will exceed grain production after several years.

2. Population exceeds grain production after about 31 years.

3. a) $(31.58, 10.07)$
  b) i) $n = 31.58$ **ii)** $0 \leq n < 31.58$ **iii)** $n > 31.58$

4. After about 31.58 years, there will not be enough grain to
  support the world population.

5. It is assumed that population growth and grain production
  will follow these growth equations.

6. Descriptions may vary.

7. a) 1 billion   **b)** About 256 billion

8. The time for the population to double should be longer;
  explanations may vary.

9. Answers may vary.

Linking Ideas: Mathematics and Technology
Solving a Two-Variable Linear System, page 330

2. Yes

4. Yes

5. Cone costs $1.75, drink costs $0.95; $2.70

6. CD costs $2.75, tape costs $1.85; $4.60

7. a) 1.17, 9976.14
  b) m is the expansion in millimetres per degree Celsius; b is
    the length in millimetres of the girder at $0°$ C

8. 0.148, 0.22

9. a) Answers may vary. $2x + 2y = 4$, $x + y = 2$; $x + y = 2$,
    $x + y = 6$; $0x + 3y = 10$, $2x + 5y = 12$
  b) Explanations may vary.

10. a) Answers may vary. $x + y = 2$, $2x + 2y = 4$;
    $2x + 3y = 10$, $4x + 6y = 15$; $0x + 3y = 10$,
    $2x + 5y = 12$
  b) Explanations may vary.

5.5 Exercises, page 334

1. $16, $28

2. $16, $12

3. $45.00

4. $150.00, $6.00

5. $35.00, $50.00

6. 380 adult tickets and 520 student tickets

7. 100

8. 150 adult tickets and 300 student tickets

9. $15.00 for between the goal lines, $10.00 for end-zone

10. 40 g, 15 g

11. 1 kg, 83 g

12. 150 of World Oil and 300 of Zinco Mines

13. $200 at 7%, $300 at 10%

14. $550 at 9%, $250 at 12%

15. $1500 at 8%, $1000 at 12%

16. 97 B737-300s and 25 B737-500s

17. $175 at 9%, $325 at 11%

18. $15/day, $0.10/km

19. 400 students attended on the first night, 450 students attended on the second night

20. 1188 km

21. Wind speed: 48.5 km/h, speed of aircraft in still air: 727.5 km/h

22. $6.90/kg for tea, $3.85/kg for coffee

23. $6.40, $9.60

Investigate, page 337

1. (1, 2, 3)

3. Explanations may vary. Every situation is different, but using addition, subtraction, multiplication and/or substitution will work. Linear systems in 3 variables can be solved like linear systems in 2 variables.

5.6 Exercises, page 341

1. $x = -1$, $y = 3$, $z = -5$; explanations may vary.

2. $x + y + z = -4$, $x - y + 2z = -13$, $2x + y - 3z = 15$; explanations may vary.

3. a) (3, 4, 1) **b)** (2, 1, −1) **c)** (5, −2, 3)

4. Explanations may vary.

5. a) (1, 2, 3) **b)** (−4, 3, −1) **c)** (−1, 2, 3)

6. a) (2, −1, 4) **b)** (0, 2, 1) **c)** (3, 5, −1)

7. a) (2, 1, −3) **b)** $\left(-6, 3, -\frac{1}{2}\right)$ **c)** $\left(\frac{1}{3}, -2, \frac{3}{5}\right)$

8. Answers may vary.
For part a: $2x + y + z = 2$, $x - y + 3z = -7$, $3x + 2y + 2z = 3$
For part b: $x + y + z = 2$, $2x + 3y + 2z = 3$, $x + 2y + 3z = 1$

9. 8, 2, 5

10. −2, 3, 1

11. a) −5, 350, 0 **b)** $r = -5p^2 + 350p$
c) Answers may vary. $10 \le p \le 70$

12. a) −500, 500, 35 000 **b)** $35 125
c) Answers may vary. $0.50 \le x \le 5$

13. a) Answers may vary. $3x - 2y + 4z = 7$, $x - y + z = 1$, $x + y - z = -3$
b) Yes, explanations may vary.

14. a) 0, 1, 3 **b)** No, explanations may vary.

15. Answers may vary.
a) Three planes can be situated as parallel planes, much like the three levels in an apartment building. The floors are not connected. The three planes could coincide, so that any point in one plane is also in the other two planes. Two planes could be parallel, with the third plane intersecting the other two planes, like the sides of a box.

b) Answers may vary. $x + y + z = 3$, $x + y + z = 4$, $x + y + z = 5$; parallel
c) Answers may vary. $x + y + z = 3$, $2x + 2y + 2z = 6$, $3x + 3y + 3z = 9$; coincident

Linking Ideas: Mathematics and Technology
Solving a Three-Variable Linear System, page 344

2. Yes

4. Yes

5. Poster costs $3.95, pennant costs $4.50, cap costs $5.75; $14.20

6. Shirt costs $18.35, sweater costs $32.15, coat costs $47.75; $98.25

7. 1.69, −125.15, 4182.53

8. a) i) 1 **ii)** 5 **iii)** 14
b) $a = \frac{1}{3}$, $b = \frac{1}{2}$, $c = \frac{1}{6}$

9. Answers may vary.
a) $x + y + z = 1$, $2x + 2y + 2z = 2$, $3x + 3y + 3z = 3$;
$x + y + z = 1$, $x + y + z = 2$, $x + y + z = 3$;
$0x + 3y + z = 4$, $5x + 6y + 3z = 5$, $4x - y + 3z = -1$
b) Explanations may vary.

10. a) $x + y + z = 1$, $2x + 2y + 2z = 2$, $3x + 3y + 3z = 3$;
$x + y + z = 1$, $x + y + z = 2$, $x + y + z = 3$;
$0x + 3y + z = 4$, $5x + 6y + 3z = 5$, $4x - y + 3z = -1$
b) Explanations may vary.

5.7 Exercises, page 350

1. Answers may vary.
a) (1, 1) **b)** (10, 0) **c)** (0, −2)
d) (2, 1) **e)** (0, 5) **f)** (1, 1)

2. Answers may vary.
a) (−1, 0) below **b)** (1, 1) below **c)** (−2, 2) above

3. a) $x + 3y < -7$ **b)** $7x - 2y > 10$
c) $3x + y < -4$ **d)** $x - 3y > -6$

5. Explanations may vary.

5.8 Exercises, page 354

1. a) $x \ge 0$, $y \ge 0$ **b)** $x \ge -4$, $y \ge -2$
c) $x \ge 0$, $x + y \le 5$ **d)** $x + y \le 5$, $x - y \le -1$
e) $x - 2y \le -4$, $5x + 3y \ge 15$
f) $2x + 5y \le 100$, $2x - 5y \ge 100$

2. a) $y > 2x + 3$, $y > 0.5x + 1$ **b)** $3x + 2y < -5$, $2x - 3y > 3$

3. a) Square; 9 square units **b)** Rectangle; 28 square units
c) Trapezoid; 6 square units

4. a) 48 square units **b)** 56 square units
c) 161 square units

5. Explanations may vary.

6. a) $3x + 2y \le 6$, $3x + 2y \ge -6$
b) $x + 2y \le 4$, $x - y \ge -2$, $x - y \le 4$
c) $3x - y \ge -3$, $x - 2y \le 2$, $3x + 2y \ge 6$

Linking Ideas: Mathematics and Technology

Solving Linear Systems, page 357

1. The solutions are shown in the square brackets on the right side of the screen.

2. **a)** $x = \dfrac{-(bf - ce)}{ae - bd}$, $y = \dfrac{af - cd}{ae - bd}$

   b) Explanations of patterns may vary.

   c) Yes, explanations may vary.

   d) $ae \neq bd$; if $ae = bd$, then the denominator is 0, which would produce a result that is not defined.

3. Formulas are practical when they are simple to use. In some cases, solving the problem through other methods may be more straightforward, as there would be fewer complicated calculations.

Problem Solving: Maximum-Minimum Problems, page 358

1. 20 footballs and 10 soccer balls

2. **a)** 30 footballs and 0 soccer balls

   b) 0 footballs and 15 soccer balls

3. **a)** 20 motorcycles and 20 bicycles

   b) 10 motorcycles and 30 bicycles

4. **a)** 0 ha of corn and 80 ha of wheat

   b) 20 ha of corn and 60 ha of wheat

   c) 40 ha of corn and 0 ha of wheat

5. **a)** 20 ha of wheat, 20 ha of barley

   b) 0 ha of wheat, $33\frac{1}{3}$ ha of barley

Mathematical Modelling: Will There Be Enough Food? (Part II), page 360

1. **a)** 120 960 000   **b)** 120 960 000

   c) 80 640 000   **d)** 0

2. **a)** 10 billion   **b)** 0 or 20 billion

3. **a)** 80 years   **b)** 150 years   **c)** 220 years   **d)** 330 years

4. Predictions may vary.

5. Prediction is not confirmed; explanations may vary.

6. **c)** **i)** $n = 51$ and $n = 110$

   ii) $n < 51$, $n > 110$

   iii) $51 < n < 110$

7. Compared to the logistic model, the linear model predicts higher populations for less than 51 years or more than 110 years, and lower populations for numbers of years between 51 and 110.

5 Review, page 364

1. **a)** $(-3, -5)$ **b)** $(3.5, 1)$ **c)** $(-2.8, 0.4)$

   d) $(-0.6, 2.2)$ **e)** $(4, 1)$ **f)** $(2.5, -1.5)$

2. Explanations may vary.

3. Answers may vary. $3x + 2y = 1$, $x + y = 1$; $x - y = -3$, $2x + y = 0$

4. Answers may vary.

   a) $x + y = 3$, $x - y = -7$ **b)** $x + y = 3$, $x - y = 5$

   c) $2x + y = 16$, $x - 2y = 3$ **d)** $x + 2y = -17$, $x - y = 1$

5. **a)** $(-1, -1)$ **b)** $\left(\dfrac{5}{3}, 0\right)$ **c)** $(-2, -1)$

   d) $(0, 2)$ **e)** $(3, -2)$ **f)** $(-4, 3)$

6. **a)** $(-2, -3)$ **b)** $(4, -3)$ **c)** $(20, -6)$

   d) $(4, -3)$ **e)** $(0, 1)$ **f)** $\left(-\dfrac{1}{6}, \dfrac{7}{12}\right)$

7. Explanations may vary.

8. **a)** $(1, 2)$, $\left(-\dfrac{1}{2}, -4\right)$

   b) $(-4.77, 22.77)$, $(3.77, 14.23)$

   c) $\left(\dfrac{3}{2}, \dfrac{3}{2}\right)$

   d) $(6.56, 2.56)$

9. $2.75/L, $0.63/L

10. $28.00/day, $0.25/km

11. $1500.00 at 7% and $3000.00 at 8.5%

12. **a)** $(1, 3, 1)$ **b)** $(2, -3, -1)$ **c)** $(-4, -3, 2)$

13. **a)** **i)** $x + y < 3$ **ii)** $x + 2y \geq -2$ **iii)** $x - y \leq -2$

    b) **i)** $x + y > 3$ **ii)** $x + 2y \leq -2$ **iii)** $x - y \geq -2$

16. **a)** $4x + 5y \leq 20$, $3x - 4y \leq -3$

    b) $2x + 3y \geq 12$, $x - 2y \leq 6$

    c) $x - 2y \geq -2$, $x + 4y \leq 20$

5 Cumulative Review, page 366

1. **a)** $1562.40   **b)** $1558.41

2. $0.60

3. **a)** Explanations may vary.   **b)** $16.12

5. $0.90

6. Answers may vary.

   a) $f(x) = (x + 4)(x - 3)(x + 2)(x - 1)$

   b) $f(x) = (x - 1)^2(x - 3)(x - 5)$

   c) $f(x) = (x + 2)(x - 6)(x + 5)(x - 4)$

   d) $f(x) = (x + 2)^2(x - 3)^2$

7. **a)** $h = 9.3 - \dfrac{540}{\pi(4.3 + r)^2}$

   c) **i)** 1.84 cm   **ii)** 3.18 cm   **iii)** 4.19 cm

   d) **i)** 0.12 cm   **ii)** 0.25 cm   **iii)** 0.39 cm

8. $3, -1$

9. **a)** $d = \dfrac{v^2}{202.86}$

   b) **i)** 17.8 m   **ii)** 39.9 m   **iii)** 71.0 m

   c) **i)** 45 km/h   **ii)** 71 km/h   **iii)** 101 km/h

10. **a)** Package A: $S = x + 1000y$

    Package B: $S = x - 10\,000 + 1500y$

    b) The salaries are equal when $y = 20$.

    c) Based on the 1995-96 results, Package A would be better for the 24 players who scored fewer than 20 goals. Package B would be better for the 38 players who scored more than 20 goals. The 4 players who scored exactly 20 goals would earn the same salary with both packages. To make a recommendation for these players, you could look at several seasons of data. If the player's goals scored have increased, you might recommend Package B.

11. $555.56 at 8% and $444.44 at 10%

Chapter 6

6.1 Exercises, page 371

1. **b)** The products of two 2-digit numbers are 3-digit or 4-digit numbers.

2. **b)** The sums of consecutive odd numbers starting at 1 are perfect squares.

3. **d)** The sum of the angles in a triangle is 180°.

4. Explanations may vary.

5. **a)** Each angle is 60°.
   b) The sum of the two acute angles is 90°.

6. **d)** Alternate angles are equal; corresponding angles are equal; interior angles are supplementary.

7. **a)** $MN = \frac{1}{2}BC$; MN is parallel to BC
   b) For any $\triangle ABC$, if M and N are the midpoints of AB and AC, respectively, $MN = \frac{1}{2}BC$ and MN is parallel to BC.

8. **b)** Parallelogram
   c) The figure formed by joining the midpoints of the four sides of any quadrilateral is a parallelogram.

6.2 Exercises, page 375

1. **a)** Teenager   **b)** Long hair   **c)** Wet
   d) Melt   **e)** Four right angles   **f)** Never meet

2. **a)** Manuel will be fit.
   b) Jennifer's writing will improve.
   c) $\angle B + \angle C = 90°$
   d) The measure of $\angle Q$ is 45°.
   e) Some rectangles have four equal sides.

3. **a)** Yes   **b)** No   **c)** Yes
   d) No   **e)** Yes   **f)** No

4. **a)** Barbados, Trinidad and Tobago, Bahamas, Jamaica
   b) Exercises may vary.

5. **c)** They are equal since they are alternate angles.
   d) They are equal since they are alternate angles.
   e) Explanations may vary.

6. The sum of the three angles in a triangle is 180°. It follows that the sum of two angle measures must be less than 180°.

7. Explanations may vary.

8. Explanations may vary.

9. Proofs may vary.

10. Proofs may vary.

11. Elise, Brittany, Alice, Dini, Carol

12. Proofs may vary.

Mathematical Modelling: Incredible Journeys, page 378

1. **a)** Answers may vary; using a ruler to draw a line, edge of a piece of paper, a crease in the palm of your hand.
   b) If the model is a crease in the palm of your hand, the diagrams would match those on pages 378, 379.

2. Answers may vary; the back of your hand. The sequence of diagrams would be the same as those on pages 378, 379.

Mathematics File: A Famous Example of Intuitive Reasoning, page 380

1. 5¢

2. No

3. 68.6 km/h

4. **a)** Suggestions may vary; ships seem to disappear below horizon, the sun rises in the East and sets in the West
   b) Lists may vary; fly into space.

Investigate, page 381

2. Natural numbers may vary; $3 = 1 + 2$; $5 = 2 + 3$; $7 = 3 + 4$

3. Natural numbers may vary; 2, 4, 8; there are no consecutive integers that add to each number.

6.3 Exercises, page 382

1. Answers may vary.
   a) 25 cannot be written as the sum of 3 perfect squares.
   b) A triangle with angles 120°, 30°, 30° is possible
   c) A baseball diamond viewed from any base is a square that does not have two horizontal and two vertical sides.

2. Parts a and b

3. **a)** False; 0
   b) True; explanations may vary
   c) False; $1^2 = 1$
   d) True; explanations may vary
   e) False; an altitude of an obtuse triangle lies outside it
   f) False; any rectangle in which its length is not equal to its width is not a square
   g) True; explanations may vary

4. **a)** Answers may vary; triangle with sides 7, $4\sqrt{2}$, 9
   b) Explanations may vary.

5. **a)** **i)** 117 shots, 16 goals   **ii)** 160 shots, 23 goals
   iii) 217 shots, 30 goals   **iv)** 267 shots, 40 goals
   v) 294 shots, 47 goals   **vi)** 403 shots, 62 goals
   The more shots a player takes, the more goals he scores.
   b) Counterexamples may vary. For example, in the 1995-96 season, Keith Tkachuk scored more goals (50) than Lindros or Selanne but took fewer shots (249).

6. **a)** Answers may vary; $20 = 17 + 3$; $28 = 23 + 5$; $30 = 23 + 7$
   b) Find an even number that cannot be written as the sum of two primes.

7. **a)** No; 1   **b)** No; 2

8. The boiling temperature of water varies with altitude, air pressure, and other factors.

9. **a)** x^x is defined when $x = -2$, since $(-2)^{-2} = \frac{1}{4}$
   b) x^x is defined when $x = -5$

10. a) For negative x-axis; the values of x are: $-0.2, -0.4, -0.6, \ldots$
   For positive x-axis; the values of x are: $0.1, 0.2, 0.3, \ldots$
   d) Explanations may vary.

6.4 Exercises, page 391

1. There is a treasure in chest #2.

2. a) There is a treasure in chest #1.
   b) There is not a treasure in chest #1 or chest #2.

3. a) $x \neq 7$ b) $x \geq 10$
   c) x is not a multiple of 2.   d) x is a factor of 24.
   e) $\angle B$ is not a right angle.   f) $\angle B$ is not an acute angle.

4. a) 11   b) 4   c) 10

5. It contains sugar or glucose or both.

7. a) Multiples of 20: 20, 40, 60, 80, 100, …
   b) Multiples of 10: 10, 20, 30, 40, 50, …
   c) 2, 8, 32

8. Explanations may vary.

9. a) 1, 2, 3, 4, 6, 9, 12, 18   b) 1, 3, 5, 7, 21, 25
   c) 4, 8, 10, 12, 16, 20, 24, 28, 30, 32, …

16. Part d

17. Parts a, b and d

18. Andrea, red; Betty, blue; Carol, white

19. 29

20. 5

21. a) 3   b) 14   c) 24   d) 31

22. 218

23. a) Titov, centre; Mellanby, left wing; LeClair, right wing
   b) Exercises may vary.

24. Part c

25. The answer would become 3, 5, 6, 10, 12, and 20. The overlapping region in the Venn diagram would not be shaded.

26. Both are truth-tellers.

27. "I will be hanged."; explanations may vary

6.5 Exercises, page 398

1. The treasure in chest #1 is worth at least $1 000 000.

2. a) Nothing
   b) The treasure is in chest #2, and is worth only $100.

3. a) True; if you live in Alberta, then you live in Red Deer; false
   b) False; if it is raining, then the ground is wet; true
   c) False; if you pass the test, then you studied; false
   d) False; if you don't win the lottery, then you didn't buy a ticket; false
   e) False; if you can't go out in a canoe, then you can't swim; false

4. a) If you don't live in Alberta, then you don't live in Red Deer; true
   b) If it is not raining, then the ground is not wet; false
   c) If you don't pass the test, then you didn't study; false
   d) If you win the lottery, then you bought a ticket; true

e) If you go out in a canoe, then you can swim; false

5. a) False; If two rectangles have the same length, then they have the same area; false
   b) False; If two positive numbers are both greater than 10, then their product is greater than 10; true
   c) True; If a right triangle contains a 60° angle, then it also contains a 30° angle; true
   d) True; If a triangle is not equilateral, then it does not have three equal sides; true
   e) True; All parallelograms are rectangles; false

6. a) If two rectangles do not have the same length, then they do not have the same area; false
   b) If two numbers are not greater than 10, then their product is not greater than 10; false
   c) If a right triangle does not contain a 60° angle, it does not contain a 30° angle; true
   d) If a triangle is equilateral, then it has three equal sides; true
   e) If a quadrilateral is not a parallelogram, it is not a rectangle; true

7. a) If I don't win the lottery, then I can start work on Monday.
   b) If I start work on Monday, then I didn't win the lottery.
   c) If I don't start work on Monday, then I won the lottery.

8. a) If anything can go wrong, then it will go wrong.
   b) If there's a will, then there's a way.
   c) If it isn't over, then it isn't over.
   d) If you eat one potato chip, then it's impossible to eat only one potato chip.

9. Parts a, b, e, and h

10. a) True; multiples of 3 are always multiples of 6; false
   b) True; if $2x$ is not a prime number, then x is a prime number; false
   c) True; a multiple of 2 is a multiple of 4; false
   d) False; a number greater than 1 is the square of a positive number; false
   e) False; any positive number is the square root of a number; true

11. a) If a number is not a multiple of 3, then it is not a multiple of 6; true
   b) If $2x$ is a prime number, then x is not a prime number; false
   c) If a number is not a multiple of 2, then it is not a multiple of 4; true
   d) If the square of a number is not greater than 1, the number is not positive; false
   e) If the number is not positive, it is not the square root of any number; false

12. Part c

13. The server

14. a) If you are not in a designated area, then do not smoke.
   b) Explanations may vary.

15. a) 3   b) 1   c) 2

16. The one you ask is a truth-teller, the other is a liar.

17. Part b or c; for part b: choose the opposite road of either response; for part c: if answer is yes, then choose the right road, if answer is no, choose the left road

Linking Ideas: Mathematics and Sports
Logical Connectives in Sports Rules, page 401

1. Sports chosen may vary.
   If the ball hits the net on a serve and falls into the correct service court, then a let is called.
   If the home team is ahead after $8\frac{1}{2}$ innings, then the game is over.
   If a team gains control of the ball, then it must try for a basket within 30 s.

2. Answers may vary.

6.6 Exercises, page 406

1. Explanations may vary.
   a) $\angle BAC = \angle CAD$, $AC = AC$, $\angle ACB = \angle ACD$ (ASA); $AC = AC$, $AB = AD$, $BC = DC$; $\angle BAC = \angle DAC$, $\angle BCA = \angle DCA$, $\angle B = \angle D$
   b) $PQ = PS$, $\angle QPR = \angle SPR$, $PR = PR$ (SAS); $PQ = PS$, $PR = PR$, $QR = SR$, $\angle QPR = \angle SPR$, $\angle PRQ = \angle PRS$, $\angle Q = \angle S$
   c) $XZ = XY$, $\angle ZXW = \angle YXW$, $XW = XW$ (SAS); $XZ = XY$, $XW = XW$, $ZW = YW$; $\angle YXW = \angle ZXW$, $\angle XWY = \angle XWZ$, $\angle Z = \angle Y$

2. a) $\triangle ABC$, $\triangle GHJ$; ASA   b) $\triangle KLM$, $\triangle NQP$; SAS

3. a) $\angle J = \angle B$ or $CR = YP$ b) $SM = PX$ or $\angle D = \angle Q$

5.–10. Explanations may vary.

11.–30. Proofs may vary.

Problem Solving: Logic Puzzles, page 410

1. $12.50
2. 3 socks
3. The green chest
4. The red chest
5. Explanations may vary.
6. Max King, Louise Port, Ken Laird
7. Ms. Vrentzos is the mother of the woman in the picture.
8. The green card and the card with a square
9. A liar
10. She's a liar and he's a truth-teller.
11. a) No; explanations may vary
    b) Yes; explanations may vary
    c) Explanations may vary.
    d) Yes; explanations may vary
    e) No; explanations may vary

6.7 Exercises, page 414

For all exercises, proofs may vary.

Linking Ideas: Mathematics and History
Measuring Earth's Circumference, page 416

1. Alternate angles are equal.
2. They are equivalent fractions.
3. a) 240 000 stades   b) 37 680 000 m; 37 680 km
4. a) 40 000 km   b) Within 2320 km
5. Explanations may vary.

6 Review, page 418

1. a) Products may vary.
   b) The product of two odd numbers is always an odd number.
2. b) $360°$
   d) The sum of the angles in a quadrilateral is $360°$.
3. Explanations may vary.
4. Explanations may vary.
5. a) $2^4 - 1 = 15$ is not a prime number
   b) Triangles may vary.
6. a) 5   b) 13   c) 12
7. a) 2, 6, 18, 54   b) 1
   c) 1, 2, 4, 5, 7, 8, 14, 28, 35, 56
   d) 7, 8, 14, 16, 21, 24,…
8. Explanations may vary.
9. a) True   b) True   c) False   d) False
10. a) If a quadrilateral is a rhombus, then it is a square; false
    b) If a quadrilateral is a parallelogram, then it has two pairs of parallel sides; true
    c) If a plant grows, then you have watered it; false
    d) If it is light outside, then it is noon; false
11. a) If a quadrilateral is not a rhombus, then it is not a square; true
    b) If a quadrilateral is not a parallelogram, then it does not have two pairs of parallel sides; true
    c) If a plant does not grow, then you have not watered it; false
    d) If it is not light outside, then it is not noon; false
12. Proofs may vary.
13. Proofs may vary.
14. Proofs may vary.
15. Proofs may vary.

6 Cumulative Review, page 420

1. a) $8630.73   b) $3426.67
2. a) $609.50   b) $1337.55
3. Explanations may vary.
4. Plan a; explanations may vary: plan a, $1257.32; plan b, $1248.18
5. a) $723.79   b) $736.08
6. a) i) $x = -1$ ii) $(-1, -3)$ iii) $-2, 0; 0$
      iv) D: all real numbers; R: $y \geq -3$
   b) i) $x = 2$ ii) $(2, 1)$ iii) $1, 3; -3$
      iv) D: all real numbers; R: $y \leq 1$

7. a) 10 000　　　　　　　　**b)** $40

8. a) D: all real numbers; R: $y \geq -16.3$
　　b) D: all real numbers; R: $y \geq 0$
　　c) D: all real numbers; R: $y \geq -0.1$

9. a) 27.60 cm by 22.60 cm by 1.20 cm
　　b) 20.94 cm by 15.94 cm by 4.53 cm

10. a) $4x^2 + 12x + 9$, $2x^2 + 3$　　**b)** $2x, \frac{x}{2}$

11. Explanations may vary.

12. a) 8　　　　　**b)** 20　　　　　**c)** 52
　　d) 20　　　　　**e)** 32　　　　　**f)** 110

13. $x^3 - x^2 - 10x - 8$; $2x^3 - x^2 - 7x + 6$

14. a) 0.94, −2.27　**b)** 1.65, −0.15　**c)** 0.64, −1.38　**d)** 0.23, 1.31

15. Explanations may vary.

16. a) Ethiopia, Algeria, Cameroon, Benin, Djibouti
　　b) Exercises may vary.

17. Proofs may vary.

Chapter 7

7.1 Exercises, page 428

1. a) OA, OB, OC　　　　　**b)** AC
　　c) AB, AC, AD　　　　　**d)** AB, AD
　　e) AB, BC, CD, AD, BD　**f)** AB, AD, BD, CB
　　g) AB, AD　　　　　　　**h)** AOB, BOC
　　i) ∠AOB, ∠BOC

2. Answers may vary.

3. Explanations may vary.

4. Answers may vary.
　　a) 1.8 cm; 4.2 cm　　　**b)** 0.3 mm
　　c) 45.2 cm^2　　　　　**d)** 2.5 cm^2

5. a) Yes　　　　　　　　　**b)** Yes

6. 1 point, 2 points, or an infinite number of points

7. Answers may vary, depending on the interpretation of the questions.

8. The centre of the circle

9. 1 point, 2 points, or an infinite number of points

10. a) i) 51 cm　　　**ii)** $48\sqrt{2}$ cm　　　**iii)** $18\sqrt{3}$ cm
　　b) Explanations may vary.

11. $24\sqrt{2}$ cm

12. d) The longest chord in a circle is a diameter.

13. b) Rectangle　　**c)** Yes　　　　**d)** Square

14. a) An infinite number of points　　**b)** 0

15. a) $P = \frac{8\pi r}{3}$　　　　　**b)** $P = 3\pi r$

Linking Ideas: Mathematics and Science
Circles in Space, page 431

1. Answers may vary.

2. Explanations may vary.

3. Answers may vary.

Investigate, page 432

1. a) 90°　　　　　　**b)** ∠OCA = 90°

2. Answers may vary.
　　a) AC = BC　　　**b)** AC = BC

3. a) O　　　　　　　**b)** It passes through O.

7.2 Exercises, page 435

1. a) 8　　　　　**b)** 5　　　　　**c)** $6\sqrt{5}$
　　d) $2\sqrt{7}$　　**e)** $\sqrt{5}$　　　　**f)** 6

2. a) 9.3　　　　**b)** 4　　　　　**c)** $\sqrt{11}$

3. Explanations may vary.

4. 6.1 cm

5. a) 3.0 cm　　**b)** 8.0 cm　　**c)** $4\sqrt{5}$ cm　　**d)** $\sqrt{5}$ cm

6. d) They lie on a semicircle.

7. b) A larger arc. The radius of the circle would be greater.
　　c) A smaller arc. The radius of the the circle would be smaller.

8. a) 8　　　　**b)** $5\sqrt{2} - 5$　　**c)** $\sqrt{17}$

9. a) i) 6　　　　**ii)** $\sqrt{85}$　　　**iii)** $\frac{5\sqrt{3}}{2}$
　　b) Explanations may vary.

10. Explanations may vary.

11. Explanations may vary.

12. c) 8 cm

13. $\sqrt{39}$ cm

14. Parts a, b, and c
　　a) $3\sqrt{11}$ cm　　**b)** $5\sqrt{3}$ cm　　**c)** $\sqrt{19}$ cm

15. $2\sqrt{20}$ cm

16. Part a　**a)** $2\sqrt{96}$ cm

17. 13.3 cm

18. 34 cm

19. Explanations may vary.

20. 77 m

21. 10 cm

22. a) 10 cm　　**b)** 14 cm　　**c)** Not enough information
　　d) 24 cm　　**e)** $6\sqrt{5}$ cm

23. The fuel line, at 0.014 cm, must be replaced.

24. a) Equal　　　　　　　　**b)** Equal
　　c) Equal chords are equidistant from the centre.

25. a) No　　　　**b)** No　　　　**c)** No

26. b) Chords of different lengths are different distances from the centre.

27. a) 12　　　　**b)** 5　　　　**c)** $6\sqrt{2}$

28. Explanations may vary.

29. Explanations may vary.

7.3 Exercises, page 444

1. Explanations may vary.

2. Explanations may vary.

3. If two chords of a circle are equidistant from the centre, they have the same length.

Mathematical Modelling: How Can We Map Earth's Surface?, page 446

1. No, explanations may vary.

2. 23 vertical and 11 horizontal lines

3. The North and South poles

4. The polar regions. Explanations may vary.

5. **b)** Winter. The southern polar region is illuminated.
   c) The northern polar region would be illuminated.

6. Answers may vary.

7. The illuminated area would have shifted to the left.

Investigate, page 448

1. **a)** $\angle C = 90°$ **b)** $\angle C = 90°$

2. **a)** $\angle C = \angle D$ **b)** $\angle C = \angle D$ **c)** $\angle C = \angle D$

3. **a)** $\angle AOB = 2\angle AOC$ **b)** $\angle AOB = 2\angle AOC$
   c) Reflex $\angle AOB = 2\angle ACB$

7.4 Exercises, page 452

1. **a)** $60°$ **b)** $90°$ **c)** $35°$
   d) $64°$ **e)** $30°$ **f)** $220°$

2. **a)** $x = 65°$, $y = 65°$ **b)** $x = 80°$, $y = 40°$
   c) $x = 50°$, $y = 55°$ **d)** $x = 110°$, $y = 35°$
   e) $x = 25°$, $y = 35°$ **f)** $x = 90°$, $y = 38°$

3. Explanations may vary.

4. **a)** Explanations may vary.
   b) The net must be a chord of a circle, where the players shoot from the circumference on the major (or minor) arc.

5. $65°$

6. $\angle DAB = 105°$, $\angle ADC = 105°$

7. **a)** $30°$ **b)** No, explanations may vary.

8. **a)** **i)** $x = 90°$, $y = 50°$, $z = 40°$
   ii) $x = 28°$, $y = 62°$, $z = 90°$
   iii) $x = 45°$, $y = 90°$, $z = 45°$
   iv) $x = 90°$, $y = 50°$, $z = 45°$
   v) $x = 55°$, $y = 110°$, $z = 35°$
   vi) $x = 30°$, $y = 30°$, $z = 50°$
   b) Explanations may vary.

9. 8.5 cm

10. **a)** Rectangle   **b)** 12 cm   **c)** 192 cm^2

11. Explanations may vary.

12. Circle; explanations may vary.

13. **c)** Answers may vary.

14. **a)** $\angle ROS$ is the central angle for $\angle RQS$.
   b) **i)** $\triangle QRO$ is a right isosceles triangle
   ii) $\triangle QOS$ is right isosceles.
   iii) $\triangle ROS$ is isosceles.   **iv)** $\triangle QRS$ is isosceles.
   c) Explanations may vary.

15. **a)** $\triangle PAC \sim \triangle PDB$ **b)** Similar triangles
   d) $\triangle PAC$ and $\triangle PBD$ are
   i) Isosceles   **ii)** Right   **iii)** Right isosceles

16. **b)** Isosceles; $\angle OAC = \angle OCA$
   c) Isosceles; $\angle OBC = \angle OCB$ **d)** $180°$
   e) Explanations may vary.

Problem Solving: Paradoxes and Proof, page 456

1. Explanations may vary.

2. Explanations may vary.

3. Neither. Self-reference is involved.

4. Neither. Self-reference is involved.

5. Undecidable

6. Undecidable

7. Explanations may vary.

7.5 Exercises, page 460

6. **b)** $\triangle PBQ$ is isosceles.   **c)** $\triangle PBQ$ is equilateral.

7. P can be anywhere on the major arc AB of one circle and Q anywhere on the major arc AB of the other circle.

Linking Ideas: Mathematics and Technology
Dynamic Circle Designs, page 462

2. **a)** 2.3 cm; explanations may vary
   b) Answers may vary. 1.2 cm, 3.5 cm; 0 cm, 4.6 cm; 1.8 cm, 6.3 cm
   c) Smallest diameter $= |2.3 - 2x|$, largest diameter $= 2.3 + 2x$

3. Explanations may vary.

7 Review, page 464

1. **a)** 24   **b)** $2\sqrt{6}$ **c)** $\sqrt{41}$

2. **a)** 6   **b)** $\dfrac{\sqrt{7}}{2}$ **c)** $4\sqrt{2}$

3. **a)** $4\sqrt{6}$ **b)** $4\sqrt{2}$ **c)** $45°$

4. Explanations may vary.

6. 7.5 cm

7. $\dfrac{\sqrt{5}}{2}$ cm

8. $9\sqrt{2}$ cm

11. 24 cm

14. **a)** Chords are concurrent.
   b) The circles are congruent and the chords are the same length.

7 Cumulative Review, page 466

1. $3724.86

2. a) $2329.92   b) $993.06

3. Explanations may vary.

4. $6\frac{3}{4}\%$ compounded annually

5. $14 000

6. $756.80, $181 632.00

7. a) $f(x) = 7x^2 - 20x - 3$ b) $f(x) = 8x^2 + 6x + 1$
   c) $f(x) = 3x^2 - 4x - 4$

8. a) 5   b) Maximum   c) 0
   d) D: all real numbers, R: $y \le 5$

9. a) $-4, 1, 3$ b) $-2, 2, 3$ c) -3

10. Explanations may vary.

11. a) $y = \dfrac{1}{\sqrt{x^2 - 5}}$ b) $y = \dfrac{1}{-4x^3 + 2x - 5}$ c) $y = \dfrac{1}{x^4 + 1}$

12. Explanations may vary.

13. a) $-3.67, 4.33$ b) $0.67, 1.75$ c) $-0.8, 0.75$

14. a) 2 different real roots   b) 2 equal real roots
    c) No real roots   d) 2 different real roots

15. Explanations may vary.

16. a) 3, 7   b) $-1, 4$ c) $-2, 3$

17. a) $x < -4$ or $x > 0$ b) $x \le -6$ or $x > -1$
    c) $-\frac{1}{2} < x \le \frac{5}{2}$

18. b) $x > -1$

19. a) $(-3, -3)$ b) $\left(-\frac{7}{3}, \frac{3}{2}\right)$ c) $(3, 1)$

20. Adult ticket costs $6.00, child ticket costs $2.50

21. a) Right triangle, area $= \frac{25}{4}$ units^2
    b) Isosceles triangle, area $= \frac{25}{12}$ units^2

22. a) Fluffy has four paws.
    b) February has more than 27 days.

23. a) 2, 6, 18, 54   b) 1, 2, 4, 5, 7, 8, 14, 28, 56

24. Explanations may vary.

Chapter 8

Investigate, page 470

1. a) $\angle$B , $\angle$D   b) They are supplementary.

2. b) $\angle$CBE $= \angle$CDA

8.1 Exercises, page 472

1. a) $x = 70°$, $y = 80°$ b) $x = 105°$, $y = 90°$
   c) $x = 115°$, $y = 100°$

2. Explanations may vary.

3. a) $x = 50°$, $y = 60°$ b) $x = 100°$, $y = 140°$
   c) $x = 70°$, $y = 40°$

4. Explanations may vary.

5. a) $x + y = 180°$ b) $x = y$

6. a) Explanations may vary.   b) No   e) Yes

7. a) $\angle$DAB $= \frac{1}{2}(\angle$DOB$)$
   b) $\angle$DCB $= \frac{1}{2}$(reflex $\angle$DOB$)$

8.2 Exercises, page 476

3. b) PQRS is a square.

4. b) APBQCR is a regular hexagon.

6. a) No special case   b) No special case
   c) ABCD is a rectangle.

Linking Ideas: Mathematics and Science
Eclipses, page 478

1. 2144 km/h

2. 8576 km

3. Explanations may vary.

5. a) 5720 km   b) 3432 km/h

6. Explanations may vary.

Mathematics File: What Is a Tangent?, page 480

1. a) No   b) No

2. a) Yes   b) Yes

Investigate, page 482

1. a) 90°   b) 90°

2. a) PA = PB   b) PA = PB

3. Answers may vary.

4. Answers may vary.

8.3 Exercises, page 485

1. a) 5   b) 18°   c) 15

2. a) $x = y = 10\sqrt{2}$ b) $x = \sqrt{377}$, $y = 19$
   c) $x = 12$, $y = 7$ d) $x = y = 4\sqrt{21}$

3. Explanations may vary.

4. a) $x = 30°$, $y = 60°$ b) $x = 40°$, $y = 25°$

5. 21.4 cm

6. 3.4 cm

7. 19.2 cm

8. 10.6 cm

9. a) $x = 90°$, $y = 135°$ b) $x = 80°$, $y = 100°$
   c) $x = 50°$, $y = 130°$ d) $x = 110°$, $y = 70°$

10. a) $y = x$, $0 < x < 90$ b) $y = 180 - 2x$, $0 < x < 90$

11. 111.40 cm

12. 19.4 cm

13. a) Supplementary   b) Yes
    c) AOBP is a cyclic quadrilateral.

14. a) EF + GH = HE + FG   **b)** Yes
  c) The sum of the lengths of opposite sides of the quadrilateral formed by the tangents at four points on a circle are equal.
15. b) Right triangles; explanations may vary.
  c) They satisfy the Pythagorean Theorem; explanations may vary.
16. 36.4 cm

8.4 Exercises, page 492

For all exercises, proofs may vary.

Mathematical Modelling: How Long Is the Chain?, page 494

1. a) 261.96 cm   **b)** 1.16 cm   **c)** 0.44%
2. a) Explanations may vary.
  b) Estimate: 249.37 cm; difference: 2.68 cm; percent: 1.06%
3. Answers may vary.
4. Answers may vary.
  d) 251.76 cm   **e)** 0.11%
5. Answers may vary.
6. Answers may vary
7. Explanations may vary.
8. a) Right triangle; PE = 12 cm, QE = 52.65 cm
  b) Explanations may vary.
9. a) $77.16°$
  b) $154.32°$, $154.32°$; explanations may vary.
  c) 0.4287, 44.44 cm   **d)** 0.5713, 102.31 cm
10. 252.05 cm
11. It is chosen so that EQ is parallel to AB.

Investigate, page 497

1. a) Answers may vary. $67°$ **b)** $67°$, $\angle QAB = \angle QRA$
  c) Equal

8.5 Exercises, page 500

1. a) $65°$ **b)** $130°$ **c)** $115°$
2. a) $30°$ **b)** $40°$ **c)** $70°$
3. Explanations may vary.
4. a) $70°$, $70°$ **b)** $115°$, $115°$ **c)** $70°$, $70°$
5. Explanations may vary.
6. Explanations may vary.
7. No, opposite angles are not supplementary.
8. a) $60°$ **b)** $60°$ **c)** 7 cm
9. $a = 70°$, $b = 50°$, $c = 50°$, $d = 60°$, $e = 60°$
10. $40°$, $60°$, $80°$
11. $50°$, $60°$, $70°$
12. a) $y = 90 - \frac{x}{2}$, D: $0 < x < 180$
  b) $y = 90 + \frac{x}{2}$, D: $0 < x < 180$

16. b) Explanations may vary.

Problem Solving: Sweeping a Circle with Lines, page 504

1. b) $PQ^2 = PA \times PB$ **c)** $PA_1 \times PB_1 = PA_2 \times PB_2$
2. $PQ^2 = PA \times PB = PR^2$
3. Answers may vary.
4. Answers may vary.
5. Explanations may vary.
6. a) Rational   **b)** Answers may vary. $y = \frac{16}{x}$
  c) D: $x > 0$, R: $y \geq x$

Investigate, page 507

1. a) $108°$ **b)** $135°$ **c)** $157.5°$
2. $180° - \frac{360°}{n}$
3. a) $360°$ **b)** $540°$ **c)** $900°$
  d) $1080°$ **e)** $1080°$ **f)** $1080°$
4. $180°n - 360°$

8.6 Exercises, page 511

1. $360°$, 4
2. a) $720°$ **b)** $1800°$ **c)** $3240°$
3. a) $540°$ **b)** $1080°$ **c)** $1440°$ **d)** $17\,640°$
4. a) $108°$ **b)** $135°$ **c)** $144°$ **d)** $176.4°$
5. Explanations may vary.
6. a) Linear
  b) Answers may vary. Sum = $180n - 360$
    i) D: positive integers greater than or equal to 3
7. a) Rational
  b) D: positive integers greater than or equal to 3
8. a) i) $720°$ **ii)** $1800°$
  b) Explanations may vary.
9. a) $1440°$ **b)** $2340°$
10. $1980°$
11. $2880°$
12. $3960°$
13. Explanations may vary.
14. a) Explanations may vary.
  b) 3, 12, 48, 192
  c) $60°$; $60°$ and $240°$; $60°$ and $240°$; $60°$ and $240°$
  d) $180°$, $1800°$, $8280°$, $34\,200°$
  e) No

Linking Ideas: Mathematics and History

A Bestseller from Way Back, page 514

1. Lists may vary.

8 Review, page 515

1. a) $x = 125°$, $y = 85°$ **b)** $x = 50°$, $y = 70°$
   c) $x = 80°$, $y = 100°$ **d)** $x = y = 90°$
   e) $x = 120°$, $y = 60°$ **f)** $x = 25°$, $y = 50°$

2. a) Parallelogram. Explanations may vary.
   b) Rectangle. Explanations may vary.
   c) The only cyclic parallelogram is a rectangle.

10. a) Explanations may vary.   **b)** $15°$
    c) Answers may vary.

11. $130°$

8 Cumulative Review, page 517

1. a) 2 years   **b)** 5 years   **c)** 3 years

2. a) $1219.50   **b)** $880.39

3. a) $4029.19   **b)** $7116.89

4. 9%

6. $478.42

7. $708, $127 440

8. a) $y = 12(x + 1)^2 + 4$ **b)** $y = \frac{5}{4}(x - 3)^2 - \frac{5}{4}$

9. Explanations may vary.

11. a) $-3.6, 1.6, 2$ **b)** ± 2.1

12. b) For part i:
   i) D: all real numbers, R: $-\frac{1}{6} \leq y \leq \frac{1}{6}$
   ii) 0   **iii)** $y = 0$
   iv) Point symmetry; descriptions may vary.
   v) Maximum: $\left(3, \frac{1}{6}\right)$, minimum: $\left(-3, -\frac{1}{6}\right)$

13. a) 4   **b)** 14   **c)** 8   **d)** 10

14. Part c

15. a) $x < -3$, $x > 5$ **b)** $-6 \leq x \leq 3$ **c)** $-1 < x < 1$

16. a) $x \leq -7$, $2 \leq x < 3$ **b)** $-6 < x < -5$, $x > 2$
   c) $x < -3$, $6 < x < 7$

17. Explanations may vary.

18. a) 3   **b)** -14

19. a) $(0, 0), (2, 4)$ **b)** $(-2, 0), (2, 0)$
   c) $(-1, -2)$ **d)** $(1, 2)$

20. Explanations may vary.

21. a) $(3, -4)$ **b)** $(-6, 10)$ **c)** $(12, -9)$

22. a) $(-2, 3)$ **b)** $(4, 1)$ **c)** $(-4, -3)$

23. 150 adult tickets, 300 student tickets.

24. a) Square; 16 square units   **b)** Triangle; 12.5 square units

25. a) $x = 4$ **b)** $x \leq 9$ **c)** x is not a factor of 25.

26. a) 10   **b)** 6   **c)** 6

27. a) $30°$ **b)** $55°$ **c)** $20°$

29. a) 4.1 cm   **b)** 7.2 cm, 6.3 cm, 4.7 cm

Chapter 9

9.1 Exercises, page 525

1. a) Yes   **b)** No   **c)** Yes   **d)** No

2. a) 9, $(3, -4)$ **b)** $\sqrt{5}$, $(-2, 1)$
   c) $\sqrt{15}$, $(-4, 0)$ **d)** $4\sqrt{3}$, $(0, 6)$
   e) 8, $(0, 0)$ **f)** $2\sqrt{3}$, $(0, 0)$

3. a) i) $x^2 + y^2 = 9$ **ii)** $x^2 + y^2 = 49$
   iii) $(x - 5)^2 + (y - 3)^2 = 16$ **iv)** $(x + 2)^2 + (y - 6)^2 = 25$
   v) $(x - 4)^2 + y^2 = 36$ **vi)** $x^2 + (y + 3)^2 = 81$
   vii) $(x - 2)^2 + y^2 = 5$ **viii)** $(x - 3)^2 + (y + 5)^2 = 10$
   b) Explanations may vary.

4. a) $(x - 1)^2 + (y - 2)^2 = 16$ **b)** $(x + 2)^2 + (y - 1)^2 = 9$
   c) $(x - 1)^2 + (y + 2)^2 = 4$

5. a) Circle with centre $(0, 0)$ and radius 1
   b) Circle with centre $(3, 2)$ and radius $\sqrt{2}$

6. a) i, ii, iii, iv   **b)** iv, vi   **c)** i, v, vi

7. Explanations may vary.

8. b) S

9. a) $(x - 2)^2 + (y + 3)^2 = 25$ **b)** $(x - 4)^2 + (y - 2)^2 = 73$
   c) $(x - 5)^2 + (y - 4)^2 = 16$

10. a) $(x - 3)^2 + (y + 2)^2 = 25$ **b)** $(x - 1)^2 + (y - 2)^2 = 17$
   c) $(x - 3)^2 + (y - 4)^2 = 5$

11. $(x - 3)^2 + y^2 = 25$ or $(x + 3)^2 + y^2 = 97$

12. a) $x^2 + y^2 = 36$ **b)** $\pm 2\sqrt{5}$

13. a) $y = \pm\sqrt{16 - x^2}$ **b)** A circle
   c) i) ± 3.46 **ii)** ± 2.83

14. $(-3, 0)$

15. a) i) $3 \pm 2\sqrt{6}$ **ii)** $2 \pm 2\sqrt{6}$
   b) Explanations may vary.

16. $(x - 2)^2 + (y - 3)^2 = 13$

17. a) Point $(0, 0)$ **b)** No points
   c) Semicircle above the x-axis with centre $(0, 0)$ and
     radius 5
   d) Semicircle to the right of the y-axis with centre $(0, 0)$ and
     radius 5

18. a) No   **b)** No

19. $(x - 1)^2 + (y - 5)^2 = 25$

20. $x^2 + y^2 = 25(3 - 2\sqrt{2})$

21. a) Hyperbola   **b)** Ellipse
   c) A "square" with rounded corners
   d) A curve in the first quadrant that approaches the y-axis as
     x approaches 0, and the x-axis as y approaches 0

Mathematics File: Circles and Regular Polygons, page 528

1. a) $(0.93, -2.85)$ **b)** $\angle CON = 36°$
   c) C$(-2.43, 1.76)$, D$(-2.43, -1.76)$

2. a) A$(3, 0)$, B$(1.50, 2.60)$, C$(-1.50, 2.60)$, D$(-3, 0)$,
     E$(-1.50, -2.60)$, F$(1.50, -2.60)$
   b) A$(3, 0)$, B$(1.87, 2.35)$, C$(-0.67, 2.92)$, D$(-2.70, 1.30)$,
     E$(-2.70, -1.30)$, F$(-0.67, -2.92)$, G$(1.87, -2.35)$

c) A(3, 0), B(2.30, 1.93), C(0.52, 2.95), D(−1.50, 2.60), E(−2.82, 1.03), F(−2.82, −1.03), G(−1.50, −2.60), H(0.52, −2.95), I(2.30, −1.93)

3. 10, 20, 30, …

9.2 Exercises, page 531

1. a) (0, 0)   b) 4 units   c) 8 units

2. ±10

3. b) $(x − 3)^2 + (y − 3)^2 = 5$

4. a) i) (1, 2), (−1, −2)   ii) (2, −1)
 iii) (3, 1), (1, −3)   iv) No points of intersection
 b) Explanations may vary.

5. b) i) (1, 0), $\left(\frac{13}{5}, \frac{16}{5}\right)$ ii) (1, 2)
 iii) No points of intersection
 c) Explanations may vary.

6. d) $y = −\frac{1}{2}x + 7$

7. b) (0, −1), (6, −9)   c) $3x − 4y = 4$, $3x − 4y = 54$

8. b) $3x + 4y = 34$, $3x + 4y = −16$

9. P(6, 8) or P$\left(\frac{48}{5}, −\frac{14}{5}\right)$

10. (−40, 0) and (−24, 32)

11. a) i) 2.1 cm   ii) 14.0 cm
 b) 3.9 cm

12. 9.5 cm

13. a) 10   b) 36.9°
 c) (14.7, 0) and (1.29, 0)   d) (14.7, 0) and (1.29, 0)

14. a) 166.7 m   b) 121.5 m

15. 10.3 m

16. $(x − 4.5)^2 + (y − 3.5)^2 = 4.5$, $(x − 6)^2 + (y − 2)^2 = 9$, or $(x − 3)^2 + (y − 5)^2 = 9$

17. b) (0, −5)

18. a) B(4, −3), D(4, 3)

19. 26 m or 14 m

Linking Ideas: Mathematics and Art
Crop Circles, page 535

4. Sets of instructions may vary.

Investigate, page 536

1. a) ∠OAB = 33.7°, ∠OBA = 56.3°, ∠MOB = 33.7°, ∠MOA = 56.3°, ∠OMA = 90°, ∠BOA = 90°, ∠BMO = 90°, BO = 6, OA = 9, BA ≐ 10.8, OM ≐ 5, BM ≐ 3.3, MA ≐ 7.5

2. 3.05

3. Descriptions may vary.

9.3 Exercises, page 540

1. a) $−\frac{3}{2}$ b) $\frac{7}{5}$ c) $−\frac{2}{3}$ d) $−\frac{1}{2}$

2. a) $2x − y − 5 = 0$ b) $4x − 5y + 31 = 0$
 c) $2x + 3y + 7 = 0$ d) $10x + 7y − 43 = 0$

3. a) (4, 2)   b) $\left(−1, \frac{7}{2}\right)$ c) $\left(\frac{3}{2}, 0\right)$ d) $\left(2, \frac{1}{2}\right)$

4. a) (2, 3), (4, 4), (6, 5)
 b) (−1, −1.5), (1, −3), (3, −4.5)
 c) (−1, 1), (−2, 2), (−3, 3)
 d) (0, 4), (2, 2), (4, 0)

5. a) i) 6   ii) 3   iii) 3
 b) Explanations may vary.

6. a) 6   b) 8   c) 24   d) 10   e) 4.8

7. b) 9   c) 3   d) 2.85   e) 2.85

8. a) 3   b) 3.64   c) 2

9. a) $\left(2, \frac{8}{3}\right)$, $\left(4, \frac{4}{3}\right)$ b) $\left(4, \frac{8}{3}\right)$, $\left(6, \frac{13}{3}\right)$

10. Explanations may vary.

11. b) $y = 7$, $y = 1$, $(x − 2)^2 + (y − 4)^2 = 9$

12. b) $y = 4$, $(x − 2)^2 + (y − 1)^2 = 36$, $(x − 2)^2 + (y − 7)^2 = 36$

13. $4x + 3y − 30 = 0$ or $4x − 3y + 30 = 0$

14. $(y − 7)(y − 1)((x − 2)^2 + (y − 4)^2 − 9) = 0$

Mathematical Modelling: What Comes after the Cube?, page 542

1. (0, 0, 0, 0), (1, 0, 0, 0), (0, 1, 0, 0), (0, 0, 1, 0), (0, 0, 0, 1), (1, 1, 0, 0), (1, 0, 1, 0), (1, 0, 0, 1), (0, 1, 1, 0), (0, 1, 0, 1), (0, 0, 1, 1), (1, 1, 1, 0), (1, 1, 0, 1), (1, 0, 1, 1), (0, 1, 1, 1), (1, 1, 1, 1)

2. 16; each branch of the tree for 3 dimensions branches into two.

3. a) $\sqrt{3}$ b) $\sqrt{4}$

4. a) 32
 b),c)

Figure	point	segment	square	cube	hypercube	5-cube	6-cube
Dimensions	0	1	2	3	4	5	6
Number of vertices	1	2	4	8	16	32	64
Number of edges	0	1	4	12	32	80	192

5. Explanations may vary.

6. a) 7   b) 17

7. Answers may vary.

Investigate, page 545

1. a) $\frac{1}{8}$, $\frac{1}{8}$, 3, 3   b) Opposite sides are parallel.

2. a) AB = $\sqrt{65}$, BC = $2\sqrt{10}$, CD = $\sqrt{65}$, DA = $2\sqrt{10}$
 b) Opposite sides are equal in length.
 c) Opposite sides of a parallelogram are equal in length.

3. a) AC: (7, 4.5); BD: (7, 4.5)
 b) Midpoints of the diagonals are the same point.
 c) The diagonals of a parallelogram bisect each other.

4. The diagonals of a rectangle are equal in length.

5. The diagonals of a rhombus bisect each other at right angles.

9.4 Exercises, page 547

1. b) Yes
  d) i) D(3, 1)   **ii)** D(−5, 5) or D(7, 9)

5. a) Answers may vary. (8, 5) and (−2, 5)
  b) 0; (3, 5)   **c)** $x = 3$
  d) Yes; explanations may vary.

6. b) $y = -\frac{1}{2}x$
  c) The centre (0, 0) satisfies the equation of the line.

7. b) $y = 7x$
  c) The product of the slopes of the lines is −1, so they are perpendicular.

8. b) $y = \frac{3}{7}x$

12. a) 16   **b)** $AB = 8$, $BC = \sqrt{41}$, $CA = 5$
  c) $s = \frac{13 + \sqrt{41}}{2}$

13. b) $AB = 2\sqrt{5}$, $BC = \sqrt{17}$, $AC = \sqrt{13}$
  c) $\angle ABC \doteq 49.4°$ **d)** $CD = \sqrt{45}$, $BD = \sqrt{26}$

14. b) $OA = \sqrt{65}$, $OB = \sqrt{65}$, $AB = 6\sqrt{5}$
  c) $\angle AOB \doteq 112.6°$ **d)** $\angle ACB \doteq 56.3°$

9.5 Exercises, page 553

1. – 9. Proofs may vary.

10. a) $y = \frac{c}{a}x - \frac{cb}{a}$ **b)** $-\frac{cb}{a}$
  c) $-\frac{cb}{a}$ **d)** Equal.

11. – 14. Proofs may vary.

Problem Solving: Creative Problem Posing, page 555

1. Answers may vary.
2. Answers may vary.
3. Answers may vary.
4. Problems may vary.
5. Problems may vary.
6. Problems may vary.
7. Explanations may vary.

9 Review, page 558

1. a) 5, (1, 0)   **b)** $2\sqrt{10}$, (−2, 3)
  c) 6, (−3, 4)   **d)** $\sqrt{55}$, (0, −5)
  e) $\sqrt{42}$, (0, 0)   **f)** 8, (0, −1)

2. a) $x^2 + y^2 = 25$ **b)** $(x - 1)^2 + (y - 3)^2 = 9$
  c) $(x - 2)^2 + (y + 2)^2 = 1$ **d)** $(x + 4)^2 + (y - 2)^2 = 3$
  e) $(x + 1)^2 + (y + 3)^2 = 8$ **f)** $(x - 2)^2 + y^2 = 36$

3. Explanations may vary.

4. 3.3 cm

5. ±6

6. b) (−7, −9) and (9, 3)
  c) $4x + 3y - 1 = 0$, $4x + 3y - 45 = 0$

7. a) 4.24   **b)** 6.01   **c)** 1.79

8. Explanations may vary.

9. a) 2.68   **b)** 8   **c)** 0.55

10. Explanations may vary.

11. a) (3, 2), (6, 1)   **b)** $\left(-\frac{5}{3}, \frac{19}{3}\right)$, $\left(\frac{1}{3}, \frac{20}{3}\right)$
  c) (−0.4, 0.4), (1.2, 0.8), (2.8, 1.2), (4.4, 1.6)
  d) (−1.6, 7.8), (−0.2, 7.6), (1.2, 7.4), (2.6, 7.2)

12. a) The shortest distance from the centre of the circle to the line, minus the radius.
  b) 2.82

13. b) No

9 Cumulative Review, page 560

1. a) $2760.26   **b)** $2738.64

2. a) $675 to $825   **b)** $150 to $250
  c) $300 to $350   **d)** $150 to $200

3. a) No   **b)** Yes

4. a) $y = \pm\frac{1}{2}\sqrt{x + 9}$ **b)** $y = 5 \pm \sqrt{x - 9}$ **c)** $y = 7 \pm \sqrt{x + 2}$

5. Explanations may vary.

8. Answers may vary.
  a) $f(x) = \frac{1}{\sqrt{x}}$, $g(x) = 3x - 4$, $k(x) = f(g(x))$
  b) $f(x) = x - 1$, $g(x) = 2x^2 + 4x + 3$, $k(x) = g(f(x))$
  c) $f(x) = x + 2$, $g(x) = 4^x$, $k(x) = g(f(x))$

9. a) −3, 0.8   **b)** −5.33, 2   **c)** 1.72, 1.28

10. Explanations may vary.

11. a) 4   **b)** 12.5   **c)** $\frac{25}{7}$
  d) $0 \leq x < 25$ **e)** $x \geq 4$ **f)** $0 \leq x \leq 18$

12. (5, 5) or (−5, −5)

13. a) ±3   **b)** 0   **c)** −1, 9

14. a) $(x - 3)(x + 1)(x - 2)$ **b)** $(x + 2)(x + 3)(2x + 1)$
  c) $(x - 1)(x - 4)(x + 2)$ **d)** $(x - 2)(2x - 3)(x + 2)$

15. a) (4, −2)   **b)** (3, 0)   **c)** (1, 2)

16. a) (3.6, 0.6)   **b)** (5, −3)   **c)** (−3, 6)

17. a) Julie has at least one pet at home.
  b) A parallelogram has four vertices.

18. a) $x = 50°$, $y = 60°$ **b)** $x = 20°$, $y = 60°$
  c) $x = 115°$, $y = 20°$

21. a) $\left(\frac{3}{\sqrt{10}}, \frac{9}{\sqrt{10}}\right)$, $\left(-\frac{3}{\sqrt{10}}, -\frac{9}{\sqrt{10}}\right)$
  b) $(1 + \sqrt{17}, -1 + \sqrt{17})$, $(1 - \sqrt{17}, -1 - \sqrt{17})$
  c) (−1.8, −2.6), (1, 3)

GLOSSARY

absolute maximum point: a point on a graph whose *y*-coordinate is greater than those of all other points on the graph

absolute minimum point: a point on a graph whose *y*-coordinate is less than those of all other points on the graph

absolute value: the distance between any real number and 0 on a number line; for example, $|-3| = 3$, $|3| = 3$

absolute value equation: an equation containing the variable inside the absolute value sign

absolute value inequality: an inequality containing the variable inside the absolute value sign

accumulated amount: the value of the principal plus interest

acute angle: an angle measuring less than 90°

acute triangle: a triangle with three acute angles

additive inverse: a number and its opposite; the sum of additive inverses is 0; for example, $+3 + (-3) = 0$

algebraic expression: a mathematical expression containing a variable: for example, $6x - 4$ is an algebraic expression

alternate angles: angles that are between two lines and are on opposite sides of a transversal that cuts the two lines

Angles 1 and 3 are alternate angles.
Angles 2 and 4 are alternate angles.

altitude: the perpendicular distance from the base of a figure to the opposite side or vertex; also the height of an aircraft above the ground

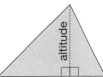

amortization: the repayment of the principal and interest of a loan by equal payments over a fixed period of time

angle: the figure formed by two rays from the same endpoint

angle bisector: the line that divides an angle into two equal angles

annuity: a series of regular, equal payments paid into, or out of, an account

approximation: a number close to the exact value of an expression; the symbol $\doteq$ means "is approximately equal to"

area: the number of square units needed to cover a region

arithmetic sequence: a sequence of numbers in which each term after the first term is calculated by adding the same number to the preceding term; for example, in the sequence 1, 4, 7, 10, ..., each number is calculated by adding 3 to the previous number

arithmetic series: the indicated sum of the terms of an arithmetic sequence

array: an arrangement in rows and columns

assessed value of property: a percent of the market value of the property used to calculate property taxes.

asymptote: a line that a curve approaches, but never reaches, as one of the variables approaches some particular value

average: a single number that represents a set of numbers; see *mean, median,* and *mode*

axis of symmetry: see *line symmetry*

balance: the result when money is added to or subtracted from an original amount

bar notation: the use of a horizontal bar over decimal digits to indicate that they repeat; for example, $1.\overline{3}$ means 1.333 333 ...

base: the side of a polygon or the face of a solid from which the height is measured; the factor repeated in a power

binomial: a polynomial with two terms; for example, $3x - 8$

bisector: a line that divides a line segment in two equal parts
The broken line is a bisector of AB.

boxplot: see *box-and-whisker plot*

broken-line graph: a graph that displays data by using points joined by line segments

budget: a written plan outlining how money is to be spent

Canada pension plan contributions (CPP): payments made by all employed Canadians between 18 and 70 years of age, and their employers, into a government-run pension scheme

centroid: the point where the three medians of a triangle intersect

chord: a line segment whose endpoints lie on a circle

circle: the set of points in a plane that are a given distance from a fixed point (the centre)

circumcentre: the point where the perpendicular bisectors of the sides of a triangle intersect

circumcircle: a circle drawn through each of the vertices of a triangle and with its centre at the circumcentre of the triangle

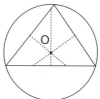

circumference: the distance around a circle, and sometimes the circle itself

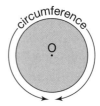

coefficient: the numerical factor of a term; for example, in the terms $3x$ and $3x^2$, the coefficient is 3

collinear points: points that lie on the same line

commission: a fee or payment given to a salesperson, usually a specified percent of the person's sales

common denominator: a number that is a multiple of each of the given denominators; for example, 12 is a common denominator for the fractions $\frac{1}{3}$, $\frac{5}{4}$, $\frac{7}{12}$

common difference: the number obtained by subtracting any term from the next term in an arithmetic sequence

common factor: a number that is a factor of each of the given numbers; for example, 3 is a common factor of 15, 9, and 21

common ratio: the ratio formed by dividing any term after the first one in a geometric sequence by the preceding term

commutative property: the property stating that two numbers can be added or multiplied in any order; for example, $6 + 8 = 8 + 6$ and $4 \times 7 = 7 \times 4$

complementary angles: two angles whose sum is 90°

∠ABC and ∠CBD are complementary angles.

complete the square: add or subtract constants to rewrite a quadratic expression $ax^2 + bx + c$ as $a(x - p)^2 + q$

complex fraction: a fraction that has other fractions in the numerator and/or denominator; for example, $\dfrac{2 + \frac{1}{3}}{3 - \frac{1}{4}}$

composite function: a function that is the result of two functions being applied successively

composite number: a number with three or more factors; for example, 8 is a composite number because its factors are 1, 2, 4, and 8

compound interest: see *interest*; if the interest due is added to the principal and thereafter earns interest, the interest earned is compound interest

compounding period: the time interval for which interest is calculated

concentric circles: circles with the same centre

cone: a solid formed by a region and all line segments joining points on the boundary of the region to a point not in the region

congruent: figures that have the same size and shape, but not necessarily the same orientation

conjecture: a conclusion based on examples

connectives: words used to combine simple statements into complex statements, such as "and," "or," or "if ...then"

consecutive integers: integers that come one after the other without any integers missing; for example, 34, 35, 36 are consecutive integers, so are −2, −1, 0, and 1

consistent system of equations: a system of equations with at least one solution

constant: a particular number

constant of proportionality: the constant k in a direct variation of the form $y = kx$; the slope of the graph of this equation

constant slope property: the slopes of all segments of a line are equal

constant term: a number

continuous function: a function whose graph can be drawn without lifting the pencil from the paper

contrapositive: *see negation;* a statement formed from an "If ... then" statement by reversing the parts and replacing each part with its negation. For example, the contrapositive of "If a triangle has three equal sides then it has three equal angles" is "If a triangle does not have three equal angles then it does not have three equal sides."

converse: a statement formed by reversing the parts of an "If ... then" statement. For example, the converse of "If a triangle has three equal sides then it has three equal angles" is "If a triangle has three equal angles then it has three equal sides."

coordinate axes: the *x*- and *y*-axes on a grid that represents a plane

coordinate plane: a two-dimensional surface on which a coordinate system has been set up

coordinates: the numbers in an ordered pair that locate a point in the plane

corresponding angles: angles that are on the same side of a transversal that cuts two lines and on the same side of each line

Angles 1 and 3 are corresponding angles.
Angles 2 and 4 are corresponding angles.
Angles 5 and 7 are corresponding angles.
Angles 6 and 8 are corresponding angles.

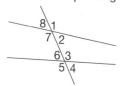

corresponding angles in similar triangles: two angles, one in each triangle, that are equal

cosine: for an acute ∠A in a right triangle, the ratio of the length of the side adjacent to ∠A, to the length of the hypotenuse

counterexample: an example that shows a conjecture to be false

credit: an arrangement where goods or services are provided on the understanding that they will be paid for at a later date

cube: a solid with six congruent, square faces

cube root: a number which, when raised to the power 3, results in a given number; for example, 3 is the cube root of 27, and −3 is the cube root of −27

cubic units: units that measure volume

cyclic quadrilateral: a quadrilateral whose vertices lie on a circle

cylinder: a solid with two parallel, congruent, circular bases

data: facts or information

database: facts or information supplied by computer software

debenture: a certificate acknowledging debt of principal and interest

deductive reasoning: reasoning based on statements that are accepted as true

denominator: the term below the line in a fraction

density: the mass of a unit volume of a substance

diagonal: a line segment that joins two vertices of a figure, but is not a side

diameter: the distance across a circle, measured through the centre; a line segment through the centre of the circle with its endpoints on the circle

difference of squares: a polynomial that can be expressed in the form $x^2 - y^2$; the product of two monomials that are the sum and difference of the same two quantities, $(x + y)(x - y)$

digit: any of the symbols used to write numerals; for example, in the base-ten system the digits are 0, 1, 2, 3, 4, 5, 6, 7, 8, and 9

discontinuous function: a function whose graph cannot be drawn without lifting the pencil from the paper

discriminant: the discriminant of the quadratic equation $ax^2 + bx + c = 0$ is $b^2 - 4ac$

Distributive Law: the property stating that a product can be written as a sum or difference of two products; for example, for all real numbers a, b, and c: $a(b + c) = ab + ac$ and $a(b - c) = ab - ac$

domain of a function: the set of x-values (or valid input numbers) represented by the graph or the equation of a function

effective rate: the annual interest rate that produces the same amount of interest per year as the given rate compounded several times per year

employment insurance contributions (EI): contributions made by all employed Canadians and their employers into a government-run insurance scheme providing insurance against some unemployment situations

entire radical: an expression of the form $\sqrt{x}$; for example, $\sqrt{20}$ is an entire radical

equation: a mathematical statement that two expressions are equal

equidistant: the same distance apart

equilateral triangle: a triangle with three equal sides

equivalent rates: rates of interest with different compounding periods that have the same effect on a given amount of money over the same time

equivalent systems: systems of equations with the same solution(s)

evaluate: to substitute a value for each variable in an expression and simplify the result

even number: an integer that has 2 as a factor; for example, 2, 4, −6

expanding: multiplying a polynomial by a polynomial

experiment: an operation, carried out under controlled conditions, that is used to test or establish a hypothesis

exponent: a number, shown in a smaller size and raised, that tells how many times the number before it is used as a factor; for example, 2 is the exponent in 6^2

expression: a meaningful combination of mathematical symbols, such as a polynomial

extraneous root: a root of an equation, obtained algebraically from the original equation, that is not a root of the original equation

extrapolate: to estimate a value beyond known values

extremes: the highest and lowest values in a set of numbers

factor: to factor means to write as a product; to factor a given integer means to write it as a product of integers, the integers in the product are the factors of the given integer; to factor a polynomial with integer coefficients means to write it as a product of polynomials with integer coefficients

factor property: if a polynomial in x has a factor $x - k$, then k is a factor of the constant term of the polynomial

fifth root: a number which, when raised to the power 5, results in a given number; for example, 3 is the fifth root of 243 since $3^5 = 243$, and -3 is the fifth root of -243 since $(-3)^5 = -243$

finance charge: the difference between the cash price and the sum of the payments made for an item

formula: a rule that is expressed as an equation

fourth root: a number which, when raised to the power 4, results in a given number; for example, -2 and 2 are the fourth roots of 16

fraction: an indicated quotient of two quantities

frequency: the number of times a particular number occurs in a set of data

function: a rule that gives a single output number for every valid input number

future value: see *accumulated amount*

general term of an arithmetic sequence: used to determine any term by substitution; for example, t_n is the nth term in the sequence, and $t_n = a + (n - 1)d$, where a is the first term, n the number of terms, and d the common difference

geometric sequence: a sequence of numbers in which each term after the first term is calculated by multiplying the preceding term by the same number; for example, in the sequence 1, 3, 9, 27, 81, …, each number is calculated by multiplying the preceding term by 3

geometric series: the indicated sum of the terms of a geometric sequence

graduated commission: see *commission*; the rate of commission increases when certain sales goals are reached

greatest common factor: the greatest factor that 2 or more monomials have in common; $4x^2$ is the greatest common factor of $8x^3 + 16x^2y - 64x^4$

gross income: the amount of money earned before deductions

guaranteed investment certificate (GIC): see *term deposit;* an account with a guaranteed rate of interest for a specified term where the deposit cannot be withdrawn before the end of the term

hectare: a unit of area that is equal to 10 000 m^2

hemisphere: half a sphere

hexagon: a six-sided polygon

horizontal intercept: the horizontal coordinate of the point where the graph of a line or a function intersects the horizontal axis

hypotenuse: the side that is opposite the right angle in a right triangle

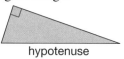

hypotenuse

incentre: the point at which the three angle bisectors of a triangle intersect

incircle: a circle drawn inside a triangle, with its centre at the incentre and with the radius the shortest distance from the incentre to one of the sides of the triangle

inconsistent system of equations: a system with no solution

inductive reasoning: reasoning based on the results of experiment

inequality: a statement that one quantity is greater than (or less than) another quantity

inscribed angle: the angle between two chords of a circle that have a common endpoint

integers: the set of numbers… −3, −2, −1, 0, +1, +2, +3, …

interest: money that is paid for the use of money, usually according to a predetermined percent

interpolate: to estimate a value between two known values

intersecting lines: lines that meet or cross; lines that have one point in common

interval: a regular distance or space between values

inverse: see *additive inverse* and *multiplicative inverse*

inverse of a function: a relation whose rule is obtained from that of a function by interchanging x and y

irrational number: a number that cannot be written in the form $\frac{m}{n}$, where m and n are integers $(n \neq 0)$

isosceles acute triangle: a triangle with two equal sides and all angles less than 90°

isosceles obtuse triangle: a triangle with two equal sides and one angle greater than 90°

isosceles right triangle: a triangle with two equal sides and a 90° angle

isosceles triangle: a triangle with at least two equal sides

kite: a quadrilateral with two pairs of equal adjacent sides

lattice point: on a coordinate grid, a point at the intersection of two grid lines

leading coefficient: the coefficient of the highest power of a polynomial expression

legs: the sides of a right triangle that form the right angle

light-year: a unit for measuring astronomical distances; one light-year is the distance light travels in one year

like radicals: radicals that have the same radical part; for example, $\sqrt{5}$, $3\sqrt{5}$, and $-7\sqrt{5}$

like terms: terms that have the same variables; for example, $4x$ and $-3x$ are like terms

line of best fit: a line that passes as close as possible to a set of plotted points

line segment: the part of a line between two points on the line, including the two points

line symmetry: a figure that maps onto itself when it is reflected in a line is said to have line symmetry; for example, line l is the line of symmetry for figure ABCD

linear equation: an equation that represents a straight line

linear equation (in two or three variables): an equation that can be written in the form $Ax + By + C = 0$ or $Ax + By + Cz + D = 0$

linear function: a function whose defining equation can be written in the form $y = mx + b$, where m and b are constants

linear system of equations: two or more linear equations in the same variables

major arc: a part of a circle that is greater than half a circle

mass: the amount of matter in an object

maturity value: see *accumulated amount;* the value of the principal plus interest at the end of the term

mean: the sum of a set of numbers divided by the number of numbers in the set

median: the middle number when data are arranged in numerical order

median of a triangle: a line from one vertex to the midpoint of the opposite side

midpoint: the point that divides a line segment into two equal parts

mill rate: the rate (in thousandths of a dollar) at which property tax is to be paid on the assessed value of property

minor arc: a part of a circle that is less than half a circle

mixed radical: an expression of the form $a\sqrt{x}$; for example, $2\sqrt{5}$ is a mixed radical

mode: the number that occurs most often in a set of numbers

monomial: a polynomial with one term; for example, 14 and $5x^2$ are each a monomial

mortgage: a long-term loan on real estate that gives the person or firm providing the money a claim on the property if the loan is not repaid

multiple: the product of a given number and a natural number; for example, some multiples of 8 are 8, 16, 24, …

multiplicative inverse: a number and its reciprocal; the product of multiplicative inverses is 1; for example, $3 \times \frac{1}{3} = 1$

natural numbers: the set of numbers 1, 2, 3, 4, 5, …

negation: a statement that has the opposite meaning of a given statement; for example, the negation of "The triangle has three equal sides" is "The triangle does not have three equal sides"

negative number: a number less than 0

negative reciprocals: two numbers that have a product of -1; for example, $\frac{3}{4}$ is the negative reciprocal of $-\frac{4}{3}$, and vice versa

net earnings: take home pay

non-linear system of equations: two or more equations in which at least one of the equations is not a linear equation

numeracy: the ability to read, understand, and use numbers

numerator: the term above the line in a fraction

obtuse angle: an angle greater than 90° and less than 180°

obtuse triangle: a triangle with one angle greater than 90°

octagon: an eight-sided polygon

odd number: an integer that does not have 2 as a factor; for example, 1, 3, -7

operation: a mathematical process or action such as addition, subtraction, multiplication, or division

opposite angles: the equal angles that are formed by two intersecting lines

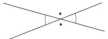

opposite number: a number whose sum with a given number is 0; for example, 3 and −3 are opposites

opposites: two numbers whose sum is zero; each number is the opposite of the other

order of operations: the rules that are followed when simplifying or evaluating an expression

ordered pair: a pair of numbers, written as (*x*, *y*), that represents a point on a coordinate grid

orthocentre: the point at which the altitudes of a triangle intersect

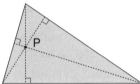

outcome: a possible result of an experiment or a possible answer to a survey question

parabola: the name given to the shape of the graph of a quadratic function

parallel lines: lines in the same plane that do not intersect

parallelogram: a quadrilateral with both pairs of opposite sides parallel

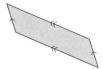

payroll deductions: deductions made by the employer from gross pay

pentagon: a five-sided polygon

per capita: for each person

percent: the number of parts per 100; the numerator of a fraction with denominator 100

perfect square: a number that is the square of a whole number; a polynomial that is the square of another polynomial

perimeter: the distance around a closed figure

period of a loan: the time it takes to pay back the loan

perpendicular: intersecting at right angles

perpendicular bisector: the line that is perpendicular to a line segment and divides it in two equal parts
The broken line is the perpendicular bisector of AB.

pi (π): the ratio of the circumference of a circle to its diameter; π ≐ 3.1416

plane geometry: the study of two-dimensional figures; that is, figures drawn or visualized on a plane

point of intersection: a point that lies on two or more figures

point of tangency: see *tangent;* the point at which a tangent to a circle intersects the circle

point symmetry: a figure that maps onto itself after a rotation of 180° about a point is said to have point symmetry

polygon: a closed figure that consists of line segments; for example, triangles and quadrilaterals are polygons

polynomial: a mathematical expression with one or more terms, in which the exponents are whole numbers and the coefficients are real numbers

polynomial function: a function where $f(x)$ is a polynomial expression

population density: the average number of people for each square unit of land

positive number: a number greater than 0

power: an expression of the form a^n, where a is called the base and n is called the exponent; it represents a product of equal factors; for example, $4 \times 4 \times 4$ can be expressed as 4^3

present value: the starting principal required to obtain a specific amount in the future

prime number: a whole number with exactly two factors, itself and 1; for example, 3, 5, 7, 11, 29, 31, and 43

principal: the amount of a loan or an investment

prism: a solid that has two congruent and parallel faces (the *bases*), and other faces that are parallelograms

probability: if the outcomes of an experiment are equally likely, then the probability of an event is the ratio of the number of outcomes favourable to the event to the total number of outcomes

proportion: a statement that two ratios are equal

pyramid: a solid that has one face that is a polygon (the *base*), and other faces that are triangles with a common vertex

Pythagorean Theorem: for any right triangle, the area of the square on the hypotenuse is equal to the sum of the areas of the squares on the other two sides

quadrant: one of the four regions into which coordinate axes divide a plane

quadratic equation: an equation in which the variable is squared; for example, $x^2 + 5x + 6 = 0$ is a quadratic equation

quadratic function: a function with defining equation $f(x) = ax^2 + bx + c$ where a, b, c are constants and a cannot be zero

quadratic inequality: an inequality where the variable is squared

quadrilateral: a four-sided polygon

radical: the root of a number; for example, $\sqrt{400}$, $\sqrt[3]{8}$, and so on

radical equation: an equation where the variable occurs under a radical sign

radical inequality: an inequality where the variable occurs under a radical sign

radical sign: the symbol $\sqrt{}$ that denotes the positive square root of a number

radius (plural, **radii**): the distance from the centre of a circle to any point on the circumference, or a line segment joining the centre of a circle to any point on the circumference

range: the difference between the highest and lowest values (the *extremes*) in a set of data

range of a function: the set of y-values (or output numbers) represented by the graph or the equation of a function

rate: a certain quantity or amount of one thing considered in relation to a unit of another thing

ratio: a comparison of two or more quantities with the same unit

rational equation: an equation where the variable occurs in the denominator of a rational expression

rational expression: an algebraic expression that can be written as the quotient of two polynomials; for example, $\frac{3x^2 + 5}{2x + 7}$

rational function: a function where $f(x)$ is a rational expression

rational inequality: an inequality where the variable occurs in the denominator of a rational expression

rational number: a number that can be written in the form $\frac{m}{n}$, where m and n are integers ($n \neq 0$)

rationalize the denominator: write the denominator as a rational number, to replace the irrational number; for example, $\frac{6}{\sqrt{2}}$ is written $\frac{6\sqrt{2}}{2}$, or $3\sqrt{2}$

real numbers: the set of rational numbers and the set of irrational numbers; that is, all numbers that can be expressed as decimals

reciprocal of a function $f(x)$: the function $\frac{1}{f(x)}$, where $f(x)$ is not zero

reciprocals: two numbers whose product is 1; for example, $\frac{3}{4}$ and $\frac{4}{3}$ are reciprocals, 2 and $\frac{1}{2}$ are reciprocals

rectangle: a quadrilateral that has four right angles

rectangular prism: a prism that has rectangular faces

rectangular pyramid: a pyramid with a rectangular base

reductio ad absurdum: an indirect form of proof where the result to be proved is assumed false and a conclusion reached that contradicts known fact(s)

reflex angle: an angle between 180° and 360°

registered retirement savings plan (RRSP): a savings plan, for people who earn income, where funds contributed and interest earned are not taxed until the funds are withdrawn

regular decagon: a polygon that has ten equal sides and ten equal angles

regular hexagon: a polygon that has six equal sides and six equal angles

regular octagon: a polygon that has eight equal sides and eight equal angles

regular polygon: a polygon that has all sides equal and all angles equal

relation: a rule that produces one or more output numbers for every valid input number

relative maximum point: a point on a graph whose y-coordinate is greater than those of the neighbouring points on the graph

relative minimum point: a point on a graph whose y-coordinate is less than those of the neighbouring points on the graph

rhombus: a parallelogram with four equal sides

right angle: a 90° angle

right circular cone: a cone in which a line segment from the centre of the circular base to the vertex is perpendicular to the base

right triangle: a triangle that has one right angle

rise: the vertical distance between 2 points

root of an equation: a value of the variable that satisfies the equation

rounding: approximating a number to the next highest (or lowest) number; for example, rounding 3.46 to the tenth is 3.5, and rounding 4.34 is 4.3

run: the horizontal distance between 2 points

scale: the ratio of the distance between two points on a map, model, or diagram to the distance between the actual locations; the numbers on the axes of a graph

scale factor: the ratio of corresponding lengths on two similar figures

scalene triangle: a triangle with no two sides equal

scatterplot: a graph of data that is a series of points

scientific notation: a number expressed as the product of a number greater than -10 and less than -1 or greater than 1 and less than 10, and a power of 10; for example, 4700 is written as 4.7×10^3

secant line: a line that intersects a curve

sector: the figure formed by an arc of a circle, the radii at the ends of the arc, and all enclosed points

sector angle: see *sector;* the angle at the centre of the circle between two radii

segment of a circle: the figure formed by an arc of a circle, the chord joining the endpoints of the arc, and all enclosed points

semi-annual: every six months

semicircle: half a circle

significant digits: the meaningful digits of a number representing a measurement

similar figures: figures with the same shape, but not necessarily the same size

simple interest: see *interest;* interest calculated according to the formula $I = Prt$

sine: for an acute $\angle A$ in a right triangle, the ratio of the length of the side opposite $\angle A$, to the length of the hypotenuse

slope: a measure of the steepness of a line; calculated as slope $= \frac{\text{rise}}{\text{run}}$

slope y-intercept form: the equation of a line in the form $y = mx + b$

sphere: the set of points in space that are a given distance from a fixed point (the centre)

spreadsheet: a computer-generated arrangement of data in rows and columns, where a change in one value results in appropriate calculated changes in the other values

square: a rectangle with four equal sides

square of a number: the product of a number multiplied by itself; for example, 25 is the square of 5

square root: a number which, when multiplied by itself, results in a given number; for example, 5 and -5 are the square roots of 25

standard form of the equation of a line: the equation of a line in the form $Ax + By + C = 0$

statistics: the branch of mathematics that deals with the collection, organization, and interpretation of data

straight angle: an angle measuring $180°$

straightedge: a strip of wood, metal, or plastic with a straight edge, but no markings

sum of an arithmetic series: the formula for the sum of the first n terms is $S_n = \left(\frac{a + t_n}{2}\right) \times n$, where a is the first term, t_n the nth term, and n the number of terms

supplementary angles: two angles whose sum is 180°

∠RST and ∠TSU are supplementary angles.

survey: an investigation of a topic to find out people's views

symmetrical: possessing symmetry; see *line symmetry;* see *point symmetry*

30–60–90 property: in a right triangle with angles 30°, 60°, and 90°, the hypotenuse is twice as long as the shorter leg, and the larger leg is $\sqrt{3}$ times as long as the shorter leg

tangent: for an acute ∠A in a right triangle, the ratio of the length of the side opposite ∠A, to the length of the side adjacent to ∠A

tangent segment: a line segment formed by a point P outside a circle and the point of tangency of a tangent from P to the circle

tangent to a circle: a line that intersects the circle in only one point

taxable income: the part of the gross earnings on which income tax must be paid

term: of a fraction is the numerator or the denominator of the fraction; when an expression is written as the sum of several quantities, each quantity is called a term of the expression

term deposit: an account whose rate of interest is guaranteed for a specified term where withdrawal before the end of the term may result in loss of interest.

tessellation: a tiling pattern

tetrahedron: a solid with four triangular faces

theorem: a conclusion reached by deductive reasoning

three-dimensional: having length, width, and depth or height

trajectory: the curved path of an object moving through space

translation: a transformation that moves all points in the plane in a given direction through a given distance

transversal: a line crossing two or more lines

trapezoid: a quadrilateral that has only one pair of parallel sides

tree diagram: a branching diagram used to show all possible outcomes of an experiment

triangle: a three-sided polygon

triangular number: a natural number that can be represented by arranging objects in a triangle; for example, 1, 3, 6, 10, 15, …

10

trinomial: a polynomial with three terms; for example, $3x^2 + 6x + 9$

truncating: approximating a number by cutting off the end digits; for example, 3.141 8 is truncated to 3.141, or to 3.14, or to 3.1, and so on

two-dimensional: having length and width, but no thickness, height, or depth

unit fraction: a fraction that has a numerator of 1

unit price: the price of one item, or the price for a particular mass or volume of an item

unit rate: the quantity associated with a single unit of another quantity; for example, 6 m in 1 s is a unit rate

unlike radicals: mixed radicals that have different radical parts; for example, $3\sqrt{5}$ and $7\sqrt{11}$

unlike terms: terms that have different variables, or the same variable but different exponents; for example, $3x$, $-4y$ and $3x^2$, $-3x$

variable: a letter or symbol representing a quantity that can vary

Venn diagram: a diagram where the elements of sets are represented by points within closed loops

vertex (plural, **vertices**): the corner of a figure or a solid

vertex of a parabola: the point where the axis of symmetry of a parabola intersects the parabola

vertical intercept: the vertical coordinate of the point where the graph of a line or a function intersects the vertical axis

vertical line test: if no two points on a graph can be joined by a vertical line, then the graph represents a function

volume: the amount of space occupied by an object

whole numbers: the set of numbers 0, 1, 2, 3,…

x-axis: the horizontal number line on a coordinate grid

x-intercept: the x-coordinate where the graph of a line or a function intersects the x-axis

y-axis: the vertical number line on a coordinate grid

y-intercept: the y-coordinate where the graph of a line or a function intersects the y-axis

zero exponent: any number, a, that has the exponent 0 $(a \neq 0)$, is equal to 1; for example, $(-6)^0 = 1$

zero of a function: a value of the variable for which the function has value zero

Zero principle: the sum of opposites is zero

INDEX

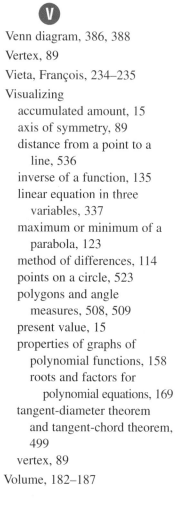

Stein Valley Nlakapamux School
PO Bag 300, Lytton, BC, V0K 1Z0
Phone:(250) 455-2522 Fax:(250) 455-2512

PHOTO CREDITS AND ACKNOWLEDGMENTS

The publisher wishes to thank the following sources for photographs, illustrations, articles, and other materials used in this book. Care has been taken to determine and locate ownership of copyright material used in this text. We will gladly receive information enabling us to rectify any errors or omissions in credits.

pp. 2-3 Ian Crysler; **p. 27** Ian Crysler; **p. 35** Ian Crysler; **p. 61** (left) Tony Stone Images, (centre) Bruce Ayres/Tony Stone Images, (right) George Hunter/Tony Stone Images; **p. 82** Canapress/Amy Sancetta; **p. 83** (bottom) Agence France Presse/Corbis-Bettmann; **p. 84** John Edwards/Tony Stone Images; **p. 98** Canapress/Frank Gunn; **p. 100** Dave Starrett; **p. 104** Copyright © Corel Corporation; **p. 108** SuperStock; **p. 142** © Harold & Esther Edgerton Foundation, 1998, Courtesy of Palm Press, Inc.; **p. 152** Ron Tanaka; **p. 187** From ONE MILLION by Hendrick Hertzberg copyright © 1993 by Hendrick Hertzberg. Reprinted by permission of Times Books, a division of Random House, Inc.; **p. 210** Bob Torrez/Tony Stone Images; **p. 214** Jon Riley/Tony Stone Images; **p. 224** (top) Marc Chamberlain/Tony Stone Images, (bottom) SuperStock; **p. 225** (centre) Mark Tomalty/Masterfile, (bottom) John Lund/Tony Stone Images; **p. 229** Canapress/Paul Chiasson; **p. 246** (left) Hulton Getty/Tony Stone Images, (centre) Corbis, (right) Margaret Gowan/Tony Stone Images; **p. 247** (left) Hulton Getty/Tony Stone Images, (centre) Sean Ellis/Tony Stone Images, (right) Loren Santow/Tony Stone Images; **p. 250** Ross Hamilton/Tony Stone Images; **p. 278** SuperStock; **p. 285** SuperStock; **p. 300** (top) Simon Jauncey/Tony Stone Images, (centre) Trevor Wood/Tony Stone Images, (bottom) Jacques Jangoux/Tony Stone Images; **p. 301** Penny Gentieu/Tony Stone Images; **p. 302** Ian Crysler; **p. 327** Ken Fisher/Tony Stone Images; **p. 335** Ian Crysler; **p. 353** (top) SuperStock, (bottom) SuperStock; **pp. 368-69** SuperStock; **p. 370** Gerben Oppermans/Tony Stone Images; **p. 373** Ian Crysler; **pp. 378-79** All of the Powers of Ten™ products are available through the Eames Office websites at www.eamesoffice.com and www.powersof10.com. The Powers of Ten™ film and Powers of Ten™ CD-ROM are also available from Pyramid Media (800.421.2304). The Powers of Ten™ book is also available from W.H. Freeman (212.576.9400, 41 Madison Ave. New York, NY 10010). The Powers of Ten™ poster and Powers of Ten™ flip book are available from the Eames Office (310.459.6703) P.O. Box 268, Venice, CA 90294 or www.eamesoffice.com.; **p. 384** Ian Crysler; **p. 385** (top) Canapress/Alain Charest, (bottom) Reuters/Corbis-Bettmann; **p. 395** David Madison/Tony Stone Images; **p. 399** Herman cartoon reprinted with permission by LaughingStock Licensing Inc.; **p. 401** (from the top) Canapress/Elise Amendola, Canapress/Kathy Willens, Reuters/Corbis-Bettmann, Canapress/Steve Russell; **p. 402** Courtesy of the United States Air Force Academy; **pp. 422-23** Satellite data provided by The Living Earth® Inc./Earth Imaging copyright © 1998 The Living Earth®, Inc. All rights reserved.; **p. 426** David Starrett; **p. 428** Susan Ashukian; **p. 431** (top) SuperStock., (centre) NASA, (bottom left) Tony Stone Images, (bottom right) Gary Braasch/Tony Stone Images; **p. 447** Satellite data provided by The Living Earth® Inc./Earth Imaging copyright © 1998 The Living Earth®, Inc. All rights reserved.; **p. 448** David Starrett; **p. 454** (top) David Starrett; **pp. 468-69** Ron Tanaka; **p. 470** Tony Wiles/Tony Stone Images; **p. 478** (top) Joel Simon/Tony Stone Images; **p. 482** David Jeffrey/Image Bank; **p. 514** (left) Hulton Getty/Tony Stone Images, (right) Hulton Getty/Tony Stone Images; **p. 522** Mark Wagner/Tony Stone Images; **p. 535** © Colin Andrews, Circles Phenomenon Research International

ILLUSTRATIONS

David Bathurst: 153
Julian Cleva: 55, 203
David Day: 369
Dayle Dodwell: 411
Kevin Ghiglione: 520-21

Dave McKay: 128, 142, 391, 398, 410, 454 (bottom), 512
Jack McMaster: 416
Jun Park: 432, 478, 479
Pronk&Associates: 454 (centre)